NanoScience and Technology

NanoScience and Technology

Series Editors:
P. Avouris B. Bhushan D. Bimberg K. von Klitzing H. Sakaki R. Wiesendanger

The series NanoScience and Technology is focused on the fascinating nano-world, mesoscopic physics, analysis with atomic resolution, nano and quantum-effect devices, nanomechanics and atomic-scale processes. All the basic aspects and technology-oriented developments in this emerging discipline are covered by comprehensive and timely books. The series constitutes a survey of the relevant special topics, which are presented by leading experts in the field. These books will appeal to researchers, engineers, and advanced students.

Please view available titles in *NanoScience and Technology* on series homepage
http://www.springer.com/series/3705/

Victor Moshchalkov
Roger Wördenweber
Wolfgang Lang
(Editors)

Nanoscience and Engineering in Superconductivity

With 186 Figures

Springer

Editors

Professor Dr. Victor Moshchalkov
Katholieke Universiteit Leuven
INPAC
Institute for Nanoscale Physics
Celestijnenlaan 200D
3001 Leuven, Belgium
E-mail: victor.moshchalkov@fys.kuleuven.be

Professor Dr. Roger Wördenweber
Forschungszentrum Jülich
Institute of Bio- and Nanosystems
52425 Jülich, Germany
E-mail: r.woerdenweber@fz-juelich.de

Professor Dr. Wolfgang Lang
University of Vienna, Faculty of Physics
Boltzmanngasse 5, A-1090 Vienna, Austria
E-mail: wolfgang.lang@univie.ac.at

Series Editors:

Professor Dr. Phaedon Avouris
IBM Research Division
Nanometer Scale Science & Technology
Thomas J. Watson Research Center
P.O. Box 218
Yorktown Heights, NY 10598, USA

Professor Dr. Bharat Bhushan
Ohio State University
Nanotribology Laboratory
for Information Storage
and MEMS/NEMS (NLIM)
Suite 255, Ackerman Road 650
Columbus, Ohio 43210, USA

Professor Dr. Dieter Bimberg
TU Berlin, Fakutät Mathematik/
Naturwissenschaften
Institut für Festkörperphyisk
Hardenbergstr. 36
10623 Berlin, Germany

Professor Dr., Dres. h.c. Klaus von Klitzing
Max-Planck-Institut
für Festkörperforschung
Heisenbergstr. 1
70569 Stuttgart, Germany

Professor Hiroyuki Sakaki
University of Tokyo
Institute of Industrial Science
4-6-1 Komaba, Meguro-ku
Tokyo 153-8505, Japan

Professor Dr. Roland Wiesendanger
Institut für Angewandte Physik
Universität Hamburg
Jungiusstr. 11
20355 Hamburg, Germany

NanoScience and Technology ISSN 1434-4904
ISBN 978-3-642-15136-1 e-ISBN 978-3-642-15137-8
DOI 10.1007/978-3-642-15137-8
Springer Heidelberg Dordrecht London New York

Library of Congress Control Number: 2010938947

Cover design: eStudio Calamar Steinen

Printed on acid-free paper

Springer is part of Springer Science+Business Media (www.springer.com)

Preface

The key technologies of the twenty-first century have been recently identified: energy, transport, nanotechnology, information/communication technology, health care, and environment. Remarkably, for several of them, superconductivity is of special interest, since it deals with the flow of charged particles (electron pairs) without dissipation, thus enabling superior performance needed for new energy-saving technologies. This unique property is a very valuable asset for developing a variety of fascinating technologies belonging to *grand challenges*. To name a few, these technologies range from next generation self-healing superconducting electricity grids, saving up to 15% of the transported electricity, superconducting generators with improved performance deployable among others in off-shore wind turbines or for ship propulsion, superconducting induction heater for power-saving metal processing, International Thermonuclear Reactor (ITER) with the superconducting magnet holding a very hot plasma making possible the fusion reaction, the levitating train operating at speeds exceeding 500 km/h, superconducting elements for quantum computing, to medical diagnostics tomography and magnetoencephalography with highest resolution.

To enable the emerging new technologies, the superconducting materials with a superior performance can be developed by manipulating the appropriate "elementary building blocks" through nanostructuring. For superconductivity, such "elementary blocks" are *Cooper pairs* and *fluxons*.

To support this research, in 2007 the European Science Foundation (ESF) had launched the 5-year Programme *Nanoscience and Engineering in Superconductivity – NES* (http://www.kuleuven.be/inpac/nes/).

Integration and efficient use of the NES experimental and theoretical techniques of the teams from the 15 EU countries participating in this program is achieved through the following activities:

- Implementation of coordinated joint scientific research
- Sharing complimentary equipment
- Joint applications for EU, ESF, and other international projects
- Joint supervision of the PhD work
- Keeping intense electronic exchange of reprints, preprints, data bases, etc
- Joint publication of original and review articles

- Exchange of lecturers, researchers, samples, software, and databases
- Creating and regularly updating dedicated NES website

The main objective of this program is to investigate the effects of the nanoscale confinement of condensate and flux on superconductivity to reveal its nanoscale evolution and to determine the fundamental relations between quantized confined states and the physical properties of these systems. Along the line of the main objective, the ESF-NES research activities are focused on the following topics

- Evolution of superconductivity at the nanoscale

The correlation between the nanograin size and the superconducting gap and the critical temperature T_c is investigated theoretically and experimentally. By systematically reducing the characteristic size of the grains and nanocells the crossover between the bulk superconducting regime and fluctuation-dominated superconductivity regime will be revealed

- Superconductivity in hybrid superconducting – normal (SN) and superconducting – ferromagnet (SF) nanosystems with tuneable boundary condition

Confined condensate is studied in superconducting nano-islands surrounded by normal metallic or ferromagnetic material. The role of proximity effects and the Andreev reflection in modifying the transparency of the sample boundaries will be revealed. The variation of the superfluid density near the boundary is mapped using the local scanning tunneling spectroscopy (STS) techniques. Different vortex configurations, including those with symmetry induced antivortices, and their dynamics are investigated in individual nanostructures of different geometries. Here strong effects of the specific boundary conditions on confined flux and condensate are expected.

- Confined flux in nanostructured superconductors and hybrid SN and SF nanosystems

Nanostructured superconductors and individual nanocells are investigated using local probe techniques, such as scanning tunneling microscopy (STM) and scanning Hall-probe microscopy, the distributions of the order parameter density and local magnetic fields are determined and then compared with the calculations of these parameters based on the solution of the Ginzburg–Landau (GL) equations with the realistic boundary conditions imposed though nanostructuring. Hybrid SN and SF arrays are also studied. Magnetic dots are used to generate local vortex–antivortex loops, which are strongly interacting with the flux lines in superconductors, creating a tunable magnetic periodic confinement. Here we can anticipate a very interesting interplay between flux generated by an applied field and magnetic dipoles, which can substantially enhance flux pinning.

- Josephson effects and tunneling in weakly coupled condensates

Josephson phenomena and phase-shifting effects are investigated in coupled superconducting condensates, where nanoscale coupling can be provided to tune the coupling strength. These phenomena are compared with Josephson effects in superfluids, mostly based on ^{3}He.

- Fundamentals of fluxonics, superconducting devices

Different devices that control the motion of flux quanta in superconductors are designed and studied. One of the focal points is on the removal of trapped magnetic flux that produces noise. The controllable vortex motion is used in nanostructured superconductors for making pumps, diodes and lenses of quantized magnetic flux. Vortex ratchets effects are investigated and then used to achieve vortex manipulation.

This book highlights the recent advances achieved along these research lines in the framework of the ESF-NES Program and presents the new ways superconductivity and vortex matter can be modified through nanostructuring and the use of the nanoscale magnetic templates. The basic nano-effects visualized by the STM/STS techniques, vortex and vortex–antivortex patterns, vortex dynamics, guided vortex motion and vortex ratchets, Josephson phenomena, critical currents, interplay between superconductivity and ferromagnetism at the nanoscale, and potential applications of nanostructured superconductors are presented. The book targets researchers and graduate students working on and/or actively interested in superconductivity and nanosciences.

Leuven *Victor Moshchalkov*
Jülich *Roger Wördenweber*
Vienna *Wolfgang Lang*
September 2010

Contents

Contributors

J. Aarts Kamerlingh Onnes Laboratory, Leiden University, P.O. Box 9504, 2300 RA Leiden, The Netherlands, aarts@physics.leidenuniv.nl

C. Attanasio Laboratorio Regionale SuperMat, CNR/INFM Salerno and Dipartimento di Fisica "E.R. Caianello", Università degli Studi di Salerno, 84081 Baronissi (Sa), Italy, attanasio@sa.infn.it

A. Barone Dipartimento di Scienze Fisiche, Università di Napoli Federico II and CNR-SPIN, Piazzale Tecchio 80 80125 Napoli, Italy, antonio.barone@na.infn.it

C. Bell Department of Advanced Materials Science, University of Tokyo, Kashiwa, Chiba 277-8651, Japan, cbell@k.u-tokyo.ac.jp

S.J. Bending Department of Physics, University of Bath, Claverton Down, Bath BA2 7AY, B&NES, UK, s.bending@bath.ac.uk

C. Berthod DPMC-MaNEP, University of Geneva, 24 quai E.-Ansermet, 1211 Geneva 4, Switzerland, christophe.berthod@unige.ch

P.L. Christiansen Department of Informatics and Mathematical Modelling, Technical University of Denmark, Fysikvej, Bldg. 309, 2800 Kgs. Lyngby, Denmark, plc@imm.dtu.dk
and
Department of Physics, Technical University of Denmark, 2800 Kgs. Lyngby, Denmark

C. Cirillo Laboratorio Regionale SuperMat, CNR/INFM Salerno and Dipartimento di Fisica "E.R. Caianello", Università degli Studi di Salerno, 84081 Baronissi (Sa), Italy, carla@sa.infn.it

Y. Fasano Instituto Balseiro and Centro Atómico Bariloche, Av. Bustillo km 9500, R8402AGP San Carlos de Bariloche, Argentine, yanina.fasano@cab.cnea.gov.ar

Ø. Fischer DPMC-MaNEP, University of Geneva, 24 quai E.-Ansermet, 1211 Geneva 4, Switzerland, oystein.fischer@unige.ch

M. Flokstra Kamerlingh Onnes Laboratory, Leiden University, P.O. Box 9504, 2300 RA Leiden, The Netherlands, flokstra@physics.leidenuniv.nl

N. Jenkins DPMC-MaNEP, University of Geneva, 24 quai E.-Ansermet, 1211 Geneva 4, Switzerland, nathan.jenkins@unige.ch

J. Kačmarčík Centre of Very Low Temperature Physics Košice at the Institute of Experimental Physics, Slovak Academy of Sciences, Watsonova 47, 04001 Košice, Slovakia, kacmarci@saske.sk

J. Koláček Institute of Physics, Academy of Sciences, Cukrovarnická 10, 162 53 Prague 6, Czech Republic, kolacek@fzu.cz
and
Faculty of Mathematics and Physics, Charles University, Ke Karlovu 3, 12116 Prague 2, Czech Republic

F.V. Kusmartsev Department of Physics, Loughborough University, Loughborough, LE11 3TU, UK, F.Kusmartsev@lboro.ac.uk

W. Lang Faculty of Physics, Electronic Properties of Materials, University of Vienna, Boltzmanngasse 5, 1090 Wien, Austria, wolfgang.lang@univie.ac.at

J. Linder Department of Physics, Norwegian University of Science and Technology, 7491 Trondheim, Norway, jacob.linder@ntnu.no

P. Lipavský Institute of Physics, Academy of Sciences, Cukrovarnická 10, 162 53 Prague 6, Czech Republic
and
Faculty of Mathematics and Physics, Charles University, Ke Karlovu 3, 12116 Prague 2, Czech Republic, lipavsky@fzu.cz

F. Lombardi Quantum Device Physics Laboratory, Department of Microtechnology and Nanoscience, Chalmers University of Technology, SE-412 96 Göteborg, Sweden, Floriana.lombardi@chalmers.se

S. Madsen Mads Clausen Institute, University of Southern Denmark, Alsion 2, 6400 Sønderborg, Denmark, soeren@madsen2themax.dk

I. Maggio-Aprile DPMC-MaNEP, University of Geneva, 24 quai E.-Ansermet, 1211 Geneva 4, Switzerland, ivan.maggio-aprile@unige.ch

M.V. Milošević Department of Physics, Universiteit Antwerpen, Groenenborgerlaan 171, 2020 Antwerpen , Belgium, milorad.milosevic@ua.ac.be

K. Morawetz University of Applied Science Münster, Stegerwaldstrasse 39, 48565 Steinfurt, Germany
and
International Center for Condensed Matter Physics, Universidade de Brasilia, 70904-910, Brasília-DF, Brazil, klaus.morawetz@physik.tu-chemnitz.de

V.V. Moshchalkov Department of Physics and Astronomy, Institute of Nanoscale Physics and Chemistry, Katholieke Universiteit Leuven, Celestijnenlaan 200 D, 3001 Leuven, Belgium, victor.moshchalkov@fys.kuleuven.be

J.D. Pedarnig Institute of Applied Physics, Johannes Kepler University, Altenbergerstrasse 69, 4040 Linz, Austria, johannes.pedarnig@jku.at

N.F. Pedersen Department of Mathematics, Technical University of Denmark, Matematiktorvet 303 S, 2800 Kgs. Lyngby, Denmark, nfp@elektro.dtu.dk

A. Piriou DPMC-MaNEP, University of Geneva, 24 quai E.-Ansermet, 1211 Geneva 4, Switzerland, alexandre.piriou@unige.ch

Z. Pribulová Centre of Very Low Temperature Physics Košice at the Institute of Experimental Physics, Slovak Academy of Sciences, Watsonova 47, 04001 Košice, Slovakia, pribulov@saske.sk

J.G. Rodrigo Laboratorio de Bajas Temperaturas, Departamento de Física de la Materia Condensada, Instituto de Ciencia de Materiales Nicolás Cabrera, Facultad de Ciencias, Universidad Autónoma de Madrid, Módulo 03, 28049 Madrid, Spain, jose.rodrigo@uam.es

M. Saarela Department of Physical Sciences, University of Oulu, P. O. Box 3000, 90014 Oulu, Finland, Mikko.Saarela@oulu.fi

P. Samuely Centre of Very Low Temperature Physics Košice at the Institute of Experimental Physics, Slovak Academy of Sciences, Watsonova 47, 04001 Košice, Slovakia, samuely@saske.sk

A.V. Silhanek Department of Physics and Astronomy, Institute of Nanoscale Physics and Chemistry, Katholieke Universiteit Leuven, Celestijnenlaan 200 D, 3001 Leuven, Belgium, alejandro.silhanek@fys.kuleuven.be

A. Sudbø Department of Physics, Norwegian University of Science and Technology, 7491 Trondheim, Norway, asle.sudbo@ntnu.no

H. Suderow Laboratorio de Bajas Temperaturas, Departamento de Física de la Materia Condensada, Instituto de Ciencia de Materiales Nicolás Cabrera, Facultad de Ciencias, Universidad Autónoma de Madrid, Módulo 03, Universidad Autónoma de Madrid, 28049 Madrid, Spain, hermann.suderow@uam.es

P. Szabó Centre of Very Low Temperature Physics Košice sice at the Institute of Experimental Physics, Slovak Academy of Sciences, Watsonova 47, 04001 Košice, Slovakia, pszabo@saske.sk

F. Tafuri Dipartimento Ingegneria dell'Informazione, Seconda Università di Napoli and CNR-SPIN, Via Roma 29, 81031 Aversa (CE), Italy, francesco.tafuri@na.infn.it

J. Van de Vondel Department of Physics and Astronomy, Institute of Nanoscale Physics and Chemistry, Katholieke Universiteit Leuven, Celestijnenlaan 200 D, 3001 Leuven, Belgium, joris.vandevondel@fys.kuleuven.be

J.M.v.d. Knaap Kamerlingh Onnes Laboratory, Leiden University, P.O. Box 9504, 2300 RA Leiden, The Netherlands, knaap@physics.leidenuniv.nl

S. Vieira Laboratorio de Bajas Temperaturas, Departamento de Física de la Materia Condensada, Instituto de Ciencia de Materiales Nicolás Cabrera, Facultad de Ciencias, Universidad Autónoma de Madrid, Módulo 03, Universidad Autónoma de Madrid, 28049 Madrid, Spain, sebastian.vieira@uam.es

D. Winkler Department of Microtechnology and Nanoscience – MC2, Chalmers University of Technology, 41296 Göteborg, Sweden, dag.winkler@chalmers.se

R. Wördenweber Institute of Bio- and Nanosystems (IBN) and JARA-Fundamentals of Future Information Technology, Forschungszentrum Jülich GmbH, 52425 Jülich, Germany, r.woerdenweber@fz-juelich.de

A. Yurgens Department of Microtechnology and Nanoscience – MC2, Chalmers University of Technology, 41296 Göteborg, Sweden, yurgens@chalmers.se

Chapter 1
Guided Vortex Motion and Vortex Ratchets in Nanostructured Superconductors

Alejandro V. Silhanek, Joris Van de Vondel, and Victor V. Moshchalkov

Abstract In type II superconductors, an external magnetic field can partially penetrate into the superconducting phase in the form of magnetic flux lines or vortices. The repulsive interaction between vortices makes them to arrange in a triangular lattice, known as Abrikosov vortex lattice. This periodic vortex distribution is very fragile and can be easily distorted by introducing pinning centers such as local alterations of the superconducting condensate density. The dominant role of the vortex-pinning site interaction not only permits to control the static vortex patterns and to enhance the maximum dissipationless current sustainable by the superconducting material but also allows one to gain control on the dynamics of vortices. Among the ultimate motivations behind the manipulation of the vortex motion are the better performance of superconductor-based devices by reducing the noise in superconducting quantum interference-based systems, development of superconducting terahertz emitters, reversible manipulation of local field distribution through flux lenses, or even providing a way to predefine the optical transmission through the system. In this chapter, we discuss two relevant mechanisms used in most envisaged fluxonics devices, namely the guidance of vortices through predefined paths and the rectification of the average vortex motion. The former can be achieved with any sort of confinement potential such as local depletion of the order parameter or local enhancements of the current density. In contrast, rectification effects result from the lack of inversion symmetry of the pinning landscape which tends to favor the vortex flow in one particular direction. We also discuss a new route for further flexibility and tunability of these fluxonics components by introducing ferromagnetic pinning centers interacting with vortices via their magnetic stray field.

1.1 Introduction

Almost a century has past since superconductivity was discovered by K. Onnes in 1911. For many years, this unique state of matter attributed to a net attraction between electrons was just a rare curiosity occurring at temperatures which seemed too low to be of any practical importance. Vigorous efforts have nevertheless inspired several important applications such as stable magnetic fields crucial for magnetic resonance imaging, detection of very small magnetic fields like those generated by brain activity using Superconducting QUantum Intereformeter Devices (SQUIDs), or levitating trains that can speed up to 500 km/h and, more recently, the high temperature superconductors motivated a new generation of self-healing superconducting electricity grids, which would save up to 15% of the transported electricity.

To ensure successful superconductor-based technologies, materials with a superior performance can be developed by manipulating the appropriate elementary building blocks through nanostructuring. For superconductivity, such elementary blocks are quantized units of magnetic flux or fluxons and Cooper pairs. Most of the electromagnetic response of superconductors to low frequency excitations (much lower than the superconducting gap) is dominated by the dissipative motion of fluxons. This is the fundamental reason for the continuous strive to comprehend the ultimate mechanisms ruling the vortex dynamics, which has made this particular topic an active theoretical and experimental line of research.

The manipulation and control of fluxons is nowadays known as fluxonics, in analogy with the control of electrons in electronics. In this chapter, we focus our attention on two particular fluxonic effects, vortex channeling and vortex ratchets, which allow a directional and orientational degree of control of the net vortex motion, respectively. In the next section, we introduce the equation of motion for a vortex in a pinning potential landscape. Section III is aimed to highlight the most relevant characteristics and fingerprint of the guided vortex motion. This phenomenon has been already addressed in the reviews of Vélez et al. [1] and Aladyshkin et al. [2] mainly for the case of magnetic pinning centers. In Sect. IV, we discuss rectification of the vortices. To guide the reader, we will start with a general introduction into the concept of particle rectification. A more comprehensive study of ratchet phenomena in general can be found in the reviews of Reimann [3] and Hänggi [4]. In the second part of this section, we will discuss in more detail the relevant parameters determining the rectification of vortices and their tunability, making these type of ratchets of particular interest for the study of ratchet systems in general.

We would like to note that the literature and references used by the authors in this chapter by no means can be considered as a complete set. Due to the dynamic and rather complex character of the subject and also to the limited space, inevitably important and interesting contributions could have been missed and, therefore, in a way, the references used in this chapter reflect the working list of publications the authors of this article are dealing with.

1.2 Equation of Motion

The pioneer work of Giaever [5] provided the first experimental evidence that dc electrical resistance in type II superconductors is a direct consequence of the motion of Abrikosov vortices. To some extent, a vortex can be imagined as a rigid entity which reacts as a whole under external excitation, due to the quantization condition. This argument is the foundation stone for justifying the description of the vortex dynamics using particle-like equations of motion [6–8]. In this approximation, we can write the equation of motion of a rigid vortex as

$$m\ddot{r} = F_{\mathrm{v-v}} + F_{\mathrm{v-p}} + F_{\mathrm{L}} + F_{\mathrm{M}} - \eta\dot{r} + F_{\mathrm{Th}} \tag{1.1}$$

where m is effective mass of a vortex per unit length, $\mathbf{r}$ is the vortex position, $F_{\mathrm{v-v}}$ is the interaction with neighboring vortices, $F_{\mathrm{v-p}}$ is vortex-pinning force, $F_{\mathrm{L}} = j \times \mathbf{\Phi_0}$ is the "Lorentz force" [9] produced by an external current $\mathbf{j} = en_s\mathbf{v_s}$, $\mathbf{v_s}$ and n_s are the superfluid velocity and the density of superconducting electrons, respectively, $F_{\mathrm{M}} = \alpha\mathbf{\Phi_0} \times \dot{\mathbf{r}}$, with α the Magnus force coefficient (the Magnus coefficient has different values in different models [10]), η is the viscosity experienced by the vortex line when moving through the superconductor, and F_{Th} represents the random thermal noise. There are several mechanisms that may contribute to the inertial mass of a vortex such as variations of the order parameter [11], time-dependent strain fields induced when a vortex moves through a deformable superconductor [12], transverse displacement of the crystaline lattice [13], or discretization of the electronic states inside the vortex core [14, 15]. In general, the vortex mass per unit length amounts to several thousands of the electron masses and represents a small contribution usually neglected in the calculations.

The ultimate mechanism for the damping coefficient η is a controversial issue. The most popular explanation is that related to the electric field needed to maintain a cycloidal motion of electrons when a vortex moves [16]. In addition, Maxwell's equations tell us that a moving flux line with velocity v should induce locally an electric field $\mathbf{E} = -\mathbf{v} \times \mathbf{B}$, where $\mathbf{B}$ is the magnetic field. As discussed by Volger [17], this field generates normal eddy currents giving a contribution far smaller than that observed experimentally [18]. Tinkham [19] proposed an additional source of damping associated with the relaxation of the order parameter as a flux line pass by. Yet another cause of flow viscosity was later on proposed by Clem [20] and is associated with the local temperature gradients generated by normal-like moving regions. In this case, as the vortex moves the leading edge of the vortex drives the superconducting material to normal and therefore the entropy increases, whereas the opposite process occurs at the trailing edge of the vortex, the resulting heat flow is the cause of vortex dissipation. As already pointed out by Suhl [11], the relaxation time of a vortex $\tau \sim \eta/m$ is of the order of the picoseconds, and therefore the dc dynamics of vortices can be well described ignoring the vortex mass. It is worth noticing that vortices not only carry flux but can also give rise to local charge separation as a consequence of the centrifugal Lorentz forces acting on the carriers circulating around the vortex core [21, 22].

The existence of a hydrodynamic force F_{M} experienced by a vortex moving through a medium of free electrons (Magnus force) was first considered by de Gennes and Matricon [23]. Later on, Bardeen [24] pointed out that the theoretical arguments used for justifying such a force were incorrect, and therefore the Magnus force should not exist. More recently, a microscopic derivation of the Magnus force has been put forward by Ao and Thouless [25–27]. A direct comparison between Magnus and Lorentz force gives $F_M/F_L = \alpha v_{vortex}/nev_s$; since α is of the order of ne [10] for low enough vortex velocities $v_{vortex} \ll v_s$, we can neglect the effects of the Magnus force. The thermal force F_{Th} can be assumed to be a Gaussian white noise of stochastic correlation relations $\langle F_{Th}(t)\rangle_t = 0$ and $\langle F^i_{Th}(t)F^j_{Th}(t')\rangle_t = 2\eta k_B T\delta_{ij}\delta(t-t')$, where k_B is the Boltzmann constant and $\langle ...\rangle_t$ denotes the average in time.

The interaction of vortex i with the rest of the vortices is $F^i_{v-v} = \sum^N_{j\neq i} \nabla_i U(r_{ij})$, where $U(r_{ij})$ is the effective vortex–vortex interaction potential. For a thin film superconductor, the intervortex interaction occurs mainly through the stray field outside the film and has a long range potential $U(r) \propto 1/r$ for $r \gg \Lambda$ and $U(r) \propto \ln(r)$ for $r \leq \Lambda$, where $\Lambda = \lambda^2/d$, λ is the bulk penetration depth and d is the film thickness [28, 29]. The interaction of a flux line with a pinning potential $U_p(r)$ is simply $F_{v-p} = -\nabla U_p(r)$. As we will discuss in the following sections, not only the depth and the typical length scale of the pinning potential but also its symmetry plays an important role in determining the vortex dynamics.

It is important to emphasize that the simplified model of a vortex as a point-like classical particle, which ignores the internal structure of the vortices and their elastic nature, fails in the limit of high vortex velocities where more realistic approaches such as time-dependent Ginzburg–Landau formalism become necessary [30]. Other effects not taken into account in the above approach are self-heating effects, self-field effects, depletion of the order parameter due to the external current, just to name a few.

The rapid development of lithographic techniques during the last decades has facilitated the fabrication of predefined pinning patterns at the submicrometer scales which has attracted a considerable attention mainly due to the flexibility of their designs [31]. A review of the typical fabrication procedures using these lithographic techniques can be found in [32]. When these nanoengineered pinning centers are arranged periodically, clear matching effects emerge as a consequence of commensurate vortex states in the periodic pinning potential. Although the formation of static vortex phases in periodic potential landscapes has been an active topic of investigation both theoretically [33] and experimentally [34], less attention has been devoted to the vortex dynamics in periodic pinning arrays. In the following section, we will demonstrate that periodic arrays of pinning sites can guide the motion of vortices in such a way that their average velocity may not be collinear with the Lorentz force. As we will discuss in detail below, this guided vortex motion leads to a very unusual field polarity independent transverse (Hall) signal in transport measurements.

1.3 Guided Vortex Motion

1.3.1 Transverse Electric Field and Guided Vortex Motion

1.3.1.1 Pinning-Free Superconductors

Let us now consider the steady state-solution (i.e. $m = 0$) to (1.1) in the simplest case of one single vortex line in a superconductor without pinning at $T = 0$ under an external current $\mathbf{j}$. The system of coordinates and the arrangement of transport current and vortex velocity are illustrated in Fig. 1.1. In this limit, the equation of motion becomes $0 = F_L + F_M - \eta\dot{r}$ and its solution is [35],

$$\dot{\mathbf{r}} = \frac{\eta}{\alpha^2\Phi_0^2 + \eta^2}\mathbf{j} \times \mathbf{\Phi}_0 + \frac{\alpha\Phi_0}{\alpha^2\Phi_0^2 + \eta^2}\mathbf{j} \tag{1.2}$$

The resulting induced electric field $\mathbf{E} = \mathbf{B} \times \dot{\mathbf{r}}$ has components,

$$E_{||} = \frac{\eta\Phi_0 B}{\alpha^2\Phi_0^2 + \eta^2}j \qquad E_{\perp} = -\frac{\alpha\Phi_0^2 B}{\alpha^2\Phi_0^2 + \eta^2}j \tag{1.3}$$

parallel and perpendicular to the current direction, respectively. In this equations, both ϕ_0 and B carry the same sign indicating the polarity of the applied magnetic field, B. Note that the longitudinal component $E_{||}(\sim\Phi_0 B)$ does not change sign when inverting the external field polarity, whereas the transversal component (or Hall signal) $E_{\perp}(\sim\Phi_0^2 B)$ is an odd function of the external field polarity like the standard Hall effect. As we will see below, the antisymmetry of $E_{\perp}$ with respect to B arising from these sort of phenomena is a crucial feature to separate their contribution from other effects such as the guided vortex motion.

1.3.1.2 Superconductors with One-Dimensional Pinning

The simplest situation to consider when introducing pinning forces is that corresponding to a random distribution of pinning centers. In this case, the pinning force is isotropic and only modifies the previous results by providing a threshold current below which no vortex motion takes place. A more interesting vortex dynamics is obtained when considering anisotropic pinning potentials such as the one described by $\mathbf{F_p} \cdot \mathbf{n} = 0$ and $|\mathbf{F_p} \times \mathbf{n}| = F_0 \neq 0$, where $\mathbf{n}$ is a unit vector (Fig. 1.1). This sort of potential can be imagined as one-dimensional channels of easy vortex flow (i.e. zero pinning force) separated by potential barriers which give rise to a finite critical current.

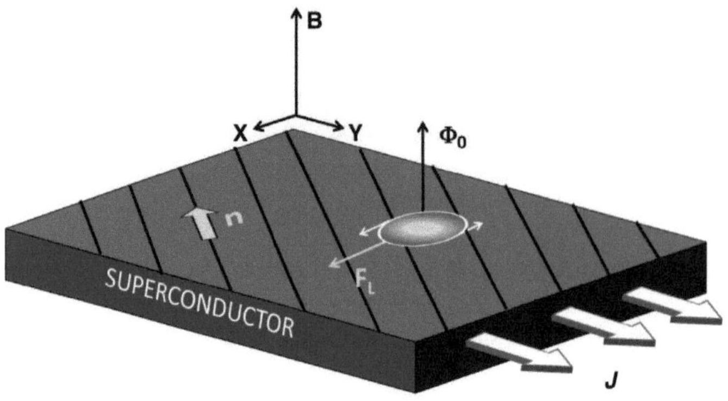

Fig. 1.1 Schematic representation of the Lorentz force F_L on a vortex in a thin-film superconductor carrying a superconducting current J. The magnetic field B is applied perpendicular to the plane of the film; the unit vector **n** indicates the direction of easy flow produced by linear modulations of the order parameter (*black stripes*)

For applied currents such that $|(\mathbf{j} \times \Phi_0) \times \mathbf{n}| < F_0$, the equation of motion has to be projected along the direction of **n** and the components of the electric field become [35],

$$E_{||} = \frac{\Phi_0 B \sin^2\theta}{\eta} j \qquad E_{\perp} = -\frac{\Phi_0 B \sin 2\theta}{2\eta} j \tag{1.4}$$

where θ is the angle between the external current and **n**. Note that in this regime of total confinement where vortices are forced to flow parallel to **n**, the Magnus force plays no role. More importantly, now the transverse component of the electric field is an even function $(\Phi_0 B)$ of the external magnetic field. *This field polarity independence of the transverse electric field is a unique property of vortex channeling* that allows one to discriminate it from other effects. For high enough applied currents such that $|(\mathbf{j} \times \Phi_0) \times \mathbf{n}| > F_0$, vortices can leak through the barriers and the complete equation of motion must be solved.

1.3.2 Experimental Results and Theoretical Investigations

1.3.2.1 Superconductors with One-Dimensional Pinning

A nanoengineered one-dimensional array of pinning centers obtained by patterning a periodic distribution of submicrometer lines made of a magnetic material (Ni) in direct contact with a superconducting Nb film was fabricated by Jaque et al. [36]. Here, the pinning potential is produced by the local depletion of the superconducting order parameter as a result of the exchange interaction or proximity effect [37]. In the mixed state, these lines induce a strong anisotropy of the vortex motion, with the hard-axis of the vortex motion perpendicular to the lines. More recently, the effect of easy vortex flow along channels of depleted superconducting order parameter has been also achieved in bilayer superconductor/ferromagnet systems [38–40]. In this case, the stray field emanating from the magnetic domain walls act as an effective pinning center and also favor the guidance of vortices. As we anticipated above and originally discussed by Daldini et al. [41], irrespective of the ultimate nature of the one-dimensional periodic arrays of pinning centers, there is a local enhancement of the critical current every time that the vortex lattice is commensurate with the pinning landscape, i.e. when the vortex lattice parameter matches the period of the pinning array. These commensurability effects become much more relevant in two-dimensional arrays [42].

In a more historical context, the first experimental investigations of guided vortex motion were carried out by Niessen et al. [43–45] in cold rolled Nb–Ta alloys. More than 20 years later, Kopelevich et al. [46] reported on the presence of a even-Hall effect, i.e., a field polarity-independent transverse voltage, in high-T_c superconductors. Similar occurrence of a transverse voltage peak at the onset of the superconductor-normal transition was later observed by Francavilla and co-workers [47, 48] in several superconductors such as Nb, NbN, NbCN, PbBi, and $YBa_2Cu_3O_7$. The original interpretation of this effect [47, 49] was based on Glazman's model [50] according to which the penetration of vortices through one border of the superconductor and antivortices on the opposite border, due to the self field produced by a bias current, tends to annihilate following curved trajectories which give rise to a transverse voltage. This effect is schematically illustrated in Fig. 1.2.

One-dimensional periodic modulations of the order parameter are also naturally produced by the layered structure due to weakly coupled CuO planes of oxide superconductors. The current-voltage characteristics in these systems were theoretically investigated by Chen and Dong [51] considering the particular case of a thermally activated single-vortex moving perpendicular to a one-dimensional

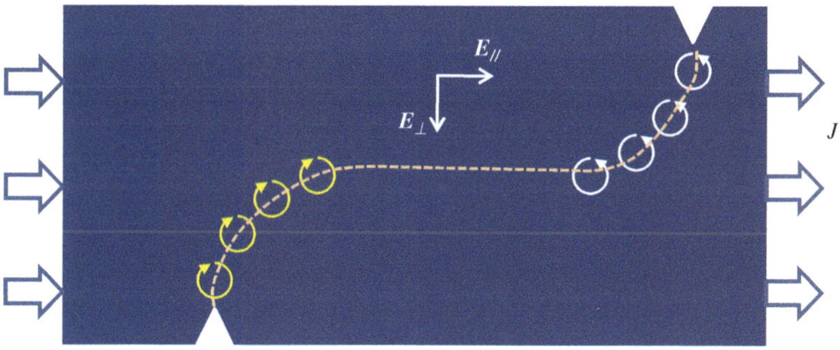

Fig. 1.2 Vortex and antivortex trajectories generated by the self-field of a bias current J and penetrating into the superconductor through small indentations at the sample's borders. According to Glazman model, the attraction between vortices and antivortices gives rise to a curved trajectory toward the annihilation point which produces a transverse electric field $E_\perp$

pinning potential with sinusoidal dependence. These authors demonstrated that at high temperatures the particles can reach a steady flux flow regime where the potential landscape plays no role, whereas at low temperatures the dominant dissipation mechanism is due to thermal activation over the potential barriers (flux creep).

There is another source of vortex confinement in layered superconductors acting perpendicularly to the CuO planes. Indeed, since the high-T_c superconductor $YBa_2Cu_3O_7$ crystalizes in an orthorhombic structure with slightly different a and b axis, the presence of crystallographic defects such as twin-boundaries separating regions where a and b are interchanged is common. Because impurities accumulate in these planes, the superconducting order parameter is locally depleted along the planes acting as efficient pinning centers. In principle, the separation between planes does not follow any periodicity; however, it is possible to prepare single crystals with only one family of planes, all of them parallel and running from side to side of the sample. Interestingly, these unidirectionally, twinned samples exhibit strong ab-anisotropy in the transport properties not only in the superconducting state but also in the normal state [52]. Since the viscosity coefficient η is inversely proportional to the normal state resistivity, it is expected that the viscosity of the vortex flow becomes also anisotropic.

It has been experimentally demonstrated [53–55] that this sort of one dimensional pinning potential gives rise to an anisotropic dissipative response due to the fact that the vortex motion deviates from the direction of the applied force. These results can be explained using an anisotropic resistivity tensor with the principal axis determined by the direction of the twin boundaries. Note that as long as vortex motion is not perfectly guided along the planes, both components of the resistivity tensor will be finite. These results were later extended by D'Anna et al. [56] to demonstrate that the guided vortex motion starts deep in the vortex liquid phase, without sharp changes at the Bose-glass transition. In addition, these authors showed that tilting away the magnetic field from the twin planes by 4° tends to wash out the guidance effect.

Early theoretical calculations of the vortex dynamics in samples with one-dimensional pinning structures with thermal fluctuations were done by Mawatari [57, 58] using the Fokker–Planck equation for a stochastic probability of vortices. This model is applicable for the case of negligible vortex–vortex interaction, i.e., fields close to H_{c1}, and assuming a sinusoidal pinning potential. Mawatari showed that the transversal component of the electric field can exhibit a non-monotonic dependence as a function of the bias current. This effect results from the competition between guided vortex motion (at low currents) and the thermal hopping of depinned vortices (at high currents). In addition, both $E_{//}$ and $E_{\perp}$ do not always decrease with increasing pinning strength, but actually increase in certain angular range of the bias current with respect to the planes. This approximation was later on extended by Shklovskii et al. [59] to investigate the nonlinear regimes of vortex motion. Interestingly, in [59], it is shown that there is also a nontrivial contribution of the guided vortex motion to the Hall-like signal (i.e., odd signal respect to field polarity inversion). Similar effects were found via molecular dynamics calculation by Zhu et al. [60], Groth et al. [61], and within the time-dependent

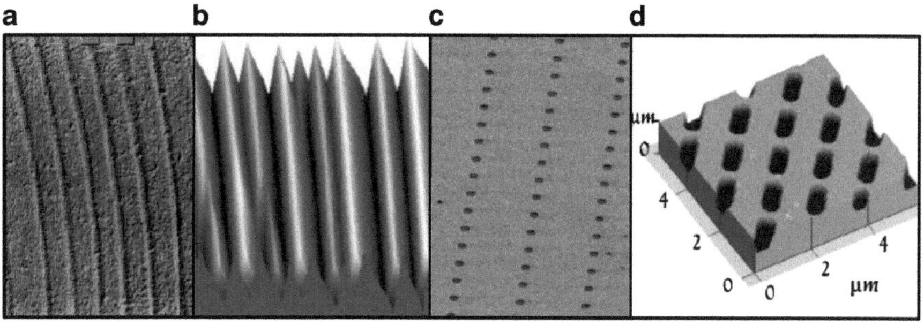

Fig. 1.3 Several examples of confining pinning potentials used for vortex channeling. (**a**) Optical photograph of mica irradiated with heavy ions through a kapton-gold handcraft mask. The period of the structure is about 20 μm. After Pastoriza and Kes [54]. (**b**) Scanning tunneling microscopy image of a Nb film deposited on top of a faceted Al_2O_3 substrate. The average periodicity is 400 nm. After Soroka et al. [65]. (**c**) Scanning electron microscopy image of a rectangular array of antidots with unit cell of 10 μm × 20 μm made on a high-T_c superconductor. After Wördenweber and Dymashevsky [71]. (**d**) Atomic force microscopy image of a square array with 1.5 μm period of rectangular antidots made on Pb. After Van Look et al. [72]

Ginzburg–Landau formalism by Crabtree et al. [62]. It is worth mentioning that the guided vortex motion produced by twin planes have implications not only on the electrical properties but also on the thermal properties of the system as demonstrated by Ghamlouch and Aubin [63].

Besides twin boundaries and intrinsic layers in superconducting CuO-based compounds, alternative ways of obtaining one-dimensional pinning centers can be achieved by modulations of the superconducting thickness using faceted substrates [64, 65], perfectly periodic thickness modulations [41], disordered paths produced by grain boundaries [35,66], or weak pinning channels separated by strong pinning enviroment [67–70]. Some examples of channeling potential landscapes are shown in Fig.1.3

1.3.2.2 Superconductors with Two-Dimensional Pinning

Guidance of vortices can also be obtained in two-dimensional structures. In the limiting case of a periodic array of pinning centers with a rectangular unit cell [73–75], it has been experimentally demonstrated that the in-plane transport properties are highly anisotropic. These findings are in agreement with earlier theoretical studies by Reichhardt et al. [76]. The experimental properties found in these systems were: (1) when the vortices move along the short side of the rectangular array, the vortex lattice matches the rectangular unit cell, whereas when they move along the long side of the rectangular array only a triangular lattice is observed [73]; (2) in a square pinning array, the vortices follow the direction of the principal axis of the array as long as the Lorentz force is within 15° away from these axis [77]; (3) for a rectangular array, this angle can amount to 85° beyond which the vortex velocity acquires a component away from the principal axis [74, 78]. Villegas et al. [78] showed that when the vortex lattice is moving along the channels formed by a rectangular array of pinning sites, there is an enhancement of the transverse pinning with respect to the static situation. Strikingly, in [76], it was theoretically predicted that a periodic pinning with a rectangular unit cell with a moderate aspect radio of 2 can lead to a huge current anisotropy with almost two orders of magnitude difference between maximum and minimum values, at fields equal to the second matching field. To the best of our knowledge, an experimental confirmation of this prediction is still lacking.

Numerical simulations of the vortex motion in square arrays of pinning centers predict that even in four-fold symmetric arrays, a guided motion of vortices along the symmetry axis of the square array should take place [79, 80]. Related investigations for two orthogonal systems of washboard potential (i.e., square array of pinning centers) was done by Shklovskij and Soroka [81] using a planar stochastic model. (It is worth mentioning that not only the symmetry of the unit cell is relevant for guiding vortices but also the symmetry of the individual pinning centers). Indeed, Van Look et al. [72] showed that a square array of rectangular antidots or blind holes gives rise also to a distinct anisotropic vortex dynamics.

In principle, to achieve an effective guidance of vortices, it is necessary that the distance between pinning centers along the line of guidance is of the same order as the superconducting coherence length (in the case of core pinning) or the penetration depth (in the case of electromagnetic pinning). It has been demonstrated by Wördenweber et al. [71, 82] that antidots, as big as $\sim 1\,\mu m$ in diameter separated by several micrometers in $YBa_2Cu_3O_7$ where $\xi \sim 1.5$ nm and the effective penetration depth is $\lambda \sim 1\,\mu m$, can indeed guide the motion of vortices. Moreover, in this high-T_c superconductor, the vortex hopping from antidot to antidot happens for all values of current down to zero external bias. This absence of critical current might be indicative of a new mechanism of flux propagation such as flux nucleation within the antidots due to the redistribution of screening currents. More recently, the deep control on the guided motion of vortices has allowed to manipulate and gain further understanding in the long standing puzzling phenomena of anomalous Hall effect [83, 84]. Direct visualization of vortex channeling in periodic arrays of pinning centers has been achieved via Lorentz microscopy [34], magneto-optical effect [85, 86], or laser scanning microscopy [87]. The resulting images are summarized in Fig. 1.4.

As we have shown above, the nanopatterning of the superconducting samples has demonstrated to be capable of modifying the macroscopic response of the system. However, the internal properties of these devices at the microscopic level are predefined, leaving little margin for subsequent adjustments. A new, alternative route toward more flexible quantum-designed hybrid nanomaterials can be obtained by exploiting the interplay of two competing collective phenomena, namely superconductivity (S) and ferromagnetism (F).

This idea was already envisaged and investigated theoretically by Carneiro [88] who considered an array of magnetic dipoles in close proximity to a superconducting film. Carneiro showed that the average vortex drift velocity depends on the

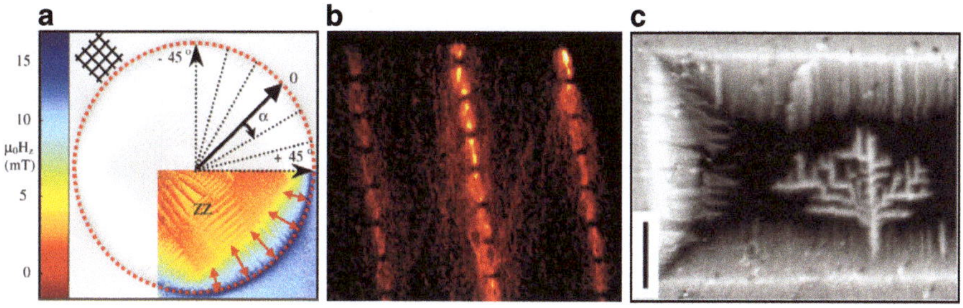

Fig. 1.4 (**a**) Vortex channeling imaged by magneto-optical Faraday effect in a disk-shaped $YBa_2Cu_3O_7$ film (diameter of 2 mm) with a *square array* of antidots as indicated by the grid at the top left. The *color code* corresponds to the out-of-plane component of the local field. After Pannetier et al. [85] (**b**) Scanning laser microscopy image taken at T = 86 K in an area of $180 \times 180\,\mu m^2$ of a $YBa_2Cu_3O_7$ with a *rectangular array* of antidots. Here, a bias current above the critical current flows from left to right. After Lukashenko et al. [87]. (**c**) Vortex avalanches observed by magneto optical imaging at $H = 1.5$ mT and $T = 6$ K in a Pb film with square array of antidots. The observed tree-like dendritic outburst follows the main directions of the *square pattern* and coexists with smooth flux penetration. The *vertical scale bar* corresponds to 0.5 mm. After Menghini et al. [86]

orientation of the magnetic moments and can be tuned by rotating the dipoles. The first experimental realization of this effect was done by Verellen et al. [89,90] using a checkerboard array of square magnetic loops underneath a superconducting film. It was shown that this system provides a high degree of flexibility since the guided vortex motion can be rerouted in two perpendicular orientation or even turned off by simply changing the magnetic state of the rings (Fig.1.5). More recently, it has been shown that if these magnetic dipoles are brought to distances shorter than the superconducting penetration depth, a very effective vortex channeling is obtained [91–93]. In this case, it might also occur that the vortex dynamics becomes fully dominated by the internal vortices generated by the magnetic template rather than by the vortices induced by the external magnetic field [92, 93].

1.4 Ratchets

A general definition of a ratchet can be given as follows: *Directed transport in spatially periodic systems far from equilibrium under alternating excitation, without the need of a non-zero applied force and/or temperature gradients.* Rectification of nanoparticles has attracted recently a lot of attention in biology and physics [3, 4, 94, 95]. In biology, typical examples are ion channels in cell membranes [96] and biological motors, whereas in physics one can find particle [97], charge [98], and vortex ratchets. In spite of the broad variety of ratchet systems, the underlying principles of these effects have many remarkable similarities which make it possible to use new findings, discovered for a given type of the devices,

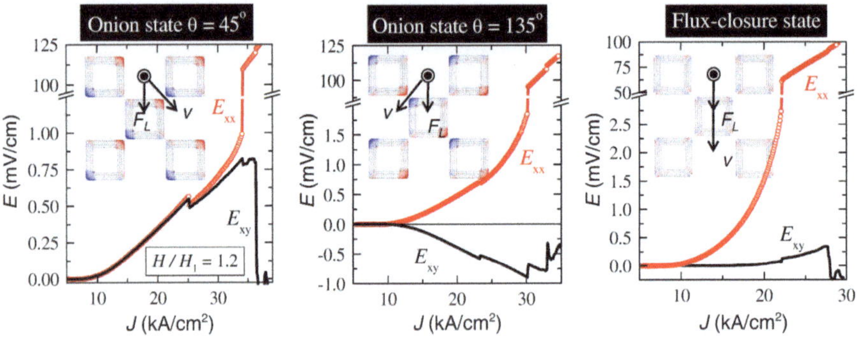

Fig. 1.5 Parallel (E_{xx}) and transverse (E_{xy}) electric field as a function of current density J at $T/T_c = 0.89$. The *leftmost panel* corresponds to the square rings magnetized at 45° as indicated in its inset. For low enough currents $J < 8$ MA/cm^2 the vortex lattice remains pinned as no dissipation is detected in either direction $E_{xx} = E_{xy} \sim 0$. Surprisingly, for currents higher than the critical current, the direction of vortex motion does not coincide with the Lorentz force. More specifically since $E_{xx} \approx E_{xy}$ vortices move at an angle of 45° as indicated in the corresponding inset. This is a clear evidence that the net displacement of the vortices is along the high symmetry axis of the magnetic pinning landscape. This situation persists up to $J_{c2} \sim 25$ MA/cm^2 where E_{xy} and E_{xx} gradually separate from each other, which means that now the component of the Lorentz force perpendicular to the channels is large enough to overcome the caging potential. The *middle panel* presents the data for the square rings magnetized at 135°, as indicated in the inset. The *rightmost* frame corresponds to the squares in the flux-closure state, as indicated in the inset. After Verellen et al. [89, 90]

for controlling the motion of nanoparticles in other devices. The rectification of magnetic flux quanta (vortices) in superconductors was proposed by Lee et al. [99]. The particular interest of this kind of rectification is the removal of unwanted magnetic flux in superconducting devices. These kind of trapped flux can be induced by fields as small as the earth magnetic field and is responsible for inherent noise levels which drastically decrease the performance of several superconducting devices. As pointed out in Sect. 1.1 recently developed lithographic techniques opened the possibility of shaping the vortex pinning potential landscape. This allows us to create, for example, specific asymmetric pinning potentials necessary for the rectification of vortices. Besides the broad variety of possible vortex rectifiers, the tunability of this particular system of interacting particles makes it an ideal candidate to study ratchets in general. The particle density can easily be adjusted by simply changing the magnetic field, while the interaction between particles can be changed continuously by sweeping temperature. These benefits have led to a very active theoretical and experimental investigation of superconducting vortex rectifiers. In the following paragraphs, we will give a brief overview of the different systems explored in this fascinating field of research.

1.4.1 Basic Ingredients

The starting point of the theoretical description of vortex rectifiers is the equation of motion given by (1.1), in which two restrictions have to be satisfied. First of all, the *symmetry* of the system has to be *broken*. In most types of vortex ratchets, a pinning potential, which lacks centrosymmetry is introduced. As a result, the reflection symmetry of the pinning force is broken along certain direction ($f_{pos} \neq f_{neg}$). In addition, the second law of thermodynamics states that if the system is kept in thermal equilibrium, no directed transport occurs, $\langle x(t) \rangle = 0$. Therefore, a second ingredient has to be implemented in (1.1) to bring the vortex system *out of equilibrium*. A possible realization of this effect is the thermal ratchet. In this type of ratchet, a periodic oscillation of temperature between two values drives the system out of equilibrium and as a result leads to directed transport of the Brownian particles. Many other possibilities exist. Nevertheless, a classification of all ratchet systems, depending on their working principle, is beyond the scope of this text. Therefore, we will focus on the ratchet type mainly realized in vortex physics: *Rocking ratchets*. In this type of ratchet systems, a time periodic ac force of zero mean value drives the system out of equilibrium, allowing it to have a net motion of particles. In the pioneering work of Lee et al. [99], the authors first described the motion of vortices in an asymmetric sawtooth potential (Fig.1.6a) by solving (1.1), neglecting any temperature fluctuations ($F_{Th} = 0$).

In the case of a single particle, meaning no vortex–vortex interaction ($F_{v-v} = 0$), an analytical result can be obtained [99]. These results are shown in Fig.1.6b and give some of the basic features common to ratchet systems. The shape of the sawtooth potential introduces two distinct pinning forces for both directions ($f_{pos} < f_{neg}$). If the particle is submitted to a square-shaped ac driving force, three different dynamical regimes can be obtained depending on the amplitude of the driving force with respect to the pinning forces. If $f < f_{pos} < f_{neg}$, all vortices are pinned and no motion occurs (Pinned vortex lattice). If $f_{pos} < f < f_{neg}$, the particles can move to the easy direction but cannot overcome the hard direction. This region of forces is called the rectification window, and a strong increase in the average velocity is observed with increased applied current. The width of the rectification window is a direct indication of the asymmetry of the underlying pinning potential. Above f_{neg}, particles start to move in both directions resulting in a strong decrease in the rectification signal.

1.4.2 Experimental Considerations

Before summarizing the theoretical and experimental results on vortex ratchets, we will give a brief summary of the exact direction and sign of all present physical quantities for currents applied in the positive and negative x direction. For both cases, the interaction of a positive magnetic field, inducing vortices in the type-II superconducting film, with the applied current results in a Lorentz force as shown in

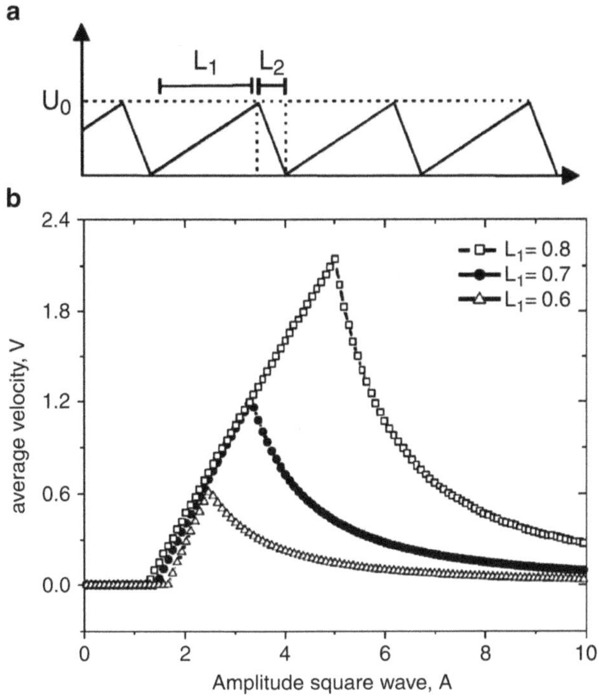

Fig. 1.6 (**a**) Schematic drawing of the sawtooth potential and its dependence on L_1, L_2 and U_0. (**b**) The average velocity as a function of the amplitude of the applied square wave, A, for different values of L_1 in the case of a normalized potential ($U_0 = 1$) and normalized periodicity ($L = 1$). After Lee et al. [99]

Fig.1.7a, b. Since the I_x and I_{-x} dc currents induce a Lorentz force in the negative and positive y directions, respectively, the vortices will probe the asymmetry of the pinning sites along the y-direction. If the Lorentz Force exceeds the pinning force, the present vortices start to move in the negative (positive) y-direction for positive (negative) applied currents. A consequence of the movement of vortices is the generation of an electric field, $\mathbf{E} = -\mathbf{v} \times \mathbf{B}$, with $\mathbf{B}$ the flux density and $\mathbf{v}$ the vortex velocity. A voltage drop, in the direction of the applied current, can be measured over the superconductor:

$$V = \mathbf{L} \cdot \mathbf{E} = \mathbf{L} \cdot (-\mathbf{v} \times \mathbf{B}) \tag{1.5}$$

with $\boldsymbol{L}$ a vector in the positive direction, with an amplitude equal to the distance between the voltage contacts. The direction of $\boldsymbol{L}$ is determined by the way the voltage drop is measured. In standard transport measurements, a voltage drop in the positive x direction will be read as a positive voltage readout. Thus, the voltage readout induced by the vortex motion is positive [$V(I_x)$] for positive applied current and negative [$V(I_{-x})$] for negative current. The difference between $V(I_x)$ and

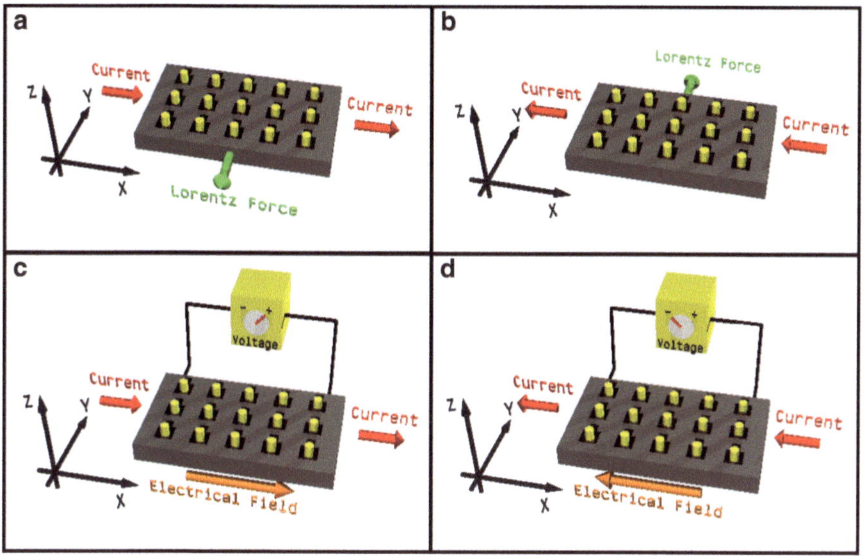

Fig. 1.7 Schematic drawings of all important physical quantities, when transport measurements are performed. Direction of the Lorentz force when a positive (**a**) and negative (**b**) current is applied. Direction of the electrical field and sign of the measured voltage drop when a positive (**c**) and negative (**d**) current is applied

$V(I_{-x})$ gives the necessary information about the asymmetry of the pinning centers in the y direction. Note that, in the case of an ac current, the voltage drop due to vortex motion is given by:

$$V = \mathbf{L} \cdot (-\langle \mathbf{v} \rangle \times \mathbf{B}), \tag{1.6}$$

similar to (1.5) except that $\langle v \rangle$ is now the average vortex velocity in one cycle. If the vortex density in the superconductor is known, we can calculate the average vortex velocity and its direction using (1.6). The dc voltage V_{dc} changes sign as H (~B) changes its polarity, meaning that vortices and antivortices are rectified to the same direction.

It is important to note that the frequencies, used in the experiments (~10 kHz), are well below the resonant frequencies due to vortices moving in a periodic potential (~100 Mhz). Therefore, in all discussed experiments each vortex travels thousands of unitcells per ac cycle, and any features in the rectification window due to discrete motion of vortices are not present.

1.4.3 Experimental Results and Theoretical Investigations

In theoretical investigations, the asymmetric pinning landscape was introduced in different ways. Ranging from a sawtooth potential (invariant in the y-direction) [99] to asymmetric vortex channels [100] and finally two-dimensional arrays of asymmetric pinning potentials [101, 102]. To obtain these types of pinning landscapes, a wide variety of experimental tools is now available, which allows researchers to shape different materials at the length scale of vortices (see Fig. 1.8). As a result, two-dimensional asymmetric pinning centers can be made by introducing a 2D array of asymmetric (blind)holes [e.g., big and small antidots placed close to each other [103, 104], triangles [105], asymmetric magnetic islands (Ni-triangles [106]) or asymmetric overdamped Josephson junction arrays [107], to name just a few]. An alternative route is the creation of an asymmetrical repulsive potential for the mobile vortices by the repulsive interaction with strongly pinned vortices [108] or the combination of amorphous-NbGe, an extremely weak-pinning superconductor, and NbN, with relatively strong pinning. In this case, the etched asymmetric channels inside the NbN act as repulsive walls for the vortices inside the NbGe [109]. The spatial symmetry of the underlying potential can also be broken by the asymmetrical distribution of symmetrical pinning sites. This was obtained in high T_c superconductors with a special arrangement of antidots [71] or exposed to an inhomogenous radiation dose [110] and in Al films with a distribution of Co/Pt multilayer dots with a linearly increasing size [111].

In all the aforementioned systems, the asymmetrical pinning potential rectifies the motion of vortices under an external ac excitation. In the following paragraphs, we will discuss some of the more interesting properties of a vortex ratchet, related to the particular nature of these systems.

1.4.3.1 Temperature Dependence

The temperature dependence of vortex ratchets is determined by a very subtle balance of different factors. The influence of temperature on a ratchet system is directly related to thermal excitations of the vortex lattice (i.e. $F_{\mathrm{Th}} \neq 0$). In Brownian

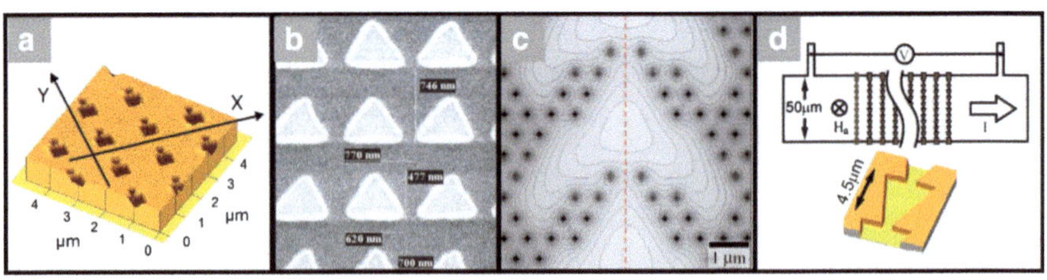

Fig. 1.8 Several designs used as vortex rectifiers. (**a**) A composite square array of pinning sites, with its unit cell consisting of a small and a big square antidot separated by a narrow superconducting wall. After Van de Vondel et al. [103]. (**b**) 2D array of submicron Ni triangles covered with a 100 nm thick Nb film. After Villegas et al. [106]. (**c**) Pattern can be regarded as a sequence of unirradiated *arrow-shaped* wedged cages, that work as microscopic "funnels" for vortex motion. After Togawa et al. [108]. (**d**) Channels fabricated from bilayer films of amorphous-NbGe, an extremely weak-pinning superconductor, and NbN, with relatively strong pinning. After Yu et al. [109]

ratchets, a decrease in the dc response as the temperature increases is expected [112]. The reason is that when the Brownian motion of particles is enhanced by thermal noise, more energy is needed to move them in a given direction. Indirectly, temperature has a threefold influence. First of all the size of a vortex, determined by $\lambda(T)$ and $\xi(T)$, increases with temperature. Therefore, an optimal temperature regime exists in which the vortex ratchet geometry can be fully explored. This effect was theoretically predicted in the work of Wambaugh et al. [100] and experimentally corroborated by de Souza Silva et al. [113]. Second, the vortex–vortex interaction is dependent on temperature, e.g., in the paper of Lee et al.:

$$F_{\mathrm{v-v}} = \frac{\phi_0^2 d}{8\eta^2\lambda^3} K_1 \left(\frac{(r_i - r_j)}{\lambda}\right) \bar{r_{ij}}, \tag{1.7}$$

with K_1 the Bessel function and d the film thickens. As a result, the interaction between vortices becomes long range at temperatures close to T_c. Finally, the pinning landscape itself will be altered by changing the temperature. In superconductors, random pinning at defects becomes more important at lower temperatures and the rectification efficiency will drastically decrease.

1.4.3.2 Sign Reversals at High Particle Densities

A clear advantage of vortex ratchets as an object of investigation compared to other systems is related with the electromagnetic nature of vortices. This gives us the possibility to vary continuously their density by changing the applied external magnetic field. Therefore, a lot of the research on vortex ratchet has been focused on their density dependence.

In a first approach, increasing the number of particles will lower their average velocity, due to smearing caused by the repulsive nature of vortices [99]. More interestingly, in the case of 2D pinning centers, the repulsive character of vortices creates so-called interstitial vortices above a certain particle density (i.e., magnetic field strength). As a result, a binary mixture exists of two types of vortices “feeling” an opposite asymmetry. Figure 1.9a shows the $V_{\mathrm{dc}}(I_{\mathrm{ac}})$ for a magnetic field able to generate three-pinned vortices and three interstitial vortices (Their distribution is shown in Fig.1.9b). At lower driving forces, the weakly pinned interstitial vortices determine the negative average velocity. At higher driving forces, the pinned vortices start to move and they play a dominant role in the rectification process, resulting in a positive rectification velocity.

It has been demonstrated that the inter-particle interactions in a chain of repelling particles captured by a ratchet potential can, in a controllable way, lead to multiple drift reversals, with the drift sign alternating from positive to negative as a function of vortex density [113, 114] . The validity of this very general prediction was demonstrated experimentally by performing transport measurements on a.c.-driven vortices trapped in a superconductor by an array of nanometer-scale asymmetric

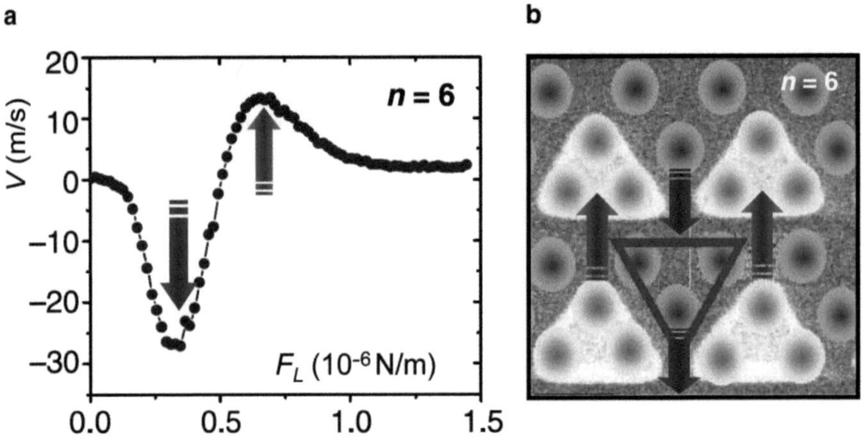

Fig. 1.9 Experimental evidence of rectifaction reversals. (**a**) Net velocity of vortices vs. the ac Lorentz force amplitude ($f = 10$ kHz), for $H = 6H_1$ at $T = 0.98T_c$; *Red* and *black arrows* show the direction of the net flow of these vortices. (**b**) Sketch of the positions of the vortices for several matching fields. Vortices pinned on the *triangles* are shown in *red* and interstitial vortices in *blue*. After Villegas et al. [106]

traps [113] (Fig.1.10). This drastic change in the drift behaviour between single- and multiparticle systems can shed some light on the different behaviour of ratchets and biomembranes in two drift regimes: diluted (single particles) and concentrated (interacting particles).

1.4.3.3 Single Vortex Rectifiers

Another possibility toward rectification in nano-engineered superconductive structures was introduced by asymmetric shapes of single superconducting cells, as it has been shown for asymmetric superconducting loops and mesoscopic triangles [115–117]. This may be understood as a result of the interplay between the geometrically induced asymmetry in the local critical current density and the Little-Parks-like quantum oscillations in the field (H)-temperature (T) phase boundary of mesoscopic superconductors [118, 119]. The underlying principle for these kind of rectifiers is fundamentally different from the vortex ratchet discussed, in the previous sections, and provides a much lower working temperature (e.g., in the work of Morelle et al. [116], rectification was sustained down to 0.7 T/T_{c0}). In the work of Van de Vondel et al. [120], a crossover was found between both the aforementioned rectifiers in a thin superconducting film with a periodic array of triangular holes. A "collective" vortex rectifier was located at high temperatures in the vortex phase. The "single" rectification regime was found very close to the phase boundary, and experiments revealed the presence of quantum oscillations in this region due to surface superconductivity appearing at the antidot edges.

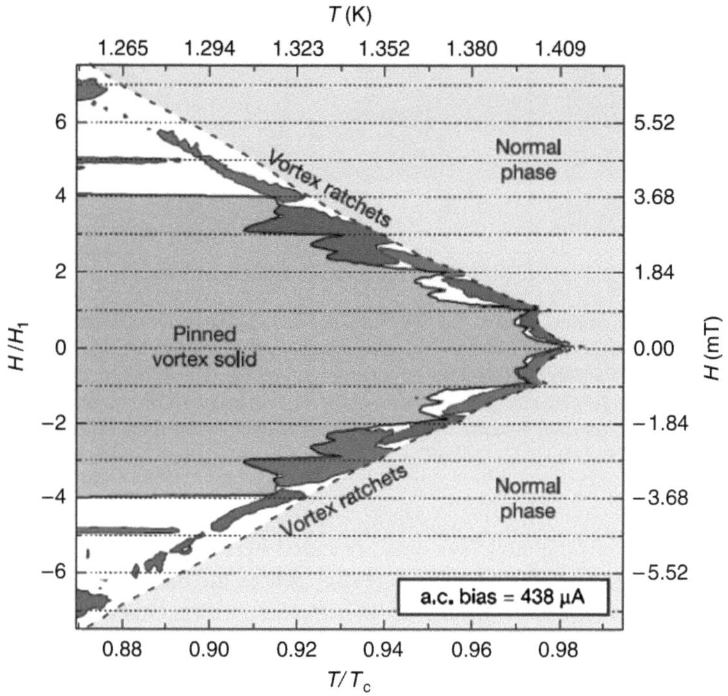

Fig. 1.10 Experimental evidence of rectifaction reversals. $H - T$ dynamical phase diagram at an a.c. current $I(t) = I_{ac}\sin(2\pi t)$, with I_{ac} = 438 mA and $f = 1$ kHz. Between the pinned vortex solid and normal phases, the voltage is dominated by vortex motion. The *green* and *red* areas correspond to positive and negative V_{dc}, respectively. In the *white* areas, vortex motion is symmetric ($V_{dc} < 0$) within the experiment accuracy. After de Souza Silva et al. [104]

1.4.3.4 Magnetic Vortex Rectifiers

As proposed by Carneiro [121], a different way to create vortex ratchets can be realized using magnetic dipoles with in-plane magnetic moment. Here, the spatial inversion symmetry is broken not by the shape of the pinning sites but rather by the vortex–magnetic–dipole interaction. This dipole-induced ratchet motion depends on the orientation and strength of the local magnetic moments, thus allowing one to control the direction of the vortex drift. It is particularly this flexibility to manipulate the vortex motion which makes this kind of pinning potentials attractive for practical applications. In [122, 123], it is shown that in-plane magnetized dipoles can indeed rectify the vortex motion (a summary of these results in shown in Fig. 1.11). Two major differences occur in comparison to a non-magnetic ratchet: (1) due to the nature of the pinning potential, vortices and antivortices feel an opposite asymmetry, which leads to a polarity-independent rectification voltage; (2) due to the presence of self-induced vortex–antivortex pairs, a rectification voltage is present at zero applied field. Such vortex–antivortex mixtures in hybrid

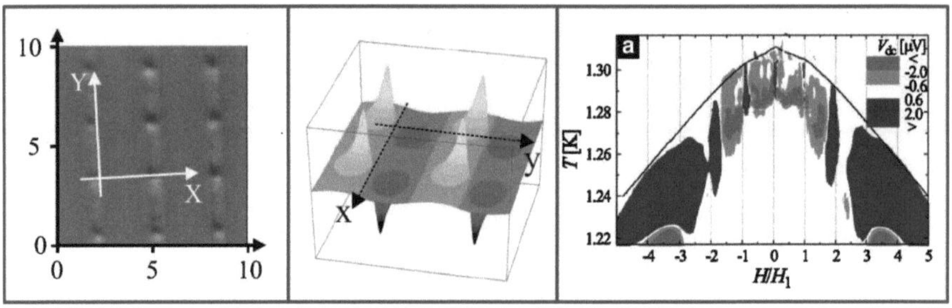

Fig. 1.11 Magnetic vortex rectifier. (**a**) MFM images and (**b**) calculated pinning potential of an array of in-plane magnetic bars on top of a superconducting Al bridge, the bars are polarized in the y direction. x indicates the direction of positive applied current. Rectified voltage V_{dc} in the temperature-field (T-H) plane for the same sample for a 1 kHz and 1.1 kHz sinusoidal ac current of amplitude 0.8 mA. *Full lines* indicate the normal-superconductor transition. After de Souza Silva et al. [122]

superconductor-ferromagnetic systems presented here are an interesting example of the ratchet effect of binary mixtures [124], where the drift direction is assisted by the interaction between the particles of different species. Other examples can be found in different physical systems such as ion mixtures in cell membranes [96] and Abrikosov–Josephson vortex mixtures in cuprate high-T_c superconductors [125].

1.4.3.5 Breaking the Time Reversal Symmetry

All the aforementioned types of vortex ratchets introduced asymmetry by means of an asymmetric potential landscape. Savel'ev and Nori proposed a ratchet system in strongly anisotropic superconductors in which asymmetry is introduced in time instead of space [126]. An externally applied field creates two types of vortices. One type of vortices, Josephson vortices, confined between the ab-plane and the other type consists of stacks of pancake vortices aligned along the c-axes. These two types of vortex lattices attract each other, and by externally moving the Josephson vortices the pancake vortices can be dragged. It is experimentally shown by Cole et al. that the asymmetrical dragging of the Josephson vortices by an external magnetic field results in lensing and antilensing of the pancake vortices [125].

1.5 Conclusion

This chapter provides a broad overview of the physical mechanisms involved in flux channeling and rectification effects in superconductors. The presented results demonstrate the enormous progress achieved to control the motion of vortices, greatly benefited by the recent advances in nanolithographic techniques. The tunability of these systems makes them ideal candidates to study these dynamical effects in general and extrapolate results toward other related systems in nature. Indeed vortices in superconductors are a unique system in which the particle (vortex) density can be easily varied by the turn of a knob (the external magnetic field), while the interparticle interactions can be fine-tuned by the temperature of the superconductor. This flexibility is even more pronounced in superconductor/ferromagnet hybrids, which introduced another degree of tunability, as the pinning landscape can be readily switched in different states. Besides this interdisciplinary aspect, manipulation of single flux quanta in superconductors is of crucial importance for controlling the flux motion in flux qubits for quantum computing and for Superconducting Quantum Interference Devices (SQUIDS) and passive elements, such as superconducting filters for telecommunication. In this chapter we give an idea about the progress made toward the practical realization of different fluxonics devices, such as flux diodes and lenses. These devices hopefully will pave the way toward more sensitive SQUID sensors and improved performance of active and passive superconducting devices.

Acknowledgements This work was supported by the Methusalem Funding of the Flemish Government, the NES-ESF program, the Belgian IAP, the Fund for Scientific Research-Flanders (FWO-Vlaanderen). A.V.S. and J.V.d.V. are grateful for the support from the FWOVlaanderen. The authors are grateful to Clecio Souza Silva and Jo Cuppens for the critical reading and valuable comments.

References

1. M. Vélez, J.I. Martín, J.E. Villegas, A. Hoffmann, E.M. Gonzalez, J.L. Vicent, Ivan K. Schuller, J. Magn. Magn. Mater. 320, 2547–62 (2008)
2. A. Aladyshkin, A.V. Silhanek, W. Gillijns, V.V. Moshchalkov, Supercond. Sci. Technol. 22, 053001 (2009)
3. P. Reimann, Phys. Rep. 361, 57 (2002)
4. P. Hänggi, F. Marchesoni, Rev. Mod. Phys. 81, 387 (2009)
5. I. Giaever, Phys. Rev. Lett. 15, 825 (1966)
6. E.H. Brandt, Phys. Rev. Lett. 50, 1599 (1983)
7. E.H. Brandt, J. Low Temp. Phys. 53, 41 (1983)
8. E.H. Brandt, J. Low Temp. Phys. 53, 71 (1983)
9. D.X. Chen, J.J. Moreno, A. Hernando, A. Sanchez, B.Z. Li, Phys. Rev. B 57, 5059 (1998)
10. C.P. Poole, H.A. Farach, R.J. Creswick, R. Prozorov, Superconductivity (Elsevier, London, 2007)
11. H. Suhl, Phys. Rev. Lett. 14, 226 (1965)
12. E. Simanek, Phys. Lett. A 154, 309 (1991)
13. E.M. Chudnovsky, A.B. Kuklov, Phys. Rev. Lett. 91, 067004 (2003)
14. A. van Otterlo, M. Feigel'man, V. Geshkenbein, G. Blatter, Phys. Rev. Lett. 75, 3736 (1995)
15. N.B. Kopnin, V.M. Vinokur, Phys. Rev. Lett. 81, 3952 (1998)
16. M.J. Stephen, J. Bardeen, Phys. Rev. Lett. 14, 112 (1965)
17. J. Volger, F.A. Staas, A.G. van Vijfeijken, Phys. Lett. 9, 303 (1964)
18. Y.B. Kim, C.F. Hempstead, A.R. Strnad, Phys. Rev. 16, A1163 (1965)
19. M. Tinkham, Phys. Rev. Lett. 13, 804 (1964)
20. J.R. Clem, Phys. Rev. Lett. 20, 735 (1968)
21. M.A.R. LeBlanc, Supercond. Sci. Technol. 10, 929 (1997)

22. J. Kolacek, P. Lipavsky, E.H. Brandt, Phys. Rev. Lett. 86, 312 (2001)
23. P.G. de Gennes, J. Matricon, Rev. Mod. Phys. 36, 45 (1964)
24. J. Bardeen, Phys. Rev. Lett. 13, 747 (1964)
25. P. Ao, D.J. Thouless, Phys. Rev. Lett. 70, 2158 (1993)
26. E.B. Sonin, Phys. Rev. Lett. 81, 4276 (1998)
27. E.B. Sonin, Phys. Rev. B 55, 485 (1997)
28. J. Pearl, Appl. Phys. Lett. 5, 65 (1964)
29. H. Brandt, Phys. Rev. B 79, 134526 (2009)
30. A. Andronov, I. Gordion, V. Kurin, I. Nefedov, I. Shereshevsky, Physica C 213, 193 (1993)
31. V.V. Moshchalkov, V. Bruyndoncx, L. Van Look, M.J. Van Bael, Y. Bruynseraede, A. Tonomura, Chapter 9 in Handbook of Nanostructured Materials and Nanotechnology, ed. by H.S. Nalwa, Electrical Properties, vol. 3 (Academic Press, New York, 2000)
32. J.I. Martín, J. Nogués, K. Liu, J.L. Vicent, I.K. Schuller, J. Magn. Magn. Mater. 256, 449 (2003)
33. see for instance C. Reichhardt, C.J. Olson, S.B. Field, F. Nori, Phys. Rev. B 57, 7937 (1998)
34. see for instance K. Harada, O. Kamimura, H. Kasai, T. Matsuda, A. Tonomura, V.V. Moshchalkov, Science 274, 1167 (1996)
35. P. Vašek, I. Janeček, V. Plecháček, Physica C 247, 381 (1995)
36. D. Jaque, E.M. Gonzalez, J.I. Martín, J.V. Anguita, J.L. Vicent, Appl. Phys. Lett. 81, 2851 (2002)
37. A.I. Buzdin, Rev. Mod. Phys. 77, 935 (2005)
38. A. Belkin, V. Novosad, M. Iavarone, J. Fedor, J.E. Pearson, A. Petrean-Troncalli, G. Karapetrov, Appl. Phys. Lett. 93, 072510 (2008)
39. A. Belkin, V. Novosad, M. Iavarone, J. Pearson, G. Karapetrov, Phys. Rev. B 77, 180506 (2008)
40. V. Vlasko-Vlasov, U. Welp, G. Karapetrov, V. Novosad, D. Rosenmann, M. Iavarone, A. Belkin, W-K. Kwok, Phys. Rev. B 77, 134518 (2008)
41. O. Daldini, C. Leemann, P. Martinoli, Solid State Comm. 16, 509 (1975)
42. M. Baert, V.V. Metlushko, R. Jonckheere, V.V. Moshchalkov, Y. Bruynseraede, Phys. Rev. Lett. 74, 3269 (1995)
43. F.A. Staas, A.K. Niessen, W.F. Druyvesteyn, J. Vonsuchtelen, Phys. Lett. 13, 293 (1964)
44. A.K. Niessen, J. Vansuchtelen, F.A. Staas, W.F. Druyvesteyn, Philips Res. Rep. 20, 226 (1965)
45. A.K. Niessen, C.H. Weijsenfeld, J. Appl. Phys. 40, 384 (1969)
46. Ya.V. Kopelevich, V.V. Lemanov, E.B. Sonin, A.L. Kholkin, JETP Lett. 50, 212 (1989)
47. T.L. Francavilla, R.A. Hein, IEEE Trans. Mag. 27, 1039 (1991)
48. T.L. Francavilla, E.J. Cukauskas, L.H. Allen, P.R. Broussard, IEEE Trans. Appl. Supercond. 5, 1717 (1995)
49. I.Yu. Antonova, V.M. Zakosarenko, E.V. Il'ichev, V.I. Kuznetsov, V.A. Tullin, JETP Lett. 54, 505 (1991)
50. L.I. Glazman, Sov. J. Low Temp. Phys. 12, 389 (1986)
51. B. Chen, J. Dong, Phys. Rev. B 44, 10206 (1991)
52. C. Villard, G. Koren, D. Cohen, E. Polturak, B. Thrane, D. Chateignier, Phys. Rev. Lett. 77, 3913 (1996)
53. A.A. Prodan, V.A. Shklovskij, V.V. Chabanenko, A.V. Bondarenko, M.O. Obolenskii, H. Szymczak, S. Piechota, Physica C 302, 271 (1998)
54. H. Pastoriza, S. Candia, G. Nieva, Phys. Rev. Lett. 83, 1026 (1999)
55. V.V. Chabanenko, A.A. Prodan, V.A. Shklovskij, A.V. Bodarenko, M.A. Obolenskii, H. Szymczak, S. Piechota, Physica C 314, 133 (1999)
56. G. D'Anna, V. Berseth, L. Forr'o, A. Erb, E. Walker, Phys. Rev. B 61, 10206 (2000)
57. Y. Mawatari, Phys. Rev. B 56, 3433 (1997)
58. Y. Mawatari, Phys. Rev. B 59, 12033 (1999)
59. V.A. Shklovskii, A.K. Soroka, A.A. Soroka, JETP Lett. 89, 1138 (1999)
60. B.Y. Zhu, J. Dong, D.Y. Xing, Z.D. Wang, Phys. Rev. B 57, 5075 (1998)
61. J. Groth, C. Reichhardt, C.J. Olson, et al., Phys. Rev. Lett. 77, 3625 (1996)

62. G.W. Crabtree, G.K. Leaf, H.G. Kaper, V.M. Vinokur, A.E. Koshelev, D.W. Braun, D.M. Levine, W.K. Kwok, J.A. Fendrich, Physica C 263, 401 (1996)
63. H. Ghamlouch, M. Aubin, Physica C 269, 163 (1996)
64. O.K. Soroka, M. Huth, V.A. Shklovskij, J. Oster, H. Adrian, Physica C 388, 773 (2003)
65. O.K. Soroka, V.A. Shklovskij, M. Huth, Phys. Rev. B 76, 014504 (2007)
66. M.J. Hogg, F. Kahlmann, E.J. Tarte, Z.H. Barber, J.E. Evetts, Appl. Phys. Lett. 78, 1433 (2001)
67. H. Pastoriza, P.H. Kes, Phys. Rev. Lett. 75, 3525 (1995)
68. N. Kokubo, R. Besseling, V.M. Vinokur, P.H. Kes, Phys. Rev. Lett. 88, 247004 (2002)
69. R. Besseling, N. Kokubo, P.H. Kes, Phys. Rev. Lett. 91, 177002 (2003)
70. K. Yu, T.W. Heitmann, C. Song, M.P. DeFeo, B.L.T. Plourde, M.B.S. Hesselberth, P.H. Kes, Phys. Rev. B 76, 220507(R) (2007)
71. R. Wördenweber, P. Dymashevski, Physica C 404, 421 (2004)
72. L. Van Look, B.Y. Zhu, R. Jonckheere, B.R. Zhao, Z.X. Zhao, V.V. Moshchalkov, Phys. Rev. B 66, 214511 (2002)
73. M. Vélez, D. Jaque, J.I. Martín, M.I. Montero, I.K. Schuller, J.L. Vicent, Phys. Rev. B 65, 104511 (2002)
74. J.E. Villegas, E.M. Gonzalez, M.I. Montero, I.K. Schuller, J.L. Vicent, Phys. Rev. B 68, 224504 (2003)
75. C.E. Chialvo, J. Guimpel, S. Reparaz, H. Pastoriza, Physica C 422, 112 (2005)
76. C. Reichhardt, G. Zimanyi, N. Gronbech-Jensen, Phys. Rev. B 64, 014501 (2001)
77. A.V. Silhanek, L. Van Look, S. Raedts, R. Jonkheere, V.V. Moshchalkov, Phys. Rev. B 68, 214504 (2003)
78. J.E. Villegas, E.M. Gonzalez, M.I. Montero, I.K. Schuller, J.L. Vicent, Phys. Rev. B 72, 064507 (2005)
79. C.C. de Souza Silva, G. Carneiro, Phys. Rev. B 66, 054514 (2002)
80. G. Carneiro, Phys. Rev. B 66, 054523 (2002)
81. V.A. Shklovskij, A.A. Soroka, Low Temp. Phys. 28, 254 (2002)
82. R. Wördenweber, P. Dymashevski, V.R. Misko, Phys. Rev. B 69, 184504 (2004)
83. R. Wördenweber, J.S.K. Sankarraj, P. Dymashevski, E. Hollmann, Physica C 434, 101 (2006)
84. R. Wördenweber, E. Hollmann, J. Schubert, R. Kutzer, A.K. Ghosh, Appl. Phys. Lett. 94, 202501 (2009)
85. M. Pannetier, R.J. Wijngaarden, I. Floan, J. Rector, B. Dam, R. Griessen, P. Lahl, R. Wördenweber, Phys. Rev. B 67, 212501 (2003)
86. M. Menghini, R.J. Wijngaarden, A.V. Silhanek, S. Raedts, V.V. Moshchalkov, Phys. Rev. B 71, 104506 (2005)
87. A. Lukashenko, A.V. Ustinov, A.P. Zhuravel, E. Hollmann, R. Wördenweber, J. Appl. Phys. 100, 023913 (2006)
88. G. Carneiro, Physica C 432, 206 (2005)
89. N. Verellen, A.V. Silhanek, W. Gillijns, V.V. Moshchalkov, V. Metlushko, F. Gozzini, B. Ilic Appl. Phys. Lett. 93, 022507 (2008)
90. N. Verellen, A.V. Silhanek, W. Gillijns, V.V. Moshchalkov, V. Metlushko, F. Gozzini, B. Ilic, Appl. Phys. Lett. 94, 219903 (2009)
91. A.V. Silhanek, J. Van de Vondel, A. Leo, G.W. Ataklti, W. Gillijns, V.V. Moshchalkov, Supercond. Sci. Technol. 22, 034002 (2009)
92. J. Van de Vondel, A.V. Silhanek, V. Metlushko, P. Vavassori, B. Ilic, V.V. Moshchalkov, Phys. Rev. B 79, 054527 (2009)
93. C.L.S. Lima, C.C. de Souza Silva, Phys. Rev. B 80, 054514 (2009)
94. P. Hänggi, F. Marchesoni, F. Nori, Ann. Phys. (Leipz.) 14, 51–70 (2005)
95. H. Linke, Appl. Phys. A 75 (special issue), 167 (2002)
96. J.H. Morais-Cabral, Y. Zhou, H. Mackinnon, Nature 414, 37 (2001)
97. S. Matthias, F. Müller, Nature 424, 53 (2003)
98. H. Linke, T.E. Humphrey, A. Löfgren, A.O. Sushkov, R. Newbury, R.P. Taylor, P. Omling, Science 286, 2314 (1999)

99. C.-S. Lee, B. Jankó, I. Derényi, A.-L. Barabási, Nature 400, 337 (1999)
100. J.F. Wambaugh, C. Reichhardt, C.J. Olson, F. Marchesoni, F. Nori, Phys. Rev. Lett. 83, 5106 (1999)
101. B.Y. Zhu, F. Marchesoni, V.V. Moshchalkov, F. Nori, Phys. Rev. B 68, 014514 (2003)
102. B.Y. Zhu, F. Marchesoni, F. Nori, Phys. Rev. Lett. 92, 180602 (2004)
103. J. Van de Vondel, C.C. de Souza Silva, B.Y. Zhu, M. Morelle, V.V. Moshchalkov, Phys. Rev. Lett. 94 (2005)
104. C.C. de Souza Silva, J. Van de Vondel, M. Morelle, V.V. Moshchalkov, Nature, 400, 651 (2006)
105. S. Ooi, T. Mochikua, K. Hirataa, Physica C 468 (2008)
106. J.E. Villegas, S. Savelev, F. Nori, E.M. Gonzalez, J.V. Anguita, R. Garcia, J.L. Vicent, Science, 302, 1188 (2003)
107. D.E. Shalóm, H. Pastoriza, Phys. Rev. Lett. 94, 177001 (2005)
108. Y. Togawa, K. Harada, T. Akashi, H. Kasai, T. Matsuda, F. Nori, A. Maeda, A. Tonomura, Phys. Rev. Lett. 95, 087002 (2005)
109. K. Yu, T.W. Heitmann, C. Song, M.P. DeFeo, B.L.T. Plourde, M.B.S. Hesselberth, P.H. Kes, Phys. Rev. B 76, 220507(R) (2007)
110. W.K. Kwok, R.J. Olsson, G. Karapetrov, U. Welp, V. Vlasko-Vlasov, K. Kadowaki, G.W. Crabtree, Physica C 382, 137 (2002)
111. W. Gillijns, A.V. Silhanek, V.V. Moshchalkov, C.J. Olson Reichhardt, C. Reichhardt, Phys. Rev. Lett. 99, 247002 (2007)
112. R. Bartussek, P. Hänggi, J.G. Kissner, Europhys. Lett. 28, 459 (1994)
113. C.C. de Souza Silva, J. Van de Vondel, B.Y. Zhu, M. Morelle, V.V. Moshchalkov, Phys. Rev. B 73, 014507 (2006)
114. Q. Lu, C.J. Olson Reichhardt, C. Reichhardt, Phys. Rev. B 75, 054502 (2007)
115. S.V. Dubonos, V.I. Kuznetsov, I.N. Zhilyaev, A.V. Nikulov, A.A. Firsov, JETP Lett. 77, 371 (2003)
116. M. Morelle, N. Schildermans, V.V. Moshchalkov, Appl. Phys. Lett. 89, 1121512 (2006)
117. N. Schildermans, A.B. Kolton, R. Salenbien, V.I. Marconi, A.V. Silhanek, V.V. Moshchalkov, Phys. Rev. B 76, 224501 (2007)
118. W.A. Little, R.D. Parks, Phys. Rev. Lett. 9, 9 (1964)
119. L.F. Chibotaru, A. Ceulemans, V. Bruyndoncx, V.V. Moshchalkov, Phys. Rev. Lett. 86, 1323 (2001)
120. J. Van de Vondel, C.C. de Souza Silva, V.V. Moshchalkov, Euro. Phys. Lett. 80, 17006 (2007)
121. G. Carneiro, Phys. Rev. B 72, 144514 (2005)
122. C.C. de Souza Silva, A.V. Silhanek, J. Van de Vondel, W. Gillijns, V. Metlushko, B. Ilic, V.V. Moshchalkov, Phys. Rev. Lett. 98, 117005 (2007)
123. A.V. Silhanek, N. Verellen, V. Metlushko, W. Gillijns, F. Gozzini, B. Ilic V.V. Moshchalkov, Physica C 468, 563 (2008)
124. S. Savel'ev, F. Marchesoni, F. Nori, Phys. Rev. Lett. 91, 010601 (2003)
125. D. Cole, S. Bending, S. Savel'ev, A. Grigorenko, T. Tamegai, F. Nori, Nat. Mater. 5, 305 (2006)
126. S. Savel'ev, F. Nori, Nat. Mater. 1, 179 (2002)

Chapter 2
High-T_c Films: From Natural Defects to Nanostructure Engineering of Vortex Matter

Roger Wördenweber

Abstract Due to their unique properties, ceramic high-T_c materials offer new perspectives for cryoelectronic applications. However, the 2D-layered structure in combination with the extremely small coherence length imposes extreme demands on the preparation of virtually single crystalline high-T_c films and the handling and manipulation of magnetic flux in devices made from these films. In this chapter, we will give an overview on vortex matter, pinning mechanism, vortex mobility, and vortex manipulation in high-T_c thin films. Vortex manipulation via artificially introduced structures will be demonstrated, and their potential for application in superconducting magnetometer and microwave devices will be sketched.

2.1 Introduction

Since the discovery of the first high-temperature superconductors (HTSs) [1], significant effort has been put into the research and realization of HTS thin film devices. On the one hand, this effort is motivated by the unique properties of HTS materials (e.g. the 2D nature or the elevated temperature up to which superconductivity persists lead to new phenomena); on the other hand, these superconductors offer new perspectives for application of cryoelectronic devices at elevated temperature.

The technology of devices and integrated circuits fabricated from low-temperature superconductors capable of operating at or near the temperature of liquid helium (4.2 K) is by now well established. A number of technical applications of HTS thin films [mainly passive (linear) and active (nonlinear) devices or electronic circuits] seem to have already a firm implementation base within the next years, and others represent tentative "dreams" for a more distant future. Typical application areas included: metrology and electronic instrumentation, radio astronomy and environmental spectroscopy (atmosphere and space), neurology and medical diagnostics, electronic warfare (radar, electronic countermeasures – ECM, magnetic anomaly detection), non-destructive materials evaluation (NDE/NDT), geological and environmental prospecting, telecommunication, ultrafast digital signal and data processing.

Nevertheless, the future application of HTS in cryoelectronic devices that are capable of operating at temperatures between typically 20 and 77 K will strongly depend on the solution of a number of problems. The major issues are:

1. The development of reproducible deposition technologies for high-quality single and multilayer HTS thin films and
2. The understanding and engineering of the superconducting properties (especially the critical properties)

Both issues are complicated by the complexity and the layered, quasi two-dimensional (2D) nature of the high-T_c superconductors.

In high-T_c material, the improvement of the superconducting properties (e.g. increase of T_c and increase of the energy gap ΔE) is accompanied by a significant enhancement of (Figs. 2.1 and 2.2):

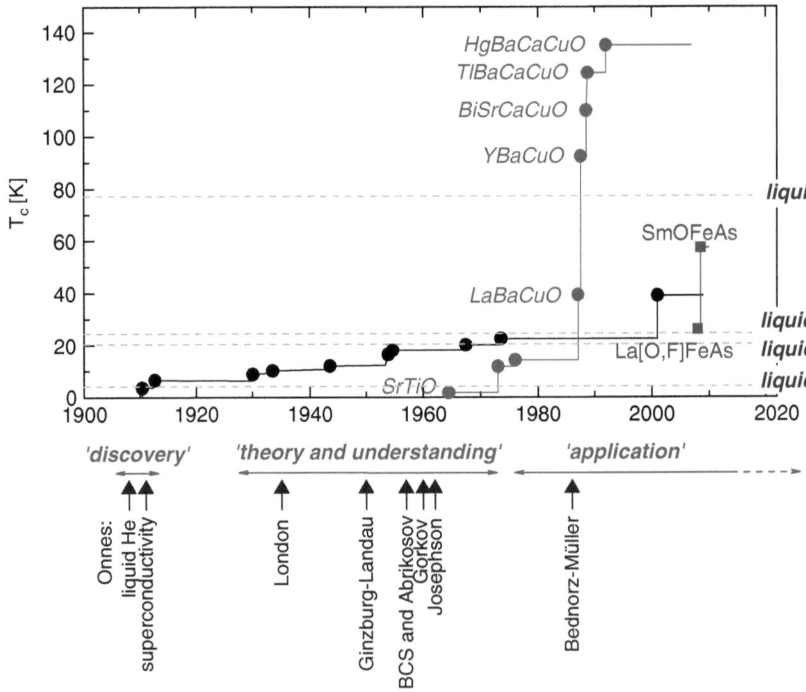

Fig. 2.1 Development of superconductivity: starting with the discovery of superconductivity in 1911 [2], the highest transition temperature T_c has been improved in time for metallic and metal compound superconductors (*black*), high-T_c ceramic superconductors (*red*) and, recently, ion arsenide-based layered superconductors (so-called pnictide superconductors, *green*). Organic superconductor (highest T_c present at 33 K for alkali-doped fullerene $RbCs_2C_{60}$ [3]) is not shown. Additionally, the normal boiling points of liquid cooling media are given, which indicate possible temperature regimes for application and major developments in discovery, theory, understanding and application are indicated

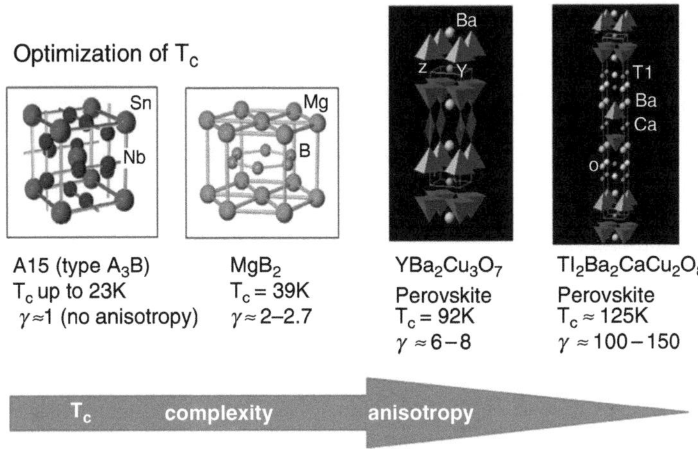

Fig. 2.2 The improvement of the superconducting properties (e.g. increase of T_c and energy gap ΔE) is achieved by an enhancement of complexity (structural as well as stoichiometric) and anisotropy of the superconductor. The HTS represents a quasi 2D-layered superconductor

1. The number of elements
2. The structural complexity
3. The anisotropy

The quasi 2D nature of the oxide superconductors related to their layered structure leads to a large anisotropy in nearly all parameters. It is characterized by the anisotropy factor

$$\gamma = \frac{\xi_{ab}}{\xi_c} = \frac{\lambda_c}{\lambda_{ab}}, \tag{2.1}$$

defined by the ratio of the coherence lengths ξ or penetration lengths λ measured for magnetic field applied in crystallographic *a–b* or *c* direction, respectively. Values of $\gamma \approx$ 6–8 for $YBa_2Cu_3O_{7-\delta}$ (YBCO), $\gamma \approx$ 55–122 for Bi compounds, and $\gamma \approx$ 100–150 for Tl compounds are reported [4–7]. Furthermore, extremely small coherence lengths are determined for these superconductors. Values of $\xi_c =$ 0.02–0.04 nm and $\xi_{ab} =$ 2–2.5 nm ($Bi_2Sr_2CaCu_2O_{8+\delta}$), $\xi_c =$ 0.3–0.5 nm, and $\xi_{ab} =$ 2–3 nm ($YBa_2Cu_3O_{7-\delta}$) are measured, which, furthermore, strongly depend on sample quality and oxygen content [8]. Therefore, local variations in the sample properties on the scale of ξ will automatically result in a spatial variation of the superconducting properties. Due to the extremely small coherence length along the crystallographic *c*-axis, only the current directed along the crystallographic *a–b* direction (i.e. along the CuO_2 planes) represents a "superconducting current" and can be used for applications. Since misalignments in the *a–b* plane lead to grain boundaries that strongly affect the critical properties, perfect biaxial *c*-axis orientated epitaxial HTS films including the avoidance of anti-phase boundaries are necessary for most applications. This requires thin film growth with perfection on nanometre scale.

The electronic properties of superconducting devices are usually strongly determined or affected by the presence or motion of quantized magnetic flux (so-called flux lines or vortices) in the type-II superconductor. Most superconducting devices will (or have to) operate in magnetic fields strong enough to create large densities of vortices within the device. Generally flux penetrates the superconductor for magnetic fields $B > B_{c1}$, with the lower critical field B_{c1}. The demagnetization effect will lead to a considerable enhancement of the magnetic field at the edge of a superconductor sample.

Moreover, calculation of the Gibb's free energy for vortices in thin film devices using an approximation of the sample geometry by a rectangular cross-section with film thickness d (parallel to the magnetic field direction) much smaller than the lateral dimension w (in contrast to the elliptic approximation of the cross-section used in the classical picture) indicates [9] that (1) the first tunnelling of single vortices (e.g. via thermal-activated penetration) through the geometrical barrier at the edge of the superconductor is expected to occur already at extremely small fields

$$B_{\mathrm{T}} = \frac{d}{2w} B_{c1} \tag{2.2}$$

and (2) that collective penetration of vortices takes place at the penetration field

$$B_{\mathrm{P}} = \sqrt{\frac{d}{w}} B_{c1}. \tag{2.3}$$

These predictions are valid for most HTS devices and patterns, i.e. as long as $w \gg d > \lambda$ is valid. As a consequence, HTS films, devices and patterns are usually strongly penetrated by magnetic flux lines (Fig. 2.3).

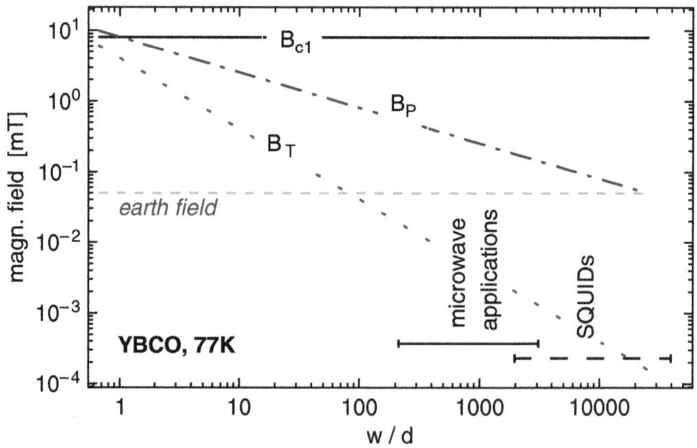

Fig. 2.3 Critical field B_{c1}, tunnelling field B_{T} and penetration field B_{P} as function of the reduced width for patterned YBCO films at 77 K. Typical regimes for cryogenic applications are indicated

The presence of vortices or its motion will lead to dissipative processes, flux noise or local modification of the superconducting properties of these devices. The performance of the device will be diminished. Due to the large anisotropy, the small coherence length and, last but not least, the elevated temperature at which the ceramic superconductors will be operated, these effects are even more severe for HTS material. Ways to avoid the impact of vortices are:

- Operation in perfectly shielded environments
- Reduction of the structures to nm size that does not allow for vortex formation
- "Manipulation" of vortex matter

Since the first two options are in most cases (extremely) costly and/or usually technically difficult or impossible, the manipulation of vortex matter appears to be an ideal solution for this problem. This is one of the motivations for scientific work on vortex manipulation in HTS thin films. Furthermore, the manipulation of vortices is of interest for the development of (novel) fluxon concepts and the understanding of vortex properties in general. In this chapter, an overview on different aspects of vortex manipulation and vortex matter in micro- and nanopatterned HTS films will be given.

2.2 Vortex Matter in High-T_c Superconductors

In order to understand vortex matter and vortex manipulation in micro- and nanopatterned HTS, one should first consider vortex matter in unpatterned type-II superconductors. This is done in the following sections. First, the impact of vortex motion on the electric properties of superconductors is discussed; then basic concepts of elementary vortex–defect interactions are sketched and, finally, different mechanisms of volume flux–flux pinning and flux motion are discussed. The intention of these sections is to provide the basic understanding of the mechanism of vortex motion and flux pinning in HTS material and to motivate different strategies to manipulate vortex motion and pinning in these complex materials.

2.2.1 Vortex Motion in Ideal Superconductors

If a vortex in an ideal superconductor is exposed to a driving force, it will start to move under the action of the force. The driving force can be provided for instance by an applied current (Lorentz force),

$$\vec{f}_{\text{vortex}} = \vec{J} \times \vec{\Phi}_{\text{o}}, \tag{2.4}$$

a temperature gradient or a magnetic field gradient. Furthermore, vortex motion can be assisted by a finite temperature (thermal activation or thermally assisted vortex motion). Within a perfectly homogeneous system, the driving force is counteracted only by a friction force:

$$\vec{f}_{\text{vortex}} = -\eta\vec{v}, \tag{2.5}$$

with v denoting the steady-state velocity of the vortex. As a consequence of the flux motion, a finite electric field

$$\vec{E} = n_{\text{v}}\vec{\Phi}_o \times \vec{v} = \vec{B} \times \vec{v}, \tag{2.6}$$

is generated in the superconductor (n_{v} represents the sheet density of vortices). Thus, a steady flow of vortices in a device will result in a finite flux-flow resistance:

$$\rho_{\text{ff}} = \frac{\partial E}{\partial J} = \frac{\Phi_{\text{o}} B}{\eta}. \tag{2.7}$$

The electric field caused by vortex motion is one of the major problems of superconducting electronic devices operating in magnetic fields. An ideal (i.e. defect-free) superconducting wire would expose a resistance to an applied current (Fig. 2.4). Thermal activation of flux lines would lead to statistic fluctuations (i.e. statistic motion) of vortices even at low temperatures that would manifest itself by (flux or voltage) noise signals, e.g. telegraph noise in superconducting magnetometers [e.g. superconducting quantum interference device (SQUID), see Fig. 2.4].

2.2.2 Flux Pinning and Summation Theories

In order to retain the dissipation-free dc current or reduce the voltage noise due to vortex motion, flux lines have to be pinned by defects in the superconducting material (so-called pinning sites). The vortex–pin interaction of the defects compensates the driving force up to a critical value. In case of current driven vortex motion, the Lorentz force defines the maximum dissipation-free current density. The experimentally determined volume pinning force F_{p}, Lorentz force and critical current density J_{c} are correlated by:

$$\vec{F}_{\text{p}} = -\vec{F}_{\text{L}} = \vec{B} \times \vec{J}_{\text{c}}. \tag{2.8}$$

The volume pinning force is either obtained via summation of the elementary pinning force f_{p} (*pin-breaking mechanism*) or defined by the stability of the strongly pinned flux-line lattice (*flux-line shear mechanism*). Both mechanisms will be sketched below.

In general, mechanisms of volume flux pinning and flux motion in real type-II superconductors have to be discussed in four steps:

1. The *interaction between individual pinning centres and vortices* represents the basis of the vortex pinning.
2. The *interaction between individual vortices* defines the formation and properties of the vortex lattice as well as its reaction on driving forces and elementary pinning.

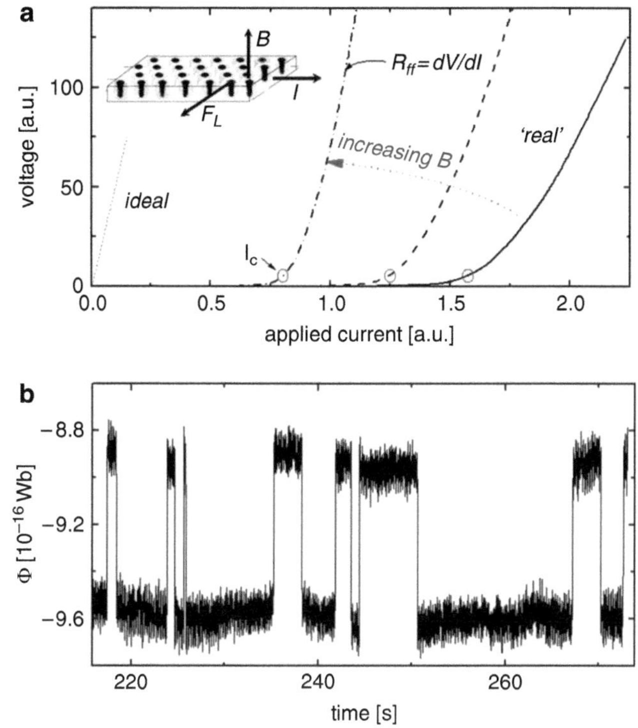

Fig. 2.4 (**a**) Schematic drawing of current–voltage characteristics for an ideal superconductor and a real superconductor at different applied fields. The critical current I_c (usually defined by a voltage criterion of 1 μV/cm) and the flux-flow resistance are indicated. The inset shows the vortex lattice and the directions of current, applied field and resulting Lorentz force for a superconducting thin film with B normal to the film surface. (**b**) Thermally activated flux noise in a HTS SQUID

3. Adequate summation of the effects of many pins usually at random position taking into account the vortex–pin interaction and the vortex–vortex interaction leads to the prediction of the maximum volume pinning force $F_{p,max}$ of the system.
4. Finally, the *homogeneity* of the superconducting material in terms of amplitude and length scale of the variation of the superconducting properties defines the resulting mechanisms of vortex motion and volume vortex pinning F_p.

Ad (1): The elementary vortex–pin interaction depends on the type of defect. It originates from the interaction between a vortex and an inhomogeneity or defects in the superconducting material. Real superconductors always posses "natural" defects (e.g. grain boundaries, dislocations, voids, precipitates) resulting in a finite critical current. Vortex–pin interactions can be classified in the following categories:

- *Geometrical* or morphological interaction (e.g. thickness variation and surfaces)
- *Magnetic* interaction arising from the interaction of superconducting and non-superconducting material (e.g. vacuum and extended precipitates) parallel to the

applied magnetic field. This interaction is determined by the field gradient in the superconductor, i.e. the penetration depth λ

- *Core* interaction arises from the coupling of locally distorted superconducting properties with the periodic variation of the superconducting order parameter. Neglecting the complicated case of pinning by spatial variation of the Fermi-surface or of magnetic properties, there exist only two predominant mechanisms for core pinning: δT_c *– pinning* (spatial variation of the density, elasticity or pairing interaction of the material) and $\delta\kappa$*-pinning* (variation of the electron mean free path)

The vortex–pin interaction depends on the gradient of the magnetic field or superconducting order parameter. In technical type-II superconductors, the magnetic penetration length λ is typically much larger than the superconducting coherence length ξ. Therefore, magnetic interaction does not play an important role for vortex pinning in these materials. In unpatterned type-II superconductors, *core pinning* and, especially, $\delta\kappa$*-pinning* dominate. The situation is slightly different for artificially induced defects. Typical candidates of artificial defects for thin film applications are *irradiation defects*, *magnetic* or *non-magnetic dots*, completely or partly etched holes (*antidots* or *blind holes*) [14, 15], specially patterned defects such as *moats*, *channels* [13,25] and many more structures. Depending on the size (most artificial defects are larger than ξ and often even larger than λ), choice of material (vacuum, non-superconducting, metal, magnetic) and shape, different types of interactions (geometric, magnetic, core) or combinations of these interactions are present. In some cases, alternative mechanisms of interactions are present. For instance, magnetic field compensation defines the major interaction of magnetic dots and vortices. As an example, the elementary vortex–pin interaction of vortex–antidot interaction will be discussed in Sect. 2.3.2.

Ad (2): Vortices–vortex interaction leads to the formation of vortex lattices. With a driving force smaller than the volume pinning force (static vortex lattice and small pinning forces), the vortex lattice will be deformed due to the vortex–pin interaction. The deformation can be elastic or plastic, or instabilities can be formed [16, 17]. The different mechanisms are comparable to the reaction of solids upon internal stress. As long as the strain is small, the vortex lattice can reach its equilibrium position with respect to the pin distribution without plastic shear taking place in the lattice. In case of larger strain, plastic shear will create a significant number of flux-line defects. The combination of (1) and (2) leads to the deformation of the vortex lattice described by the displacement field. The deformation can be two-dimensional (transversal displacement) [16] or three-dimensional [17].

Ad (3): The adequate summation of the effects of many pins usually at random position leads to the prediction of the maximum volume pinning force, $F_{p,max}$, taking into account the elementary vortex interaction, the distribution and density of pinning sites and the kind of deformation in the vortex lattice. Note that this volume pinning force, $F_{p,max}$, is not automatically identical with the force $F_p = J_c B$, which is defined by the onset of vortex motion. The summation problem can be solved in some ideal or model systems. In the easiest case, every pinning centre is able to exert its maximum pinning force, f_p, on the vortex lattice, and the net volume pinning force would be given by the *direct summation* [18]:

$$F_{p,DS} = -F_{p,max} = \sum \left(\frac{f_p}{V} \right). \quad (2.9)$$

This case is usually not observed in real (i.e. non-artificial) systems at moderate or large applied magnetic induction. In very small magnetic fields, for which the vortex lattice parameter $a_o = 1.075(\Phi_o/B)^{1/2} \gg \lambda$, direct summation is appropriate, since the vortex–vortex interaction is negligible. At higher fields, a totally flexible vortex lattice or a distribution of pin sites which is matched to the vortex lattice [14, 15] would be necessary to apply direct summation. The latter is for instance in case of commensurability of the vortex lattice with an artificial patterns (see Sect. 2.3.3.1).

The realistic scenario for real type-II superconducting material in moderate or large magnetic fields is different. The volume pinning force has to be evaluated for the case of larger correlated regions in the vortex lattice via adequate summation of the elementary vortex–pin interactions. This summation of the elementary pinning forces of a realistic defect system has been a long standing problem which was first solved in the form of the theory of collective pinning [19]. Other approaches of the evaluation of correlation lengths due to small disorder in the vortex lattice caused by the vortex–pin interaction are discussed in the literature [20]. Modifications of the collective pinning theory have been introduced especially for HTS, assuming a system of point pins with typical dimensions smaller than ξ [21–23].

Ad (4): Finally, the mechanism of flux motion determines the onset of dissipation due to a driving force. This mechanism defines the irreversible properties, i.e. the experimentally determined volume pinning force F_p given by the critical current density J_c:

$$\vec{F}_p = \vec{J}_c \times \vec{B}, \quad \text{with } F_p \leq F_{p,max}. \quad (2.10)$$

The volume critical force F_p can differ strongly from the maximum volume pinning force $F_{p,max}$. It depends upon (1) the relation between vortex–vortex interaction and vortex–pin interactions and (2) the homogeneity of the superconductor on a length scale larger than the coherence length [24–27]. Thus, the mechanism of vortex motion strongly depends upon strength, density and distribution of pinning centres, on the elastic or plastic response of the vortex lattice, and the homogeneity of the superconducting material. The two different kinds of interactions (vortex–vortex and vortex–pin) in combination with the homogeneity of the superconductor automatically introduce two different mechanisms of vortex dynamic: *pin breaking* and *flux-line shear mechanisms*.

If the differences between depinning forces of neighbouring vortices are small compared to the vortex–vortex interaction, the complete vortex lattice will be pinned or depinned. This situation is referred to as *pin breaking*. The volume pinning is given by statistical summation of the elementary interactions in the correlation volume $V_c = L_c\, R_c^2$ [19]:

$$F_p = F_{p,PB} = F_{p,max} = \sqrt{\frac{n\left\langle f_p^2\right\rangle}{V_c}} = \sqrt{\frac{W(0)}{V_c}}. \tag{2.11}$$

Here n denotes the density of pinning sites, $W(0)$ represents the collective pinning parameter, and L_c and R_c are the correlation lengths perpendicular and parallel to the magnetic field direction, respectively. The resulting field dependence is shown in Fig. 2.5a for a weak-pinning type II superconductor (a-Nb_4Ge). Up to a given field, the elastic deformation of the vortex lattice sustains and the field dependence of the volume pinning force is nicely described by the collective pinning theory (2.11). At high fields, plastic deformations set in leading to an increase in the pinning force with respect to the predictions of the collective pinning theory. The so-called peak effect at high fields (Fig. 2.5a) is a characteristic feature of the collective pinning behaviour in weak pinning material.

When the local pinning force strongly varies over length scales comparable to or larger than the vortex–vortex distance, vortices or bunches of vortices will start to move independently as soon as the driving force exceeds the flow stress of the vortex

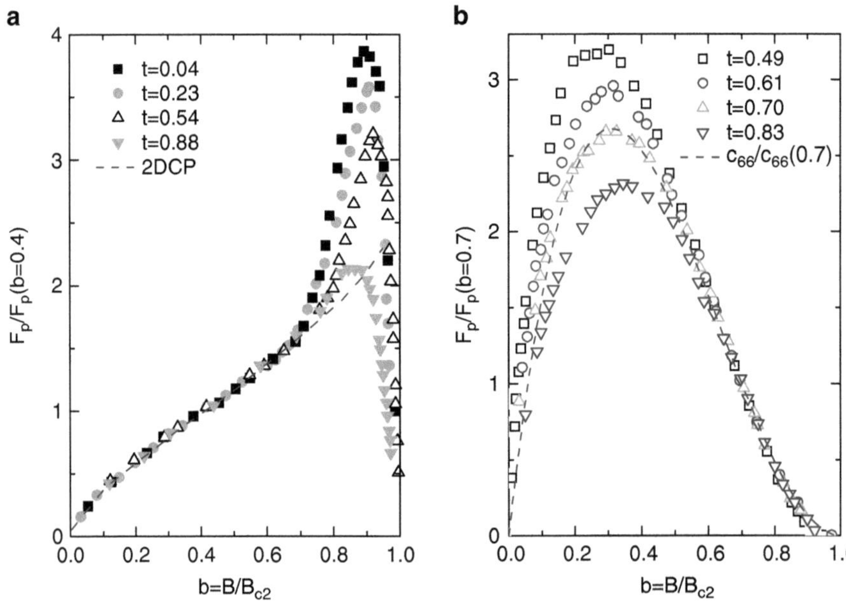

Fig. 2.5 Normalized volume pinning force as a function of applied magnetic induction for (**a**) *pin breaking mechanism* in weak pinning amorphous Nb_4Ge thin films with $F_p(b = 0.4)$ typically of the order of 10^5–10^6 N/m^3 at 2.2 K [16] and (**b**) *flux-line shear mechanism* in strong-pinning NbN thin films with $F_p(b = 0.7)$ typically of the order of 10^8–10^9 N/m^3 at 4.2 K [25]. The *dashed lines* represent the field dependence of 2D collective pinning theory with $L_c = d$ (**a**) and the shear module (**b**), respectively; d represents the film thickness, and t denotes the reduced temperature $t = T/T_c$

lattice. In the so-called flux-line shear mechanisms, the volume pinning force is determined by the vortex–vortex interaction. It is not given by the maximum volume pinning force $F_{p,max}$ of the weak or strong pinning areas, but it is determined by the plastic shear properties of the vortex lattice, since areas, that are weakly pinned, shear away from strongly pinned regimes. The resulting volume pinning force for the *flux-line shear mechanisms* is given by [24–27]:

$$F_p = F_{p,FLS} = G\, c_{66} \propto \frac{B_{c2}^2}{w} b(1-b)^2, \tag{2.12}$$

with $c_{66} \propto B_{c2}^2 b(1-b)^2$ representing the shear modulus of the vortex lattice, G a geometrical factor, $b = B/B_{c2}$ the reduced field, and B_{c2} the upper critical field. The typical field dependence obtained for a strong pinning superconductor is shown in Fig. 2.5b. It is characterized by a broad peak at small magnetic field typically at $B \approx B_{c2}/3$. The flux-line shear mechanism is usually encountered in strong pinning (i.e. technical) systems, whereas weak pinning superconductors often show collective pinning behaviour.

2.2.3 *Pinning Mechanism in HTS*

Bearing in mind the small coherence length and anisotropy, high-T_c material clearly shows a flux pinning according to the *flux-line shear mechanisms*. Vortices move in weak pinning "channels" when the critical current is exceeded. The field dependence of F_p shows the typical peak at small field, and it can be described by the field dependence of the flux-line shear mechanism (2.12), additionally taking into account the impact of thermal activation and/or a distribution in the channel properties (width, orientation, etc.). This is discussed in detail in [28]. The mechanism of vortex motion can also be visualized. For instance, low-temperature SEM experiments or magneto-optic recording of flux motion and penetration demonstrates the presence of channels along which vortices preferentially move (Fig. 2.6).

2.3 Vortex Manipulation in HTS Films

The *flux-line shear* mechanism automatically affects the conditions for vortex manipulation in HTS. In contrast to the *collective pinning* mechanism that is demonstrated for weak-pinning superconductors (e.g. amorphous conventional superconductors such as a-Nb_xGe, a-Mo_xSi or a-Pb), the properties of weak-pinning channels determine vortex motion and, thus, generation of flux noise in HTS devices. Vortex pinning and motion in HTS films are defined by the film's inhomogeneity and the vortex–vortex interaction. Additional (artificial) pinning sites are not expected to have a large impact on the critical properties of these materials. As a consequence, the strategy for vortex manipulation has to be different compared to vortex manipulation in weak pinning (e.g. amorphous) superconductors.

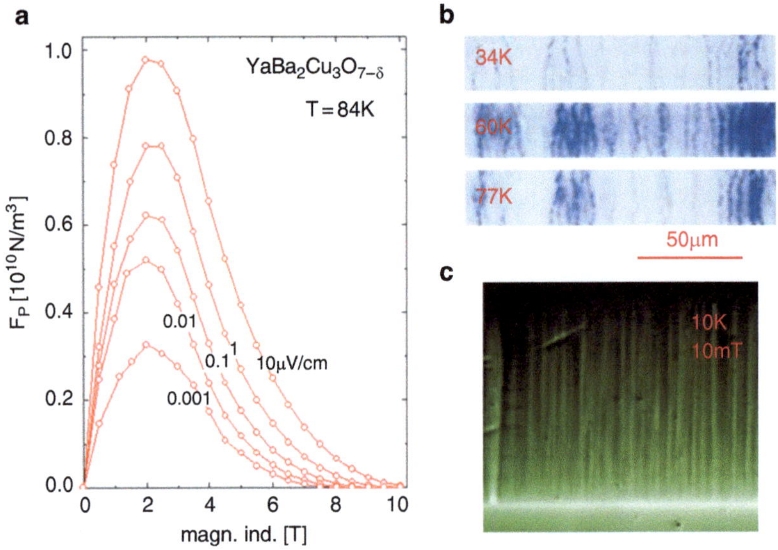

Fig. 2.6 Electronic and visual demonstration of the flux-line shear mechanism in high-T_c films: (**a**) magnetic field dependence of the volume pinning force for a YBCO thin film stripline at 84 K for various voltage criteria [28]; (**b**) low-temperature scanning electron microscope (LTSEM) voltage image of vortex motion in a 40-μm-wide YBCO stripline at various temperatures (R. Gross and D. Koelle, private communication) [29]; (**c**) Magneto-optical image of the flux distribution in a YBCO film on a 14° miscut $NdGaO_3$ substrate taken at 10 K and 10 mT. [12]. Darker areas correspond to larger contributions to flux motion (**b**) or smaller magnetic fields (**c**)

As a consequence, vortex manipulation has to be achieved via *guidance of vortices* rather than by vortex pinning or trapping for HTS films. The guidance of vortices could lead to a reduction in the driving force (e.g. guidance under an angle with respect to the Lorentz force, Sect. 2.3.3.2), guidance towards extended pinning centres (Sect. 2.3.3.1), guidance along special paths that reduce the impact of the motion of vortices (see for instance noise reduction in SQUIDs, Sect. 2.4.1), or guidance of vortices with asymmetric driving potential (see for instance vortex ratchets, Sect. 2.4.2.2). The different types of defects for guidance of vortices and the different strategies are sketched in the following sections.

2.3.1 Vortex Manipulation via Artificial Structures

Vortices can be manipulated by various modifications in the superconducting film.

1. Connected superconducting areas with locally tailored electrical properties (mainly "channels for vortex motion") are mainly aiming at vortex guidance. They can be created among others by:

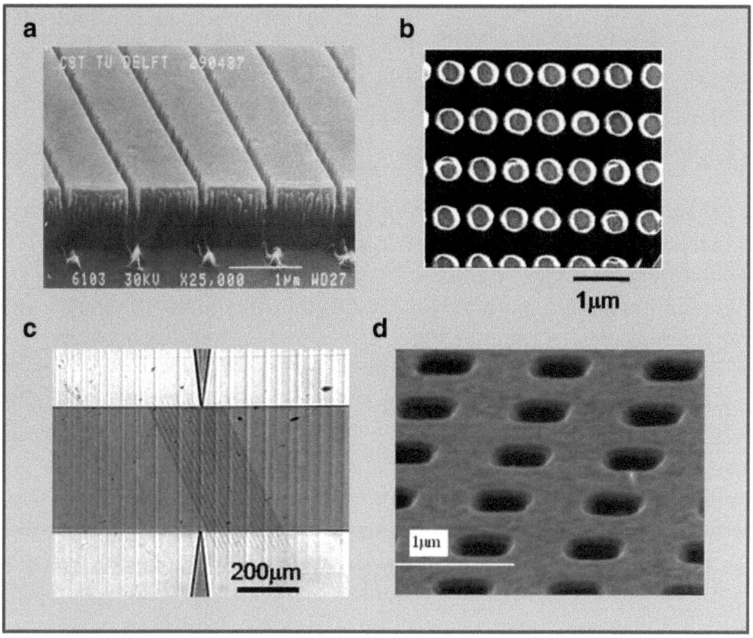

Fig. 2.7 SEM images of artificial structures for vortex guidance in NbN/Nb_3Ge bilayer [10,13,25] (**a**), magnetic dots (Ni) on Nb [30] (**b**), heavy-ion patterned channels in $YBa_2Cu_3O_{7-\delta}$ [31] (**c**) and antidots in $YBa_2Cu_3O_{7-\delta}$ [15, 32, 33] (**d**)

(a) Combination of weak and strong pinning material (Fig. 2.7a) [10, 13, 25]
(b) Thickness variation (see for instance [11])
(c) Grain boundaries and other extended natural defects (e.g. films on vicinal cut substrates, Fig. 2.6c) [12, 34]
(d) Modification of superconducting properties (e.g. ion beam patterning, Fig. 2.7c) [31]

2. Non-simply connected micro- or nanostructures will cause a long-range electronic or magnetic interaction between vortex and artificial structures. Arrangements (rows or arrays) of various structures are feasible and will lead to vortex pinning or trapping as well as vortex guidance. Different micro- and nano-objects are feasible:

(a) Dots (magnetic or non-magnetic) [30, 35–37]
(b) Completely etched holes (antidots) [14, 15, 32, 33, 38–43]
(c) Partially etched holes (blind holes) [44].

Not all of these structures are suitable for HTS films. Some of the different ways to control vortex pinning and vortex motion are sketched below.

Artificial channels: An obvious method to provide controlled vortex motion (guided vortex motion) is given by the pattering of narrow channels into superconducting material. This can be done by etching channel structures into a single layer leading to modification of the pinning force due to thickness variation, local modification of the superconducting properties or by combining two layers of superconducting material with different pinning properties.

An intriguing example of easy vortex flow channels consists of a weak-pinning amorphous Nb_3Ge bottom layer combined with a strong-pinning NbN top layer, into which the small channels are etched (Fig. 2.7a). A detailed report of experimental results obtained from resistive measurements on these channels is given in [25], and simulation of the vortex motion in the channels is given in [13]. The vortices in the channels predominantly experience the interaction with the row of pinned vortices at the edge of the channel within the NbN. The commensurability between the vortex lattice with field dependent lattice parameter $a_o = b_o(\phi_o/B)^{1/2}$ and the channel size leads to periodic oscillations of the volume pinning force. It is demonstrated that vortices move within the channels if the total driving force on the vortices in the channel exceeds the shear forces at the channel edges.

Similar structures are possible for combinations of HTS films with weak pinning conventional superconductors. However, to preserve the high temperature operation, all high-T_c combinations have to be chosen. Since combinations of different HTS films prove to be difficult to obtain, modification of the superconducting properties are the easiest way to pattern channels into HTS films.

Heavy ion lithography: Ion irradiation of HTS films offers a unique possibility to create a wide range of different defects and to tailor the electrical and superconducting properties. Depending on the species of ions used during the irradiation, their energy and fluence, nanoscale columnar pinning centres can be created, which locally enhance or diminish the pinning properties. A review of this effect is given in this book [45]. While, the irradiation with relatively low fluence of high energy heavy ions leads to an enhancement of the critical current due to the strong vortex pinning at columnar defects, relatively high fluence leads to a reduction of the critical properties. Thus, the superconducting properties can be controlled and modulated locally. Recently, the preparation of artificial channels for flux motion has been demonstrated with this technology [31].

Dots and antidots: In contrast to simply connected structures, dots and holes (antidots) offer a number of advantages for the manipulation of vortices:

1. Dots and/or antidots of various size, properties and shapes can be used or even combined (see Sect. 2.3.3)
2. They can be positioned more or less at wish, fancy structures are possible (see Sect. 2.4.2.2)
3. They can be used as pinning or trapping sites as well as for guidance of flux
4. The superconductivity is not modified over the entire cross-section like in case of the formation of channels in HTS films

The latter is important for a number of possible applications, e.g. when a reduction in the transition temperature is not tolerable.

While (magnetic) dots have not been successfully been tested for HTS films, the preparation (of arrays) of micron- or nanosize antidots has been proven to be a very effective way to manipulate flux in HTS films and devices. In contrast to other pinning defects, which have to be of the size of the superconducting coherence length ξ (core interaction [33, 46–51]), antidots with sizes much larger than ξ will trap magnetic flux very effectively [15]. The advances in lithography techniques and the possible use of antidots in applications (e.g. SQUIDs [52,53], vortex diodes [33,48–51] and microwave devices [57]) have led to a renewed interest in the research of superconducting films containing antidots or antidot lattices. Antidots are successfully prepared in films of conventional superconductors (typically weak pinning Pb, V, or a-WGe thin films, Pb/Ge multilayers, Pb/Cu bilayers or Nb foils) [14, 33, 39–41, 43, 58–63] as well as HTS material (YBCO) [15, 32, 33, 52, 53, 64]. Intensive studies have been performed of:

1. Commensurability effects (matching effects) between the antidot lattice and vortex lattice
2. Multiquanta formation in the antidots
3. Guided motion along rows of antidots (e.g. via resistive Hall-type measurements)
4. Visualization of vortex motion (e.g. magneto-optical imaging, Hall-probe scanning microscopy, Lorentz microscopy)
5. Implementation of antidots in cryoelectronic devices (e.g. noise reduction in SQUIDs or improvement of the microwave properties of HTS resonators and filters)

2.3.2 Theoretical Considerations of Vortex Manipulation via Antidots

There are a number of theoretical methods to analyse vortex matter in patterned superconducting thin films ranging from electrodynamics, free energy considerations and Ginzburg–Landau formalism to numerical computer simulation. A few basic ideas are sketched in this section.

Computer simulations of vortex motion in thin films are usually based on 1D or 2D (depending on the examined problem) versions of a molecular dynamics code solving the overdamped equation of motion for individual vortices that are subject to an external driving force (e.g. Lorentz force f_L), vortex–vortex repulsion f_{VV} and additional forces (e.g. due to thermal fluctuations f_T). The net force f_i acting on an individual vortex and that leads to a motion with velocity v_i that is typically given by [46, 65–68]:

$$\vec{f}_i = \vec{f}_L + \vec{f}_{VV}(\vec{r}_i) + \vec{f}_p(\vec{r}_i) + \vec{f}_T(\vec{r}_i) = \eta \vec{v}_i. \tag{2.13}$$

With adequate expressions for the different forces and the viscosity η molecular dynamics simulation of vortex motion has proven to be a powerful tool for analysing, predicting and visualizing vortex dynamics in complex thin film systems including studies of vortex phases and phase transitions, vortex mobility and

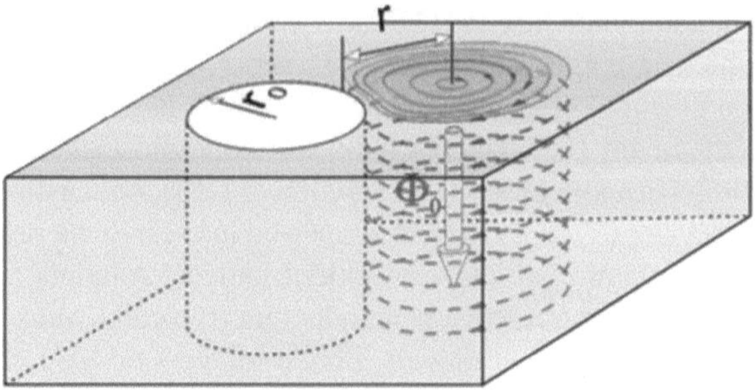

Fig. 2.8 Sketch of a vortex interacting with a cylindrical cavity (e.g. an antidot with radius r_o) [54]

dynamics up to the development of novel "fluxonic" devices. The major handicap of this approach is given by the fact that vortices are treated as "molecules" in this model, i.e. modifications of vortices for instance at boundaries – especially, vortex nucleation or annihilation – are not properly described (Fig. 2.8). The latter is of importance for the description of vortex mater in micro- or nanostructured superconductors.

Vortex–antidot interaction: The interaction energy between a vortex and a small insulating cylindrical cavity (analogue to an antidot) in high-κ superconductors has been calculated using the London approximation [55] and in an alternative approach (using the analogy between a vortex close to an antidot and a charge line in an infinite dielectric close to a cylindrical cavity of different dielectric permittivity) in [56]. Later, the calculations have been extended to arbitrarily large cavities [69]. It is demonstrated that the interaction energy becomes the same as the one between a vortex and the straight edge of a superconductor (Bean–Livingston barrier [70]) when the radius of the antidot goes to infinity. Although the precise form of the interaction potential between a vortex and a cylindrical antidot is slightly different in these studies [55,56,69], the main conclusions are identical.

Based on a series expansion of Bessel functions K_o of the second kind, the free energy of a vortex at a radial distance r of an antidot with radius r_o is given by [55,56]:

$$F(r) = \frac{\Phi_o^2}{4\pi\mu_o\lambda^2}\left[K_o\left(\frac{\xi}{\lambda}\right) + n^2 K_o\left(\frac{r_o}{\lambda}\right) + 2nK_0\left(\frac{r}{\lambda}\right) + \ln\left(1 - \frac{r_o^2}{r^2}\right)\right] \quad (2.14)$$

for $\xi < (r,\ r_o) << \lambda$ and n representing the number of flux quanta that are already trapped in the antidot. The elementary vortex–antidot interaction is given by $f_{VA} = -\partial F/\partial r$.

The vortex–antidot interaction depends upon the magnetic flux $\Phi = n\Phi_o$ trapped in the antidot (Fig. 2.9). The interaction between a vortex and an "empty" antidot is attractive for all vortex–antidot distances. As soon as one flux quantum is trapped in the antidot, the interaction potential changes and a surface barrier emerges at the edge of the antidot. The height of the barrier increases with increasing number of

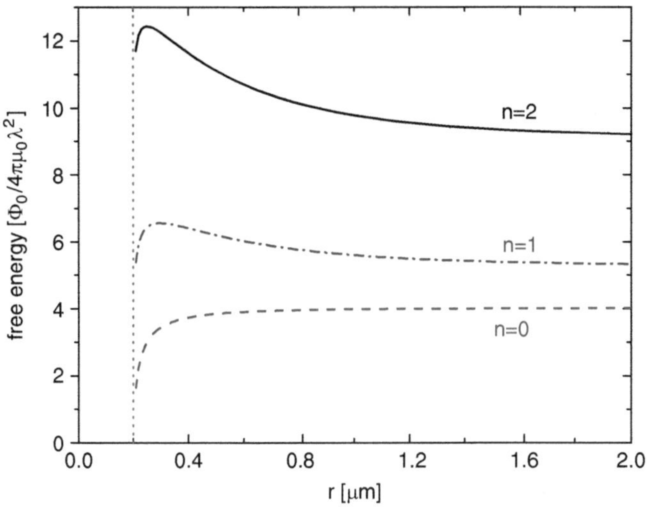

Fig. 2.9 Free energy of a vortex as function of the radial distance from an antidot for different occupation numbers n of the antidot with $\lambda = 600\,\text{nm}$, $\kappa = 50$ and $r_o = 200\,\text{nm}$

trapped flux quanta. This automatically defines a saturation number n_s. For $n \geq n_s$, the interaction becomes repulsive, and no additional flux can be trapped by the antidot. The saturation number depends upon the size of the antidot. For small antidots (single antidot with $r_o \ll \lambda$), the saturation number can be approximated by [55]

$$n_s \cong \frac{r_o}{2\xi(T)}. \tag{2.15}$$

The pinning force per unit length depends on the radial distance between vortex and antidot and has a maximum value close to the antidot edge:

$$f_{p,\max} \approx \frac{\Phi_o^2}{4\sqrt{2}\pi\mu_o\lambda^2\xi}\left(1 - \frac{n}{n_s}\right), \tag{2.16}$$

i.e. the maximum vortex–antidot interaction decreases with increasing occupation number and becomes zero for $n = n_s$.

Generally, the saturation number depends upon temperature, i.e. it increases with decreasing temperature. Furthermore, a consequence of a saturation number $n_s > 1$ is the formation of so-called multi-quanta vortices. Experimental proofs of the existence of multi-quanta vortices will be given in Sect. 2.3.3.1.

Ginzburg–Landau formalism: Vortex matter in nanostructured superconductors is generally very complicated. For instance, flux motion across a superconducting microbridge (e.g. formed by closely spaced antidots) of the size of the characteristic length of the superconductor will not lead to vortex formation. A phase slip line will develop along the path of the flux motion leading to Josephson-like behaviour of the bridge [71]. According to theory, the characteristic length of the superconductor for this effect should be the coherence length ξ. However, it has been shown that phase slip lines are present for bridge dimensions w up to the penetration length λ [71]. The effective penetration length $\lambda_{\text{eff}}(T) = \lambda_L(T)\coth[d/2\lambda_L(T)]$ in thin $YBa_2Cu_3O_{7-\delta}$ films with thickness of typically $d = 80$–200 nm range between 400 and 1000 nm. Furthermore, it has been shown that phase slip can be present even in bridges with larger dimensions, i.e. up to $w = 1$–$1.5\,\mu\text{m} > \lambda$ [72,73].

Another interesting effect that might be relevant for the nucleation of vortices in mesoscopic systems is known from observations of classical von Karman vortices. After nucleation the vortex structure develops in time. Due to the motion, a given space is necessary for a vortex to fully develop. If this space is not provided, vortices will not be originated. A similar situation might apply for Abrikosov vortices in mesoscopic superconductors. The nucleation of flux in the superconductor and its shuttling between adjacent antidots might not lead to vortex formation due to the restricted distance between the antidots. This is among others described by the term "kinematic" vortices [74].

As a consequence, flux transport in HTS microstructures (e.g. closely spaced antidots) is not properly described by molecular dynamics or energy considerations that are based on "pristine" Abrikosov vortices. For this problem, the Ginzburg–Landau formalism might be more adequate.

Depending on the considered problem, different approaches are chosen. In the *stationary case*, a self-consistent set of differential equations has to be solved for the order parameter ψ and the vector potential $\mathbf{A}$ of a magnetic field $\mathbf{H} = \nabla \times \mathbf{A}$ [75,76] which can be presented in the dimensionless form as follows:

$$\begin{aligned} &(-i\nabla - \mathbf{A})^2\psi - \psi\left[\left(1 - \tfrac{T}{T_c}\right) - |\psi|^2\right] = 0 \\ &-\kappa^2\Delta\mathbf{A} = -\tfrac{i}{2}(\psi * \nabla\psi - \psi\nabla\psi *) + \mathbf{A}\,|\psi|^2 = \mathbf{j}, \end{aligned} \tag{2.17}$$

with the boundary condition $\text{n}(-i\nabla\psi - \mathbf{A}\psi)|_{\text{boundary}} = 0$, $\mathbf{n}$ representing the unit vector normal to the boundary and $\kappa = \lambda(T)/\xi(T)$. For the analysis of *dynamic problems*, the time-dependent Ginzburg–Landau equation has to be used [77–79] where the time τ_{in} of the nonequilibrium quasiparticle distribution function is explicitly included.

Solutions based on the Ginzburg–Landau formalism exist among others for the vortex–antidot interaction [54], the current and magnetic field distribution in nanostructured films (e.g. for superconducting films with antidot arrays Fig. 2.10), the vortex and flux transfer between antidots (e.g. see Fig. 2.11) and the vortex dynamics up to high velocities [81].

Critical vortex velocity: Studies of the energy dissipation due to vortex motion is predominantly focussed on the limit of small driving forces and small frequencies, i.e. the onset of vortex dynamics in the limit of small vortex velocities. However, a number of HTS applications such as coated conductors, fault-current limiters or microwave devices operate at high power levels or frequencies and, therefore, potentially in the regime of high vortex velocities. Based on Eliashberg's

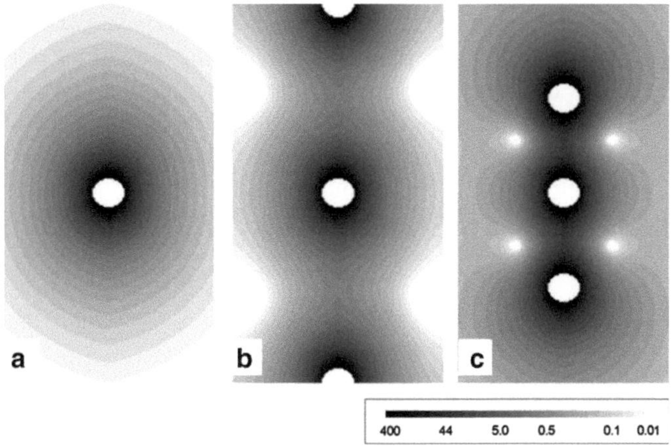

Fig. 2.10 Density distribution of the screening current for two-dimensional superconductor (e.g. thin film) with a row of antidots placed at different distances d_v (measured in units of the antidot radius) from each other obtained from simulations on the basis of the Ginzburg–Landau equations in the high-κ limit: (**a**) $d_v = 66.7$, (**b**) $d_v = 13.3$, (**c**) $d_v = 6.7$. High local current densities are indicated with dark grey; the scale is given in arbitrary units [33]

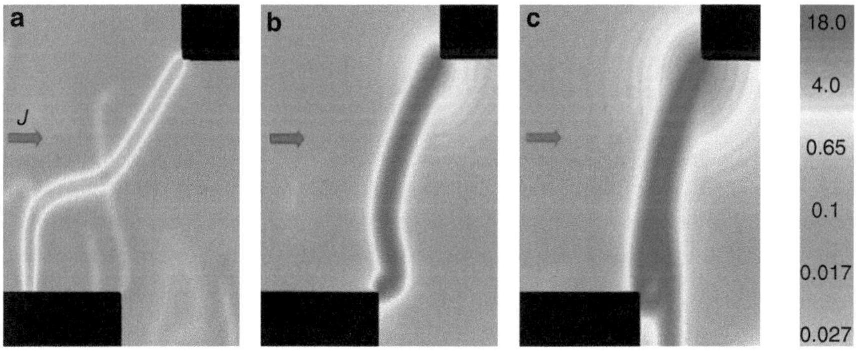

Fig. 2.11 Accumulated trajectories of vortices in a thin YBCO film with two indentations (antidots) for different applied current densities $0.333 \times 10^{10}\,A/m^2$ (**a**), $0.666 \times 10^{10}\,A/m^2$ (**b**) and $1.332 \times 10^{10}\,A/m^2$ (**c**). The sizes of the areas are $0.24\,\mu m \times 0.24\,\mu m$, the *black* areas represent parts of the square size antidots and the *arrow* indicates the nominal direction of the applied current [80]

ideas on non-equilibrium effects in superconductors, Larkin and Ovchinnikov predicted that a non-equilibrium distribution and relaxation rate of the normal charge carriers (treated as quasiparticles) develop during the motion of vortices at high velocities [82]. As a result, they expected a discontinuity in the current–voltage characteristic (IVC).

According to the Larkin and Ovchinnikov (LO) theory, the viscous damping coefficient at a vortex velocity v is given by [82]:

$$\eta(v) = \eta(0)\left[1 + \left(\frac{v}{v^*}\right)^2\right]^{-1}, \tag{2.18}$$

with a critical vortex-velocity v^*:

$$v^* = \left[\frac{D\sqrt{14\zeta(3)}}{\pi\tau_{\text{in}}}\sqrt{1-\frac{T}{T_{\text{c}}}}\right]^{1/2}, \tag{2.19}$$

τ_{in} denoting the inelastic quasiparticle scattering time, $\zeta(x)$ the Riemann–Zeta function, $D = v_{\text{F}}l_{\text{o}}/3$ the quasiparticle diffusion coefficient, v_{F} the Fermi velocity and l_{o} the electron mean free path. According to LO, a nonlinear IVC is expected for a critical electric field $E^* = v^*B$ according to:

$$J = \frac{E}{\rho_{\text{ff}}}\left\{\left[1 + \left(\frac{E}{E^*}\right)^2\right]^{-1} + \gamma\left[1-\frac{T}{T_{\text{c}}}\right]^{1/2}\right\} + J_{\text{c}} \tag{2.20}$$

with $\gamma \approx 1$. This behaviour (see sketch in Fig. 2.12a) is observed for conventional superconductors [85, 86] as well as for high-T_{c} thin films [83, 84, 87]. It appears for instance when the current limit is exceeded in superconducting resistive fault-current limiters. In this case, the limiter shows an extremely sharp and sudden voltage peak at the critical electric field E^* during the quench of the superconductor at high power [88].

Large vortex velocities are also expected in fluxon devices working at high frequencies. For instance, a net flux transport in an impedance-matched microwave stripline would require a vortex motion across in a stripline of a typical width of $w = 400$–$500\,\mu\text{m}$ in a time $t < 1/2f$. As a result, a vortex velocity $v > 1\,\text{km/s}$ is expected for frequencies $f > 1\,\text{MHz}$. Thus, vortex velocities that clearly exceed the critical velocity for high-T_{c} films are expected for fluxon devices operating in the MHz–GHz frequency regime and temperatures not too close to T_{c} (see inset of Fig. 2.12b). One of the major questions is whether vortex properties such as vortex–antidot interaction or vortex mobility will be modified at high frequencies.

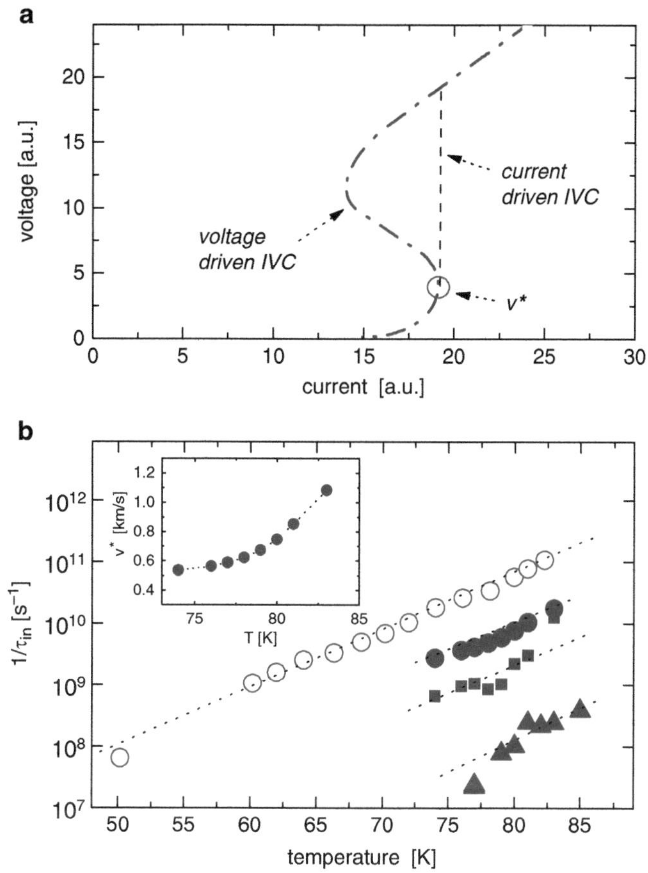

Fig. 2.12 (**a**) Schematic drawing of voltage-controlled and current-controlled IVCs measured up to voltages $V > V^*$ and (**b**) inelastic scattering rate and critical vortex velocity (*inset*) as function of temperature for YBCO films obtained from IVC measurements via rapid current ramps (*open symbol*) [83] and current pulses (*solid symbols*) [84]

2.3.3 Experimental Demonstration

Vortex manipulation has been demonstrated among others via microscopy (scanning Hall experiments, magneto-optic imaging, Lorentz microscopy or laser scanning microscopy), resistive, Hall-type, inductive or magnetization measurements or microwave experiments. Most of these experiments have been performed for weak pinning superconducting systems (e.g. Pb, V, or a-WGe thin films, Pb/Ge multilayers, Pb/Cu or Nb/Ni bilayers). However, for most applications strong-pinning superconductors and, especially, high-T_c films are of interest.

A first report of a successful preparation and a clear indication of an attractive interaction between antidots and vortices in HTS is given in [15]. Based on

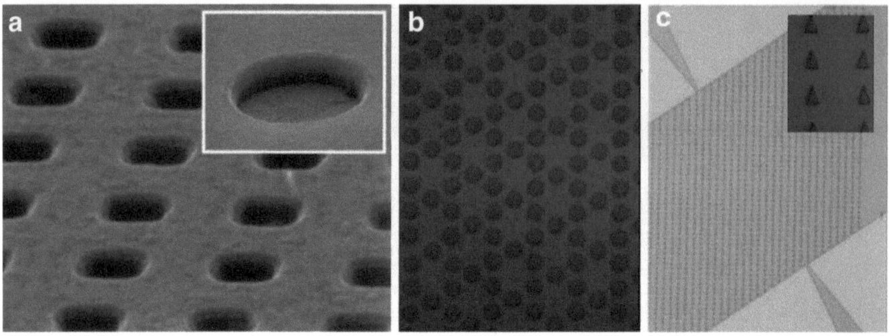

Fig. 2.13 SEM images of various antidot lattices in YBCO films: (**a**) square lattice (periodicity $d_{AA} = 1\,\mu m$, antidot radius $r_o = 220\,nm$), (**b**) Kagome lattice ($d_{AA} = 6\,\mu m$, $r_o = 1\,\mu m$), and (**c**) rows of anisotropic antidots ($d_{AA} = 10\,\mu m$ in the row)

this work, antidots in HTS thin films have been used for guidance of vortices, microwave experiments, improvement of existing HTS devices and development of novel fluxon concepts.

Generally, high structural perfection and small surface roughness are required for an optimized patterning process for micrometer and, especially, for submicrometer structures in YBCO films. Furthermore, for microwave experiments suitable substrates (e.g. $LaAlO_3$ or CeO_2 buffered sapphire) should be chosen. High quality $YBa_2Cu_3O_{7-\delta}$ (YBCO) films are typically deposited via PVD technologies, e.g. high-pressure sputter technique, pulsed laser deposition or thermal coevaporation. Typical deposition parameters are for instance:

- *Pulsed laser deposition*: substrate temperature of 700 °C, pressure of 0.5 mbar of pure oxygen, pulse rate of 10 Hz (1 J/pulse) and energy density of $5\,J/cm^2$ at the target lead to a deposition rate of $\sim 1.67\,nm/s$
- *Sputter deposition*: 6″ dc magnetron at 180 W, pressure of 0.6 mbar of Ar/O_2 ratio of 2:1, heater temperature of 870 °C yield a deposition rate of $\sim 0.8\,nm/min$ for YBCO and 6″ rf magnetron at 150 W, 0.13 mbar of Ar/O_2 ratio of 8:1, heater temperature of 870 °C yield a deposition rate of $\sim 0.3\,nm/min$ for CeO_2.

Typical images of antidots and antidot lattices in YBCO films are given in Fig. 2.13. In the following sections, experimental results are discussed that are recently obtained for vortex manipulation via antidots in high-T_c films.

2.3.3.1 Vortex–Antidot Interaction and Multi-Quanta Formation

First clear indications of an attractive interaction between antidots and vortices in HTS are given in [15]. Figure 2.14 represents a comparison between the magnetic field dependence of critical current I_c and critical current density J_c as a function of the magnetic field for a YBCO thin films with and without antidots. First, the critical current seems to be enhanced by the presence of the antidot lattice (inset

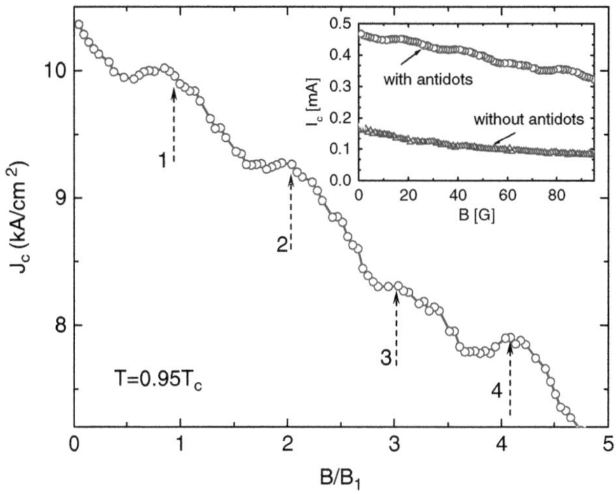

Fig. 2.14 Critical current density and critical current (*inset*) as a function of magnetic induction for YBCO without and with antidots measured at $T = 0.95\ T_c$ [15]. The antidot lattice is shown in Fig. 2.13a; it is a square lattice with periodicity $d_{AA} = 1\ \mu m$ and antidot radius $r_o = 220$ nm. The resulting first matching is expected to be $B_1 = \Phi_o/d_{AA}^2 = 2.07$ mT. Both samples (with and without antidots) are obtained from the same film and posses identical dimensions

of Fig. 2.14). It demonstrates the additional flux pinning provided by the antidots. Second, the reference sample (patterned simultaneously into the same film but without antidots) shows the usually observed monotonic field dependence of J_c. In contrast, the perforated YBCO stripline shows peaks or cusps in the magnetic field dependence of the critical current at the theoretically expected matching fields $B_n = nB_1$. Matching peaks could be resolved at integer $n = 1, 2, 3$ and 4. In Sect. 2.4.1.1, it is shown that commensurability effects are even present at rational positions (n/m [89]). Small shifts in the peak positions have to be ascribed to field gradients in the sample that develop during the field ramps in strong-pinning material [15]. The experiments shown in Fig. 2.14 are executed at temperatures close to T_c. The enhancement of the volume pinning force and the presence of matching effects [cusps in $J_c(B)$] could be resolved only down to temperatures of about 50 K. At lower temperatures, the pinning by natural defects seems to dominate the vortex–antidot interaction.

Another interesting aspect of vortex pinning by antidots is given by the formation of *multi-quanta vortices*. Already in the first theoretical considerations [55] a flux occupation of the antidot of $\Phi = n\Phi_o$ with $1 \leq n \leq n_s$ [see (2.15)] was predicted. Due to their specific character, antidots trap quantized flux and not Abrikosov vortices. By producing the so-called blind holes (not completely etched holes), the Abrikosov vortex character is preserved and the formation of multi-flux-quanta trapping by antidots can be visualized in the form of flux splitting into single-flux-quanta vortices within each blind hole. This phenomenon has been demonstrated via Bitter-decoration experiments on Nb films with regular arrays of blind holes [60,61].

The existence of multi-quanta vortices in antidots in HTS thin films has been demonstrated by inductive measurements that have been performed on YBCO films with regular square lattices of antidots (Fig. 2.15) [32]. The experimental technique is based on the superposition of an ac and dc magnetic field normal to the film surface. The amplitude of the ac field is typically much smaller than the amplitude of the dc field. The in-phase and out-of-phase components of the fundamental signal are recorded by the pick-up coil. They are proportional to the in-phase and out-of-phase components of the complex susceptibility $\chi(T,B) = \chi'(T,B) + i\chi''(T,B)$ that represent the stored energy ($W_m = \chi' {B_a}^2/2\mu_o$) and the

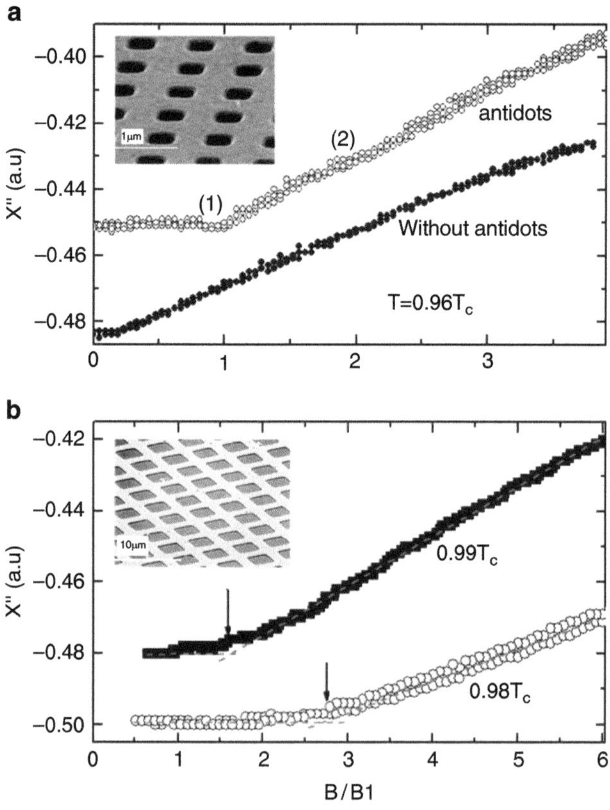

Fig. 2.15 Out-of-phase susceptibility component χ'' as a function of the dc bias field for YBCO films without and with a square lattice of antidots of different sizes [32]. B_1 refers to the first matching field of the given antidot lattice. (**a**) χ'' obtained for a reference film (without antidots) and a film with an antidot lattice ($d_{AA} = 1\,\mu$m, antidot radius $r_o \approx 0.34\,\mu$m) at $T/T_c = 0.96$ ($n_s = 1$). (**b**) χ'' is shown for a film with an antidot lattice ($d_{AA} = 5\,\mu$m, $r_o \approx 1.75\,\mu$m) at $T/T_c = 0.99$ ($n_s \approx 2$) and $T/T_c = 0.98$ ($n_s \approx 3$). The insets show SEM images of the YBCO films with different square lattices

dissipated energy ($W_q = 2\pi\chi''$), respectively. By varying the dc magnetic field B, the dissipated energy due to vortex motion in the film can be recorded as a function of temperature and field.

Figure 2.15 shows the variation of out-of-phase component χ'' as a function of the magnetic field at temperatures close to T_c (high mobility of the vortices) for YBCO thin films without antidots and with square arrays of antidots of different sizes. For the measurements on the reference films without antidots, χ'' increases monotonically (except for the fields below the magnet's remnant field of about 0.5 mT). In contrast, the out-of-phase component for the film with small antidots is constant up to the first matching field B_1 for which the density of vortices matches the density of antidots. Above B_1, χ'' increases with a slope similar to that observed for the reference sample. Small curvatures are observed at the matching fields $2B_1$ and $3B_1$. The inductive measurement of the antidot sample in Fig. 2.15a confirms the presence of matching effects and demonstrates that the saturation number $n_s = 1$ for this antidot configuration and temperature. Multi-quanta vortices are expected at lower temperatures. Unfortunately a high mobility of the vortices is needed for this experiment; hence, experiments at lower temperatures are not possible. However, by increasing the size of the antidots the presence of multi-quanta vortices can be demonstrated. This is demonstrated in Fig. 2.15b for larger antidots ($r_o \approx 1.75\,\mu m$) where the saturation number n_s changes from 2 to 3 for a small temperature reduction.

Multi-quanta formation in antidot arrays in YBCO films has also been visualized via high-resolution magneto-optic technique that produces a 2D image of the local magnetic field [90, 91]. Figure 2.16 shows a 3D reconstruction of the trapped flux in a hexagonal antidot lattice in a YBCO film, which was recorded at 7 K and zero magnetic field after a field of $B = 1$T had been removed. The height of the peaks

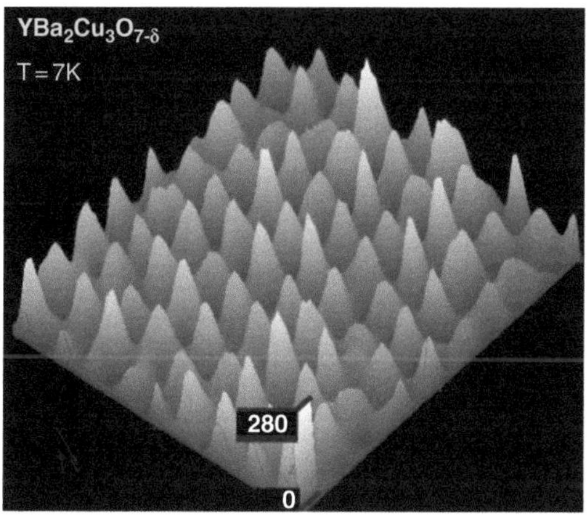

Fig. 2.16 Reconstruction of the trapped magnetic flux distribution obtained from a magneto-optic image that was recorded at 7 K in zero-field after applying at field of 1T [90]. The local field shows clear maxima at the position of the antidots (*triangular lattice*, $d_{AA} = 10\,\mu m$, $r_o \approx 1\,\mu m$). The scale for the height of the peaks is given in numbers of trapped flux quanta

represents the amount of trapped flux quanta. No flux (vortices) could be detected between the antidots which indicate the higher mobility (small pinning potential) of the interstitial vortices compared to flux in the antidots. Multi-quanta vortices with flux up to $\Phi = 280\ \Phi_o$ are trapped at this temperatures after removal of the field. Inserting $n_s \approx 280$ into expression (2.15) yields a reasonable value for the coherence length, i.e. $\xi(7\mathrm{K}) \approx 1.8$ nm. Furthermore, using the extended expressions [92] derived for the saturation number in square or hexagonal arrays leads to a reasonable value for the lower critical field of $B_{c1}(7\mathrm{K}) \approx 41$ mT [15].

2.3.3.2 Guided Vortex Motion via Antidots

Probably even more important than flux trapping is the guidance of vortices by artificial defects for the general understanding of vortex matter and applications of HTS. Guided motion has been demonstrated in conventional weak pinning superconductors using lithographically etched channels (Fig. 2.7a) [10, 13, 25]. The use of connected superconducting areas with locally tailored electrical properties that form channels for flux motion is also possible for HTS material. Guidance has been demonstrated with channels prepared by heavy ion irradiation [31] in HTS thin films. However, an intriguing and more flexible method of vortex guidance is provided by special arrangements of antidots [90, 91].

First indications for guidance of vortices by rows of antidots have been obtained from magneto-optic imaging of flux penetration in patterned HTS films [91]. Figure 2.17 represents a magnetic-field map of a disc-shaped YBCO thin film after cooling in zero field to 4.2 K and successive increase in the field up to 30 mT. There are two important features [91]:

1. The flux penetration pattern shows a very clear fourfold (D_4) symmetry, i.e. the penetration is highly anisotropic, compared to the cylindrical symmetry that is observed for non-perforated disc samples.
2. The flux penetrates preferentially along the antidot lattice vector that is closest to the direction of the Lorentz force. The Lorentz force is caused by the screening current, and its direction is normal to the edge of the sample.

A first basic explanation of the angular dependence of the guidance of vortices via rows of antidots is given by the n-channel model [33]. The sketch in Fig. 2.18 illustrates the angular dependence of V_{Hall} given in a simplified one-channel model, in which the flux is expected to drift only along rows of antidots. The orientation of the rows is given by the angle γ. The component of the Lorentz force, which compels vortices to move along the antidot rows (i.e. guided motion), is $F_{\mathrm{guid}} = F_{\mathrm{L}} \cos\gamma$, where F_{L} is the modulus of the Lorentz force: $F_{\mathrm{L}} = |\boldsymbol{F}_{\mathrm{L}}|$. The components parallel and perpendicular to the applied current are:

$$
\begin{aligned}
F_{\mathrm{x,guid}} &= F_{\mathrm{guid}} \sin\gamma = F_{\mathrm{L}} \cos\gamma \sin\gamma \\
F_{\mathrm{y,guid}} &= F_{\mathrm{guid}} \cos\gamma = F_{\mathrm{L}} \cos^2\gamma,
\end{aligned}
\tag{2.21}
$$

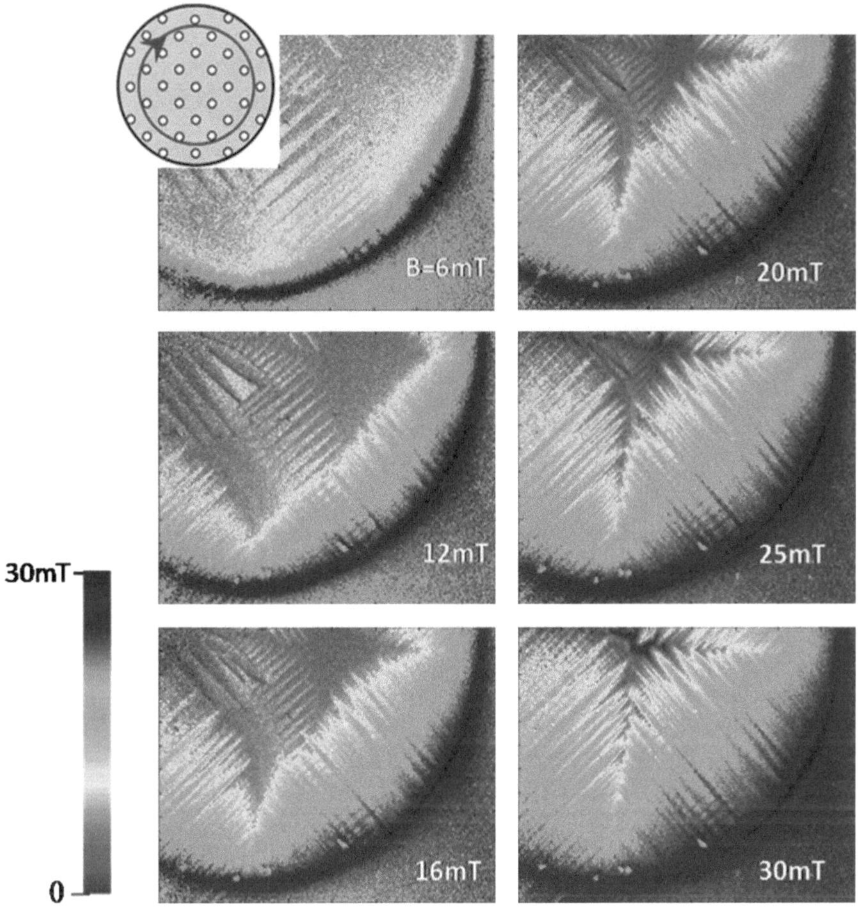

Fig. 2.17 Magneto-optic visualization of the anisotropic (guided) flux penetration in a disk-shaped YBCO film with a square lattice of antidots ($d_{AA} = 10\,\mu m$, $r_o \approx 1\,\mu m$) according to [91]. To enhance spatial resolution, only one-quarter of the disk is shown. The images are obtained at 4.2 K after zero field cooling, the field is gradually increased to $B_a = 6$, 12, 16, 20, 25 and 30 mT, respectively. The scale indicates the magnitude of the local field. The orientation of the square antidot lattice and the direction of the screening current that leads to radial oriented Lorentz force are indicated in the sketch

with $F_{x,guid}$ and $F_{y,guid}$ contributing to the Hall and longitudinal voltage signal, respectively. The experimentally determined angular dependence of the Hall voltage roughly obeys this simple relation $V_{Hall} \propto F_L \cos\gamma \sin\gamma$ obtained in this "1- channel model" [33]. Actually, it has to be considered that vortices can move with some probability also between antidots of neighbouring rows. These additional channels of vortex motion become important for large angles γ (Fig. 2.22) [93]. However, for these values of γ, the contribution of the vortex motion to V_{Hall} is small [see (2.21)]. Nevertheless, taking into account these additional channels of vortex

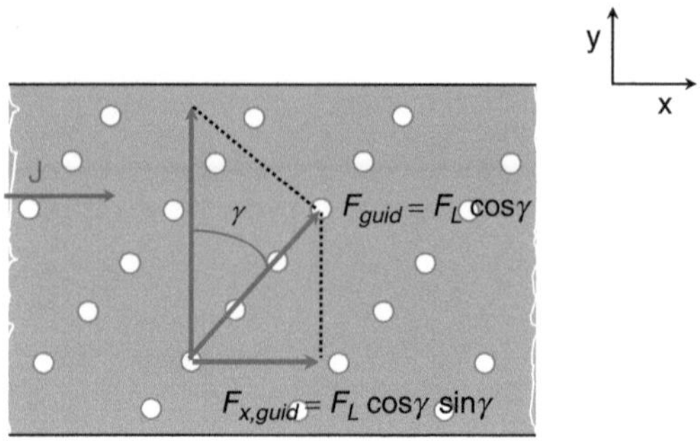

Fig. 2.18 Schematic illustration of the angular dependence of the Hall signal according to the n-channel model [33]

motion (i.e. vortices motion between antidots of adjacent rows) (2.21) should be substituted by a more general expression yielding

$$V_H \propto \sum_i P_i(\gamma) F_L \cos\gamma \ \sin\gamma, \tag{2.22}$$

where the summation is performed over all the channels of vortex motion, and $P_i(\gamma)$ is the angle-dependent probability of the motion along the i-th channel.

In order to obtain a more detailed insight into the guidance of vortices via rows of antidots, resistive six-probe dc measurements are executed on HTS films equipped with rows of antidots that simultaneously record the longitudinal voltage signal and the Hall signal [33,64]. Figure 2.19a shows schematically the typical sample design that is also suitable for microwave experiments on vortex manipulation [64].

The longitudinal voltage (contacts V_x) represents the standard parameter to characterize the critical properties of a superconducting stripline. In the normal regime, $V_{longitudinal}$ represents the normal state resistivity, whereas in superconducting regime it is generated by the component of vortex motion along the Lorentz force. In the latter case, $V_{longitudinal}$ is a measure for the flux transfer across the stripline that is collected between the longitudinal voltage contacts labelled V_x (integral signal). It characterizes the average velocity component $<v_y>$ of vortex in the stripline. In contrast, the Hall signal in superconducting regime characterizes the complementary velocity component $<v_x>$ of vortex and represents a more local analysis of the vortex motion that is restricted to the vicinity of the contacts pair (Hall contacts V_y in Fig. 2.19a).

High-T_c materials typically reveal the, so-called, anomalous Hall effect (AHE), i.e. the sign inversion of the Hall signal below T_c [94–103]. It has been shown that the AHE in high-T_c material is caused by vortex motion. Most likely the additional component of vortex motion is caused by the Magnus force affecting vortex motion close to T_c when pinning is very weak [103]. At temperatures close to T_c, the AHE competes with the guidance of vortices by rows of antidots [guided motion (GM)]. Depending on the orientation of the rows of antidots, the AHE will be suppressed. This is sketched in Fig. 2.19a for different orientations γ of rows of antidots. The experiment (Fig. 2.19c) shows the transition to the superconducting

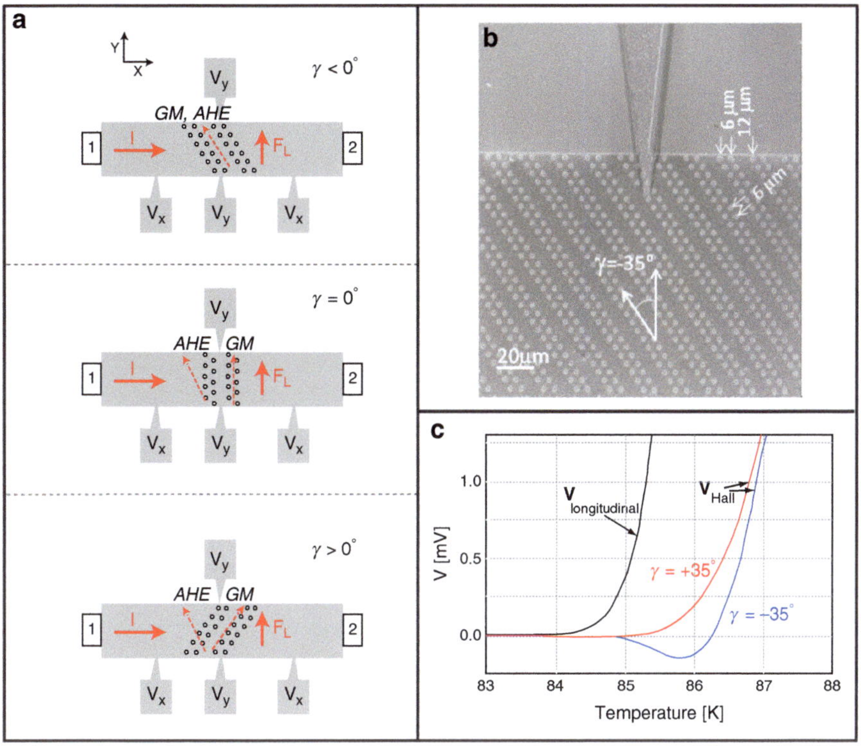

Fig. 2.19 (**a**) Sketch of a typical arrangement of antidots lattices for six-probe dc and microwave measurements of the longitudinal and Hall voltage to analyse guided vortex motion via rows of antidots and the anomalous Hall effect for negative, zero and positive angles γ, respectively. The angle γ is defined by the nominal direction of the Lorentz force due to the current and the orientation of the rows (single, pairs or triple rows are examined). The voltage probes, input and output ports for dc and microwave current (1, 2), the directions of current, Lorentz force, anomalous Hall effect (AHE) and guided motion (GM) are shown. (**b**) Microscopic image of a section of a sample depicting one of the Hall contacts and the configuration of antidot rows for $\gamma = -35°$. (**c**) Temperature dependence of the longitudinal voltage (contacts V_x) and Hall signals (contacts V_y) at the transition temperature. Depending on the direction of rows of antidots contacts, the anomalous Hall effect is present ($\gamma = -35°$) or suppressed ($\gamma = +35°$)

state determined by the measurements of the longitudinal voltage together with the temperature dependence of the Hall signals for two designs $\gamma > 0°$ and $\gamma < 0°$. While the Hall contacts positioned at antidots rows with $\gamma > 0°$ display the AHE, the Hall signal measured at antidots rows with orientation $\gamma < 0°$ do not show signs of an AHE.[1]

[1] Please note that the longitudinal voltage collects the total flux transport across the stripline registered between the voltage contacts V_x, whereas the Hall signal records the vortex motion "across the imaginary line between each pair of Hall contacts V_y". Consequently, the longitudinal voltage

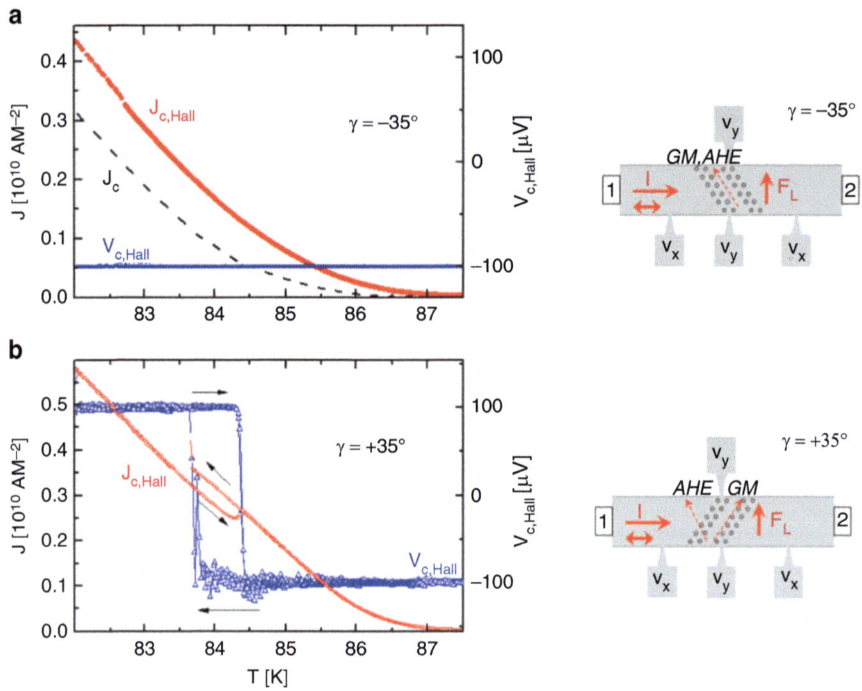

Fig. 2.20 Critical current densities $J_c(T)$ (*black dashed line*, voltage criterion $V_c = 5\,\mu V/cm$) and $J_{c,Hall}(T)$ (*red plot*) for two different orientations of rows: $\gamma = -35°$ (**a**) and $\gamma = +35°$ (**b**). $J_{c,Hall}$ is defined by a Hall voltage $|V_{c,Hall}| = 100\,\mu V$ (*blue line* and *triangles*). The *arrows* indicate the direction of temperature change during the experiment, and the magnetic field was 0.6 mT. *Right side*: the corresponding sample configurations from Fig. 2.19

Measurements of the transition curves (e.g. in Fig. 2.19c) are typically performed at small current densities. In the following, the impact of larger driving forces (i.e. large current densities) on the guidance by antidots is discussed. Additionally to the transition temperature, the critical current density J_c can be recorded via the different contacts. The standard J_c value is defined by measurements of the longitudinal voltage (voltage contacts V_x). It defines the onset of vortex motion and is determined by keeping the longitudinal voltage constant (typically at a value of a voltage criterion of a few $\mu V/cm$). The resulting current defines the critical current $I_c = J_c dw$ (d and w represent the thickness and width of the stripline). For YBCO films with and without antidot arrangements, $J_c(T)$ shows the classical behaviour, i.e. J_c decreases linearly with increasing temperature and at the transition this decrease smoothly approaches zero (Fig. 2.20a).

signal is orders of magnitudes larger compared to the Hall signal for the case of vortex motion. Therefore, a finite signal $V_{longitudinal}$ persists to lower temperatures where the Hall signal is already so small that it cannot be measured.

Similar to measurements of the classical critical current, we could keep the component of vortex motion $<v_x>$ constant by controlling the Hall signal. This way, the critical current density $J_{c,Hall}$ is defined in analogy to the classical critical current density J_c defined by the longitudinal voltage. The advantage of the analysis of $J_{c,Hall}$ is that a defined component of vortex motion perpendicular to the Lorentz force is established. This component of motion can only be caused by additional forces acting perpendicular to the Lorentz force, like the Magnus force or the guidance force. Therefore, $J_{c,Hall}$ is more suitable to characterize properties related to these forces than the classical definition of J_c.

Figure 2.20 shows the plots of critical current densities $J_{c,Hall}$ for both orientations of rows of antidots in a small applied magnetic field of 0.6 mT. To obtain a high mobility of vortices, a large voltage criterion is chosen for these measurements, $|V_{c,Hall}| = 100\,\mu V$. Due to the larger voltage criterion, $J_{c,Hall}$ is larger than J_c. Nevertheless, the overall functional dependencies of J_c and $J_{c,Hall}$ are the same for both orientations of antidot rows, except for the behaviour of the sign of the critical voltage (direction of vortex motion) and a small hysteretic jump in $J_{c,Hall}$ for one of the orientations of antidot arrows. This is discussed below.

$\gamma < 0°$: For these configurations (Fig. 2.20a), the Hall signal is negative over the whole investigated temperature range below T_c, the vortices moved in the direction of rows, indicated by the dashed red arrow in the sketch of Fig. 2.20a. The value of $J_{c,Hall}$ (corresponding to $V_{c,Hall} = -100\,\mu V$) increases monotonically with decreasing temperature. It agrees qualitatively with the behaviour of the critical current J_c in the superconducting state (black dashed line). The quantitative difference between J_c and $J_{c,Hall}$ is explained by differences in the voltage criteria ($V_c = 5\,\mu V/cm$ and $|V_{c,Hall}| = 100\,\mu V/cm$) and the different components of vortex motions (perpendicular or parallel to the Lorentz force) that are characterized.

$\gamma > 0°$: In this case (Fig. 2.20b), a different vortex behaviour is observed. The Hall voltage changes sign when cooling down from T_c. A negative voltage is observed close to T_c, when vortices do not follow the guidance of antidot rows. However, a few K below T_c, the Hall voltage changes abruptly to a positive value. The vortices move now in the direction imposed by antidot rows. This change of direction of vortex motion is hysteretic, i.e. it occurs at 83.7 and 84.4 K for decreasing and increasing temperatures, respectively. It is accompanied by a hysteretic change in the critical current $J_{c,Hall}$, seen in Fig. 2.20b, and a change of the microwave transmission coefficient S_{21} (not shown here) at exactly the same temperatures that indicate a change of the mobility of vortices [64]. The abrupt jump in the Hall voltage, critical current and microwave properties that is observed for the YBCO films with $\gamma > 0°$ suggests a change of the mechanism of vortex motion. It is discussed in the following.

Close to T_c, the negative Hall voltage is indicative of the presence of the AHE [94–103]. The vortex–antidot interaction is weak in this temperature range. Consequently, the guidance of vortex motion by the rows of antidots is negligible. The AHE dominates and vortices move in the direction defined by the AHE as indicated in sample drawings of Fig. 2.20.[2] With decreasing temperature, the vortex–antidot interaction strength increases and, finally, the guidance of vortices via the rows of antidots starts to dominate the AHE. The vortices move within the rows of antidots [93].

When the orientation angle of antidot rows coincides with the direction of $<v>$ resulting in AHE, as is the case for the design of Fig. 2.20a, AHE and guided motion (GM) by antidot rows point in the same direction. Therefore, the Hall voltage stays negative over the whole temperature range, and changes in the critical current density and microwave properties are monotonous functions of T. In contrast, a change in the direction of motion is necessary when transition from AHE to guided motion occurs in the design of Fig. 2.20b. Close to T_c vortices shuttle between the rows of antidots, while below a certain temperature they move within the rows of antidots. This implies a modification of the mechanism of vortex motion. While vortices mainly travel in the superconductor in the first case, they "hop" from antidot to antidot in the latter case. This difference in the mechanism of vortex motion is visible in the resulting jumps in the critical current density and the microwave loss. The Hall critical current is modified by 10–12% at the transition. Vortex motion within the rows of antidots seems to be energetically more favourable than it is in the superconducting matrix [64].

Generally, guidance of vortices via antidot arrangements depends in a complicated way on the relation (amplitude and direction) of the different forces and potential acting on the vortices. The major forces to be considered are the "background" pinning force of the superconductor, the vortex–antidot interaction, driving forces (e.g. Lorentz force), Magnus force and thermal activation. Some of these interactions depend on temperature, and others on their orientation direction. This can lead to transition in the direction of vortex motion as function of temperature (Fig. 2.20), current density (see Fig. 2.21) or orientation of antidot arrangement (Fig. 2.22). As a consequence, guidance of vortices strongly depends upon temperature, driving force and geometrical arrangement of the antidots. If additionally anisotropic pinning potentials are introduced (e.g. via asymmetrically shaped antidots), preferentially directed vortex motion can be induced leading to ratchet effects or vortex diodes.

2.4 Vortex Matter in Superconducting Devices

In the previous sections, it has been shown that vortices easily penetrate into HTS thin film devices leading to vortex motion that will finally affect the performance of the device. For instance, it has been shown that vortex motion in active devices (e.g. SQUIDs and flux quantum logic) leads among others to increased low-frequency noise, reduced sensitivity or increased bit error rate [104, 105]. In

[2] It should be noted that only the direction but not the angle for this motion can be measured.

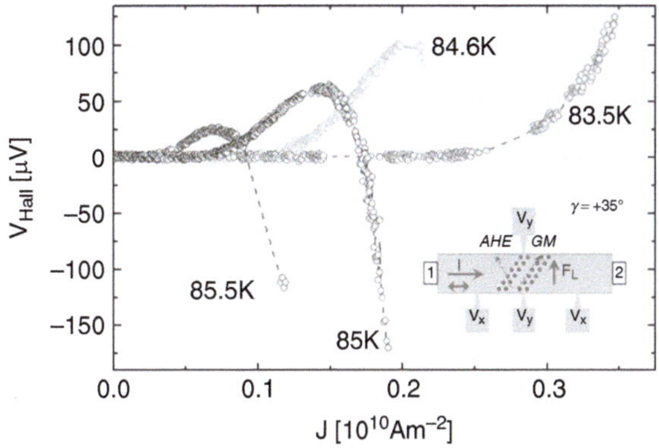

Fig. 2.21 Hall voltage of the sample shown in Fig. 2.20b as function of applied dc current density. A small magnetic field of 220 μT was applied; the arrangement of antidots is shown in the inset. At lower temperatures, a transition from guided motion (positive voltage) to AHE (negative voltage) is induced by the large driving forces, i.e. large current densities

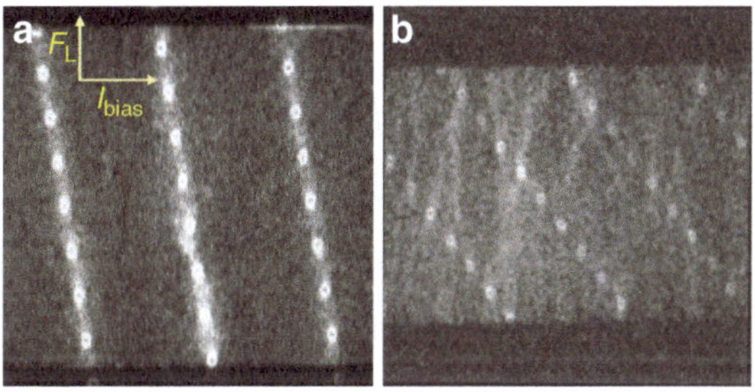

Fig. 2.22 Low temperature scanning microscopy images (area 200 × 200 μm^2) of YBCO films with rows of antidotes arranged at different angles $\gamma = -9°$ (**a**) and $\gamma = -38°$ (**b**) with respect to the Lorentz force [93]. Temperatures are 86 and 86.5 K, current density $J = 1.7\ J_c$ and $J = 1.2\ J_c$ in (**a**) and (**b**), respectively. Brighter regions correspond to higher amplitudes of the voltage response. In (**a**), the flux is guided along the rows of antidots, whereas in (**b**) flux starts to shuttle between adjacent rows (see *dashed ellipsoid*)

microwave devices (e.g. high-Q resonators, filters and antennas), vortices cause a reduction in the quality factor and power handling capability [57].

In this chapter, examples of active devices and passive microwave devices will be given that illustrate the role that vortices play in superconducting devices.

First, an efficient method for the reduction of *low-frequency noise* in *SQUIDs* will be described in which pinning of vortices by artificial defects is used. It will be demonstrated that (a) vortex motion is responsible for the low-frequency noise in SQUIDs and (b) an effective and field-independent noise reduction can be achieved by means of "strategically positioned" (thus, artificial) defects.

Second, *coplanar microwave resonators* are examined in small magnetic fields. It is shown that (a) the performance of these devices is strongly suppressed by vortices in the device and (b) a careful analysis of the local distribution of vortices in the devices can be used for the understanding and optimization of the microwave properties of these devices.

Finally, concepts for novel *fluxonic devices* working at low and high frequencies are sketched. These concepts are based on vortex manipulation using various antidots arrangements. Especially the use of non-isotropic shapes or arrangements of antidots could lead to vortex ratchet effect, vortex-based rf-dc converter, microwave filter devices and many more interesting applications.

2.4.1 Low-Frequency Noise in SQUIDs

Superconducting quantum interference devices (SQUIDs) represent the most sensitive sensors of magnetic flux or small physical quantities that can be transformed into magnetic flux. They are of special interest in applications where extremely small magnetic fields have to be detected, e.g. non-destructive testing (NDE), geomagnetic applications and, especially, medical or biomagnetic applications (Fig. 2.23). For an overview on the working principle and the application of SQUIDs, the reader should refer to [104, 105]. Here, only a brief introduction into the concept of SQUIDs and the problem of vortex motion in SQUIDs is given.

The SQUID is a deceptively simple device, consisting of (1) a small superconducting loop with (2) one (rf-SQUID) or two (dc-SQUID) damped (resistively shunted) Josephson junctions. Magnetic flux threading the superconducting loop is always quantized in units of Φ_o. Thus, the total flux in the SQUID consists of the measured flux Φ plus the flux Φ_{ind} that is induced by current in the loop to

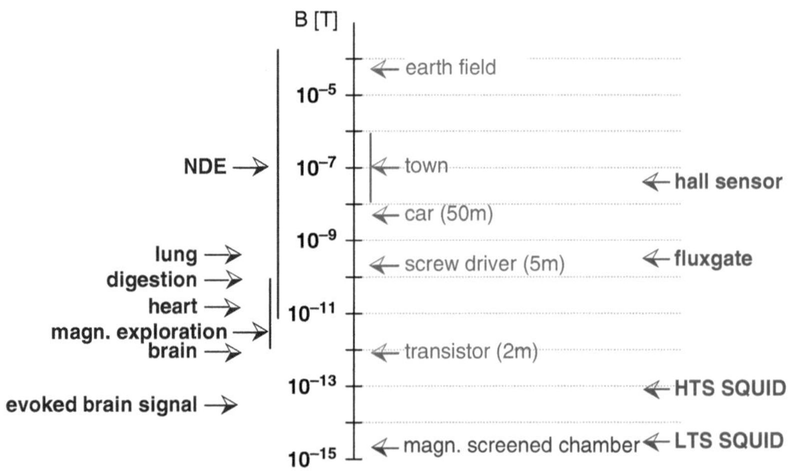

Fig. 2.23 Comparison of the magnetic-field sensitivity of conventional and superconducting field sensors (*right*), level of magnetic-field noise (*middle*) and field sensitivity required for various applications (*left*)

compensate according to: $\Phi_{tot} = n\Phi_o = \Phi + \Phi_{ind}$. The phase difference $\Delta\varphi$ across the Josephson junction is correlated with the superconducting tunnel current I_c according to the first Josephson equation [106]:

$$I_c = I_o \sin(\Delta\phi) = I_o \sin\left(\frac{\pi\Phi}{\Phi_o}\right) \tag{2.23}$$

with the tunnel current I_o at zero phase difference being a characteristic property of the contact and a phase that depends upon the magnetic flux in the contact according to $\Delta\varphi = \pi\Phi/\Phi_o$.

The SQUID makes use of the Josephson effect. By reading out the tunnel current (or related quantities), the magnetic flux is given in units of $\Phi_o = 2.068\times10^{-15}$ Wb. At a constant bias current, the voltage across the Josephson junction is periodically modified in a dc SQUID (Fig. 2.24) and the flux quanta in the SQUID loop can be counted. Moreover, in conjunction with suitable feedback electronics bucking the measured flux to maintain a stable operating point, the device can be used as a linear null detector. In this so called flux-lock-loop (FLL) mode, it becomes the most sensitive fluxmeter known, with a capability of resolving better than $10^{-6}\Phi_o$. The ranges of measured signal frequencies and intensity are wide and principally only limited by the electronics used. Typical room temperature electronics offer a frequency range extending from dc to kHz, and a dynamic range of 120 dB or more [105].

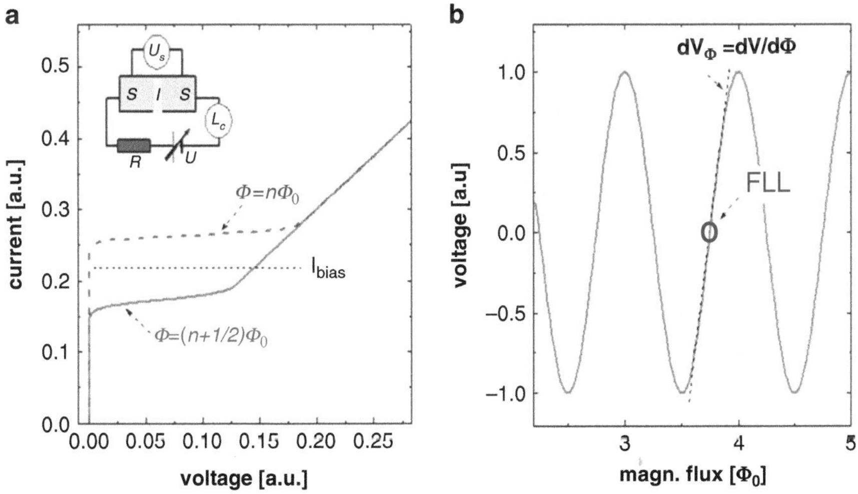

Fig. 2.24 Schematic drawing of the working principle of SQUIDs. The nonlinear IVC of a Josephson junction depends strongly on the magnetic flux (**a**). The superconducting tunnel current I_s changes periodically with the flux that threats the junction (**b**) (2.23). This periodical change is recorded in the SQUID for instance by applying an adequate bias current. The sensitivity of the SQUID can be enhanced using a feedback electronic (FLL) to fractions of the magnetic flux quantum $\Phi_o = 2.068 \times 10^{-15}$ Wb

In order to couple sufficient magnetic flux with a SQUID, either flux transformers are added or the superconducting loop (washer) is extended to provide sufficient flux focusing. Without going in detail (see e.g. [104, 105]), in both cases the inductance of the different components plays an important role and determines the final design of the device. In the first case, the large inductance of the transformer has to be matched to the SQUID inductance, i.e. large transformers with large inductance require SQUIDs with comparable inductance (i.e. SQUID of larger size). In the second case, the inductance of the extended washer has to be minimized to minimize the SQUID noise. The inductance of the SQUID is dominated by the diameter of the SQUID hole [107]. In both cases, the resulting SQUID possesses a large outer dimension (e.g. large focusing effect) and a small hole (small inductance). In conclusions, SQUIDs designed for technical applications generally posses rather extended structures (especially washers) that allow vortex to penetrate the superconductor. As a consequence, vortex motion within the superconductor plays a major role in these devices leading to a significant reduction in the SQUIDs sensitivity, especially for measurements that are performed at low frequencies.

Generally, the sensitivity of SQUIDs is limited by the frequency-dependent noise level of the device. Additional to the contribution of the electronics (usually white noise), in superconducting active devices two different sources are considered to be responsible for the noise. These are the contribution of the active part of the device, which usually consists of Josephson junctions, and the noise of the passive component, the superconducting thin film. The noise mechanisms in Josephson junctions are well understood [108–111] and a reduction in this noise contribution by simple electronic means has successfully been demonstrated [112–117]. The contributions of the superconducting thin film (mainly washer or flux transformer) to the noise of active devices are basically understood. They contribute to the low-frequency noise and are ascribed to motion of quantized flux (vortices) in the superconducting thin films. A nice illustration of this contribution is given by the so-called telegraph noise that occurs when a vortex hops between two pinning sites (Fig. 2.4b). In case of a statistical motion of many vortices, a scaling of the spectral noise density S_Φ with frequencies f and the applied magnetic field B is expected:

$$\sqrt{S_\Phi\,(f, B)} \propto \frac{B^n}{f^m} \tag{2.24}$$

with $n = m = 0.5$. This so-called $1/f$ noise-spectrum is ascribed to an incoherent superposition of many thermally activated microscopic fluctuators, which are given by moving or hopping vortices. The characteristic $1/f$ frequency dependence is evidence for a distribution of activation energies for the vortex hopping [118, 119]. Since most measurements of $1/f$ noise are sensitive only to small activation energies, usually only the low-energy part of this distribution is probed [120].

Typical examples of the spectral noise density and field dependences of the $1/f$ integral noise of rf-SQUIDs designed for applications are given in Fig. 2.25. The $1/f$ low-frequency noise is limited by the corner frequency (10–20 Hz at zero-field) above which the frequency-independent white noise sets in. The low-frequency

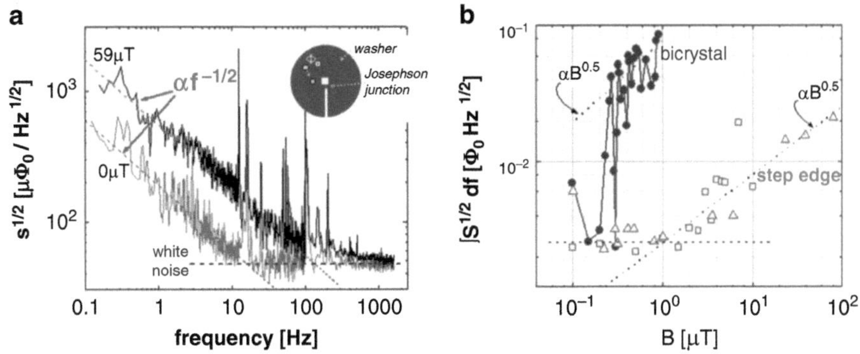

Fig. 2.25 (**a**) Spectral noise density for different magnetic fields measured in a HTS step-edge type rf-SQUID [52]. The lines idealize the different noise contributions, i.e. the field-dependent $1/f$ low-frequency noise and the field- and frequency-independent white noise. The schematic drawing sketches the geometry of the rf-SQUIDs and the problem of moving vortex in the washer. (**b**) Magnetic field dependencies of the integral noise of a step-edge and bicrystal Josephson junction according to [52]. The noise data are characterized by the normalized integration of the noise spectra in the low-frequency regime between 0.5 and 10 Hz. The *dotted lines* represent the theoretical field dependence of the low-frequency noise (2.24)

noise increases with increasing field for SQUIDs with extended washer, i.e. SQUIDs that are designed for application. This is demonstrated in Fig. 2.25b for two different types of SQUIDs, i.e. step edge and bicrystal type rf-SQUIDs [52]. For magnetic fields larger than the penetration field, the noise increases according to the theoretical expectation, i.e. $S_\Phi \propto B$ (2.24). The low-frequency noise represents a serious limitation for a number of applications of SQUIDs especially if the sensors are used in unshielded or barely shielded environment.

Various remedies to reduce the low-frequency noise have been suggested and tested that can be classified into two categories:

1. Either vortex penetration of the superconductor has to be avoided [121, 122] or
2. Vortices have to be pinned by sufficiently strong pinning sites in the superconductor [52, 89, 123]

In case of flux avoidance, extremely good screening or, alternatively, relatively small superconducting geometrical structures are required. Both solutions are technically difficult to realize and very costly. For instance, the patterning of the complete washer into a "grid" of striplines of linewidth $w < 6\,\mu\text{m}$ [124] would be necessary for application of these devices in earth field (50 μT). This turns out to be technically quite complicated considering the requirements of the patterning. An alternative is given by vortex pinning at strong pinning sites. In the following, we will describe that (i) the low-frequency noise in SQUIDs can be modified via flux pinning due to artificial pinning sites (antidots) and (ii) only a few pinning sites very effectively suppress the low-frequency noise if they are strategically positioned in the SQUID.

2.4.1.1 Manipulation of the Low-Frequency Noise via Antidot Arrays

In order to demonstrate the effect of vortex–antidot interaction upon the low-frequency noise in HTS devices, a characteristic property of bicrystal rf-SQUIDs (BS) is exploited. At extremely small magnetic fields ($B_{p,GB} \approx 200\,nT$ for the BS used in the following experiment) clearly below the penetration field ($B_{p,SE} \approx 1$–$2\,\mu T$, see Fig. 2.25b), at which flux penetrates the washer in step-edge SQUIDs (SES) of identical geometry, the grain boundary in the washer of the BS serves as a channel for flux motion. This flux motion causes a drastic increase in 1/f noise (Fig. 2.25b). Once flux has penetrated the gain boundary, the further increase in the low frequency noise with field agrees with the expected field dependence $S_\Phi \propto B$ (2.24).

In order to examine the impact of the interaction between an antidot array and vortex lattices on the flux noise in a SQUID, a YBCO film with a square lattice ($d_{AA} = 5\,\mu m$) of antidots is mounted in flip-chip configuration on top of the grain boundary of the BS washer (see sketch in Fig. 2.26). For these measurements, the square antidot lattice is oriented with one axis along the grain boundary.

The resulting low-frequency noise-spectra of the arrangement (BS with antidot lattice) are given in Fig. 2.26. As expected, the lowest flux noise is recorded for zero magnetic field. At non-zero field, the low frequency noise strongly depends upon the exact values of the applied magnetic induction. In contrast to the standard field dependence $S_\Phi \propto B$ (2.24), it varies non-monotonically and over several orders of magnitude in noise level. For example, the noise at fields of 750 or 900 nT is more than 2 orders of magnitude larger than for the matching fields M_1 and M_2 (828 and 845 nT, respectively) at which the noise level of the zero-field spectrum is recorded.

A comparison of the field dependence of the low-frequency noise of the bare SQUID and the same SQUID with the antidot lattice is given in Fig. 2.27. In contrast to the reference measurements on the bare SQUID, which show the usually observed behaviour (i.e., constant noise level below $B_{p,GB}$ and linear increase $S_\Phi \propto B$ above $B_{p,GB}$), the data display clear minima and maxima. The minima are observed exactly at matching conditions. Matching or commensurability occurs when the vortex lattice matches the antidot lattice and the grain boundary, along which the "noisy" motion of the vortices takes place. Since the vortex lattice parameter $a_o \propto B^{1/2}$ varies with magnetic field, the resulting matching conditions can easily be recorded. They are listed in the table in Fig. 2.27. It should be noted that extremely small matching fields down to 247 nT $= B_1/324$ can be observed in this experiment. The noise level at matching fields is comparable to the noise level recorded at zero field. This indicated that vortex motion within the grain boundary of the washer is completely suppressed at matching condition. In contrast, at fields between the matching fields even larger noise levels (up to ten times larger) are observed for the SQUID with antidot lattice with respect to the bare SQUID. Thus, vortex motion is even enhanced in case of incommensurability between vortex lattice, antidot array and grain boundary. We can conclude as follows:

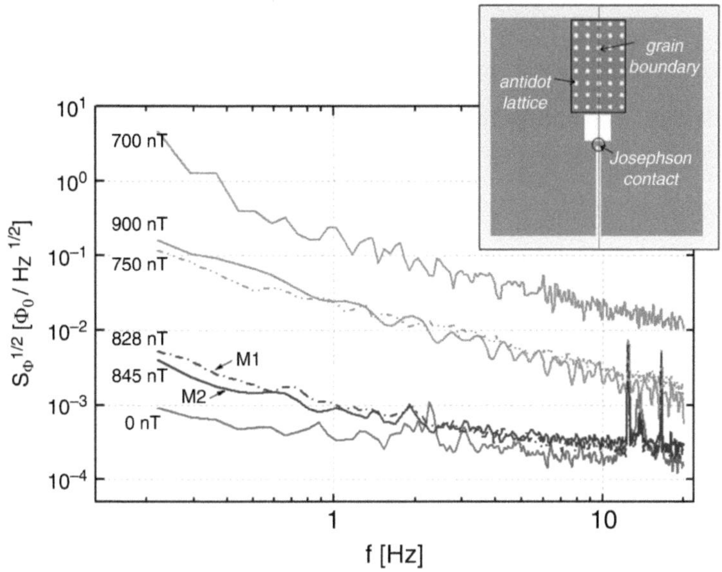

Fig. 2.26 Low-frequency noise spectra of a bicrystal rf-SQUID (BS) with a square antidots lattice on top of the grain boundary of the washer for different magnetic fields [89]. M_1 and M_2 label two matching conditions (see Table 2.1 in Fig. 2.27). The inset shows a sketch of the experimental arrangement. A YBCO film with square antidot lattice ($d_{AA} = 5\,\mu m$) is mounted in flip-chip configuration on top of the grain boundary of the BS. One of the main axes of the antidot lattice is oriented parallel to the grain boundary

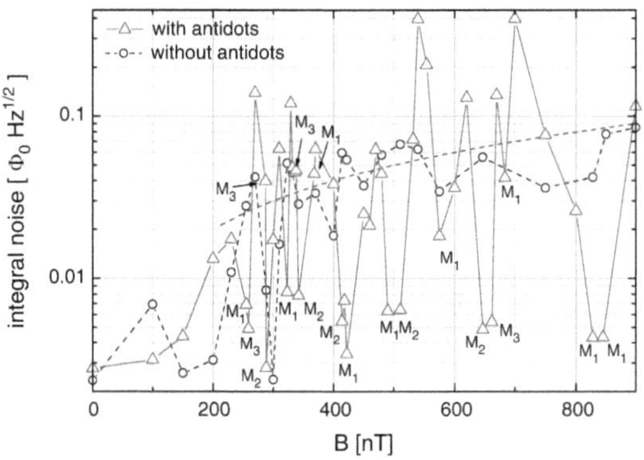

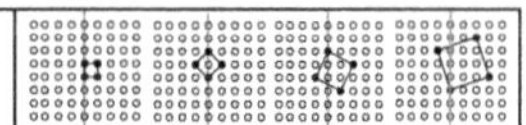

n	M_1	M_2	M_3	M_4
	a_0=d	a_0=	a_0=	a_0=
4				518 nT
5			662nT	331 nT
6			460nT	
7		845nT	338nT	
8		647nT	259nT	
9		511nT		
10	828nT	414nT		
11	684nT	342nT		
12	575nT	288nT		
13	490nT			
14	422nT			
15	368nT			
16	323nT			
17	287nT			
18	256nT			

Fig. 2.27 Magnetic field dependence of the normalized and integrated (0.5–10 Hz) low-frequency noise of a bicrystal rf-SQUID with and without an additional YBCO layer with an antidot lattice [89]. The noise data are measured several times and could be reproduced within the experimental error. The *blue dashed line* represents the theoretical field dependence $S_\Phi \propto B$ (2.24) of the SQUID without antidot lattice. The table represents the matching fields M_i with $I = 1$–4 in the field range between 255 and 900 nT. They are calculated for different vortex configurations (see schematic sketch) that match the square antidot array ($d_{AA} = 5\,\mu m$)

1. Noise manipulation via antidot lattices is demonstrated for HTS SQUIDs:
   (a) In case of commensurability between vortex and antidot lattice the flux noise is reduced to zero-field level
   (b) In case on non-commensurability, the noise level is even enhanced with respect to the noise of the SQUID without antidots lattice.
2. The noise measurements demonstrated commensurability effects between vortex and antidot lattices down to extremely small magnetic fields (i.e. fractions of the first matching field $\sim B_1/324$).
3. For noise reduction in SQUID applications working in variable magnetic fields, regular lattice of antidots are not suitable. Minima in the noise level at matching fields alternate with excessively high noise-levels for non-commensurability. Different solutions to this problem have been discussed, such as non-regular antidot lattices, extended holes or moats or the use of a few *strategically positioned* antidots. The latter is demonstrated in the next section.

2.4.1.2 Noise Reduction via Strategically Positioned Antidots

In the previous section, it was demonstrated that the low-frequency noise in SQUIDs can be manipulated via antidots. However, it can be seen in Figs. 2.26 and 2.27 and it is obvious from theoretical considerations that regular arrays of antidots lead to noise reduction only at discrete values of the magnetic field (at matching fields), whereas in case of non-commensurability between the vortex and antidot lattice even an increase in the low-frequency noise is observed. Therefore, it is better to use only a few "strategically positioned" antidots in the superconducting device, which trap only those vortices that attribute strongly to the low-frequency noise (i.e. vortices close to the SQUID hole) and leave the vortex lattice free to arrange itself within the device. Thus, the important issue is to allocate those strategic positions. A nice experimental demonstration for noise reduction via strategically positioned antidots is given in [52, 53, 89] for the case of rf-SQUIDs. It is sketched in the following.

Noise reduction by vortex trapping: Generally, the largest impact of vortex motion upon the SQUIDs flux noise is expected for vortex motion at a position close to the SQUID hole and Josephson junction. Vortex motion at this position (a) is more likely due to the large Lorentz force (caused by the read-out current) and (b) will lead to large flux changes at the Josephson contact [52]. Thus, obviously the first choice for strategically positioned antidots would be a position close to the Josephson junction and close to the SQUID hole. Figure 2.28 shows the corresponding arrangements of "strategically positioned" antidots and the resulting modification of the noise properties determined via field-cooled experiments.

The resulting noise reduction can be seen in *field-cooled* experiments in Fig. 2.28b. In these experiments, the SQUID is cooled from the normal to the superconducting state in an applied field B oriented normal to the film surface.

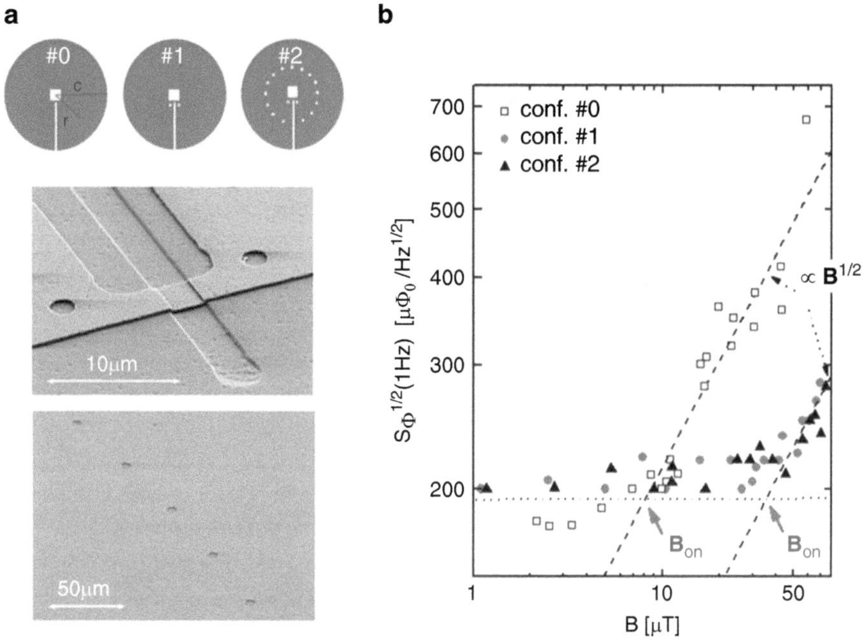

Fig. 2.28 (**a**) Sketch of a YBCO rf-SQUID with different antidot configurations [without antidots (#0), with two antidots in the vicinity of the junction (#1) and, finally, two antidots at the junction and an additional ring of antidots at radial position $r = c/2$ (#3)]; SEM images of the two antidots in the vicinity of the junction (antidots radius $r_o = 750\,\text{nm}$) and the ring of antidots ($d_{AA} = 50\,\mu\text{m}$, $r_o \sim 1.25\,\mu\text{m}$). (**b**) Spectral noise density at 1 Hz ($1/f$ noise) as a function of the magnetic field for field-cooled measurements and the different configurations of strategically positioned antidots. The onset of the increase in the low-frequency noise is characterized by the onset fields B_{on} for the different configurations. The proportionality $S_\Phi \propto B$ of the low-frequency noise for $B > B_{on}$ (*dashed lines*) agrees with the theoretical expectation in (2.24). The onset field for the configurations with antidots (#1 and #2) is comparable to the earth field of $\sim 50\,\mu\text{T}$

A comparison of the spectral noise density in the low-frequency regime ($1/f$ noise) for a SQUID with the different successively patterned antidot arrangements demonstrates[3]:

1. At low fields $B < B_{on}$, the spectral noise density is field independent and the same for all configurations.
2. At higher fields $B > B_{on}$, the spectral noise densities increase linearly with increasing field according to the theoretical expectation $S_\Phi \propto B$.
3. However, the transition from field-independent to field-dependent spectral noise density is significantly increased from $B_{on} \approx 8\,\mu\text{T}$ for configuration #0 (without antidots) to $B_{on} \approx 40\,\mu\text{T}$ for the configurations with antidots (#1 and #2).

[3] Note that always the same SQUID has been used in this experiment.

The significant increase in the onset field B_{on} by the arrangement of only two strategically positioned antidots is sufficient for most applications of SQUIDs in unshielded environment (earth field $B_{earth} \approx 50\,\mu T$).

Usually SQUIDs are used in varying field. As a consequence, flux will penetrate the SQUIDs' washer when the magnetic field is changed; this automatically leads to enhanced flux noise due to vortex motion. A detailed discussion of this effect including the possibility of determining the position of individual vortices in the washer is given in [52, 53, 89]. One solution for this problem is the operation of the SQUID in the so-called flux-lock-loop. Using a feedback electronic, the flux is kept constant at the SQUID. Nevertheless, large or sudden flux changes cannot always be avoided. As a result, the SQUID will be "noisy" for a long time or has to be warmed up to the normal state to remove unwanted flux in the washer. An alternative is given by introducing a ring of antidots at a constant radial position (see configuration #2 in Fig. 2.28). It allows the vortices to (1) easily enter the washer without and (2) modify the flux in the SQUID hole once the vortex is in the washer. As a consequence, a change in the applied magnetic field would only result in brief increase of the noise density followed by the fast reduction of the flux noise to the operation level. This effect has been demonstrated for configuration 2 (Fig. 2.28) for magnetic fields up to $24\,\mu T$ [52].

In conclusion of this section, the noise of SQUIDs proved to be extremely sensitive to vortex motion and commensurability effects of the vortex and antidot lattice. Even the position of vortices and their motion in the SQUIDs washer can be determined experimentally. As a consequence, extremely simple arrangements of only a few but well-positioned antidots lead to a considerable reduction of flux noise in HTS SQUIDs. For instance, a pair of antidots arranged in the vicinity of the Josephson junction of rf-SQUIDs strongly reduces the $1/f$ noise in ambient magnetic fields, and well-directed penetration of vortices in the washer can be achieved via rows of antidot. Thus, vortex manipulation in HTS SQUIDs is a powerful tool to improve the performance of these devices.

2.4.2 *Vortex Matter in Microwave Devices*

The behaviour of a conductor in a microwave field can generally be characterized by the complex surface impedance $Z_s = R_s + iX_s$ with surface resistance R_s and surface reactance X_s. Z_s describes the electrodynamics at the interface between vacuum and (super)conductor. The surface resistance represents the dissipated energy, whereas the surface reactance characterizes the energy stored in the conductor. The surface resistance is given by:

$$R_s = X_s = \sqrt{\frac{2\pi^2\mu_0 f}{\sigma_n}} \text{ and } R_s \approx 2\pi^2\mu_0^2 f^2\sigma_n\lambda^3 \tag{2.25}$$

for the normal conductor and superconductor (two-fluid model), respectively. Here, f represents the frequency and σ_n is the conductivity due to normal charge carriers.

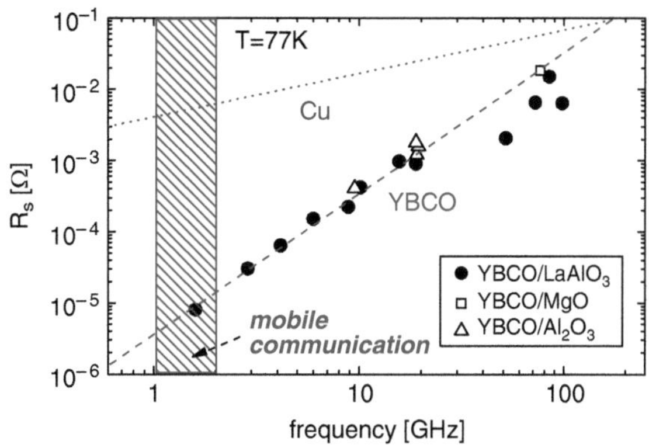

Fig. 2.29 Comparison of the frequency dependence of the microwave surface resistance R_s at $T = 77$ K for YBCO films on different substrates and Cu. The frequency regime used for mobile communication is marked

The surface resistance of a type-II superconductor and a normal conductor shows a significant difference in frequency dependence and magnitude. At frequencies up to a few GHz, technical superconductors (e.g. Nb, NbN and HTS) possess orders of magnitude smaller R_s values that automatically result in smaller microwave losses.

From this point of view, the superconductor offers a significant benefit in microwave applications. As demonstrated in Fig. 2.29, the smaller losses can be used to improve the quality factors $Q = f_0/\Delta f$ of resonators or filters (f_0 represents the centre frequency and Δf is the full width of the resonance at -3 dB), or lead to miniaturization of existing devices [125, 126]. As a consequence, usually large power densities are encountered in HTS microwave devices. However, in contrast to their normal-conducting competitors, the power handling capability of superconducting microwave devices is limited by the nonlinearity of the surface resistance. At high microwave power, R_s increases significantly (Fig. 2.30). This not only leads to an increase in dissipated energy that finally causes a heating of the superconductor, but also causes a degradation of Q values and the development of unwanted intermodulation signals.

Although various physical origins are discussed (e.g. weak links, thermal or magnetic effects), the physical mechanism, which is responsible for the nonlinear surface resistance and that represents a serious restriction for a number of microwave applications, is not fully understood. Nevertheless, it has been demonstrated that vortex matter has a significant impact on the properties of HTS microwave devices, including the quality factor, the nonlinearity and the power handling capability [57]. In this section, we will give a few examples of experimental evidence for the vortex-induced modifications of microwave properties of HTS films and HTS devices. Then we will sketch possible ways to overcome these problems. Finally, we will demonstrate that one might use the impact of vortices on microwave properties to design novel microwave fluxonic devices.

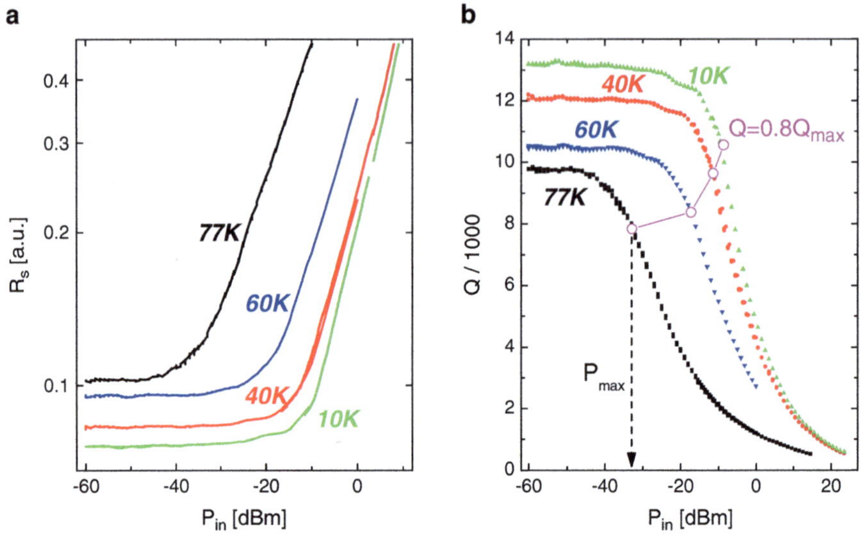

Fig. 2.30 The nonlinearity of the microwave surface resistance of HTS films leads to a degradation of the performance of superconducting devices at large microwave power. This is visible for instance in a significant increase of the surface resistance (**a**) and the degradation of the quality factor (**b**) measured in this example for a 1.6 GHz coplanar YBCO resonator

2.4.2.1 Impact of Vortices on the Microwave Properties

The basic components of microwave devices made from HTS thin films are impedance-matched stiplines. Due to the thickness and the dielectric properties of the substrate, an impedance of 50 Ω results in a typical width of the structures of $w = 300$–$600\,\mu$m. On the one hand, the microwave current is strongly peaked at the edge of the conductor. For example, the experimentally determined current densities in the central conductor of a coplanar microwave devices can be approximated by [127]

$$J_{\mathrm{rf}}(x) = \frac{I}{wK\left(\frac{w}{a}\right)} \begin{cases} \left[\left(1-\left(\frac{2x}{w}\right)^2\right)\left(1-\left(\frac{2x}{a}\right)^2\right)\right]^{-1/2} & |x| \le \frac{w}{2}-\lambda \\ \left[\left(1-\left(\frac{w}{a}\right)^2\right)\frac{\lambda}{w}\right]^{-1/2} & \frac{w}{2}-\lambda \le |x| \le \frac{w}{2} \\ -\left[\left(\left(\frac{a}{w}\right)^2-1\right)\frac{\lambda}{w}\right]^{-1/2} \quad \text{for} & \frac{a}{2} \le |x| \le \frac{a}{2}-\lambda \\ -\left[\left(\left(\frac{2x}{w}\right)^2-1\right)\left(\left(\frac{2x}{a}\right)^2-1\right)\right]^{-1/2} & |x| \ge \frac{a}{2}+\lambda \end{cases} \tag{2.26}$$

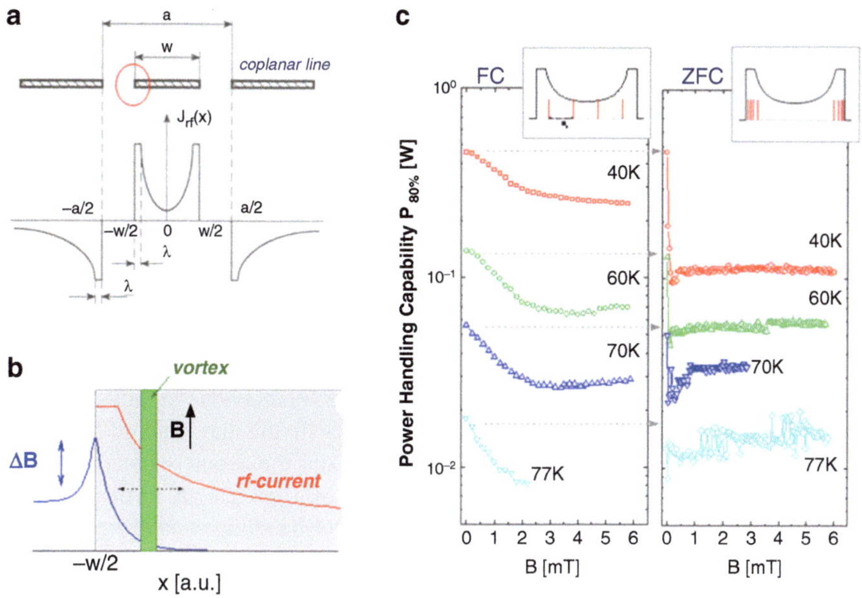

Fig. 2.31 Schematic drawing of (**a**) the microwave current distribution J_{rf} in a coplanar superconducting lines according to (31) and (**b**) the microwave current and magnetic field distribution at the edge of a superconducting stripline. (**c**) Field dependence of the power handling capability of a coplanar YBCO resonator for field-cooled and zero-field cooled experiments at different temperatures [57]. The sketches in (**c**) indicate the different vortex distribution in field cooled and zero-field cooled experiments and the microwave current distribution in the central conductor of the coplanar microwave resonator

with w representing the width of the stripline, a the distance between the ground planes and K the complete elliptical integral of the first kind. A sketch of the cross-section of the structure and the resulting microwave current distribution is given in Fig. 2.31a. On the other hand, vortices will penetrate at the edge of the superconducting stripline upon modification of the applied magnetic field. Due to the large aspect ratio, small changes of the external field will result in large changes of the magnetic field at the edge of the stripline. Depending on the position of the vortex, the microwave current will be affected differently. This situation is sketched in Fig. 2.31b.

Experimentally, it has been demonstrated by comparing *field-cooled* (fc) and *zero-field cooled* (zfc) measurements of the power handling capability of YBCO thin film microwave resonators that were exposed to small magnetic fields (Fig. 2.31c) [57]. In these experiments, the power handling is characterized by the degradation of the loaded quality factor $Q_L = f_o/\Delta f$, i.e. P_{max} is defined by the condition $Q_L(P_{max}) = 0.8\ Q_L(P_o)$ (see also Fig. 2.30b). In fc experiments, the resonator is cooled to the superconducting state in the applied magnetic field (i.e. the vortex distribution is expected to be homogeneous), whereas in zfc experiments the resonator is cooled to the superconducting state in zero-field and the magnetic field is applied in the superconducting state. In the latter case, vortices have to penetrate from the edge of the superconductor, leading to an inhomogeneous distribution of vortices. Thus, the expected vortex distribution is different for fc and zfc experiments, and it is sketched in the insets of Fig. 2.31c.

The resulting power handling capability P_{max} depends strongly upon the way the magnetic field is established. While fc measurements show a gradual decrease of P_{max}, in zfc experiments a strong decrease of the power handling capability within a few μT is followed by an almost constant P_{max} (Fig. 2.31c). The different behaviours are consequences of the vortex distribution in the resonator. For fc experiments, the moderate increase in the homogeneous vortex density with increasing field seems to lead to the moderate and gradual decrease of P_{max}. In contrast to the fc experiments, an inhomogeneous vortex distribution is expected for the zfc experiment. Due to the flux penetration at the edge of the superconductor, in the zfc case the position of the largest vortex density coincides with the maximum of the rf current at the edge of the central conductor. This explains the strong reduction of P_{max} that is observed already at extremely small fields of a few μT. A further increase of the field in the zfc case does not affect the power-handling capability seriously. This is consistent with other observations of flux penetration at small fields, where an increase in the magnetic field first leads only to a shift of the flux front into the superconductor. Since the interior of the superconductor does not carry large rf currents, this shift does not affect the power handling capability seriously. Furthermore, the fc and zfc values of P_{max} seem to merge at large fields. This is expected for the field for which the zfc flux front approaches the centre of the sample. For these fields, similar vortex densities are expected at the edge of the superconductor for the fc and zfc case.

The complete picture of the impact of vortices on the properties of microwave resonators that also confirm the interpretation of the fc and zfc data given above is illustrated by field-cooled sweep (fcs) experiments. In fcs experiments, the sample is first cooled to the superconducting state in an applied field B_{fc} (where the field-cooled state is established) and, subsequently, the field is modified by ΔB. Figure 2.32 shows a typical example of fcs measurements together with fc and zfc data recorded for a 1.4 GHz YBCO coplanar resonator at 60 K. The zfc and fc data represent lower and upper values for the power handling capability, respectively, whereas fcs measurements start with the fc value and show a strong and complex variation of P_{max} that is shown in detail in the inset of Fig. 2.32. The behaviour of P_{max} in fcs experiments can be interpreted in terms of vortex penetration and annihilation caused by magnetic fields of the order of μT (for a detailed discussion, refer to [57]). It demonstrates that not only flux and vortices have a large impact upon the performance of a microwave device, but also the field direction and, especially, the correlation between the spatial distributions of vortices and microwave current density are of importance. The largest impact of vortices upon the microwave properties is observed for vortices at the edge of the microwave device that represents the part of the device that normally carries the highest microwave current density. As a consequence, vortices have to be manipulated (e.g. guided into the centre of the

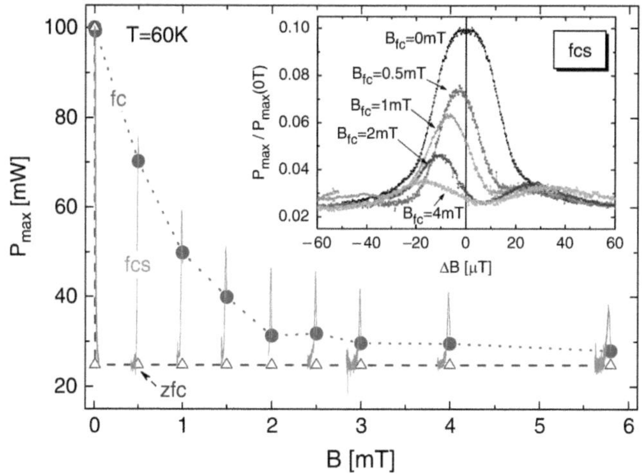

Fig. 2.32 Comparison of the magnetic field dependence of the power handling capability of a coplanar 1.4 GHz YBCO resonator for fc (*solid red circles*), zfc (*open blue triangles*) and fcs (*green lines*) experiments. The inset shows the power handling capability in fcs measurements for different starting fields B_{fc} on a normalized scale $\Delta B = B - B_{fc}$

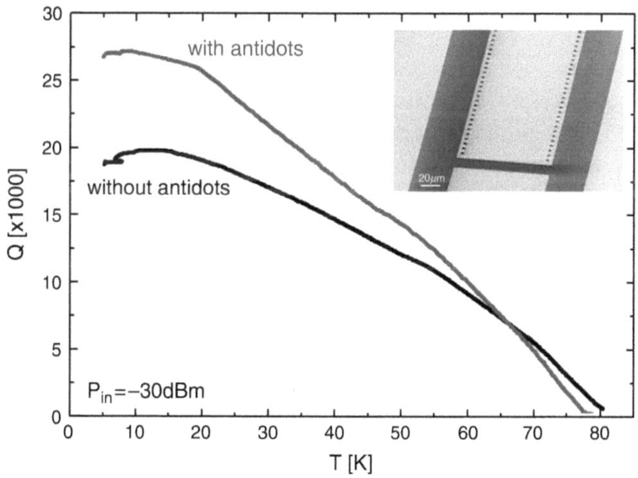

Fig. 2.33 Loaded quality factor of a 1.4 GHz YBCO resonator measured in small varying magnetic fields of the order of a few μT with and without antidots. The inset shows the arrangement of antidots that are positioned at the edge of the central line of the coplanar resonator

microwave device) to conserve the performance of microwave devices operating in magnetically unshielded environment. This situation is very similar to the situation encountered in SQUIDs where flux is guided into the washer by a row of antidots. First attempts to improve the performance of microwave resonators that are exposed to magnetically unshielded environment via antidots are demonstrated in Fig. 2.33.

2.4.2.2 Concepts for HTS Fluxonic Devices

In the previous sections, strategies to improve existing superconducting devices by manipulating (i.e. trapping and guiding) vortices were introduced. However, the field of vortex manipulation by micro- or nanostructures in HTS films is much more colourful. Different feasible concepts are sketched in Fig. 2.34. For instance, vortices might be generated, guided and trapped by slits, small and large antidots, respectively. By adequate magnetic field variations or using the current curls created by a slit at opposing edges of a stripline, even vortices and antivortices can be created and manipulated. The operation of some of these components has been demonstrated for low-T_c films (e.g. Nb, Pb or Al films) or their impact upon magnetic flux has been visualized by magneto-optic experiments [128, 129]. An interesting option is given using asymmetric antidots that could lead to the directed vortex motion, the so-called ratchet effect. This will be briefly sketched in this section.

Generally, *ratchets* are formed from spatially asymmetric confining potentials. They can rectify oscillatory driving forces and generate directed motion. Ratchet scenarios were already considered by Feynman in his lecture notes in 1963 [130] and are related to earlier problems of thermodynamic studied by Smoluchowski in 1912 [131]. Ratchets represent a major component of particle transport in nanoscale systems, both in solid-state systems and in biology. Ratchets in biological systems (e.g. biomolecular motors) can be found in nature, including the kinesin and dynein proteins that provide transport functions within the cell [132]. Ratchets can be produced by biomolecular engineering, e.g. molecular walker constructed from strands of DNA [133] and controlled motion of kinesin-driven microtubules along lithographically patterned tracks have been demonstrated recently [134]. Advances in nanofabrication made it feasible to develop and investigate ratchets formed from

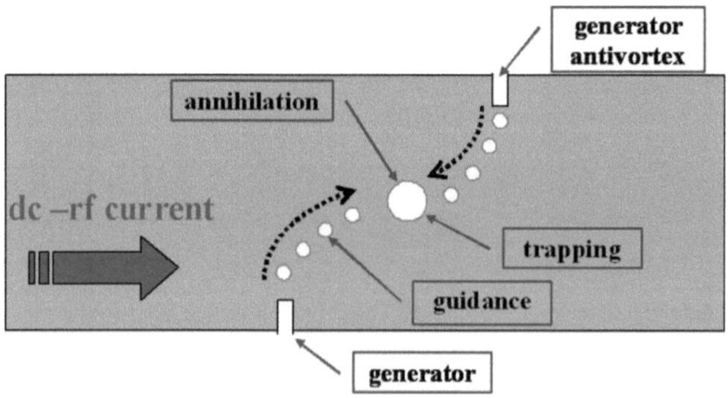

Fig. 2.34 Sketch of different nano- and microstructure components for vortex manipulation in HTS films that might lead to novel fluxonic applications. Vortices and antivortices (vortices with inverted field component) can be generated by curls in the dc or microwave current caused by slits at opposite edges of a stripline. Guidance and trapping are achieved by antidots of different size. Annihilation of vortices would be possible in case of guidance of vortex and antivortex towards the same trapping site

solid-state systems involving electronic devices or microfluidics. Such devices can be used (1) as analogue systems for modelling biomolecular motors, (2) to understand novel particle transport at the nanoscale or (3) to develop new devices for application. One advantage of solid-state nanofabrication is the possibility to tailor the ratchet potential and to control driving parameters, temperature or other process parameters. Implementations of ratchets in electronic devices have been demonstrated recently, e.g. the use of asymmetric structures of electrostatic gates above a 2D electron gas [135] and arrays of Josephson junctions with asymmetric critical currents [136].

Vortices in superconductors form an ideal system for exploring ratchet phenomena. The control of vortex dynamics via micro or nanostructures allows for the tailoring of vortex confining potentials. One approach for controlling and rectifying vortex motion in superconductors involves the use of arrays of antidots. Vortex ratchet effects obtained by various arrangements of antidots have recently been demonstrated for low-T_c films [48–51] and high-T_c films [33]. The asymmetric pinning potential is achieved by asymmetrically shaped antidots, combinations of antidots of different sizes, asymmetric arrangement of symmetric antidots or even mixing of dc and ac currents

Figure 2.35 shows an example of a HTS ratchet that also illustrates the potential use of ratchets in microwave applications. The design of the device (Fig. 2.35a) is similar to the design used for the demonstration of guided vortex motion (Fig. 2.19), except for the fact that asymmetric (here, triangular shaped) antidots are used. The

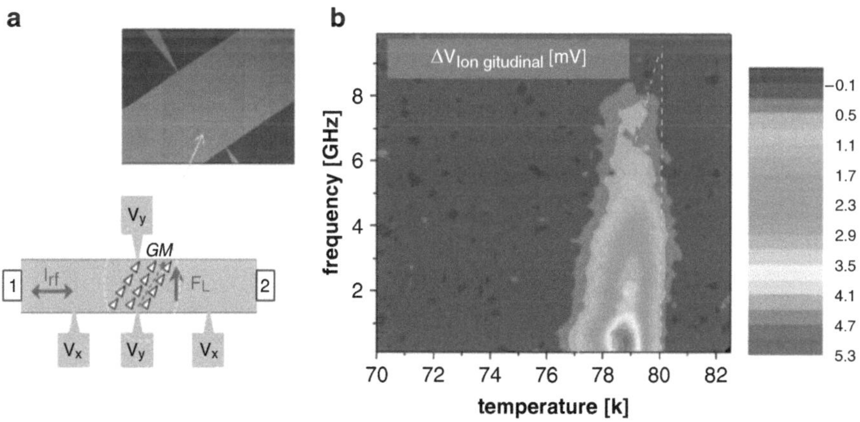

Fig. 2.35 Demonstration of the ratchet effect in HTS microwave devices. Schematic drawing and microscope image of a YBCO ratchet based on rows of triangular-shaped antidots (**a**) and contour plot of the change of the dc voltage signal $\Delta V_{\text{longitudinal}} = V_{\text{longitudinal}}(-10\,\text{dBm}) - V_{\text{longitudinal}}(-50\,\text{dBm})$ as function of temperature and applied microwave frequency (**b**). The microwave current is applied via ports 1 and 2, the vortices are rectified by rows of triangular antidots and the dc longitudinal and dc Hall voltages are recorded at contacts V_x and V_y. The microscope image shows the stripline in the vicinity of the Hall contacts. The change in the rectified longitudinal voltage signal due to the microwave power demonstrates the rectified vortex motion for temperatures close to T_c (high vortex mobility) and up to frequencies of about 10 GHz

tilting of the direction of antidot rows with respect to the Lorentz force allows for simultaneous measurements of a longitudinal and Hall signal in case of guided vortex motion. In case of a rectification of the vortex motion (ratchet effect), a microwave driving current would be transformed in a directional motion of vortices and, thus, a dc voltage at the Hall and longitudinal voltage pads. Figure 2.35b shows a contour plot of the dc voltage (only the longitudinal voltage is shown) obtained in a typical experiment. In order to improve the resolution of the measurement, a reference signal taken at low microwave power (−50 dBm) is subtracted from the signal recorded at large microwave power (−10 dBm).

The experiment demonstrates that HTS vortex ratchets operate, although in a restricted temperature regime close to T_c and a frequency regime up to about 10 GHz. An obvious restriction of the operation regime is imposed by the limited microwave power used in the experiment. Close to T_c, the pinning force in small and vortices move easily. At lower temperatures, pinning force and viscosity increase and, as a consequence, larger microwave power is needed to move vortices between the antidots. This restricts the temperature regime for operation to temperatures close to T_c. Nevertheless, the temperature regime is still quite large compared to operation regimes of low-T_c films, where ratchet effects are typically present in the regime of $0.995 < T/T_c < 1$. Whether the frequency limit ($\sim$10 GHz) is a fundamental limit for motion of Abrikosov vortices in HTS devices is not clarified till now. Experiments executed at larger microwave power or the use of smaller antidot distances might prove this in the future.

The principle of vortex ratchet is one of the most interesting components for basic analysis and maybe in the future also for application of vortex manipulation in HTS films and devices. Its operation has been demonstrated. HTS vortex ratchets might be used as a converter (microwave-to-dc), filters or as a component in more complex microwave devices that could offer interesting and novel properties.

2.5 Conclusions

Vortex matter in HTS films and devices not only is an interesting topic for basic research but also plays a substantial role for application of superconductivity in general. In most electronic applications, magnetic flux will penetrate the superconductor. Magnetic flux and flux motion affect the performance of superconducting devices. For instance, the reduction of the sensitivity in SQUIDs or the power-handling capability in microwave devices, or the increase of the error rate in logic devices is a consequence of the presence of flux and flux motion in these devices. Guidance and trapping of vortices can reduce or even prevent this effect.

Moreover, vortex manipulation not only is a useful tool to avoid degradation of HTS film and device properties, but can also be used to analyse and understand novel and interesting physical properties and to develop new concepts for application of HTS. Various concepts for vortex manipulation are sketched. The advantage of the use of micro- and nanopatterns (especially antidots) to guiding and trapping of vortices is discussed, and experimental evidence of vortex guidance and vortex trapping by various arrangements of antidots is given.

Thus, the vortex state of matter appears to be very important in applications of superconductivity that requires further investigation. A better understanding will clearly lead to an improvement in the performance of HTS components, such as reduced noise, better power handling capability or improved reliability. Furthermore, it promises deeper insight into the basic physics of vortices and vortex matter, especially at high frequencies. The use of different experimental techniques in combination with micro- or even nanopatterning of high-T_c superconducting film might pave the way towards strategic manipulation of vortices. Systematic analysis of manipulation methods could in turn pave the route towards interesting and innovative fluxonic effects and device concepts.

Acknowledgements The author likes to acknowledge the experimental work of A.M. Castellanos, P. Selders, M. Pannetier, R. Wijngaarden, A. Pruymboom, P. Dymachevski and P. Lahl. Furthermore, A. Offenhäuser, A.I. Braginski, V.R. Misko, V. Yurchenkov, E. Hollmann and R. Kutzner are acknowledged for their support, cooperation and discussion. This work was supported by the ESF program "Nanoscience and Engineering in Superconductivity – NES".

References

1. J.G. Bednorz, K.A. Müller, Z. Phys. B **64**, 189 (1986)
2. H. Kamerlingh Onnes, Leiden Comm. **120b**, **122b**, **124c** (1911)
3. A.G. Lebed (ed.), The physics of organic superconductors and conductors. Springer Ser. Mater. Sci. 110 (2008), ISBN: 978-3-540-76667-4
4. D.E. Farrell, C.M. Williams, S.A. Wolf, N.P. Bansal, V.G. Kogan, Phys. Rev. Lett. **61**, 2805 (1988)
5. D.E. Farrell, S. Bonham, J. Foster, Y.C. Chang, P.Z. Jiang, K.G. Vandervoort, D.L. Lam, V.G. Kogan, Phys. Rev. Lett. **63**, 782 (1989)
6. D.E. Farrell, R.G. Beck, M.F. Booth, C.J. Allen, E.D. Bukowski, D.M. Ginsberg, Phys. Rev. B **42**, 6758 (1990)
7. K. Okuda, S. Kawamata, S. Noguchi, N. Itoh, K. Kadowaki, J. Phys. Soc. Japan **60**, 3226 (1991)
8. J.G. Ossandon, J.R. Thompson, D.K. Christen, B.C. Sales, H.R. Kerchner, J.O. Thomson, Y.R. Sun, K.W. Lay, J.E. Tkaczyk, Phys. Rev. B **45**, 12534 (1992)
9. E. Zeldov, A.I. Larkin, V.B. Geshkenbein, M. Konczykowski, D. Majer, B. Khaykovich, V.M. Vinokur, H. Shtrikman, Phys. Rev. Lett. **73** 1428 (1994)
10. N. Kokubo, Phys. Rev. Lett. **88**, 247004 (2002)
11. K. Yu, T.W. Heitmann, C. Song, M.P. DeFeo, B.L.T. Plourde, M.B.S. Hesselberth P.H. Kes, Phys. Rev. B **76**, 220507 (2007)
12. V.V. Yurchenko, A.J. Qviller, Y. Galperin, P. Mozhaev, J.B. Hansen T.H. Johansen, Effect of self-organized periodic pinning arrays on magnetic flux dynamics in high-T_c superconductors (to be published)
13. R. Besseling, Thesis, University of Leiden (2001)
14. M. Baert, V.V. Metlusko, R. Jonckheere, V.V. Moshchalkov, Y. Bruynseraede, Phys. Rev. Lett. **74**, 3269 (1995)
15. A.M. Castellanos, R. Wördenweber, G. Ockenfuss, A. v.d. Hart, K. Keck, Appl. Phys. Lett. **71**, 962 (1997)
16. R. Wördenweber, P.H. Kes, C.C. Tsuei, Phys. Rev. B **33**, 3172 (1986)

17. R. Wördenweber, P.H. Kes, Phys. Rev. **B34**, 494 (1986)
18. D. Dew-Hughes, Philos. Mag. **30**, 293 (1974)
19. A.I. Larkin, N. Yu, J. Ovchinnikov, Low Temp. Phys. **34**, 409 (1979)
20. H. Kronmüller, M. Fähnle, M. Domann, G. Grimm, R. Grimm, B. Gröger, J. Magn. Magn. Mater. **13**, 53 (1979)
21. G. Blatter, M.V. Feigel'man, V.B. Geshkenbein, A.I. Larkin, V.M. Vinokur, Rev. Mod. Phys. **66**, 1125 (1994)
22. P.H. Kes, J. van den Berg, in *Studies of High Temperature Superconductors*, vol. 5, ed. by A. Narlikar (NOVA Science Publishers, New York, 1990), p.83
23. E.H. Brandt, U. Essmann, Phys. Status Solidi B **144**, 13 (1987)
24. E.J. Kramer, J. Nucl. Mater. **72**, 5 (1978)
25. A. Pruymboom, P.H. Kes, E. van der Drift, S. Radelaar, Appl. Phys. Lett. **52**, 662 (1988); Phys. Rev. Lett. **60**, 1430 (1988)
26. D.O. Welch, J. Adv. Sci. **4**, 81 (1992)
27. R. Wördenweber, Phys. Rev. B **46**, 3076 (1992)
28. R. Wördenweber, Rep. Prog. Phys. **62**, 187 (1999)
29. R. Gross, J. Bosch, H.-G. Wener, J. Fischer, R.P. Hübener, Cryogenics **29**, 716 (1989)
30. M.P. González, E. Hollmann R. Wördenweber, J. Appl. Phys. **102**, 063904 (2007)
31. F. Laviano, G. Ghigo, E. Mezzetti, E. Hollmann, R. Wördenweber, Engineering confined vortex flow in YBCO films patterned by heavy-ion lithography (to be published)
32. R. Wördenweber, A.M. Castellanos, P. Selders, Physica C **332**, 27 (2000)
33. R. Wördenweber, P. Dymashevski, V.R. Misko, Phys. Rev. B **69**, 184504 (2004)
34. T. Haage, J. Zegenhagen, J.Q. Li, H.-U. Habermeier, M. Cardona, C. Jooss, R. Warthmann, A. Forkl H. Kronmüller, Phys. Rev. B **56**, 8404 (1997)
35. J.I. Martín, M. Vélez, J. Nogués I.K. Schuller, Phys. Rev. Lett. **79**, 1929 (1997)
36. M. Andersson, J.C. Cuevas M. Fogelström, Physica C **367**, 117 (2002)
37. M. Lange, M.J. Van Bael, Y. Bruynseraede V.V. Moshchalkov, Phys. Rev. Lett. **90**, 197006 (2003)
38. E. Rosseel, M. van Bael, M. Baert, R. Jonckheere, V.V. Moshchalkov, Y. Bruynserade, Phys. Rev. B **53**, R2983 (1996)
39. V.V. Metlushko, M. Baert, R. Jonckheere, V.V. Moshchalkov, Y. Bruynserade, Solid State Commun. **91**, 331 (1994)
40. V.V. Moshchalkov, M. Baert, E. Rosseel, V.V. Metlushko, M.J. van Bael, Y. Bruynserade, Physica C **282**, 379 (1997)
41. V.V. Moshchalkov, M. Baert, V.V. Metlushko, E. Rosseel, M.J. van Bael, K. Temst, Y. Bruynserade, R. Jonckheere, Phys. Rev. B **57**, 3615 (1998)
42. V.V. Moshchalkov, M. Baert, V.V. Metlushko, E. Rosseel, M.J. van Bael, K. Temst, R. Jonckheere, Y. Bruynserade, Phys. Rev. B **54**, 7385 (1996)
43. L. van Look, E. Rosseel, M.J. van Bael, K. Temst, V.V. Moshchalkov, Y. Bruynseraede, Phys. Rev. B **60**, R6998 (1999)
44. A. Bezryadin, B. Pannetier, J. Low Temp. Phys. **98**, 251 (1995)
45. W. Lang, J. Pedarnig, *Ion Irradiation of High-Temperature Superconductors and Its Application for Nanopatterning* (this book)
46. J.F. Wambaugh, C. Reichhardt, C.J. Olson, F. Marchesoni, F. Nori, Phys. Rev. Lett. **83**, 5106 (1999)
47. J.E. Villegas, S. Savel'ev, F. Nori, E.M. Gonzalez, J.V. Anguita, R. García, J.L. Vicent, Science **302**, 1188 (2003)
48. B.Y. Zhu, F. Marchesoni, V.V. Moshchalkov F. Nori, Phys. Rev. B **68**, 014514 (2003)
49. J. Van de Vondel, C.C. de Souza Silva, B.Y. Zhu, M. Morelle V.V. Moshchalkov, Phys. Rev. Lett. **94**, 057003 (2005)
50. C.C. de Souza Silva, J. Van de Vondel, B.Y. Zhu, M. Morelle V.V. Moshchalkov, Phys. Rev. B **73**, 014507 (2006)
51. C.C. de Souza Silva, J. Van de Vondel, M. Morelle V.V. Moshchalkov, Nature **440**, 651 (2006)
52. P. Selders, R. Wördenweber, Appl. Phys. Lett. **76**, 3277 (2000)

53. R. Wördenweber, P. Selders, Physica C **366**, 135 (2002)
54. G.R. Berdiyorov, M.V. Milošević, F.M. Peeters, Qualitative modification of the pinning force due to vortex-cavity adherence in perforated superconducting films (unpublished)
55. G.S. Mkrtchyan, V.V. Shmidt, Sov. Phys. JETP **34**, 195 (1972)
56. A.I. Buzdin, D. Feinberg, Physica C **256**, 303 (1996)
57. P. Lahl, R. Wördenweber, Appl. Phys. Lett. **81**, 505 (2002)
58. A.N. Lykov, Solid State Commun. **86**, 531 (1993)
59. K. Harada, O. Kamimura, H. Kasai, T. Matsuda, A. Tonomura, V.V. Moshchalkov, Science **274**, 1167 (1996)
60. A. Bezryadin, Y.N. Ovchinikov, B. Pannetier, Phys. Rev. B **53**, 8553 (1996)
61. A. Bezryadin, B. Pannetier, J. Low Temp. Phys. **102**, 73 (1996)
62. V.V. Metlushko, U. Welp, G.W. Crabtree, Zhao Zhang, S.R.J. Brueck, B. Watkins, L.E. DeLong, B. Ilic, K. Chung, P.J. Hesketh, Phys. Rev. B **59**, 603 (1999)
63. V.V. Metlushko, U. Welp, G.W. Crabtree, R. Osgood, S.D. Bader, L.E. DeLong, Zhao Zhang, S.R.J. Brueck, B. Ilic, K. Chung, P.J. Hesketh, Phys. Rev. B **60**, R12585 (1999)
64. R. Wördenweber, E. Hollmann, J. Schubert, R. Kutzner, A.K. Ghosh, Appl. Phys. Lett. **94**, 202501 (2009)
65. B.Y. Zhu, L. Van Look, V.V. Moshchalkov, B.R. Zhao, Z.X. Zhao, Phys. Rev. B **64**, 012504 (2001)
66. F. Nori, Science **271**, 1373 (1996)
67. C. Reichhardt, C.J. Olson, F. Nori, Phys. Rev. B **57**, 7937 (1998)
68. C. Reichhardt, F. Nori, Phys. Rev. Lett. **82**, 414 (1999)
69. H. Nordborg, V.M. Vinokur, Phys. Rev. B **62**, 12408 (2000)
70. C.P. Bean, J.D. Livingston, Phys. Rev. Lett. **12**, 14 (1964)
71. K.K. Likharev, L.A. Yakobson, Sov. Phys. Tech. Phys. **20**, 950 (1976)
72. J. Schneider, H. Kohlstedt, R. Wördenweber, Appl. Phys. Lett. **63**, 2426 (1993)
73. J. Schneider, M. Mück, R. Wördenweber, Appl. Phys. Lett. **65**, 2475 (1994)
74. A. Andronov, I. Gordion, V. Kurin, I. Nefedov I. Shereshevsky, Physica C **213**, 193 (1993)
75. V.L. Ginzburg, L.D. Landau, Zh. Eksp. Teor. Fiz. **20**, 1064 (1950)
76. M. Tinkham, *Introduction to Superconductivity*, 2nd edn (McGraw-Hill, New York, 1996)
77. A. Weber L. Kramer, J. Low Temp. Phys. **84**, 289 (1991)
78. A. Andronov, I. Gordion, V. Kurin, I. Nefedov I. Shereshevsky, Physica C **213**, 193 (1993)
79. L.I. Glazman, Fiz. Nizk. Temp. **12**, 688 (1986); Sov. J. Low Temp. Phys. **12**, 389 (1986)
80. V.M. Fomin, R. Wördenweber, Proc. SPIE **7364**, 73640P (2009)
81. G.R. Berdiyorov, A.K. Elmurodov, F.M. Peeters D.Y. Vodolazov, Phys. Rev. B **79**, 174506 (2009)
82. A.I. Larkin, Y.N. Ovchinnikov, Zh. Eksp. Teor. Fiz. **68**, 1915 (1975); Sov. Phys. JETP **41**, 960 (1976)
83. S.G. Doettinger, R.P. Huebener, R. Gerdemann, A. Kühle, S. Anders, T.G. Träuble, J.C. Villégier, Phys. Rev. Lett. **73**, 1691 (1994); Physica C **235–240**, 3179 (1994)
84. P. Lahl, R. Wördenweber, Inst. Phys. Conf. Ser. **158**, 77 (1997)
85. L.E. Musienko, I.M. Dmitrenko, V.G. Volotskaya, Pis'ma Zh. Eksp. Teor. Fiz. **31**, 603 (1980); JETP Lett. **31**, 567 (1980)
86. W. Klein, R.P. Huebener, S. Gauss und J. Parisi, J. Low. Temp. Phys. **61**, 413 (1985)
87. S.G. Doettinger, R.P. Huebener, A. Kühle, Physica C **251**, 285 (1995)
88. R. Wördenweber, U. Krüger, J. Schneider, R. Kutzner, G. Ockenfuß, Inst. Phys. Conf. Ser. **148**, 619 (1995)
89. P. Selders, A.M. Castellanos, M. Vaupel, R. Wördenweber, Appl. Supercond. **5**, 269 (1998)
90. R. Surdeanu, R.J. Wijngaarden, R. Griessen, J. Einfeld, R. Wördenweber, Europhys. Lett. **54**, 682 (2001)
91. M. Pannetier, R.J. Wijngaarden, I. Fløan, J. Rector, R. Griessen, P. Lahl, R. Wördenweber, Phys. Rev. B **67**, 212501 (2003)
92. E.H. Brandt, Phys. Stat. Sol. B **77**, 551 (1976); J. Low Temp. Phys. **26**, 709, 735 (1977); J. Low Temp. Phys. **28**, 263, 291 (1977)

93. A. Lukashenko, A.V. Ustinov, A.P. Zhuravel, E. Hollmann, R. Wördenweber, J. Appl. Phys. **100**, 023913 (2006)
94. S.J. Hagen, C.J. Lobb, R.L. Greene, M.G. Forrester, J.H. Kang, Phys. Rev. B **41**, 11630 (1990)
95. S.J. Hagen, A.W. Smith, M. Rajeswari, J.L. Peng, Z.Y. Li, R.L. Greene, S.N. Mao, X.X. Xi, S. Bhattacharya, Qi Li, C.J. Lobb, Phys. Rev. B **47**, 1064 (1993)
96. P. Ao, J. Phys. Condens. Matter **10**, L677 (1998)
97. H.J. Jensen, P. Minnhagen, E. Sonin, H. Weber, Europhys. Lett. **20**, 463 (1992)
98. R.A. Ferrell, Phys. Rev. Lett. **68**, 2524 (1992)
99. A. van Otterlo, M. Feigel'man, V. Geshkenbein, G. Blatter, Phys. Rev. Lett. **75**, 3736 (1995)
100. D.I. Khomskii, A. Freimuth, Phys. Rev. Lett. **75**, 1384 (1995)
101. N.B. Kopnin, Phys. Rev. B **54**, 9475 (1996)
102. J. Kolacek, P. Vasek, Physica C **336**, 199 (2000)
103. R. Wördenweber, J.S.K. Sankarraj, P. Dymashevski, E. Hollmann, Physica C **434**, 101 (2006)
104. D. Koelle, R. Kleiner, F. Ludwig, E. Dantsker, J. Clarke, Rev. Mod. Phys. **71**, 631 (1999)
105. A.I. Braginski, H.J. Krause, J. Vrba, *Handbook of thin film devices, Superconducting Film Devices*, vol. 3, Academic, San Diego, pp. 149–225 (2000), ISBN: 0-12-408952-6
106. B.D. Josephson, Phys. Lett. **1**, 251 (1962)
107. M.B. Ketchen, W.J. Gallagher, A.W. Kleinsasser, S. Murphy, J.R. Clem, in *SQUID'85 – Superconducting Quantum Interference Devices and Their Applications*, ed. by H.D. Hahlbohm, H Lübbig (deGruyter, Berlin, 1985), p. 865ff
108. R. Gross, B. Mayer, Physica C **180**, 235 (1991)
109. M. Kawasaki, P. Chaudhari, A. Gupta, Phys. Rev. Lett. **68**, 1065 (1992)
110. A. Marx, R. Gross, Appl. Phys. Lett. **70**, 120 (1997)
111. A.H. Miklich, J. Clarke, M.S. Colclough, K. Char, Appl. Phys. Lett. **60**, 1899 (1992)
112. R.L. Forgacs, A.F. Warwick, Rev. Sci. Instrum. **38**, 214 (1967)
113. V. Foglietti, W.J. Gallagher, M.B. Ketchen, A.W. Kleinsasser, R.H. Koch, S.I. Raider, R.L. Sandstrom, Appl. Phys. Lett. **49**, 1393 (1986)
114. R.H. Koch, J. Clarke, W. M. Goubau, J. M. Martinis, C. M. Pegrum, D. J. van Harlingen, J. Low Temp. Phys. **51**, 207 (1983)
115. R.H. Koch, W. Eidelloth, B. Oh, R.P. Robertazzi, S.A. Andrek, W.J. Gallagher, Appl. Phys. Lett. **60**, 507 (1992)
116. A.H. Miklich, D. Koelle, E. Dantsker, D.T. Nemeth, J.J. Kingston, R.F. Kromann, J. Clarke, IEEE Trans. Appl. Supercond. **3**, 2434 (1993)
117. M. Mück, C. Heiden, J. Clarke, J. Appl. Phys. **75**, 4588 (1994)
118. P. Dutta, P. Dimon, P.M. Horn, Phys. Rev. Lett. **60,** 646 (1979)
119. P. Dutta, P.M. Horn, Rev. Mod. Phys. **53**, 497 (1981)
120. M.J. Ferrari, M. Johnson, F.C. Wellstood, J.J. Kingston, T.J. Shaw, J. Clarke, J. Low Temp. Phys. **94**, 15 (1994)
121. E. Dantsker, S. Tanaka, J. Clarke, Appl. Phys. Lett. **70**, 2037 (1997)
122. R.H. Koch, J.Z. Sun, V. Foglietta, W.J. Gallagher, Appl. Phys. Lett. **67**, 709 (1995)
123. T.J. Shaw, J. Clarke, R.B. van Dover, L.F. Schneemeyer, A.E. White, Phys. Rev. **B54**, 15411 (1996)
124. J.R. Clem, Vortex exclusion from superconducting strips and SQUIDs in weak perpendicular ambient magnetic fields (unpublished) (1998)
125. H.J. Chaloupka, in *Microwave Application of High Temperature Superconductors, NATO ASI Series*, ed. by H. Weinstock (Kluwer, Dordrecht) (1999)
126. H.J. Chaloupka, M. Jeck, B. Gurzinski, S. Kolesov, Electron. Lett. **32**, 1735 (1996)
127. I.B. Vendik, O.G. Vendik, D.I. Kaparkov, Superconductor Microwave Technology, Part II: Superconducting Microwave Circuits, Chalmers University of Technology, Report No. 24, ed. by E. Kollberg, ISSN: 1103-4599, ISRN CTH-MVT-R-24–SE (1996)
128. M. Baziljevich, T.H. Johansen, H. Bratsberg, Y. Shen P. Vase, Appl. Phys. Lett. **69** 3590 (1996)
129. V.V. Yurchenko, R. Wördenweber, Yu. M. Galperin, D.V. Shantsev, J.I. Vestgården, T.H. Johansen, Physica C **437–438**, 357 (2006)

130. R.P. Feynman, R.B. Leighton M. Sands, *The Feynman Lectures on Physics*, vol. I (Addison-Wesley, Reading, 1963)
131. M. Smoluchowski, Phys. Z. **13**, 1069 (1912)
132. E. Kay, D. Leigh F. Zerbetto, Angew. Chem. Int. **46**, 72 (2007)
133. T. Omabegho, R. Sha N.C. Seeman, Science **324**, 67 (2009)
134. Y. Hiratsuka, T. Tada, K. Oiwa, T. Kanayama T. Uyeda, Biophys. J. **81**, 1555 (2001)
135. H. Linke, T.E. Humphrey, A. Löfgren, A.O. Sushkov, R. Newbury, R.P. Taylor P. Omling, Science **286**, 2314 (1999)
136. D.E. Shalóm H. Pastoriza, Phys. Rev. Lett. **94**, 177001 (2005)

Chapter 3
Ion Irradiation of High-Temperature Superconductors and Its Application for Nanopatterning

Wolfgang Lang and Johannes D. Pedarnig

Abstract Many of the cuprate high-temperature superconductors have transition temperatures well above the boiling point of liquid nitrogen and can be operated under technically viable cooling conditions. On the other hand, they have complex and sensitive crystallographic structures that impose severe restrictions for nanopatterning by the established methods. Ion irradiation of these materials offers a unique possibility to create a wide range of different defects and to tailor the electrical and superconducting properties. Depending on the species of ions used during the irradiation, their energy and fluence, nanoscale columnar pinning centres can be created that enhance the critical current, or randomly distributed point defects that change the superconducting properties. The various effects of ion bombardment on the structural and electrical properties of a representative high-temperature superconductor are reviewed with an emphasis on He^{+} irradiation with moderate energy and the prospects discussed to create nanostructures in thin films of these superconductors.

3.1 Introduction

Nanodevices made of superconductors would offer many advantages over today's semiconductor technology. Heat dissipation is a severe obstacle for the ongoing down-scaling trend in electronic circuits and is likely to cause a significant slow-down of the progress. Although superconductive electronic circuits have been successfully demonstrated with classical superconductors [1], the cooling to a very low operating temperature is expensive and limits its use to a high-technology environment. The novel high-temperature superconductors (HTS) with their much more practical cooling requirements would result in a big step forward, but, unfortunately, the well-established lithography technologies from semiconductor manufacturing, that can be applied – with some modifications – to metallic superconductors, are not suitable for creating high-resolution structures in HTS. Several fabrication methods have been explored to produce micro- and nanostructures in HTS, for instance to manipulate vortices [2], but most of them are suitable for laboratory purposes only.

The systematic modification of the superconducting properties is a very important prerequisite for many commercial applications of HTS. Similarly, the fabrication of semiconductor devices was possible only after the discovery that a tremendous variation of the electrical conductivity, as well as the control of the carrier polarity and lifetime can be achieved by various kinds of impurities and their concentration.

One route of tailoring the electrical properties of HTS that has been explored from right after their discovery is the modification of the carrier density by changing the chemical composition to the under- or overdoped state [3]. Such investigations have been primarily attempted to shed light on the microscopic mechanisms that evoke high-temperature superconductivity. This method is not so practical for device manufacturing, since the chemical composition is best established during the material's synthesis and later patterning of structures is hardly possible.

Another way to substantially change the properties of a HTS is by means of creating defects. In early HTS research, tremendous efforts have been undertaken to grow as clean single crystals as possible. The quest for 'good' samples has somewhat overwhelmed the large potential that the systematic creation of defects provide for a thorough understanding of the cuprates. Systematic studies of ion-beam-induced disorder in HTS found a substantial decrease in the critical temperature T_c [4, 5] and, eventually, a disorder-driven superconductor to insulator transition in both in-plane and out-of-plane resistivity [6]. Such as creation of correlated defect structure is another parameter to tune the properties of HTS.

Defects that are invoked by irradiation of particles with different mass, charge and energy can be often controlled accurately and have the additional advantage that they can be formed cumulatively, i.e., one can trace the material's properties during a series of irradiation steps and, thus, better understand the effects of disorder created by the irradiation. An important consideration is the penetration range of the irradiation. For bulk materials, only neutrons and high-energy light ions can be considered [7]. Depending on the particular irradiation conditions, different kind of defects can be created. On the one hand, point-like defects are produced with electrons [8], protons [9] or other light ions, and columnar defect tracks, on the other hand, by heavy ions of GeV energy [10]. Irradiation with nearly GeV energy protons creates splayed columnar defects that result from induced fission of the constituents of the HTS [11].

High-energetic irradiation with heavy ions has been explored extensively since it provides a tool to deliberately pattern columnar defect tracks into the superconductor that enhance the pinning of vortices significantly. Consequently, the critical current is raised and the irreversibility line shifted [10]. Apart from these application-oriented aspects, the columnar defects allow for a wide variation of the complex physical properties of vortices in HTS [12, 13].

Point defects are imperfections on an atomic scale that are formed typically under conditions, where the energy transfer to the crystal is of the order of the energy necessary for vacancy formation. This is the case with light ions of energy ranging from about 0.05 MeV to several MeV. The typical penetration depth of such ions ranges from about 100 nm to few micrometers. It follows that low-energetic irradiation is applicable to thin films of HTS only, since otherwise the ions will be implanted in the material which causes additional effects apart from the defect creation. This might be a reason that low-energetic light-ion irradiation has been rarely explored so far. For superconducting electronics based on thin film devices, however, the small penetration depth of the ions is not a constraint.

3.2 Defect Creation by Ion Irradiation

3.2.1 Methods

Ion irradiation of a crystalline target results in several different effects. Depending on the ion species, their energy, the angle of incidence, the composition and thickness of the target, the ions will be scattered back, implanted in the target material or transmitted through it with a characteristic probability. The interaction with the target can lead to sputtering of atoms, secondary electron generation and displacements of atoms in the bulk of the target, eventually to the extent of a local destruction of the crystalline structure. Displaced atoms might take up so much energy to evoke further defects off the ion path through the material. Such collision cascades are fundamental for the understanding of defect creation during ion irradiation. While extended amorphous regions in the target, as they are commonly caused by heavy-ion irradiation, can be visualized by transmission electron microscopy, statistically distributed point defects evoked by the displacement of individual atoms are hardly directly experimentally accessible.

On the other hand, computer simulations of atomic collision cascades are well established and can predict the damage effects of various ion species in a wide range of energies. In linear cascades, the sequence of recoil generations are calculated including the secondary, tertiary and so on knock-on atoms, until eventually the energies of the incident ions and the knock-on atoms fall below the threshold displacement energy for damage production, or the ion leaves the target with some residual energy. We have used two well-established algorithms, the programs SRIM/TRIM [14] and MARLOWE [15] and, by running both codes on the same problem, have found a mutually good agreement of the results. Both methods are based on the binary collision approximation and include quantum-mechanical treatment. While SRIM assumes an amorphous target and a Monte Carlo method for the evaluation of the scattering cascades, MARLOWE uses the ideal crystallographic structure of the target material. For an evaluation of ion channeling effects and the angle dependence of defect creation, only MARLOWE is suitable. It should be kept in mind that such calculations do not include thermal annealing effects that occur after the damage creation and, thus, tend to overestimate the number of defects. The computer simulations offer excessive possibilities for parameter variations, so that we have oriented our simulations on the parameters that correspond to our experimental investigations discussed later, i.e., a 100-nm-thick film of

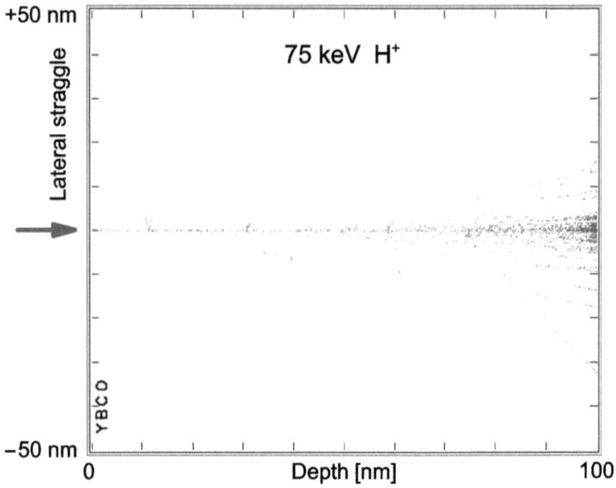

Fig. 3.1 Scattering cascades of target atoms in a 100-nm-thick YBCO film after irradiation with 100 H^+ ions. The direction of incident ion beam and the point of ion injection into the YBCO film are marked by the *arrow*. *Red dots* mark vacancies created by the incident ions, *darker shadings* (*blue*: oxygen, *magenta*: copper, *green*: yttrium, *purple*: barium) indicate vacancies that were created by recoiling target atoms

the cuprate superconductor $YBa_2Cu_3O_7$ (YBCO) on MgO substrate. Simulations on other cuprate superconductors such as $Bi_2Sr_2CaCu_2O_8$ and $HgBa_2CaCu_2O_6$ produced very similar results [16].

3.2.2 Ion Species

Variations of the key parameters for ion irradition have been investigated. First, the collision cascades of several representative ion species, such as H^+, He^+, Ne^+ and Pb^+, have been calculated with SRIM and are visualized in Figs. 3.1–3.4, respectively.

In the case of protons with 75 keV energy, very little interaction with the target atoms takes place and, consequently, a very high dose would be needed to produce a significant amount of defects in the YBCO film (Fig. 3.1). About 99.7% of the incident ions penetrate the target; the other ions are backscattered at the surface and none of them is implanted in the YBCO target. The average number of defects created is 0.5/ion. An interesting feature of proton irradiation is that the ions are little deflected from their incidence direction during traversing the target. Such a small lateral straggle is demanded for preparing nanostructures in thin HTS films by ion damage.

The impact of 75 keV He^+ ions on YBCO is shown in Fig. 3.2. The number of defects created in YBCO is significantly larger as compared to protons (9.7/ion)

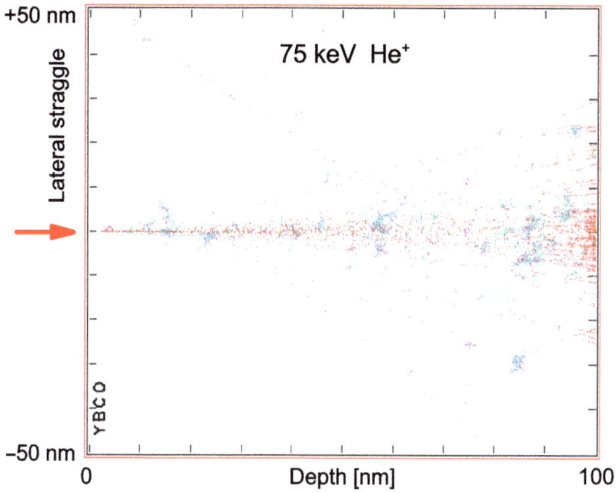

Fig. 3.2 Scattering cascades of target atoms in a 100-nm-thick YBCO film after irradiation with 100 He^{+} ions. The visualization is equivalent to that in Fig. 3.1

and the lateral straggle is about 10 nm when 99% of the ions exit at the backside of the YBCO film. This would still provide an acceptable resolution for ion patterning. The number of implanted He^{+} ions is negligible.

Another frequently used ion species is Ne^{+}. The corresponding results are displayed in Fig. 3.3. The projectiles are strongly deflected from their incidence direction and cause a large number of defects, 476/ion on the average, in particular also many displacements of the heavier atoms Y and Ba. About 37% of the ions are implanted in the target and it is obvious that such irradiation parameters are not suitable for nanopatterning with high resolution. Also, the large number of defects per crystallographic unit cell presumably leads to a breakdown of the crystal structure at moderate fluence and to amorphization of the target material.

As an archetype for heavy ions, the impact of Pb^{+} ions with 75 keV energy is shown in Fig. 3.4. The energy of the ions is exhausted after an average penetration depth of about 30 nm and they are implanted into the target material. As will be discussed later, an almost complete transmission of the ions through the superconducting film is necessary to pattern structures into the HTS film that can be used for superconducting circuits. In principle, it is possible to accelerate heavy ions to higher velocities to achieve this condition, but then the above-mentioned columnar damage tracks will be created.

3.2.3 Ion Energy Dependence

The above results indicate that He^{+} ions are most suitable for a systematic creation of point defects in YBCO thin films. In general, low-energetic ions have a very

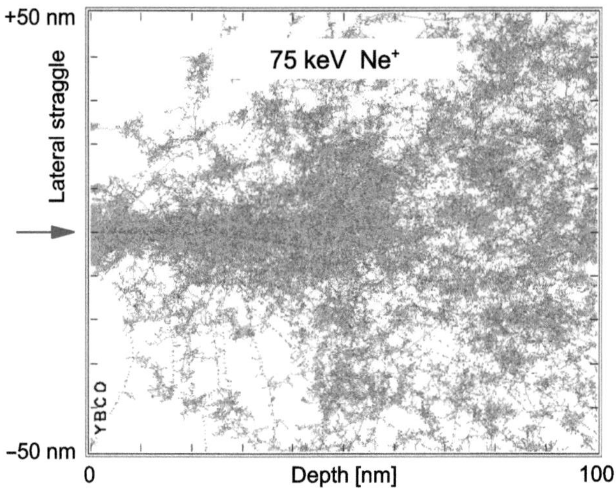

Fig. 3.3 Scattering cascades of target atoms in a 100-nm-thick YBCO film after irradiation with 100 Ne^+ ions. The visualization is equivalent to that in Fig. 3.1. Defects are created far off the line of ion beam incidence. As for H^+ and He^+, the majority of ions are not implanted in the YBCO film

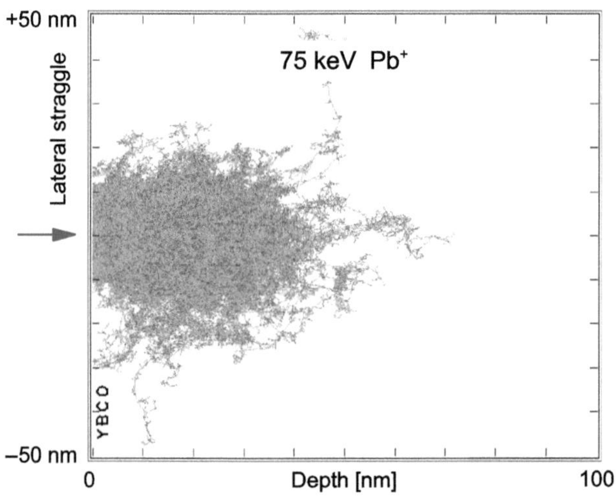

Fig. 3.4 Scattering cascades of target atoms in a 100-nm-thick YBCO film after irradiation with 100 Pb^+ ions. The visualization is equivalent to that in Fig. 3.1. The heavy Pb^+ ions cannot penetrate through the film and implant into the YBCO thin film, creating many defects

limited penetration depth into the target, whereas at higher energies the scattering cross-section diminishes and the efficiency of defect creation is reduced. In Fig. 3.5, the probability for transmission, backscattering or implantation of He^+ ions in a 100-nm-thick YBCO film is plotted as a function of the ion energy [17]. The simulation was performed with MARLOWE on the ideal crystalline structure of YBCO

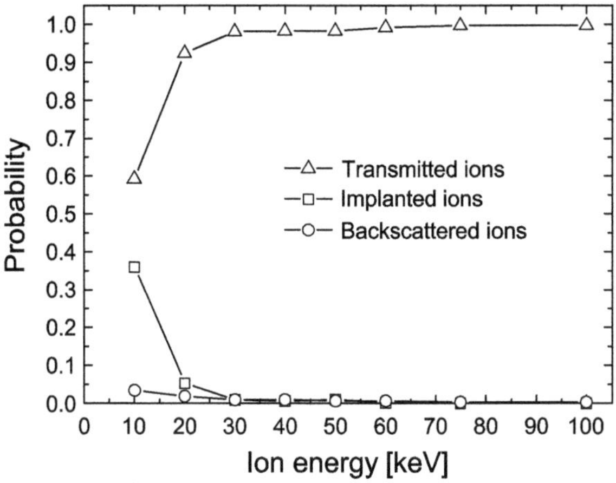

Fig. 3.5 Probability for He^+ ions to be transmitted, backscattered or implanted in a 100-nm-thick YBCO film as a function of energy. The incidence direction is slightly tilted with respect to the crystallographic axes to avoid ion channeling effects

Fig. 3.6 Histogram of the energy distribution of 75 keV He^+ ions after transmission through a 100-nm-thick YBCO film

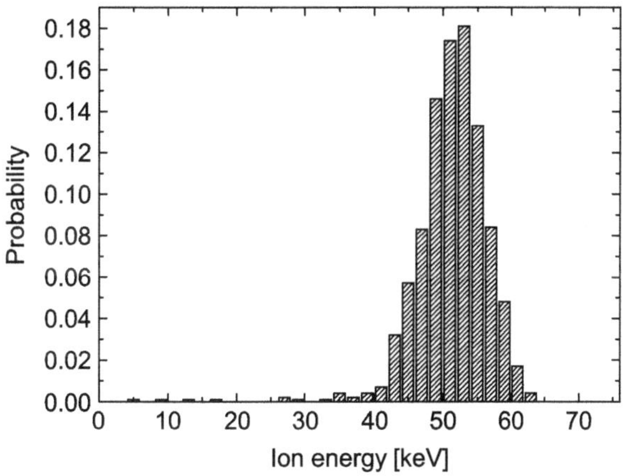

at an incidence angle 5° off the *c* axis and an azimuth angle of 5° from the *a* axis to avoid ion channeling effects. Backscattering of ions is rare and happens only at energies below 30 keV. In the same energy range, those ions that are not transmitted through the target are mostly implanted. Above about 50 keV, however, more than 99% of the ions are transmitted through a 100-nm-thick YBCO film. Thus, it appears reasonable to select an energy of 75 keV for our further investigations.

Although the 100-nm-thick YBCO film is almost transparent to 75 keV He^+ ions, they lose some energy after the passage of the target. In Fig. 3.6, a histogram of the residual energy distribution is shown that was calculated with SRIM [16]. The peak is around 53 keV, corresponding to an average loss of 22 keV after passing the target, and the FWHM is about 10 keV. It is important to note that a rather narrow distribution of the ions that exit on the back side is vital for the homogeneity of

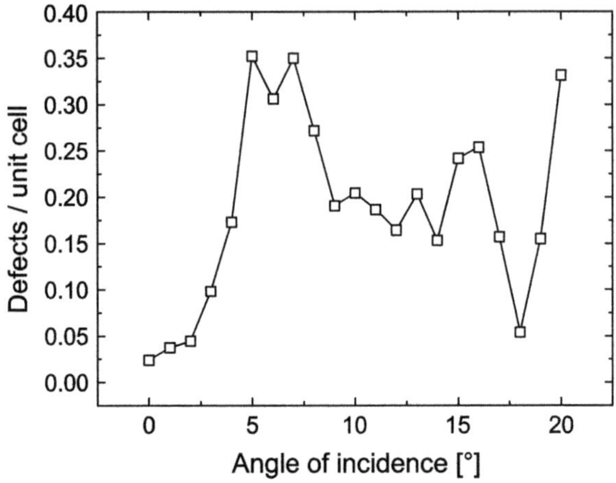

Fig. 3.7 Number of oxygen defects per unit cell in a 100-nm-thick YBCO film after irradiation with 75 keV He^+ ions at various incidence angles from the surface normal at a fluence of 1×10^{15} cm^{-2}

the depth profile of the created defects. If this were not the case, or if most of the ion energy was absorbed in the target, the concentration of defects and of implanted ions would significantly increase near the target's back side.

3.2.4 Angle Dependence

Channeling of ions through thin crystalline targets is a well-known effect that can be often used as an analytical measurement method. Although YBCO thin films do not have a crystal perfection comparable to single crystals of, e.g., semiconductors, an angle dependence of the defect generation efficiency can be envisaged. In Fig. 3.7, the number of oxygen defects per unit cell that are created in YBCO by an 75 keV He^+ ion beam under various incident angles is plotted [17]. The ion fluence was set to 1×10^{15} cm^{-2} for any incidence angle, i.e., in units of ions per *YBCO surface area*. Displacements of oxygen atoms are the most frequent defects, mainly because oxygen is the most abundant atom species in YBCO and, to lesser extent, because it is the lightest constituent. The angle dependence is pronounced with channeling minima at 0° and 18° and maxima at 5° and 7°, respectively. One has to keep in mind that the calculation was performed for an ideal crystal structure and the sharp maxima and minima will be less pronounced in real materials. Nevertheless, a conclusion is that an efficient irradiation should take place at angles of $6 \pm 1°$ from the c axis.

3.2.5 Experimental Results

From these simulation results, it is obvious that the formation of significant defects in YBCO that can lead to a modification of the electrical and superconducting properties of YBCO might be expected not only for highly energetic heavy ion irradiation, but, in a less dramatic fashion, also for the light ions of moderate energy. Several previous experimental studies have investigated the impact of ions with low or moderate mass on HTS thin films. A comparative study of the irradiation of YBCO thin films with either 300 keV protons or 600 keV Ar^+ ions indicated different defect structures that were attributed to displaced oxygen atoms in the former and the overlap of insulating regions in the latter case, respectively [18]. Proton irradiation leads to small defects that are statistically distributed in the material, but the complete suppression of superconductivity in YBCO requires a dose of the order of several 10^{16} cm^{-2} that might be prohibitive for practical applications. Such behavior confirms our results presented in Fig. 3.1. Irradiation of YBCO with 50 keV or 130 keV He^+ resulted in an orthogonal to tetragonal transition of the crystal lattice due to oxygen displacement at a fluence of about 5×10^{15} cm^{-2}, but no extended defect clusters can be detected in TEM investigations below this threshold [19]. An early work reported the creation of amorphous zones after irradiation with 500 keV O^+ ions [20]. In the isostructural compound, $GdBa_2Cu_3O_7$ stable amorphous areas that aggregate and overlap with increasing dose after irradiation with 300 keV Ne^+ and 100 keV Xe^+ ions were found in high-resolution transmission electron microscopy investigations [21].

We have investigated the influence of 75 keV He^+ ion irradiation on the structure of YBCO thin films by X-ray analysis, shown in Fig. 3.8. The (00l) peaks of YBCO exhibit only a slight decrease in intensity up to the ion dose of 5×10^{15} cm^{-2}. The

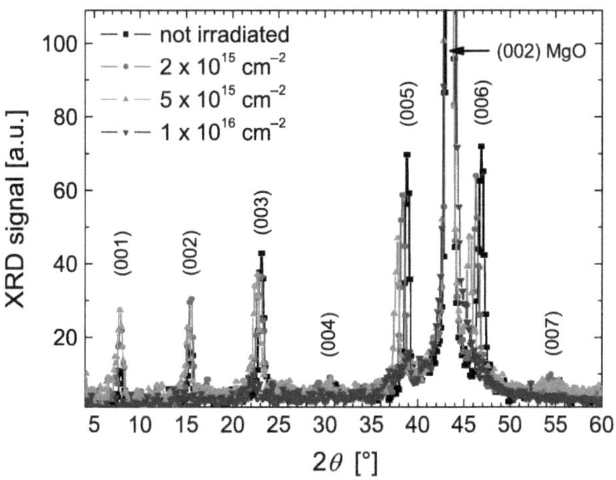

Fig. 3.8 X-ray analysis of a thin YBCO film before and after cumulative irradiation with 75 keV He^+ ions

rocking curves indicate an unchanged linewidth of the reflections. From the shift of the peaks, a small increase in the c axis lattice parameter can be inferred that we attribute to enhanced disorder in the crystal structure. A similar relation between disorder of oxygen atoms and the c-axis lattice constant has been observed during room-temperature annealing of oxygen-deficient YBCO [22]. The sample was heated to above 300°C, where the oxygen atoms are thermally disordered and then rapidly quenched to low temperature. It was demonstrated that during annealing at a timesscale in the order of 100 h, the critical temperature T_c increases and the c-axis length decreases synchronously. It was concluded that both effects are evoked by the gradually growing order of oxygen atoms.

Only at the fluence 10^{16} cm^{-2}, the characteristic peaks of the YBCO crystal structure disappear, indicating the amorphization of the sample. This result is in reasonable accordance with the results of the simulations shown in Fig. 3.7 that imply that about 50% of the oxygen atoms in a unit cell are displaced at such irradiation dose, causing a breakdown of the crystal structure.

3.3 Electrical Properties after Ion Irradiation

3.3.1 Brief Review

Previous investigations of the resistance of HTS did not indicate the particular advantages of the light-ion irradiation. For instance, irradiation with 200 keV Ne^+ ions resulted in a large broadening of the superconducting transition with an almost unchanged onset temperature of superconductivity [23]. The amorphous areas of several nanometer diameter [21] that were created resulted in a 'swiss cheese' structure and a percolative electrical transport. Similarly, irradiation with 600 keV Ar^{2+} with a dose up to 4.1×10^{13} cm^{-2} still shows an onset of superconductivity at $T \sim 50$ K and a very broad transition, although the normal state resistivity already has a semiconducting-like behavior [18]. For Be^+, Ne^+ and Ar^+ ions of energies ranging between 1 and 3.5 MeV, a pronounced increase in the resistivity and a reduction in the critical temperature T_c was observed [24]. The superconducting transition becomes significantly broader after irradiation. Ion beam-induced defects that cluster into small (<10 nm) disordered areas were detected. The size and density of these possibly insulating regions increase with the ion fluence. All these results indicate that the aforementioned conditions result in inhomogeneous electric properties of the samples that might be useful to enhance pinning but not for a systematic variation of the material's superconducting and electric properties.

Only in few studies that have used He^+ ion irradiation with moderate energy, it was reported that the room-temperature resistivity was increased and T_c reduced with the transition remaining sharp [25, 26], supporting our computer simulation results in Chap. 3.2.

3.3.2 Experimental Techniques

The typical procedure for obtaining the experimental results presented here started by the fabrication of thin films of $YBa_2Cu_3O_{7-\delta}$ on (100) MgO substrates by pulsed-laser deposition using 248-nm KrF-excimer-laser radiation and stoichiometric ceramic targets [27]. The films were highly epitaxial with the c axis oriented perpendicular to the substrate surface. The onset of superconducting transition was at about 90 K with a transition width $\Delta T_c \sim 1$ K and the critical current densities $j_c \sim 3$ MA/cm^2 at 77 K and zero magnetic field.

For the transport measurements, the films were patterned to bridges, typically with the dimensions $300 \times 20\ \mu m^2$ by photolithography and wet chemical etching. Contacts were established in a four-probe geometry using sputtered Au-pads. The bridges – but not the contact areas – were irradiated with 75 keV He^+ ions at room temperature using either a masked ion beam lithography system capable for large field structuring [28] or a commercial ion implanter with dose monitoring by Faraday cups. Beam current was kept small and the sample stage was cooled to avoid heating of the YBCO film that might have resulted in loss of oxygen. The beam direction was tilted off the sample's c-axis direction by an angle of roughly 10° to avoid channeling of the He^+ ions.

For the measurements described in this chapter, the irradiation was applied cumulatively, and before irradiation and after each step the resistivity and the Hall effect were measured in a closed-cycle refrigerator with temperature control by a calibrated platinum resistor and computer-controlled compensation for its magnetoresistance. Magnetic fields in both polarities up to $B = \pm 1.2$ T were supplied by an electromagnet. Low-current density excitation was performed with a dc constant-current source in both polarities and the longitudinal and transverse voltages measured by a digital voltmeter and nanovoltmeter, respectively.

3.3.3 Resistivity

The resistivity of YBCO exhibits an exponential increase as a function of the cumulative 75 keV He^+ ion fluence at both room temperature (300 K) and at 100 K, shown in Fig. 3.9. The resistivity at 100 K is increased by about two orders of magnitude upon irradiation with a fluence of 5×10^{15} cm^{-2}. Correspondingly, the critical temperature (defined at half-transition) is rapidly suppressed as shown in Fig. 3.10. A dose of 3×10^{15} cm^{-2} marks the crossover from superconducting material with a metallic-like temperature dependence to a semiconducting temperature behavior and the disappearance of superconductivity.

The effect of the ion irradiation on the resistivity vs temperature curves of a 100-nm-thick YBCO film is presented in Fig. 3.11 [26]. The superconducting transitions remain sharp up to a dose of 1.5×10^{15} cm^{-2} and then become somewhat broader. After irradiating with a fluence of 3×10^{15} cm^{-2}, the temperature dependence is almost flat from 20 to 300 K. With irradiation below 2×10^{15} cm^{-2}, the linear

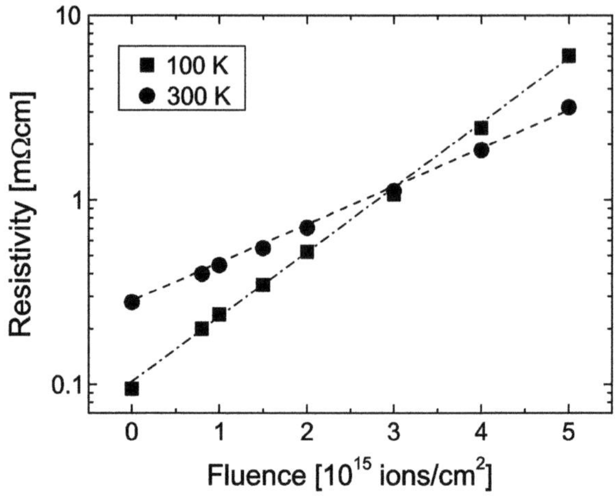

Fig. 3.9 Resistivity increase at 100 K and 300 K of a YBCO film after irradiation with 75 keV He^{+} ions as a function of the cumulative fluence. *Lines* are guides to the eye

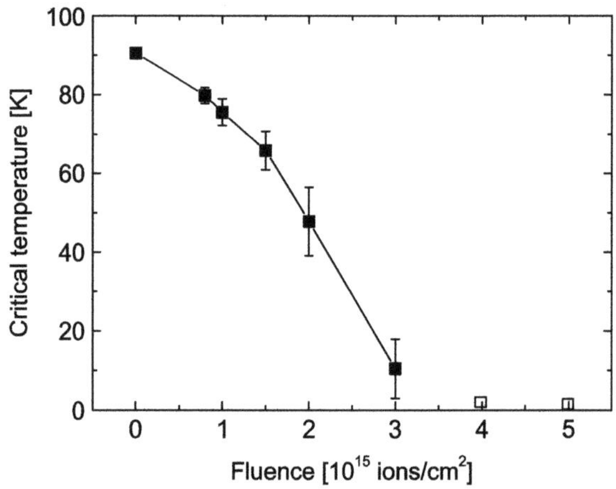

Fig. 3.10 Decrease in the critical temperature T_c of the YBCO film after ion irradiation. T_c is defined by the midpoint of the superconducting transition, the error bars indicate an estimate of the width of the superconducting transition by the 10% and 90% resistivity levels. Line is a guide to the eye. *Open symbols* indicate that no superconductivity was found down to 4.2 K

$\rho(T)$ behavior in the normal state that is characteristic for optimally doped YBCO is preserved and the slope remains the same, but the offset at $T = 0$ K increases with the ion dose, as shown in Fig. 3.12. According to the generally accepted picture of electric properties of HTS, this indicates that the number of mobile carriers remains unchanged, but the number of defects is increased with the irradiation dose and act as scattering centres [29]. No indication for an inhomogeneous critical temperature and a percolative onset of superconductivity that typically shows up as a shoulder in the transition curve can be found.

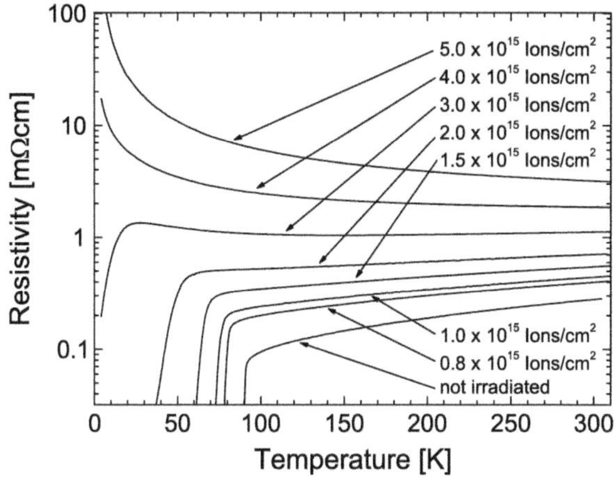

Fig. 3.11 Temperature dependence of the resistivity of an YBCO thin film after irradiation with 75 keV He^{+} ions. The indicated doses are the respective cumulative values

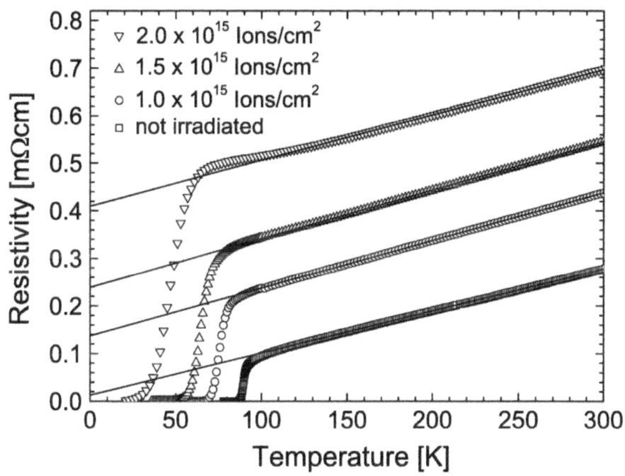

Fig. 3.12 Temperature dependence of the resistivity of YBCO before and after irradiation with 75 keV He^{+} ions at lower doses on a linear scale. The *full lines* are fits to the data in the temperature range from 120 to 300 K

3.3.4 Hall Effect

The measurement and analysis of the Hall effect provide deeper insights into the relevant electrical transport mechanisms. However, the interpretation of the Hall effect is not straightforward in HTS materials. The main problem is that the Hall coefficient R_H in the HTS shows an unusual temperature dependence. In a classical metal, R_H is more or less temperature independent and the Hall number $p_H = (R_H e)^{-1}$ closely matches the concentration of mobile charge carriers in a simple one-band model. In metallic YBCO [30], as well as in other HTS [29], p_H depends almost

Fig. 3.13 Hall coefficient of a YBCO thin film as a function of temperature, after cumulative irradiation with 75 keV He^+ ions

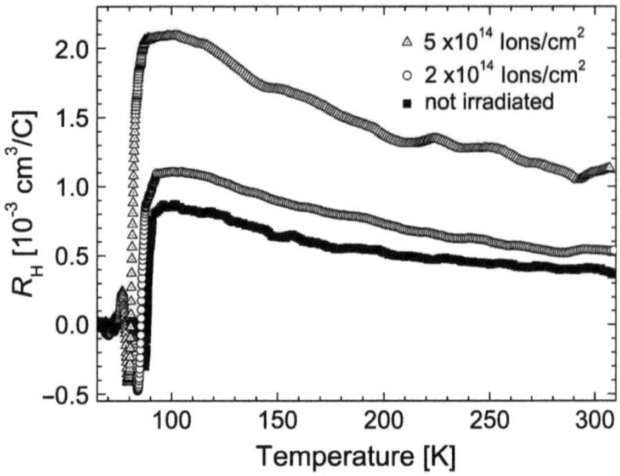

linearly on T above T_c. Thus, p_H calculated from the Hall coefficient R_H cannot be directly identified as the density of charge carriers. The Hall coefficient of YBCO after irradiation with rather low doses of 75 keV He^+ ions is shown in Fig. 3.13 [31]. The unirradiated sample exhibits the typical Hall behavior of a close-to-optimally doped HTS. It is expected that the ion irradiation does not change the mobile carrier concentration and, thus, one would expect only an insignificant change of the Hall effect. On the contrary, R_H is strongly enhanced after irradiation over the entire temperature region displayed in Fig. 3.13.

In addition, an investigation of the Hall angle $\tan\theta_H = \mu B$, where μ is the carrier mobility, provides more insight, since it obeys the semi-empirical law proposed by Anderson in HTS materials [32]

$$\cot\theta_H = \alpha T^2 + C, \tag{3.1}$$

where α and C are constants. The T^2 dependence indicates that the electron–electron (or, more generally, fermion–fermion) scattering mechanism dominates. Such mechanism plays an important role in normal metals only at low temperatures, while in HTS it appears to be dominant up to room temperature. Thus, α is proportional to the hole concentration in HTS, whereas the carrier mobility μ can be inferred directly from the Hall angle.

The carrier mobility of irradiated YBCO is displayed in Fig. 3.14 [26]. The pristine and the weakly irradiated samples show a pronounced $\sim T^{-2}$ temperature dependence that indicates the predominance of hole–hole scattering. At the highest irradiation levels, $\mu(T)$ becomes almost temperature independent since it is then dominated by defect scattering. The mobility at $T = 120$ K, shown in Fig. 3.15 [31] decreases almost linearly with the irradiation dose, which is a further indication of the rising number of defects induced by the irradiation process. A similar relation between the number of defects and the Hall mobility has been found in $YBa_2Cu_{3-x}Zn_xO_{7-\delta}$ single crystals, where defects have been systematically introduced by replacing Cu with Zn [33].

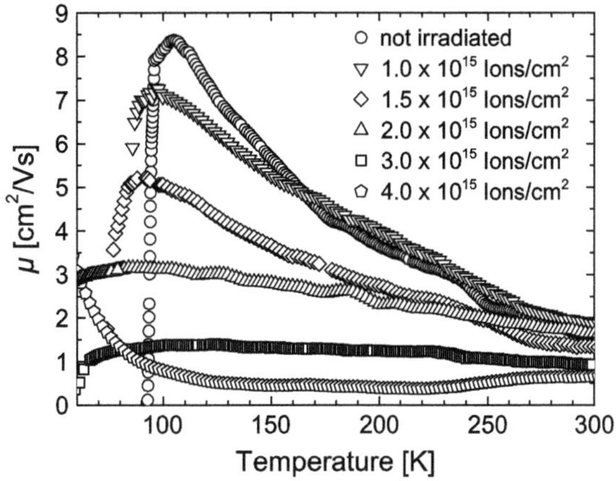

Fig. 3.14 Carrier mobility in YBCO, inferred from the Hall angle, as a function of temperature before and after irradiation with 75 keV He^+ ions at various cumulative doses

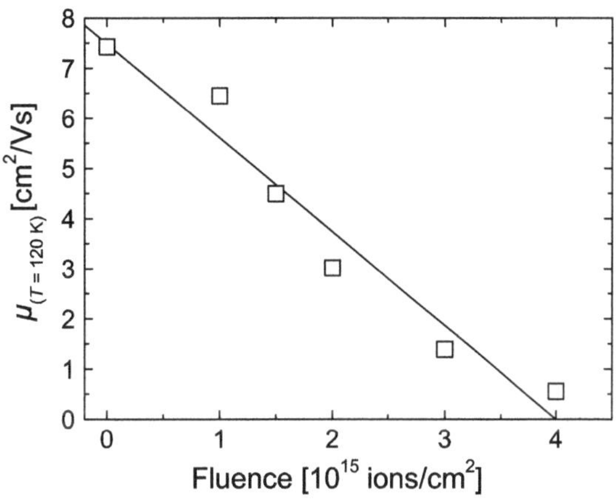

Fig. 3.15 Carrier mobility in YBCO at 120 K as a function of the irradiation dose. The *line* is a linear fit to the data

Figure 3.16 shows the inverse Hall angle of irradiated YBCO as a function of the square of the temperature [26]. The curves precisely follow Anderson's law (3.1) in the normal state even for the semiconducting samples. Following the arguments in [32], the increase in the offset C indicates enhanced defect scattering. The curves for 3×10^{15} and 4×10^{15} cm^{-2} have only a slightly smaller slope α than those at low irradiation doses, which indicates that the reduction in the carrier density upon irradiation is minor even for the highest doses. In sharp contrast to our results, a reduction in the oxygen content in YBCO that is known to reduce the density of mobile carriers results in a significantly lower slope α [30]. Thus, our observations are not compatible with a significant change of the carrier density.

In conventional BCS superconductors, T_c is determined primarily by the density of states, and thus connected to the carrier density, but almost insensitive to moderate defect concentrations. This seemingly contradiction to our results, where we observe a systematic decrease of T_c with the irradiation dose, but almost constant carrier density, can be explained as follows. Substitution of Ba by the isovalent Sr in YBCO does not change the carrier density, but induces a buckling of the CuO_2 planes in

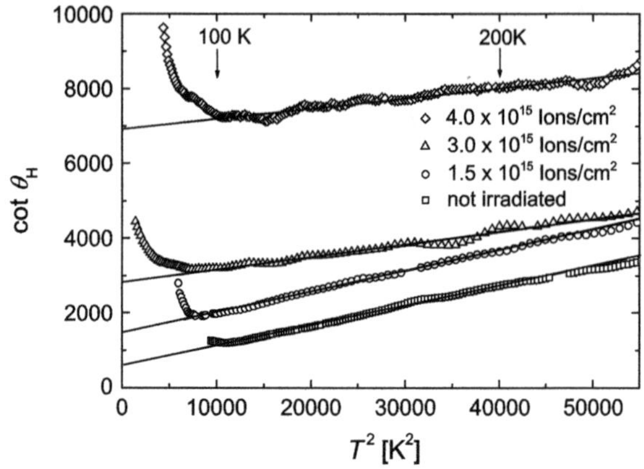

Fig. 3.16 Cotangent of the Hall angle in YBCO as a function of T^2 according to Anderson's law before and after irradiation with 75 keV He^+ ions at various cumulative doses. The *full lines* are fits to the linear part of the curves

the crystal structure that in turn acts as a scattering defect for the carriers. In this situation, also a reduction of T_c was observed [34]. In d-wave superconductors, like YBCO, T_c is a universal function of the in-plane impurity scattering rate that results from the pair-breaking theory for this particular gap symmetry [35].

We can therefore conclude that all our results consistently indicate the creation of defects upon irradiation that result in enhanced carrier scattering and reduction of their mobility, together with a constant density of mobile carriers up to a dose of 1.5×10^{15} cm^{-2} and an only slight reduction of carrier density at the highest irradiation levels.

Other authors have investigated the topological nature of the defects under similar – although lower dose – irradiation conditions, and have found that the mass anisotropy was not changed [25]. They also found strong indications that the defects created by light ion irradiation are point-like, which is in sharp contrast to most of the previous studies that have created columnar damage tracks in the HTS by heavy-ion irradiation.

3.3.5 *Long-term Stability*

With any creation of point defects, the question about possible annealing effects arises. For instance, defects created in YBCO by electron irradiation can be partially annealed at elevated temperatures [36]. For 100 keV proton irradiation at low temperature ($T = 50$ K), it was reported that most of the low energy defects disappear by bringing the sample up to room temperature, resulting in a T_c increase of more than 15 K as measured immediately after irradiation and after room temperature annealing [37]. Similarly, we have found that a significant part of the defects from He^+ irradiation can be healed at about 300°C. However, for practical purpose, it is more interesting to know what extent annealing processes are present at room temperature.

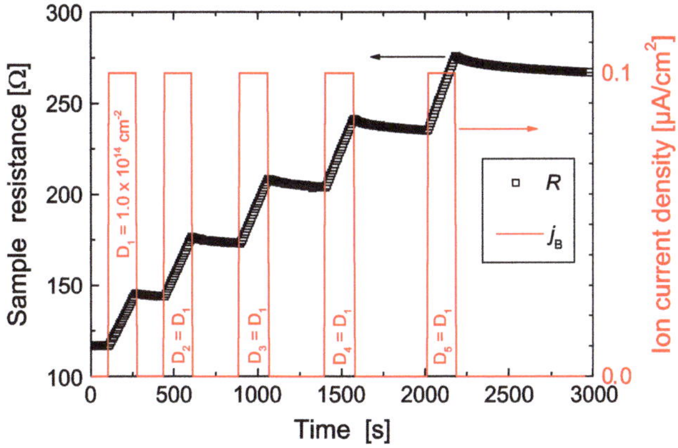

Fig. 3.17 Time evolution of the resistance of a thin YBCO film during cumulative irradiation with 75 keV He^+ ions. Each individual irradiation step was performed with a dose of 1×10^{14} cm^{-2} (ion current density $j_B = 0.1 \pm 0.0025\,\mu A/cm^2$). The resistance was monitored in situ during ion irradiation and during intervals between individual exposure steps

A c-axis-oriented YBCO film on MgO substrate with a thickness of 310 nm was patterned into bridges of 100 μm width and 1,000 μm length. The YBCO film was then irradiated with 75 keV He^+ ions at normal incidence. The ion beam current density was $j_B = 0.1 \pm 0.0025\ \mu A/cm^2$ and was monitored during sample irradiation using Faraday cups placed next to the sample. An exposure time of about 160 s corresponded to an irradiation dose of 1.0×10^{14} cm^{-2}. The time required for mechanical opening of the beam line and for deflecting the beam away from the sample after completion of irradiation was less than 1 s. A stainless steel platelet with a hole of 2.0-mm diameter was used as mask to restrict ion exposure to the YBCO film bridges and to protect the contact pads against irradiation. In situ electrical measurements were performed by the four-probe method with periodical polarity reversal of the excitation current.

In Fig. 3.17, the change of YBCO film resistance is monitored during an alternating sequence of irradiation and beam-off periods. During each irradiation step with a dose of 1×10^{14} cm^{-2}, the resistance appears to increase linearly. This is the first-order approximation in a limited range of the exponential increase displayed in Fig. 3.9. After blocking the ion beam, the resistance relaxes slightly. Obviously some of the created defects are trapped with a low energy barrier and can anneal at room temperature with time constants in the order of seconds. The relaxation behaviour as a function of the ion fluence is displayed in Fig. 3.18. The normalized sample resistance $R(t)/R_0$ relaxes by a few per cent within about 1,000 s after terminating the irradiation. The relaxation behaviour seems to vary with the accumulated dose employed to the sample, being less pronounced at the highest dose of 5.0×10^{15} cm^{-2}.

A crucial question is how many defects will anneal at long time constants. Resistance relaxation measured on an YBCO film of 100-nm thickness on a $SrTiO_3$

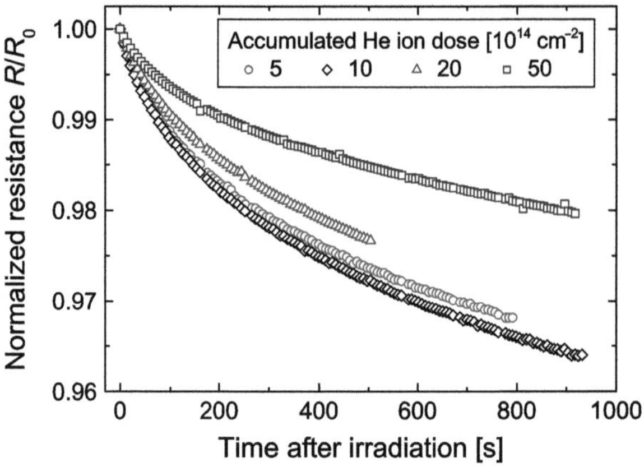

Fig. 3.18 Relaxation of YBCO thin film resistance R/R_0 after He^+ ion irradiation at different dose. The value R_0 is the film resistance measured when the indicated dose is reached and the ion beam is deflected from the sample

substrate immediately after irradiation with a fluence of 3.5×10^{15} cm^{-2} for 3 days revealed that it follows a stretched-exponential behaviour

$$R(t) = R(t = 0)\exp\left[-(t/\tau)^{\beta}\right] + R_0, \tag{3.2}$$

where $R(t)$ and $R(t = 0)$ are the sample resistance as a function of time and at start, respectively, τ is the characteristic time constant, β the stretching exponent and R_0 the equilibrium resistance at $t \to \infty$. Stretched exponential relaxation is a common process with disorder in solids [38] and allows to extract some fundamental parameters. In our case, values of $\tau \sim 5 \times 10^4$ s, $R_0/R(t = 0) \sim 0.80$ and $\beta \sim 0.3$ were obtained. This result indicates that the resistance of irradiated YBCO decreases by about 20% before reaching thermal equilibrium after storing the sample at room temperature for weeks. In our earlier experiments, we have checked the resistance of irradiated samples several months after the first measurement (that took place several days after irradiation) and found no change. From the above, we conclude that despite some annealing at room temperature, the He^+ ion bombardment yields a permanent change of the electrical properties of YBCO.

3.4 Nano-patterning by Masked Ion Beam Irradiation

3.4.1 Previous Attempts to Nanopatterning of HTS

Common techniques for the patterning of HTS films are chemical etching or ion milling, where the structures are defined by a layer of photoresist on top of the HTS.

Apart from possible degradations of the surface after removal of the photoresist, this process results in the formation of a textured surface that is a major disadvantage for growing additional epitaxial layers of HTS, a protection layer or other material. Strain on the edges of the patterned film and the wavelength of the light used for exposing the photoresist limit the minimum size of device structures. In addition, such techniques involve several processing steps even for the fabrication of a simple circuit like a superconducting connecting line between two devices.

Josephson junctions in HTS have been only demonstrated with rather complicated and time-consuming production methods [39]. Bicrystal grain boundary junctions [40], edge-type junctions [41] and ion-beam damage junctions have been reported. Also in the latter case, several production steps were necessary: either a layer of photoresist was processed by focused ion beam [42] or standard UV lithography [43] and then etched, or a metal mask was directly deposited on the YBCO film and then patterned by ion beam milling [37,44,45]. In a final step, the HTS was ion irradiated through the resulting trench.

The reproducible and systematic change of the electrical properties of HTS by irradiation with He^+ ions of moderate energy, as discussed above, can be used for the novel method **M**asked **I**on **B**eam direct **S**tructuring (MIBS) of HTS that is schematically shown in Fig. 3.19. A film of a HTS is prepared on a suitable substrate with a thickness small enough that the ions of the selected energy can penetrate through this film and implantation is avoided. A mask is placed at small distance from the film and protects selected areas of the HTS film from being irradiated. After subjecting the arrangement to an ion beam with large cross-section and very low beam divergence, selected parts of the film are converted to non-superconducting, where those regions protected by the mask pattern remain superconducting. The major advantage of such single-step technique is the direct, non-contact structuring of superconducting devices with a resolution that is essentially only limited by the masking technique. It does not require the multiple processing steps connected with the use of photoresist and avoids possible surface degradations. The large ion-beam exposure fields would enable the processing of many structures in parallel.

3.4.2 Computer Simulation Results

For a practical realization of MIBS, it is crucial to investigate the possible lateral resolution of this technique. A simulation of the lateral straggle of the collision cascades and the homogeneity of defect generation along the thickness of the film is presented in Fig. 3.20 [46]. The irradiation is applied through a mask with a $50 \times 50\ nm^2$ hole, and it is assumed that the ion beam has no divergence. The defect area laterally widens to about $60 \times 60\ nm^2$ at the bottom of the YBCO film, which indicates that irradiation with He^+ through a mask should achieve a principal resolution of about 10 nm. Close inspection of the detailed view shown at the bottom of Fig.3.20 shows that preferentially the oxygen atoms in the YBCO crystal structure are displaced, whereas the impact on the heavier atoms is significantly

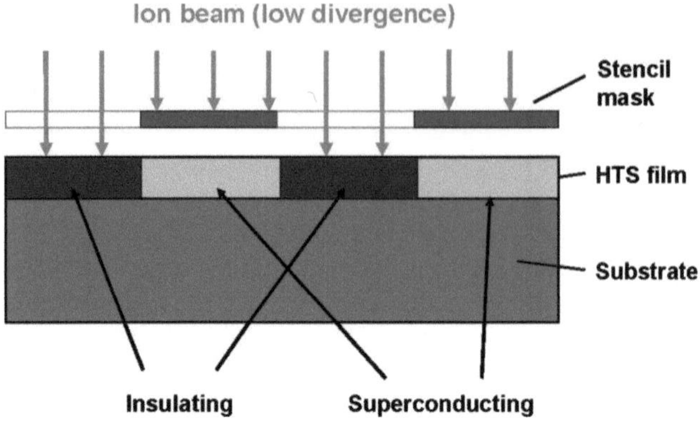

Fig. 3.19 The principle of masked ion beam direct structuring (MIBS) of HTS. A mask with the particular pattern is placed at small distance above a superconducting film prepared on a substrate. The pattern can be projected to the HTS film in a 1:1 reproduction with a large exposure field

lower, in accordance with the microscopic results [19]. The atoms are displaced by a short distance only, typically at the order of the size of a unit cell. This is a further indication that no loss of oxygen from the YBCO has to be expected.

3.4.3 Experimental Patterning Tests

For experimental patterning tests, epitaxial films of YBCO that are similar to those used for the electrical measurements were fabricated as described in Sect. 3.3.2. A silicon stencil mask with various test patterns and features in the dimensions ranging from about 50 nm to 1.5 μm was placed on top of the YBCO film according to the arrangement presented in Fig. 3.19 and irradiated with 75 keV He^+ ions. Arrays of ion-irradiated regions were created in the YBCO film, corresponding to the holes in the mask and investigated by scanning electron microscopy (SEM), where dark regions correspond to the irradiated areas. As examples, a periodic pattern of irradiated dots is shown in Fig. 3.21 and of stripe-like features in Fig. 3.22 [47]. The contrast in the SEM picture results presumably from the different electronic work functions of regions with different conductivity. The irradiated dots have a diameter of about 200 nm. The irradiated stripe pattern shows a 90° turn according to the mask structure with a distance of about 74 nm between individual stripes. Further experiments with different masks indicated that even significantly smaller features might be possible with this technique.

The nano-structured YBCO films have been performed by ion projection patterning with an 1:1 reproduction scale. With advanced ion-beam systems using 200:1 reduction optics and programmable aperture plate systems [48], the patterning of smaller structures of arbitrary size and shape over a large YBCO film area is envisaged.

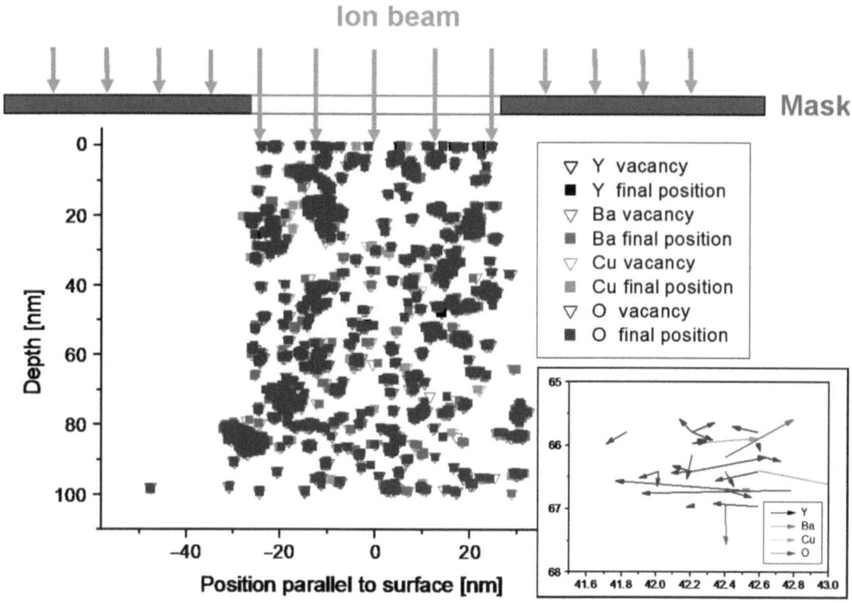

Fig. 3.20 Computer simulation of the ion-beam-induced defects with the program MARLOWE. A 100-nm-thick film of $YBa_2Cu_3O_7$ is irradiated with 75 keV He^+ ions through a mask with a 50×50 nm^2 hole, at a dose of 1×10^{14} cm^{-2}. Target atoms that are displaced are indicated by their original (*open symbols*) and stopping positions (*full symbols*). In the inset, a blow up of a small area located at the right fringe of the defect zone is shown in greater detail and the displacement vectors of atoms indicated by *arrows*

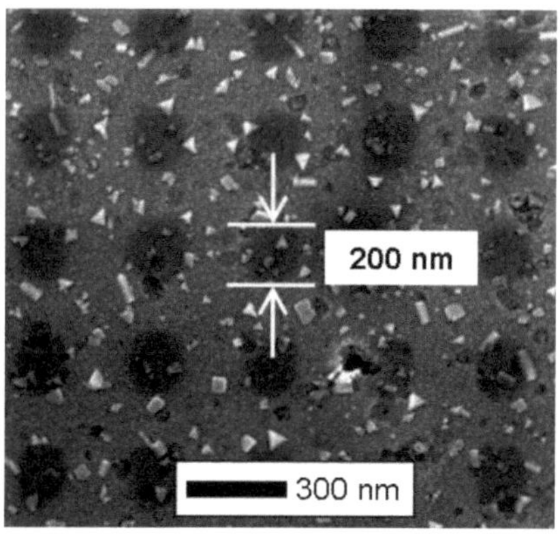

Fig. 3.21 Scanning electron microscopy (secondary electron detection) picture of a thin YBCO film after He^+ ion irradiation through a silicon stencil mask. The *dark* regions correspond to ion-irradiated arrays of nano-dots with a diameter of about 200 nm according to the Si mask structures

3.5 Conclusions and Outlook

Irradiation of HTS by He^+ ions of moderate energy provides a convenient method to systematically alter the electrical properties, increase the resistivity, reduce the carrier mobility and reduce the critical temperature, even eventually fully suppress superconductivity. These modifications can be achieved by introducing point defects,

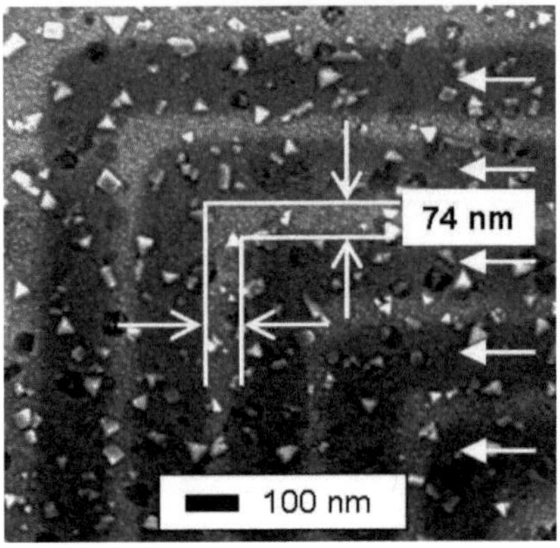

Fig. 3.22 Scanning electron microscopy picture of a thin YBCO film after He^{+} ion irradiation through a silicon stencil mask. *Arrows* indicate line-shaped patterns, separated by unirradiated material of about 74-nm width

but with leaving the framework of the crystalline structure intact. Together with appropriate masking techniques, ion irradiation allows for patterning structures that consist of both high resistive, non-superconducting regions and pristine material side-by-side with high lateral resolution. It can be expected that with this technique not only superconducting interconnects but also Josephson junctions and superconducting devices can be produced on a nano-scale in a single processing step.

Acknowledgements This work was supported by the Austrian Science Fund (FWF), the Micro @Nanofabrication Austria (MNA) Network funded by the Austrian Federal Ministry for Economic Affairs and Labour, and the European Science Foundation program Nanoscience and Engineering of Superconductors (NES). We are greatly indebted to our many collaborators who have made invaluable contributions to these results: P. Bauer, D. Bäuerle, M. A. Bodea, E. Cekan, M. Dineva-Stavreva, T. Enzenhofer, K. Haselgrübler, C. Hasenfuss, R. Kolarova, R. Korntner, H. Loeschner, W. Markowitsch, M. Marksteiner, L. Palmetshofer, M. Peruzzi, E. Platzgummer, H. Richter and K. Siraj.

References

1. T. Van Duzer, C.W. Turner, Principles of Superconductive Devices and Circuits, 2nd edn. (Prentice Hall PTR, Upper Saddle River, NJ, 1999)
2. R. Wördenweber, High-T_c Films: From Natural Defects to Nanostructure Engineering of Vortex Matter (this book)
3. J.T. Markert, Y. Dalichaouch, M.B. Maple, in Physical Properties of High Temperature Superconductors I, ed. by D. Ginsberg (World Scientific Publishing, Singapore, 1989), pp. 265–338
4. B. Roas, B. Hensel, G. Saemann-Ischenko, L. Schultz, Appl. Phys. Lett. 54, 1051 (1989)
5. A. Pomar, Z. Konstantinovic, L. Martel, Z.Z. Li, H. Raffy, Phys. Rev. Lett. 85, 2809 (2000)
6. J.Z. Wu, N. Yu, W.K. Chu, Phys. Rev. B 48, 9929 (1993)
7. H.W. Weber, in Handbook of Superconducting Materials, vol. 1, ed. by D.A. Cardwell, D.S. Ginley (IOP Publishing, London, 2003), pp. 407–418
8. J. Giapintzakis, W.C. Lee, J.P. Rice, D.M. Ginsberg, I.M. Robertson, R. Wheeler, M.A. Kirk, M.O. Ruault, Phys. Rev. B 45, 10677 (1992)

9. L. Civale, A.D. Marwick, M.W. McElfresh, T.K. Worthington, A.P. Malozemoff, F.H. Holtzberg, J.R. Thompson, M.A. Kirk, Phys. Rev. Lett. 65, 1164 (1990)
10. L. Civale, A.D. Marwick, T.K. Worthington, M.A. Kirk, J.R. Thompson, L. Krusin-Elbaum, Y. Sun, J.R. Clem, F. Holtzberg, Phys. Rev. Lett. 67, 648 (1991)
11. L. Krusin-Elbaum, J.R. Thompson, R. Wheeler, A.D. Marwick, C. Li, S. Patel, D.T. Shaw, P. Lisowski, D.T. Ullmann, Appl. Phys. Lett. 64, 3331 (1994)
12. E.H. Brandt, Rep. Prog. Phys. 58, 1465 (1995)
13. G. Blatter, V.B. Geshkenbein, in The Physics of Superconductors, vol. 1, ed. by K.H. Bennemann, J.B. Ketterson (Springer, Berlin, 2003), pp. 725–936
14. J.F. Ziegler, J.P. Biersack, U. Littmark, The Stopping and Range of Ions in Solids (Pergamon Press, New York, 1985). http://www.srim.org
15. M.T. Robinson, MARLOWE Binary Collision Cascade Program Version 15b (2002)
16. M. Dineva-Stavreva, Monte-Carlo Simulation der Defekterzeugung durch Ionenstrahlen in Hochtemperatursupraleitern. Ph.D. Thesis, University of Vienna, (2005)
17. M. Marksteiner, Einfluss winkelabhängiger Ionenbestrahlung in Hochtemperatursupraleitern. Diploma Thesis, University of Vienna, (2006)
18. O. Meyer, T. Kroener, J. Remmel, J. Geerk, G. Linker, B. Strehlau, T. Wolf, Nucl. Instr. Meth. B65, 539 (1992)
19. F. LeGoues, M. Ruault, G. Clark, H. Bernas, M. Gasgnier, Philos. Mag. A 60, 525 (1989)
20. G. Clark, F. LeGoues, A. Marwick, R. Laibowitz, R.H. Koch, Appl. Phys. Lett. 51, 1462 (1987)
21. G. Van Tendeloo, M.O. Ruault, H. Bernas, M. Gasgnier, J. Mater. Res. 6, 677 (1991)
22. J.D. Jorgensen, P. Shiyou, P. Lightfoot, S. Hao, A.P. Paulikas, B.W. Veal, Physica C 167, 571 (1990)
23. S. Matsui, H. Matsutera, T. Yoshitake, J. Fujita, T. Satoh, Nucl. Instr. Meth. B 39, 635 (1989)
24. C.H. Chen, A.E. White, K.T. Short, R.C. Dynes, J.M. Poate, D.C. Jacobson, P.M. Mankiewich, W.J. Skocpol, R.E. Howard, Appl. Phys. Lett. 54, 1178 (1989)
25. Z. Sefrioui, D. Arias, E.M. Gonzalez, C. Leon, J. Santamaria, J.L. Vicent, Phys. Rev. B 63, 064503/1 (2001)
26. W. Lang, T. Enzenhofer, M. Peruzzi, J.D. Pedarnig, D. Bäuerle, C. Horner, E. Cekan, E. Platzgummer, H. Loeschner, Inst. Phys. Conf. Ser. 181, 1549 (2004)
27. D. Bäuerle, Laser Processing and Chemistry, 3rd edn. (Springer, Berlin, 2000)
28. http://www.ims.co.at
29. Y. Iye, in Physical Properties of High Temperature Superconductors III, vol. 3, ed. by D.M. Ginsberg (World Scientific, Singapore, 1992), pp. 285–361
30. B. Wuyts, E. Osquiguil, M. Maenhoudt, S. Libbrecht, Z. Gao, Y. Bruynseraede, Phys. Rev. B 47, 5512 (1993)
31. T. Enzenhofer, Eigenschaften von Hochtemperatur-Supraleitern nach Bestrahlung mit leichten Edelgasionen. Diploma Thesis, University of Vienna, (2003)
32. P.W. Anderson, Phys. Rev. Lett. 67, 2092 (1991)
33. T.R. Chien, Z.Z. Wang, N.P. Ong, Phys. Rev. Lett. 67, 2088 (1991)
34. B.W. Veal, W.K. Kwok, A. Umezawa, G.W. Crabtree, J.D. Jorgensen, J.W. Downey, L.J. Nowicki, A.W. Mitchell, A.P. Paulikas, C.H. Sowers, Appl. Phys. Lett. 51, 279 (1987)
35. S.K. Tolpygo, J.Y. Lin, M. Gurvitch, S.Y. Hou, J.M. Phillips, Phys. Rev. B 53, 12454 (1996)
36. S.K. Tolpygo, J.Y. Lin, M. Gurvitch, S.Y. Hou, J.M. Phillips, Phys. Rev. B 53, 12462 (1996)
37. M.G. Blamire, D.J. Kang, G. Burnell, N.H. Peng, R. Webb, C. Jeynes, J.H. Yun, S.H. Moon, B. Oh, Vacuum 69, 11 (2003)
38. J.C. Phillips, Rep. Prog. Phys. 59, 1133 (1996)
39. F. Tafuri, J.R. Kirtley, Rep. Prog. Phys. 68, 2573 (2005)
40. P. Chaudhari, J. Mannhart, D. Dimos, C.C. Tsuei, J. Chi, M.M. Oprysko, M. Scheuermann, Phys. Rev. Lett. 60, 1653 (1988)
41. J. Gao, Y. Boguslavskij, B.B.G. Klopman, D. Terpstra, G.J. Gerritsma, H. Rogalla, Appl. Phys. Lett. 59, 2754 (1991)
42. A.S. Katz, S.I. Woods, R.C. Dynes, J. Appl. Phys. 87, 2978 (2000)
43. F. Kahlmann, A. Engelhardt, J. Schubert, W. Zander, C. Buchal, J. Hollkott, Appl. Phys. Lett. 73, 2354 (1998)

44. D.J. Kang, G. Burnell, S.J. Lloyd, R.S. Speaks, N.H. Peng, C. Jeynes, R. Webb, J.H. Yun, S.H. Moon, B. Oh, E.J. Tarte, D.F. Moore, M.G. Blamire, Appl. Phys. Lett. 80, 814 (2002)
45. N. Bergeal, X. Grison, J. Lesueur, G. Faini, M. Aprili, J.P. Contour, Appl. Phys. Lett. 87, 102502 (2005)
46. W. Lang, H. Richter, M. Marksteiner, K. Siraj, M.A. Bodea, J.D. Pedarnig, C. Grigoropoulos, D. Bäuerle, C. Hasenfuss, L. Palmetshofer, R. Kolorova, P. Bauer, Int. J. Nanotechnol. 6, 704 (2009)
47. J.D. Pedarnig, K. Siraj, M.A. Bodea, I. Puica, W. Lang, R. Kolarova, P. Bauer, K. Haselgrübler, C. Hasenfuss, I. Beinik, C. Teichert, Thin Solid Films 518, 7075 (2010)
48. E. Platzgummer, C. Klein, P. Joechl, H. Loeschner, M. Witt, W. Pilz, J. Butschke, M. Jurisch, F. Letzkus, H. Sailer, M. Irmscher, Proc. SPIE 7488, 74881D (2009)

Chapter 4
Frontiers Problems of the Josephson Effect: From Macroscopic Quantum Phenomena Decay to High-T_C Superconductivity

Antonio Barone, Floriana Lombardi, and Francesco Tafuri

Abstract The macroscopic quantum decay phenomena occurring in Josephson structures represent one of the most intriguing issues of the weak coupling superconductivity, due to both fundamental physics implications and stimulating potential applications. In this chapter, the main aspects of the macroscopic quantum tunneling (MQT) are discussed for both low Tc and high Tc Josephson junctions. Particular attention will be dedicated to the latter for which the unconventional character of the order parameter plays a very important role. Both experimental and theoretical aspects are reviewed.

4.1 Introduction

This contribution tries to merge two main themes: the Josephson effect [1–7]/ macroscopic quantum phenomena [8–13] and high critical temperature superconductors (*HTS*) [14, 15]. The element of conjunction is in this case provided by the recent experiments of macroscopic quantum tunneling (*MQT*) on *HTS* Josephson Junctions (*JJ*s) [16, 17] described in Sect. 4.4. On one hand, this is an occasion to revisit the progress in the realization and understanding of the properties of *HTS* junctions, stressing on elements of major impact and significance for experiments on *MQT* and nanophysics (such as d-wave order parameter symmetry and the presence of low energy quasi-particles). On the other hand, macroscopic quantum effects are a required frame for the quantum computation "golden age", for architectures which, at least partly, use superconducting components. Macroscopic quantum effects in superconducting junctions have a double function: (1) they will promote novel qubit designs, which will integrate superconductors into hybrid structures, to emphasize the functionalities of the various components; (2) they represent a powerful tool for the diagnostics of the physics of weak links and of the materials, which compose the junctions, and they inspire novel insights on macroscopic quantum physics. The focus of this contribution is more on the latter aspects.

The nature of superconductivity in oxide compounds lies in the background with its still mysterious origin. The phenomenology of *HTS* encompasses a wide range of interesting issues at the border of our understanding of solid-state systems and at the limit of current capabilities of material science and nanotechnology. The complexity of these materials, the extreme values of their characteristic lengths and energies, the possibility to tune their order, and their properties through the oxygen content are crucial ingredients for a composite and complicated phenomenology.

The Josephson junctions have been playing an irreplaceable role in defining crucial properties of *HTS*. The d-wave order parameter symmetry is probably the most remarkable example [18, 19]. Recent reviews on *HTS JJ*s [19–21] focus on the peculiar features of *HTS* and try to bring back the phenomenology of *HTS JJ*s when possible, to the well-established tracks of the Josephson effects in low critical temperature superconductor junctions [2–7].

If we imagine to be a few months before the discovery of *HTS*, who would have imagined that in a few months a supercurrent would have flown up to about 100 K? For the Josephson effect (*JE*) this turns in: who would have imagined a supercurrent between two phase coherent electrodes up to about 100 K? What about the thermal energy, the gap value, the Josephson coupling energy, the charging energy, the coherence length, the critical stoichiometry, etc.? These considerations lead to the first obvious feature, which is independent of the still mysterious origin of superconductivity in *HTS*, of their very complicate structure, and so on: oxides enlarge the occurrence of superconductivity to unexpected energy and length scales. This is obviously true also for the *JE*. Dimensionality, strong correlations, and d-wave order parameter symmetry (*OPS*) will represent the other ingredients which contribute to define the properties of *HTS*. The new energy and length scales have an enormous impact on the way Cooper pair phase coherence will manifest and will be the continuous background where to develop more advanced issues.

4.2 Grain Boundary Junctions: The Tool

Because of the difficulties in producing a traditional tri-layer structure (superconductor–insulator (normal metal)–superconductor, S-I (N)-S) composed by *HTS* electrodes [20, 21], in analogy to low critical temperature superconductor (*LTS*) junctions shown for instance in Fig. 4.1a, various alternative types of junctions have been developed. The intuitive picture of the Josephson effect is given in Fig. 4.1b, and Josephson coupling will survive as far as the two macroscopic wavefunctions of the two electrodes will overlap. The *HTS* junctions are described in extensive reviews [19–21] with all proper references. Here, we shortly outline the main concepts of those junctions employed in the experiments described in the following. Grain boundary junctions (*GB*s) in *HTS* compounds take advantage of a significant reduction in the critical current between two grains with different orientations, which generates weak coupling and Josephson-like behavior between the two electrodes.

Any *GB* can be considered as the result of three fundamental operations: one electrode can be tilted with respect to the other electrode around the c-axis, or tilted

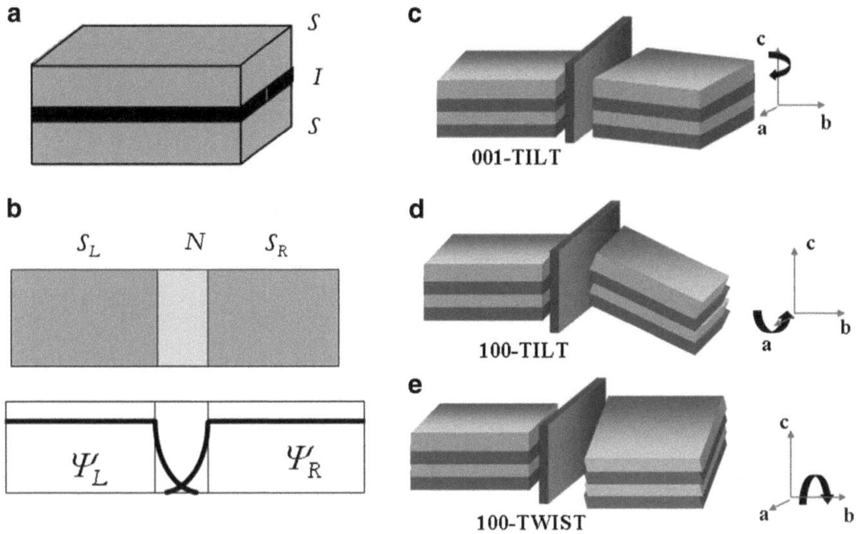

Fig. 4.1 Different configurations of weak-links: junction with (**a**) dielectric or normal metal barrier; (**b**) Josephson coupling will survive as far as the two macroscopic wavefunctions of the two electrodes will overlap; (**c**) 001 tilt, (**d**) 100 tilt and (**e**) 100 twist grain boundary junction

around the *a* or *b* direction (tilt of the *c*-axis) or twisted along the junction interface, respectively, as shown in Fig. 4.1c–e.

The bicrystal technique, based on the union of two substrates with different crystal orientations, is the most direct way to create a *GB* [22–24]. The *HTS* films, if they grow epitaxially, will reproduce the relative orientations of each of the two substrates (Fig. 4.2a). This is the only way to vary the relative orientations of the electrodes in all possible combinations. The performances of junctions with different misorientation angles have been compared in several works (see for instance, [20, 22–30]).

The step, step-edge [31–34], and biepitaxial [35–38] techniques can overcome the constraints on the location of the junctions imposed by the bi-, tri-, or tetra-crystal techniques [20, 21]. All these techniques employ photholithographic means to define the *GB* interfaces. Although they are more flexible in the location of the junctions for fundamental studies and more efficient for circuit design, the *GB*s are limited to particular misorientation angles, and the performances of some of these junctions may be less reliable [20, 21]. While on bicrystals, different *HTS* compounds have been fabricated, for the other junctions all studies basically refer to $YBa_2Cu_3O_{7-x}$ (*YBCO*).

The biepitaxial technique uses changes of the orientation of *HTS* films induced by epitaxial growth on structured template layers. In the original technique [35], an MgO template layer on a r-plane sapphire produces an in-plane rotation by 45° of a $SrTiO_3$/YBCO bilayer when compared with an identical bilayer grown directly on

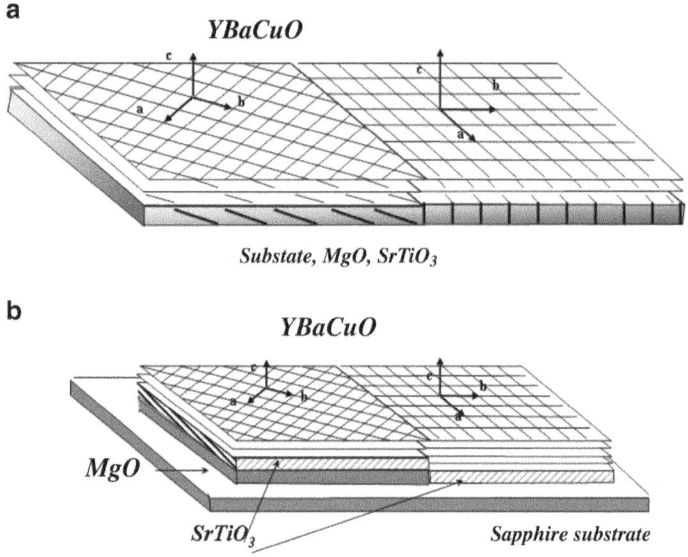

Fig. 4.2 Sketch of the grain boundary structures in bicrystal (**a**) and classical biepitaxial (**b**) junctions

the sapphire, producing a grain boundary with a 45° tilt around the [001] direction (Fig. 4.2b).

More recently, the biepitaxial technique has been extended to novel configurations, in which one of the electrodes does not grow along the *c*-axis orientation [36–38]. A specific feature of these structures is the use of a (110)-oriented MgO (Fig. 4.3a) or CeO_2 (Fig. 4.3b) buffer layer, deposited on (110) $SrTiO_3$ substrates. *YBCO* grows along the [001] direction on the MgO and on the CeO_2 seed layers, while it grows along the [103]/[013] direction on $SrTiO_3$ substrates. In order to select the [103] or [$\bar{1}$03] growth and to ensure a better structural uniformity of the *GB* interface, vicinal *STO* substrates have also been used. The presence of the CeO_2 produces an additional 45° in-plane rotation of the *YBCO* axes with respect to the in-plane directions of the substrate (Fig. 4.3b). As a consequence, the *GB*s are the product of two 45° rotations, a first one around the *c*-axis and a second one around the *b*-axis. This configuration produces a 45° misorientation between the two electrodes to enhance *d*-wave order parameter effects, by varying the interface orientation. Atomically flat interfaces (as shown in the transmission electron microscope image in Fig. 4.3c) can be achieved in appropriate conditions [36]. The possibility to have such "self-assembled" barriers is very relevant in view of the "search of the ideal junction". We will report in Sects. 4.3 and 4.5 on experiments which definitely benefit of the properties of such clean interfaces. The tilt of one of the electrodes and the in-plane 45° rotation on the CeO_2 are the main ingredients to achieve high-quality barriers. They both contribute to decrease the barrier transmission and the in-plane rotation enhances the desired d-wave features, as discussed in the next subsection.

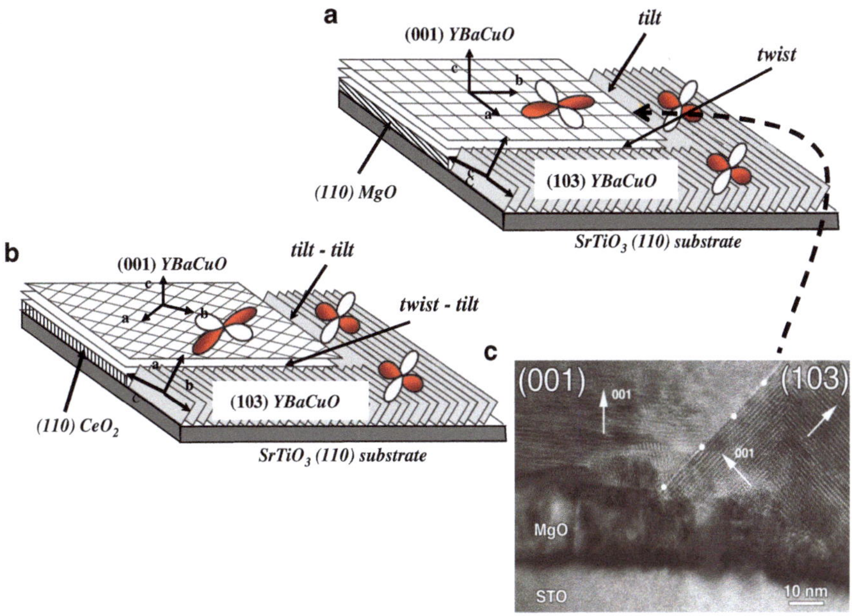

Fig. 4.3 Sketch of the grain boundary structures in biepitaxial MgO(**a**) and CeO_2(**b**)-based out-of-plane biepitaxial junctions. In (**a**) and (**b**) the two limit configurations, tilt and twist, are indicated. When compared with MgO, the presence of the CeO_2 produces an additional 45° in-plane rotation of the *YBCO* axes with respect to the in-plane directions of the substrate. In (**c**) a cross section Transmission Electron Microscope image of the atomically flat interface of the tilt limit of MgO-based out-of-plane biepitaxial junctions is reported [36]

*GB*s have been used for several corner-stone experiments on *HTS* junctions [19–21]. The possibility to span over 4 orders of magnitude in the critical current density and the normal state resistance by selecting specific misorientation angles in *GB* junctions remains as one of the most significant advantages. This is mostly ruled by the d-wave *OPS* and *GB* microstructure. We can have, on one hand, large J_C values for relatively low misorientation angles, which allow studies on static and dynamical effects associated with half flux quantization. On the other hand, J_C values of the order of 10^2 A/cm^2 lead to junctions with hysteretic behavior in the current-voltage characteristics up to 90–95% (see Sect. 4), which have permitted the macroscopic quantum experiments discussed later. It is important to recall that the tunnel limit, as commonly reported for traditional low critical temperature superconductors (*LTS*) Josephson junctions, has never been achieved for *HTS* junctions based on thin film technology. However 'tunnel-like' behavior is often reported in literature.The expression 'tunnel-like' is commonly used to indicate *HTS* junctions characterized by a significant hysteretic behavior in the $I-V$ curves, with critical currents typically not larger than a few μA and normal state resistances R_N not lower than a few hundred Ω for micron size junctions with thickness of about 100–200 nm. Such junctions represent a minority among the data available in literature.

Despite the success of these experiments, *GB*s still appear to suffer problems of reproducibility and yield. The feeling is that some of the limits can still be overcome by further technological improvements, using nanotechnology and by some deeper understanding on the transport in *HTS* junctions. These efforts will lead to understand the ultimate limit of *GB* performances. Nanoscale junctions get closer to the characteristic scale lengths of *HTS* (i.e., coherence length, charge domains) and *HTS* structures (i.e., *GB* facets, *GB* dislocation distribution, nano-channel size) and have full potentials to isolate intrinsic features and transport regimes of *HTS* systems. Once a refined control on each single fabrication step is achieved on a width scale of about 100 nm, the ultimate performances of these junctions will be probably reached. Studies on single faceted nanostructures might also inspire deeper *thinking* on the subtle equilibrium between coherence and disorder, which seem to permeate in different forms the phenomenology of the Josephson effect in *HTS*.

4.3 Retracing d-wave Order Parameter Symmetry in Josephson Structures

An example of one of the most successful achievements of the Josephson effect (*JE*) in the *HTS problem* is the demonstration that a predominant d-wave order parameter symmetry (*OPS*) [18, 19, 39–41] can explain a large part of the phenomenology of HTS junctions [19–21,42]. D-wave *OPS* implies the contribution of antinodal (high energy), and nodal (low energy) quasi-particles in the conduction across junctions and the absence of sharp gap features in the density of states of the weak link. Following [43], we first refer to the two corner-stone experiments [18,19,39–41], which first devised the d-wave *OPS*, and then focus on three examples which are particularly significant for the themes of this contribution and have been made possible by the progress in the realization of *HTS* devices.

A superconducting loop including a Josephson junction could have a π-phase shift in the absence of an externally applied field or current, as proposed by Bulaevskii et al. in 1977 [44]. Such a ring is now termed a π-ring. Such a π-phase shift could result from spin-flip-assisted tunnelling within the tunnel junction itself. A π-phase shift into a superconducting ring has been also introduced by: (1) a more 'intrinsic' mechanism, namely that related to order parameter symmetry in unconventional superconductors [45, 46], (2) by tunnelling through ferromagnetic layers [47–51], and (3) by running supercurrent through two closely spaced electrodes along the ring [52–54]. We are particularly interested in unconventional superconductors for producing π-rings. Geshkenbein, Larkin, and Barone first predicetd the existence of the half flux quantum in proposing π-rings to test for unconventional pairing symmetry in the heavy fermion superconductors [45, 46]. Sigrist and Rice [55–57] interpreted the paramagnetic Meissner effect [58–61] occurring in ceramic samples of $Bi_2Sr_2CaCu_2O_8$, in terms of naturally occurring π-rings due to Josephson contacts between the grains. The controlled geometries that have been used for observing the effects of an intrinsic π-shift in a superconducting ring can be divided into two classes. In the first, the $0-\pi$ junction, one section of a Josephson junction has an intrinsic π phase shift relative to the other. A second geometry for phase sensitive tests of the pairing symmetry in unconventional superconductors is the superconducting quantum interference device (*SQUID*), a superconducting ring with at least one Josephson junction, with in general an intrinsic phase shift π upon circling the ring [18, 62].

The first phase-sensitive pairing symmetry experiments on the cuprate superconductors were by Wollman et al. [40, 41] on π-rings and $0-\pi$-junctions between single crystals of *YBCO* and the s-wave *Pb*. A phase shift of π in the dependence of the critical currents of their π-*SQUID*s on applied flux, relative to that expected for a conventional SQUID, resulting in a minimum in the critical current of their $0-\pi$ junctions at zero applied field. More advanced technique to make high quality $0-\pi$ junctions and π-rings in bi and quad-crystals of $SrTiO_3$ using the cuprate superconductors allowed to obtain responses very close to the ideal behavior [63].

The first experiments to detect the spontaneous flux predicted for π-SQUIDs and $0-\pi$ junctions fabricated from cuprate superconductors were by Tsuei and Kirtley [19]. They formed π-rings [39] using a tricrystal geometry with grain boundary weak links. A scanning SQUID microscope (*SSM*) was used to image the spontaneous currents generated in the ring. Tsuei and Kirtley used tricrystal pairing symmetry tests to infer that the optimally hole-doped cuprates and the optimally electron-doped cuprates [19] have predominantly $d_{x^2-y^2}$ pairing symmetry. Same results were obtained in the hole-doped cuprates over a broad doping range [19]. Predominantly d-wave pairing symmetry could also be inferred from the characteristics of asymmetric $0-45°$ junctions, which have rapidly alternating 0- and π-junctions due to faceting (see end of this section). Such faceting results in unusual magnetic interference patterns [21, 64–66] and spontaneous flux generation in the grain boundaries [64–68]. Spontaneous magnetization along the grain boundaries generated by the series of π-rings have been discussed also in terms of fractional vortices [21, 64–66]. An example is given in the *SSM* image reported in Fig. 4.4 relative to a *YBCO* sample. An opportune geometry has been designed on the basis of the characteristic features and properties of the biepitaxial junctions. To some extent, spontaneous currents can be switched on and off, depending on the interface orientation, barrier transparency, the ratio between the Josephson penetration depth, and the width of the junction [21, 68, 69]. Time reversal symmetry breaking (*BTRS*) has been predicted in the absence of an imaginary component of the *OPS* [70–74] and motivated attempts to see spontaneous flux and fractional vortices also in systems without *GB*s [75, 76].

Advances in the quality of junctions have promoted additional confirmations on d-wave *OPS*. We refer to ramp-edge YBCO-Nb junctions [77] and the above-mentioned off-axis *YBCO* biepitaxial *GB* junctions (Fig. 4.3c). YBCO-Nb ramp junctions have been used to 'extend the half flux quantum vortex experiments' to very large arrays of π-rings [77–79]. The high values of the critical current density favors the spontaneous generation of half-flux quantum vortices in faceted $0-\pi$ YBCO-Nb junctions [78, 79]. A half-flux quantum vortex is generated at

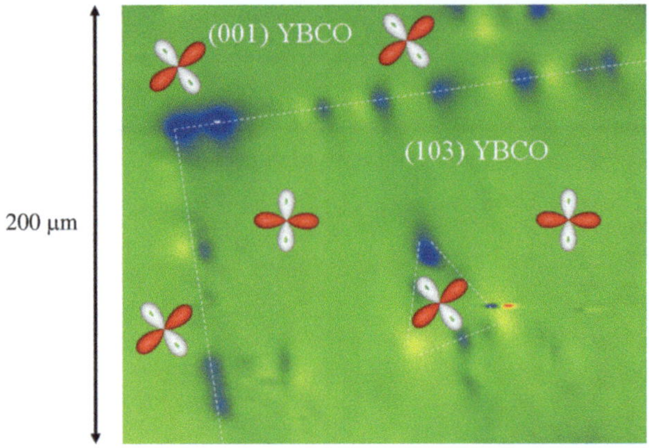

Fig. 4.4 Scanning SQUID microscope image of a $200 \times 200\ \mu m^2$ area, enclosing tilt–tilt and twist–tilt in CeO_2-based biepitaxial *GB*s.The *GB*s are marked by the presence of spontaneous currents. The sample was cooled and imaged at $T = 4.2$ K in nominally zero field. Adapted from Tafuri et al. [68]

each facet corner as the sample is cooled through the Nb superconducting transition temperature.

The off-axis *BP* junctions have been used to measure the angular dependence of the Josephson critical currents of c-axis tilt biepitaxial grain boundary YBCO junctions (Fig. 4.5) [37].

The solid symbols in Fig. 4.5 are normalized critical current densities from junctions with widths of 10 μm (triangles) and 4 μm (stars), as a function of the angle θ of the junction normal relative to the a or b axis of the (001) film. In the Sigrist–Rice phenomenological approach [55–57], in which the Josephson current is proportional to the projection of the momentum-dependent energy gap onto the junction normals, the critical current density is given by

$$J_c = J_0(n_x^2 - n_y^2)_L(n_x^2 - n_y^2)_R \sin(\phi), \tag{4.1}$$

where J_0 is the maximum Josephson current density, and n_x, n_y are the x, y components of the junction normals on the two sides of the junction (L(R) stands for left (right)), and a pure $d_{x^2-y^2}$ pairing symmetry has been assumed. In the biepitaxial grain boundary structure of Fig. 4.3c, this reduces to

$$J_c \sim \sin 2\theta(2 - \cos^2\theta)(1 - 3\sin^2\theta)/(1 + \sin^2\theta), \tag{4.2}$$

which is plotted as the dashed line in Fig. 4.5. The experimental critical current has minima at approximately 0°, 35° and 90°, as expected for a $d_{x^2-y^2}$ superconductor with this geometry. These results have been confirmed on submicron biepitaxial junctions, where also $I_C R_N$ has been found in complete agreement with the results of Sigrist and Rice [80]. These measurements also demonstrate that 'intrinsic' d-wave effects are dominant in the phenomenology of all *HTS* Josephson junctions

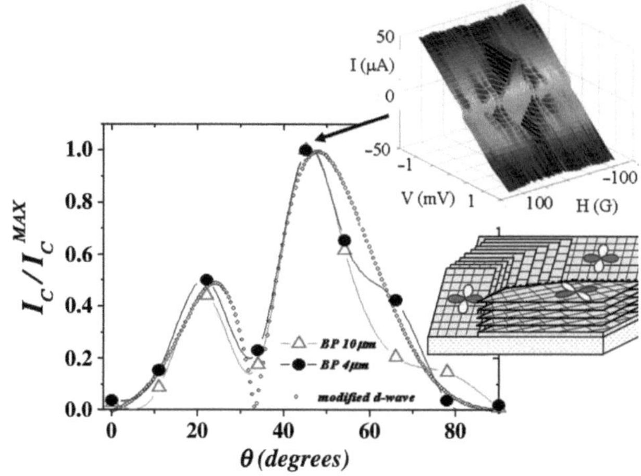

Fig. 4.5 Normalized critical current density J_C vs. angle θ for two sets of c-axis tilt biepitaxial YBCO junctions, with width 10 μm (*triangles*) and 4 μm (*stars*). The *solid lines* connecting the symbols are guides to the eye. The *dotted line* is the Sigrist–Rice-like formula assuming pure $d_{x^2-y^2}$ pairing symmetry in this geometry [37]. In the bottom inset, the scheme of the junction is reported for three different angles, along with the d-wave profiles of the two electrodes. In the top inset, the I–V characteristics of the junction specified by the interface orientation $\theta = 50°$ are reported as a function of the external magnetic field. In this limit, an ideal Fraunhofer-like pattern is obtained. Adapted from Lombardi et al. [37]

themselves independent of the interface details. In the previous corner-stone experiments [18,19,39–41], junctions were inserted in a loop or composed of a low critical temperature superconductor counter-electrode. The d-wave-induced effects in these junctions are apparently very robust, and the coherence related to the OPS seems to be preserved on a scale larger than the faceting and/or any interface impurity characteristic lengths. This may be due to the junction configuration with low barrier transmission probabilities, which preserves the directionality of the Cooper pairs.

Smilde et al., [81, 82] have used the YBCO-Nb ramp-edge technology to produce a series of junctions with varying junction normals relative to the a-axis. Slightly different outcomes have been found on junctions employing twinned and untwinned YBCO films. Junctions made with twinned YBCO films showed the fourfold symmetry, with nodes at 45°, expected for a $d_{x^2-y^2}$ symmetry. Those made with untwinned YBCO showed nodes offset by about 5°, consistent with a small s-wave component to the gap, as expected for this orthorhombic superconductor [19]. The measured angular dependence of the critical currents for YBCO using this ramp edge junction technology was consistent with a substantially in-plane $d_{x^2-y^2}$ gap with an amplitude 50% higher in the b (Cu–O chain) direction than in the a direction [81, 82].

These experiments on YBCO-Nb ramp-edge junctions also provided a clean confirmation of the effects on faceting on the transport properties of a Josephson junction, first found on high angle bicrystal GBs junctions [64–66]. The grain boundary itself represents a sequence of 0 and π junctions: faceting rotates the normal angle, producing alternating positive and negative critical currents along the grain boundary. If the sum of all π-junctions equals the sum of all 0 junctions, the total macroscopic current is 0, unless an external magnetic field provides an additional phase modulation. Only if there is an imbalance between 0 and π, there will be a macroscopic critical current [64–66].

4.4 Macroscopic Quantum Phenomena in Josephson Systems: Fundamentals and Low Critical Temperature Superconductor Junctions

4.4.1 *Resistively and Capacitively Shunted Junction Model and the "Washboard" Potential*

In this section, we briefly recall some standard arguments based on the resistively and capacitively shunted junction (*RCSJ*) model, which allows a general classification of weak links. For details, we refer to traditional textbooks and reviews [2–5]. Within the *RCSJ* model, the Josephson inductance L_J and capacitance C act as an anharmonic LC resonator (at zero voltage) with resonance frequency $\omega_\mathrm{P} = (L_\mathrm{J}C)^{-1/2}$ (plasma frequency), where $L_\mathrm{J} = \Phi_\mathrm{O}/(2\pi I_\mathrm{CO}\cos(\phi)) = \Phi_\mathrm{O}/(2\pi(I_\mathrm{CO}^2 - I^2))^{1/2}$, $\omega_\mathrm{P} = (2\pi I_\mathrm{CO}/(\Phi_\mathrm{O}C))/(1-(I/I_\mathrm{CO})^2))^{1/4}$ and I_CO, represents the maximum Josephson current. Representing the displacement current by a C capacitor and the sum of the quasi-particle and insulator leakage current by a resistance (R), we can devise an equivalent circuit for the junction:

$$I + I_N = I_\mathrm{CO}\sin(\phi) + V/R + C\mathrm{d}V/\mathrm{d}t \tag{4.3}$$

The second term contains the well-known dc Josephson equation $I_\mathrm{C} = I_\mathrm{CO}\sin(\phi)$. The noise source I_N is associated with its shunt resistance. A wide variety of I–V characteristics can be described through an opportune choice of the parameters. We can therefore pass from a tunnel-like behavior with high values of the capacitance, characterized by hysteretic behavior and by the presence of switching currents, to a regime where capacitance plays a marginal role and no hysteretic behavior is present. This situation is particularly significant for *HTS JJ*s, where most junctions present a phenomenology consistent with low capacitance barriers. 'Tunnel-like' junctions represent a minority for *HTS JJ*s [21], but are of great impact for the experiments discussed in the following sections.

Neglecting the noise term and using substantially the ac Josephson equation $V = (\hbar\partial\phi)/(2e\partial t)$, we obtain:

$$I = \frac{\Phi_o}{2\pi}C\frac{\partial^2\phi}{\partial t^2} + \frac{\Phi_o}{2\pi}\frac{1}{R}\frac{\partial\phi}{\partial t} + I_\mathrm{C}\sin(\phi), \tag{4.4}$$

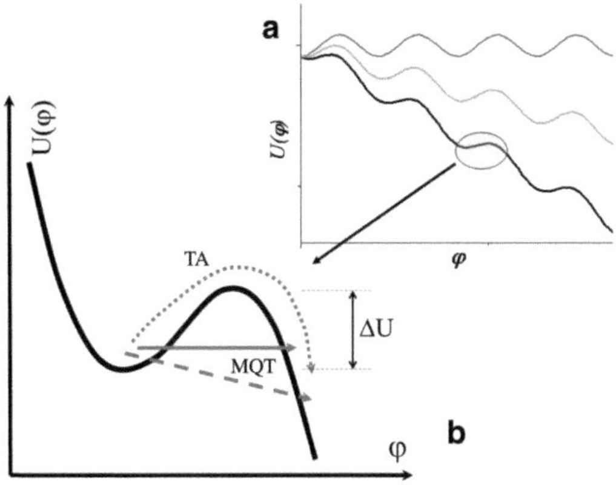

Fig. 4.6 (**a**) Wash-board potential in the RCSJ model for zero and for two different finite values of the bias current, respectively; (**b**) detail of (**a**) for finite bias current, ΔU represents the energy barrier. *TA* stands for thermal activation (*dotted line*), *MQT* processes are indicated for small (*solid line*) and large (*dashed line*) dissipation [92–94]

which can be expressed as:

$$\left(\frac{\Phi_o}{2\pi}\right)^2 C\frac{\partial^2\phi}{\partial t^2}+\left(\frac{\Phi_o}{2\pi}\right)^2\frac{1}{R}\frac{\partial\phi}{\partial t}+\frac{\partial}{\partial\phi}U=0, \tag{4.5}$$

where

$$U=-\Phi_o/(2\pi)(I_C\cos(\phi)+I\phi). \tag{4.6}$$

Considerable insights into the nonlinear dynamics of the junction can be gained by realizing that this equation describes the motion of a ball moving on the 'tilted washboard' potential U [2–6] (Fig. 4.6). The term involving C represents the mass of the particle, the $1/R$ term represents the damping of the motion, and the average 'tilt' of the washboard is proportional to $-I$. For values of $I < I_{CO}$, the particle is confined to one of the potential wells, where it oscillates back and forth at the plasma frequency ω_P. In this case, the average voltage across the junction is zero. When the current I exceeds I_{CO}, the particle rolls down the washboard; in this case, a voltage appears across the junction.

The considerations above can be extended to a more complicated 'tilted washboard' potential U, where effects related to the presence of a second harmonic [83–87] in the critical current are taken into account [16, 17, 88, 89]. This is of more interest for *HTS JJ*s ($I = I_1(\sin\varphi - a\sin 2\varphi)$); the strength of the second harmonic can be defined by the parameter $a = I_2/I_1$, where I_2 and I_1 are the second and first harmonic components in the current–phase relationship (*CPR*), respectively.

4.4.2 Macroscopic Quantum Tunnelling (MQT) and Energy Level Quantization (ELQ)

When the bias current $\gamma = I/I_{CO}$ is ramped from $\gamma = 0$ to $\gamma < 1$, the junction is in the zero voltage state in the absence of thermal and quantum fluctuations. At finite temperature, the junction may switch into a finite voltage state for a bias current $\gamma < 1$. This corresponds to the particle escaping from the well either by a thermally activated process or by tunneling through the barrier potential, macroscopic quantum tunneling (*MQT*). This only occurs if the kinetic energy gained from running "down hill" in the tilted-washboard potential is not all dissipated, but enough remains to carry the representative point over next "hill". Once the phase point gets over a hill by a fluctuation, it keeps running, provided that the damping is below some critical value. The dimensionless quality factor (Q factor) measures the damping of the plasma oscillation by the effective shunt resistance R of the junction and estimates the number of periods during which the oscillation energy will be dissipated. For pure Ohmic damping, Q is defined as $\omega_P RC$.

In the pure thermal regime, the escape rate for weak to moderate damping ($Q > 1$) is determined by [90]:

$$\Gamma_t(a) = a_t \frac{\omega_P(a)}{2\pi} \exp\left(-\frac{\Delta U(a)}{k_B T}\right), \tag{4.7}$$

where $\Delta U(a) = c(a)\left(E_J 4\sqrt{2}/3\right)(1-\gamma)^{3/2}$ is the barrier height for γ close to one, and $c(a)$ is the renormalization factor [16]. The escape rate will be dominated by *MQT* at low enough temperature [8–10, 91] (Figs. 4.6 and 4.7): for $Q > 1$ and γ close to 1, it is approximated by the expression for a cubic potential

$$\Gamma_q(a) = a_q \frac{\omega_P(a)}{2\pi} \exp\left[-7.2 \frac{\Delta U(a)}{\hbar\omega_P(a)}\left(1 + \frac{0.87}{Q}\right)\right], \tag{4.8}$$

where $a_q = (864\pi \Delta U/\hbar\omega_P)^{1/2}$.

The *MQT* rate is affected by dissipation, the irreversible energy transfer between the system and the environment, because of the damping-dependent factor [8–10, 91–94]. This is physically due to smearing of quantum levels in the washboard

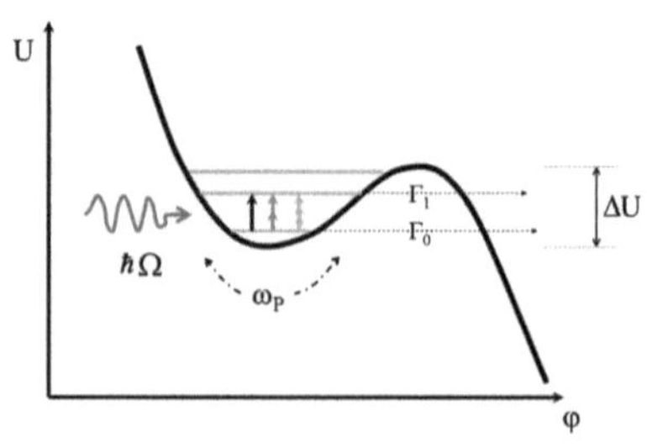

Fig. 4.7 Detail of wash-board potential in the RCSJ in the presence of microwaves. Γ_0 and Γ_1 indicate tunneling process from the ground state and the first excited state, respectively

potential with decreasing Q. The influence of dissipation on the switching statistics of the critical current of the *JJ*s, which is the technical tool to investigate the thermal and quantum activation just introduced, has been widely discussed both theoretically [8–10, 95–99] and experimentally [11–13]. It is convenient to express the thermal escape through an escape temperature $T_{\rm esc}$ defined as:

$$\Gamma_q = \frac{\omega_{\rm P}(a)}{2\pi} \exp\left[\frac{\Delta U}{k_{\rm B} T_{\rm esc}}\right], \tag{4.9}$$

This is made possible because in both the classical and quantum regimes $T_{\rm esc}$ is very nearly independent of the bias current. The crossover temperature $T_{\rm cr}$ between the thermal and quantum regimes is given by: $T_{\rm cr} = (\hbar\omega_{\rm P}/2\pi k_{\rm B})\left\{\left(1 + 1/4Q^2\right)^{1/2} -1/2Q\right\}$.

An analytical expression for the retrapping rate from the resistive to the superconducting state is known only for strongly underdamped *JJ*s:

$$\Gamma_{\rm R} = \omega_{\rm PO} \frac{I - I_{\rm RO}}{I_{\rm CO}} \sqrt{\frac{E_J}{2\pi k_{\rm B} T}} \exp\left[\frac{-E_J Q_{\rm O}^2}{2k_{\rm B}T}\left(\frac{(I - I_{\rm RO})^2}{I_{\rm CO}^2}\right)\right] \tag{4.10}$$

where $I_{\rm RO}$ is the fluctuation-free retrapping current, and $\omega_{\rm PO}$ and $Q_{\rm O}$ are the plasma frequency and quality factor for $I = 0$, respectively.

The very first experiments on *MQT* in a Josephson junction were carried out by Voss and Webb [100] and by Jackel et al. [101], while related experiments on a junction inserted in a superconducting loop were realized by de Bruyn Ouboter et al. [102–104], Prance et al. [105], and Dmitrenko et al. [106]. The temperature dependence of the effect of damping on the tunneling has been addressed by later experiments [107, 108]. The most systematic approach to *MQT* has been carried out by Devoret, Martinis, and Clarke [11–13]. They have neatly demonstrated *MQT* and energy level quantization (*ELQ*), and established almost all experimental procedures and concepts that have been used for all subsequent *MQT* experiments. It has been clearly addressed the problem of the complex impedance presented to the junction at microwave frequency by the wires directly connected to it or by any circuit in its vicinity, and for the first time classical phenomena have been used to measure all relevant parameters of the junction in situ [11–13].

The behavior of the phase difference φ was deduced from measurements of the escape rate Γ of the junctions from its zero-voltage state. To determine the escape rate, 10^4 10^5 events were typically collected for each set of parameters. The resulting distribution of the switching probability $P(I)$ is used to compute the mean escape rate out of the zero-voltage state as a function of I following Fulton and Dunkleberger [109]: $\Gamma(I) = \frac{1}{\Delta I}\frac{{\rm d}I}{{\rm d}t}\ln\left(\frac{\sum_{i \geq I} P(I)}{\sum_{i \geq I+\Delta I} P(I)}\right)$, where ${\rm d}I/{\rm d}t$ is the current ramp rate and ΔI is the channel width of the analog-to-digital converter. The relevant parameters of the junction (critical current and shunting admittance) were determined in situ in the thermal regime from the dependence of Γ on bias current and from resonant activation in the presence of microwaves. The shunting capacitance is dominated the self-capacitance of the junction, while the bias circuitry is determined the shunting conductance. In the quantum regime, Γ became independent of temperature below the crossover temperature T_{cr}. When I_C was reduced with a magnetic field to leave the junction in the thermal regime at low temperatures, Γ followed the predictions of the thermal model [11–13].

In a further series of experiments, the existence of quantized energy levels in the potential well of the junction was demonstrated spectroscopically [11–13]. The escape rate from the zero-voltage state was increased when the microwave frequency Ω corresponded to the energy difference between two adjacent energy levels (Fig. 4.7). A crucial point is that the anharmonic nature of the well, which results from the nonlinear inductance of Josephson junctions, causes the energy spacing to decrease as the quantum number progressively increases, so each transition has a distinct frequency. The discrete Lorentzian-shaped resonances observed as a function of bias current are characteristic of transitions between quantized energy levels in the well, the position of the energy levels agreed quantitatively with quantum-mechanical predictions involving junction parameters measured in the thermal regime [11–13]. The transition may involve more than one photon at once (see gray and light gray lines in Fig. 4.7), thus called multi-photon transition, which has been observed experimentally [110]. Problems related to the distinction between quantum and classical limits of the macroscopic system have also been discussed [111–113] .

The crossover temperature T_{cr} between thermally activated and *MQT* regimes and the escape time τ_{esc} have been recently found in the more general case of a finite-size Josephson junction (*JJ*) placed in magnetic field [114, 115]. Apart from the phase, the potential itself depends on space variables. Usually the phenomenon of *MQT* is considered in a "point-like" *JJ*, i.e., completely neglecting the finiteness of the junction size L. This (zero order in L) approximation is based on the assumption that the junction size is much smaller than all other related parameters of the problem such as the Josephson penetration depth, etc. [114, 115]. It was demonstrated that the effects due to the junction's size result in the appearance of a strong sensitivity of the *MQT* process on applied magnetic field, making the crossover temperature be nonmonotonic function of it [114, 115]. Since magnetic field is an easily adjustable parameter, it can become an important tool in the study of such a quantum coherent phenomenon without modification of other junction parameters.

4.4.3 Developments of Quantum Measurements for Macroscopic Quantum Coherence Experiments

The above-mentioned *MQT* experiments are among the most elegant and "dense of physical contents" studies in the field of weak superconductivity, and have clearly proved that φ is a quantum variable [116]. Although this system contains a large number of atomic constituents, it is atom-like in the sense that it has a single degree of freedom behaving quantum mechanically. Thermal energy must be sufficiently low to avoid incoherent mixing of eigenstates, and the macroscopic degree of freedom must be sufficiently decoupled from other degrees of freedom for the lifetime of the quantum states to be long on the characteristic time scale of the system [8–10]. The next step in the demonstration of macroscopic quantum physics has been macroscopic quantum coherence, i.e., to implement devices showing the superposition of two quantum states $|\Psi_1\rangle$ and $|\Psi_2\rangle$ in the form $|\Psi\rangle = \alpha|\Psi_1\rangle + \beta|\Psi_2\rangle$ [117]. We follow [116] in recalling the main ideas and experiments in this direction. The positive answer was first given in 1997 by Nakamura et al. with the first experiment on a charge qubit [118, 119], showing spectroscopically the superposition of the Cooper-pair states $|n\rangle$ and $|n+1\rangle$, where the integer n is the quantum number specifying the number of Cooper pairs. Demonstrations of the superposition of states in a flux qubit by the Stony Brook [120] and Delft [121] groups followed. A flux qubit consists of a superconducting loop interrupted by one [120] or three [121] Josephson junctions. The two quantum states consist of supercurrent flowing in an anticlockwise or clockwise direction or, equivalently, flux pointing up and flux pointing down, respectively. The Saclay group [122, 123] realized in 2002 a qubit, named 'quantronium', in which two small junctions are connected by a superconducting island, involving the superposition of the Cooper-pair states $|n\rangle$ and $|n+1\rangle$.

In the phase qubit realized by Martinis et al. in 2002 [124], the relevant quantum states are the ground state and the first excited state, and the final device is basically the same used earlier to observe quantized energy levels [11–13].

These first experiments which also classify the three main different types of qubit, i.e., charge, flux, and phase, opened the way to a vast series of studies [116]. Measurements in the time domain necessary to determine the dynamical behavior of a qubit are another example of the amazing progress registered in the field. Spectroscopy is important for establishing that a given qubit is a functional device, and it enables energy-level splitting to be measured as a function of relevant control parameters. Rabi oscillations [125] (coherent oscillations between the eigenstates of a two-level quantum system which is subject to a resonant perturbation), Ramsey fringes [126], spin-echo technique [127] measurements have been realized on several different systems. They require manipulating the state of the qubit by using appropriate microwave pulses, which are also the operative tools to implement single-qubit gates for quantum computing. Macroscopic resonant tunneling (MRT) when energy levels in each well are aligned turns out to be an "accessible" reliable tool to perform on a routine basis the task to extract all devices quantum parameters, flux noise, and for a diagnostics of fabrication processes and materials. This represents another example of how "quantum measurements" become more and more accessible, and a tool to drive material science choices on the nature and the performances of the junctions. Typical relaxation time T_1 (the time required for a qubit to relax from the first excited state to the ground state) range from values of the order of 1 ns to many microseconds [116].

Another big force in driving research in the superconducting quantum measurements/qubit field has been to engineer systems to make them as less sensitive as possible to decoherence. This work has led to remarkable increases in decoherence times compared with those of early devices [116]. The developments of more advanced charge qubits such as the transmon [128] and quantronium [122, 123] are examples of improved charge qubits, where attention is paid to protect the devices from low-frequency noise and from electrons moving among defects.

The main intrinsic limitation on the coherence of superconducting qubits results from low-frequency noise, notably '1/f noise', arising from a uniform distribution of two-state defects [129]. Each defect produces random telegraph noise, and a superposition of such uncorrelated processes leads to a 1/f power spectrum. Recognized sources of 1/f noise are [116]: (1) critical-current fluctuations, which arise from fluctuations in the transparency of the junction caused by the trapping and untrapping of electrons in the tunnel barrier [130]; (2) charge fluctuations, which arise from the hopping of electrons between traps on the surface of the superconducting film or the surface of the substrate; (3) magnetic-flux fluctuations. In particular, the considerations on decoherence has led the community to believe that a better quality junction is an important issue to improve qubit performances, and that material science aspects are a possible solution for junctions of superior quality. Bright ideas on qubit design can circumvent some intrinsic limitations due to the nature of the junctions, but at the end the quality of the junction is one of the central elements of the superconducting qubit.

4.5 Macroscopic Quantum Effects in High-T_C Josephson Junctions and in Unconventional Conditions

All the milestone experiments discussed in the previous section give a clear feeling of the amazing progress in the field and clearly show future trends of developments [116]. At the same time, they promote some more experiments aimed to clarify still very debated key issues on coherence and dissipation in solid state systems. *HTS* may be an interesting reference system for novel ideas on these topics because of their unusual properties. This section is divided into three subsections to cover unconventional aspects of *MQT* and *ELQ*.

4.5.1 Macroscopic Quantum Phenomena in High-T_C Josephson Junctions

HTS can be a reference system for studies on dissipation and coherence because of the presence of low energy quasi-particles due to nodes in the d-wave order parameter symmetry. This has represented since the very beginning a strong argument against the occurrence of macroscopic quantum effects in these materials. Quantum tunnelling of the phase leads to fluctuating voltage across the junctions which excites the low energy quasi-particles specific of d-wave junctions, causing decoherence. The demonstration of *MQT* [16] and energy level quantization (*ELQ*)

[17] in *HTS* indicates that the value of dissipation mechanisms in *HTS* is below some threshold, and their role has to be revised [16,17]. Thus, macroscopic quantum effects turn to be a useful tool to understand some aspects of the physics of *HTS* and *HTS JJ*s and give some support to proposals on d-wave "quiet" qubit functionalities [131–138].

'Qubit' proposals involving high-T_C superconductors [132–138] most often exploit Josephson junction circuits with an energy-phase relation with two minima in the absence of an external magnetic field, meaning that there is no need to apply a constant magnetic bias, unlike in systems based on low temperature superconductors. Naturally, degenerate states and violation of the time reversal symmetry become the keywords and concepts of all qubit proposals [92,11]. Subsequent investigations raised concerns about the quietness of these devices [133, 139]. This analysis led to the development of alternative designs, in which a five-junction loop (with four ordinary junctions and a π junction) takes the place of the original s-wave–d-wave superconductor junction. In this 'macroscopic analogue' of [133], the π junction removes the need for a constant magnetic bias near $\Phi_O/2$.

Contributions to dissipation due to different transport processes, such as channels due to nodal quasi-particles, midgap states, or their combination, have been identified and distinguished [140–144]. In particular cases, decoherence times and quality factors were calculated considering the system coupled to an Ohmic heat bath. It has also been argued that problems in observing quantum effects due to the presence of gapless quasi-particle excitations can be overcome by choosing the proper working phase point [141]. In particular, decoherence mechanisms can be reduced by selecting appropriate tunnelling directions because of the strong phase dependence of the quasi-particle conductance in a d-wave *GB* junction.

This was the state of art in January 2005, when the first results on *MQT* appeared in *Physical Review Letters* after 2 years of experiments on switching current probability distributions on high quality *YBCO* off-axis biepitaxial *GB* junctions [16, 17]. As a matter of fact, the search of macroscopic quantum effects become feasible once some kind of *HTS* 'tunnel-like' Josephson junctions was available. We can distinguish two classes of experiments, which are based on two different complementary types of junctions: (1) *MQT* and *ELQ* [16,17] on the just mentioned *YBCO* grain boundary biepitaxial junctions, where the experiment has been designed to study d-wave effects with a lobe of the former electrode facing the node of the latter; (2) *MQT* and *ELQ* on intrinsic junctions on single crystals of different materials [145, 146], where d-wave are expected to play a minor role [141, 142]. The experiments using *GB*s are more complicated because of the complexity of these junctions, but are complete and allow to address issues of the effects of a d-wave *OPS* on dissipation and coherence. Only *GB*s junctions can be more easily integrated into circuits.

The *GB* biepitaxial junctions used in [16, 17] had reproducible hysteretic behavior up to 90%. In addition, the possibility to tune the critical current I_C through the interface orientation θ in complete agreement with the predictions of a d-wave *OPS* (see Fig. 4.5) allows to select the junction for the *MQT* experiment knowing the *OPS* configuration exactly. This means that d-wave effects are dominant for some types of junctions, and robust in the sense that interface microstructure cannot mask them. The suitable junction can be therefore selected for the experiment. Since the interest was mostly focussed in those features that are distinct from the case of low T_C superconductor (*LTS*) junctions, namely effects due to OPS, and dissipation due to low energy quasi-particles, the junction in the tilt configuration (angle $\theta = 0°$) turns out to be the most interesting case for the *MQT* and *ELQ* experiments. This configuration (lobe to node) maximizes d-wave-induced effects and allows to explore the effects of low energy quasi-particles.

We follow [16] and [17] in reporting on the first experimental measurements on *MQT* in *HTS JJ*s. Figure 4.8 shows a set of switching current probability distributions as a function of temperature for the biepitaxial *JJ*. In the inset of Fig. 4.8, the switching current probability distribution measured at $T = 0.019$ K is "projected" on the original I–V, from where the switching current probability distribution was extracted with the standard procedure outlined in Sect. 4.3. This is meant to visualize the type of measurement, and as a term of comparison also the switching current probability distribution at $T = 0.1$ K is reported. The dependence of the distribution width σ on temperature is reported in Fig. 4.9. The measured σ (in the top inset the way σ is obtained is for instance indicated for a specific temperature) saturates

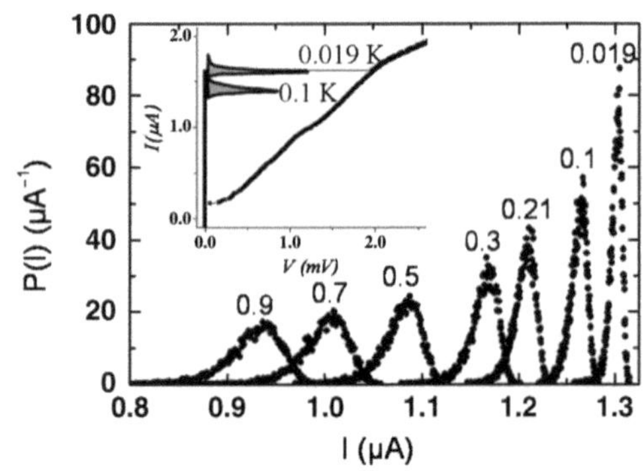

Fig. 4.8 Switching current probability distribution for $I_{C0} = 1.40\,\mu$A at $B = 0$ T for different bath temperatures T_{bath}. Adapted from Bauch et al. [16]

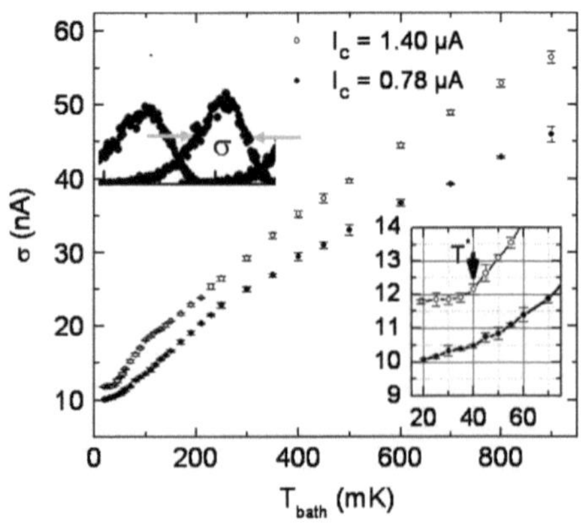

Fig. 4.9 The measured σ (see top inset to see how it is derived) saturates below 50 mK, indicating a crossover from the thermal to the quantum regime. The width σ for $B = 2$ mT and the data for $B = 0$ mT are shown in the bottom inset. Adapted from Bauch et al. [16]

below 50 mK, indicating a crossover from the thermal to the quantum regime. To rule out the possibility that the saturation of σ is due to any spurious noise or heating in the measurement setup, the switching current probability distributions were measured for a reduced critical current ($I_{CO} = 0.78$ μA) by applying an external magnetic field $B = 2$ mT. The width σ for $B = 2$ mT and the data for $B = 0$ mT are shown in the bottom inset of Fig. 4.9. The data in the presence of a magnetic field clearly show a smaller width σ, which does not saturate down to the base temperature [16].

Measurements on the same samples in the presence of microwaves followed a few months later [17]. In analogy with *LTS* systems discussed in the previous section, energy-level quantization was clearly observed, additional signature of macroscopic quantum behavior and indication that the dissipation in a d-wave *JJ* is low enough to allow the formation of the 'sharp' energy levels. Microwaves at fixed frequency μ_{mw} were transmitted to the junction via a dipole antenna at a temperature below the crossover value, T_{cr} [17]. When μ_{mw} of the incident radiation (or multiples of it) coincides with the bias current-dependent level separation of the junction, $\mu_{01}(I) = m\mu_{mw}$, the first excited state is populated [110]. Here, m is an integer number corresponding to an m-photon transition from the ground state to the first excited state (Fig. 4.10). At low power values (−20 dBm), the escape is basically from the ground state, since the occupation probability of the first excited state is negligible. When the applied power is increased (−17 dBm and −16 dBm), the first excited state starts to be populated. Then in the histogram, two peaks appear corresponding to tunneling from both the first excited (Γ_1) and ground (Γ_0) states. The escape from the first excited state is exponentially faster and dominates, and the switching current distribution is again single peaked at −14 dBm [17]. From the Lorentzian-shape of the escape rate, a Q value of the order of 40 is extracted [17],

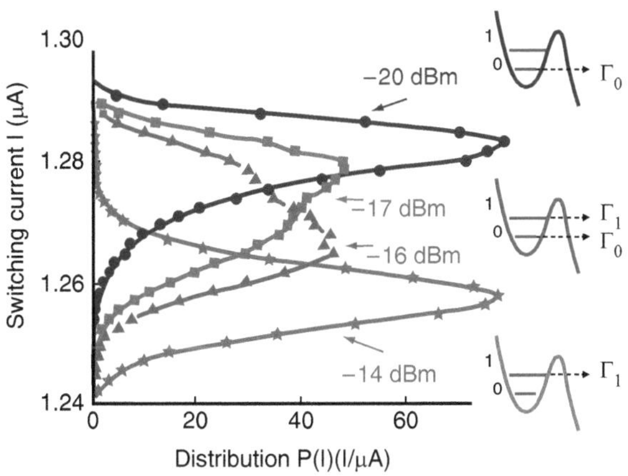

Fig. 4.10 Measured switching current probability distribution $P(I)$ in the presence of microwaves at a frequency of 850 MHz and temperature $T = 15$ mK. The applied power at the room temperature termination varies from 20 to 14 dBm. Adapted from Bauch et al. [17]

comparable with the first best results obtained in *LTS* junctions [12, 13]. Specific effects related to stray capacitance and large kinetic inductance have been discussed both in the original paper [17] and in subsequent papers [147].

The observation of quantum tunneling, narrow width of excited states, and a large Q value support the notion of "quiet" qubits based on d-wave symmetry superconductor, but the meaning of the experiments goes beyond [17, 21]. There may be some mechanism preventing the low-lying quasi-particles in the d-wave state from causing excessive dissipation. For example, the physical properties of the low-lying quasi-particles are found to resemble those in BCS theory in the SU(2) slave-boson model [148]. These results may open up the possibility for some kind of freezing mechanism for quasi-particles at very low temperature and/or the existence of a subdominant imaginary s-wave component of the order parameter inducing a gapped excitation spectrum. This last possibility has been discussed in various experiments available in literature, but there is neither a convincing reproducible proof nor a neat definition of the controllable experimental conditions which lead to this effect, especially in *GB*s [19, 21]. An imaginary component of the order parameter, even rising at the lower temperatures, would be a more orthodox explanation as opposed to freezing mechanisms and more subtle quantum protectorates [149].

Experiments on Intrinsic Josephson junctions (*IJJ*) from $Bi_2Sr_22CaCu_2O_8$ single crystals [145, 146] have been more aimed to increase the crossover temperature (T_{cr}) and to clarify the nature of *IJJ* junctions, rather than raising novel themes of coherence in d-wave systems. As a matter of fact, these junctions mostly exploit c-axis transport, i.e., perpendicular to the copper oxide layers, where the nodes of the $d_{x^2-y^2}$ order parameter should not affect *MQT*. Josephson coupling between CuO_2 double layers has been proved, and most of the materials behaved like stacks of $S-I-S$ *JJ*s with effective barriers of the order of the separation of the CuO_2 double layers (1.5 nm) (J_C typically 10^3 A/cm^2) [150–152]. *IJJ*s have a much higher Josephson coupling energy than *GB* junctions, the I–V curves exhibited large hysteresis and multiple branches, indicative of a series connection of highly capacitive junctions. Practical realizations of *IJJ* have been designed to nominally avoid heating effects [151–154]. However, at high voltages caution is required when extracting information because of possible unavoidable heating problems. The temperature T_{cr} of the crossover between thermal and quantum escape was reported to be about 800 mK, remarkably higher than those usually reported on *LTS* systems. Using microwave spectroscopy, the unique uniform array structure of intrinsic Josephson junction stacks have been considered responsible for a remarkable enhancement of the tunneling rate [146]. This enhancement adds a factor of approximately N^2 to the quantum escape rate of a single Josephson junction, also resulting in a significant increase in the crossover temperature T_{cr} between the thermal activation regime and the quantum tunneling, where N is the number of the junctions in the stack. This effect can be caused by large quantum fluctuations due to interactions among the N junctions [146].

4.5.2 Switching Current Statistics in Moderately Damped Josephson Junctions

The influence of dissipation on the switching current statistics of moderately damped Josephson junctions, especially in the thermal regime, has become an argument of great interest [153–156]. Both low-T_C and *HTS* junctions may fall in this regime, having for low-T_C junctions the additional possibility of a finer control of the damping [153, 154]. These studies basically respond, among the others, to the intriguing question whether the amplitude of fluctuations of physical properties always increase with the temperature.

The majority of the large body of previous work in the literature have concentrated on junctions in either the underdamped $Q \gg 1$ or the overdamped $Q \sim 1$ limit. As opposed to the strongly underdamped Josephson junctions case, well described by the analysis of Fulton and Dunkleberger [109] in the thermal region, overdamped junctions show nonhysteretic behavior, with a finite voltage on the supercurrent branch of the IV characteristic, associated with thermally activated phase diffusion, and thermal fluctuations leading to very much smaller variations in the switching behavior. If the damping around the junction plasma frequency is sufficiently high, at low bias the phase particle will always retrap in a local minima after escape and a finite resistance phase-diffusion branch appears on the I–V curve of the junction (the appearance of a small voltage, before switching to a voltage on the order of twice the superconducting gap Δ). Phase diffusion in junctions with hysteretic IV characteristics has been discussed in literature [157, 158] and has been associated with frequency-dependent damping, such that junctions are underdamped at low frequencies but are in the overdamped limit at high frequency.

Phase diffusion process has reappeared in a different regime in magnetometers with much larger and unshunted junctions used for qubits readout [155]. The retrapping process is affected by the frequency-dependent impedance of the environment of the dc SQUID. For $Q > 1$ with relatively small I_C, such that $E_J \sim k_B T \gg E_C$, it has been shown that a regime exists where escape does not lead to a finite-voltage state, but rather to underdamped phase diffusion [156]. In other words, in addition to the usual crossover between macroscopic quantum tunnelling and thermally activated (TA) behavior, the transition from TA behavior to underdamped phase diffusion has been also observed, resulting in (T, E_J) phase diagram with various regimes [156].

In [153, 154], a systematic analysis has been carried out on different junctions: low ohmic Nb–Pt–Nb (S-N-S) junctions, Nb–CuNi–Nb (S-F-S) junctions with a diluted ferromagnetic alloys, Nb–InAs–Nb (S-two-dimensional electron gas-S), and BiSrCaCuO (2212) *IJJ*. The damping parameter of the junctions has been tuned in a wide range by changing temperature, magnetic field, and gate voltage and introducing a ferromagnetic layer or in situ capacitive shunting. The phenomenon of an unexpected collapse of switching current fluctuations with increasing T is explained by the interplay of two counteracting consequences of thermal fluctuations [153, 154]. On one hand, thermal fluctuations assist in premature switching into the resistive state and, on the other hand, help in retrapping back to the superconducting state [153, 154]. In other words, temperature does not only provide energy for excitation of a system from equilibrium state but also enhances the rate of relaxation back to the equilibrium.

The crossover temperature T^* for this effect is defined as the temperature with positive $d\sigma/dT$ below T^* and negative $d\sigma/dT$ above T^*, being σ the width of the switching current [153,154,159]. The low-temperature behavior fits the expectations for underdamped junctions, and the high temperature behavior resembles previous observations for overdamped junctions with phase diffusion. This crossover from underdamped to overdamped behavior has been also discussed in terms of a temperature-dependent damping Q [160, 161]. A temperature-independent Q can fit experimental data as well, if the junctions are in the moderately damped regime $Q \sim 5$, and T^* is a measure of the damping [153, 154]. Monte Carlo simulations of thermal fluctuations in moderately damped Josephson junctions take into account multiple escape and retrapping, switching- and return-current distributions [159]. It is found that the characteristic temperatures T^* are not unique for a particular Josephson junction but have some dependence on the ramp rate of the applied bias current.

Multiple-retrapping processes in a hysteretic BiSrCaCuO (2212) *IJJ* with a high tunneling resistance have been reported to govern the switching from a resistive state in the phase-diffusion regime into the quasi-particle tunneling state [162]. The frequency-dependent junction quality factor, representing the energy dissipation in a phase-diffusion regime, determines the observed temperature dependence of the switching current distribution and the switching rate.

Phase dynamics has been investigated also in underdamped ferromagnetic *JJ* by measuring the switching probability in both the stationary and nonstationary regimes [163–165]. The junction is Nb–Al_2O_3–PdNi (10% Ni)-Nb. Incomplete relaxation leads to dynamical phase bifurcation. Bifurcation manifests itself as a premature switching, resulting in a bimodal switching distribution [163]. Escape rate measurements at temperatures T down to 20 mK show that the width of the switching current histogram decreases with temperature and saturates below $T = 150$ mK in Nb/ Al_2O_3/$Cu_{40\%}$ $Ni_{60\%}$/Nb heterostructures [164, 165].

4.5.3 *MQT Current Bias Modulation*

Mechanisms of barrier penetration in nonstationary field, such as phenomena belonging to photon-assisted tunnelling and Euclidean resonance, and temporally modulated barriers deserve great attention [166, 167]. Recently, possible influence of current bias modulation on the dynamics of Josephson junctions has been also discussed in connection with the quantum Zeno (QZE) and anti-Zeno (AZE) effects [168, 169], namely slowdown and speedup, respectively, of the decay of quantum states into an energy continuum due to frequent measurements. The effect of a nonstationary current through the junction on tunneling rate has been considered through an extension of the Wentzel, Kramers, and Brillouin (WKB) approach for a nonstationary case. An extremely small probability of quantum tunneling may become not very small under the action of an ac component of the bias current [166, 167]. The tunneling rate has a peak as a function of a dc component of the bias current (Euclidean resonance). This effect does not involve transitions between energy levels and, thus, is different from the well-known process of photon-assisted tunnelling. In low-photon processes, a typical time scale of the problem is $1/\Omega_{\mathrm{P}}$. In the multiphoton case, the number of participating photons $E_J/\hbar\Omega_{\mathrm{P}}$ is large and a much shorter time scale of the order of $\hbar/E_J$ controls the dynamical process [166, 167]. The conventional Josephson dynamics, in terms of φ, works well if the intrinsic dynamic time $\tilde{\tau}$ of a superconductor is shorter than the time scale of dynamical processes. The intrinsic time $\tilde{\tau}$ is determined by the branch imbalance relaxation and the order parameter dynamics [166, 167]. The reduced dynamical description holds at sufficiently long characteristic time scale of the phase $\varphi(t)$ (adiabatic condition) when the condensate follows a time-dependent voltage. When φ(t) varies faster than the electron-hole imbalance relaxes (relaxation time $\tilde{\tau}$), the condensate does not follow the voltage. In this limit, the Josephson relation breaks down. The electron-phonon mechanism of the charge imbalance motion at low temperature is characterized by the relaxation time $\tilde{\tau} \approx \omega_{\mathrm{D}}^2(\hbar/\Delta)^3$, where ω_{D} is the Debye frequency and Δ is the superconducting energy gap. For Nb, one can estimate $\tilde{\tau} \approx 10^{11}$s. The adiabatic condition $\tilde{\tau} < \hbar/E_J$ becomes $I_{\mathrm{C}} < 2e/\tilde{\tau}$. For Nb or Pb, this restriction implies approximately $I_{\mathrm{C}} <10$ nA. For high-T_{C} materials, $\tilde{\tau}$ is much shorter and values of $I_{\mathrm{C}} \approx 1\,\mu$A [166, 167] are accessible for experiments.

4.6 Mesoscsopic Effects and Coherence in *HTS* Nanostructures

Nanotechnology can provide another path to study coherence in *HTS* and relaxation processes for quasi-particles.

Nanoscale junctions get closer to the characteristic scale lengths of *HTS* (i.e., coherence length, charge domains) and *HTS* structures (i.e., *GB* facets, *GB* dislocation distribution, nano-channel size) and have full potentials to isolate intrinsic features and transport regimes of *HTS* systems. These efforts will lead to understand the ultimate limit of *GB* performances and to valuable progress in the study of coherence in *HTS* systems. Looking at the great amount of data available in literature, to the apparent contrast of several results, or to the fact that some predicted phenomena derived from well-established effects have not yet been clearly detected or only at intermittence (such as 2φ component, time reversal symmetry breaking, all deriving from d-wave *OPS* [19,21]), the only reasonable explanation is to assume that several different tunneling and diffusive processes are active in the transport, but only the morphology of the barrier and of the GB will discriminate the role of each transport mechanism. The microstructure acts as a filter determining additional constraints on the various transport modes, and nanostructures may help to isolate the various contributions [80, 83–87, 173–177].

The first studies on bicrystal submicron *JJ*s gave encouraging results as the reduction of decoherence [170], the presence of the 2φ component [83–87] b, or of Andreev bound states [172, 173].

Recently, submicron biepitaxial junctions have been realized down to about 500 nm using both e-beam lithography and C and Ti masking [80, 174]. Yield and reproducibility have been improved on this width scale, while the early successful attempts on fabricating a junction on a single facet on a (103) grain have been made. This classical controllable "top-down" approach is going to be accompanied by some sort of "bottom-up" techniques, which fund on the intrinsic nature of *GB*. The complex growth process may determine self-assembled nanochannels of variable dimensions, ranging typically from 20 to 200 nm. These nano-contacts can be considered self-protected as far as they are "enclosed" in macroscopic impurities. Even if this very last technique is not ideal on the long range for applications, since it needs an additional critical step to locate the nanobridges and etch the *HTS* thin film, it can be really helpful to understand the ultimate limit of the junction performances and to understand the transport mechanisms.

An example of how to use the natural self-assembly to extract information on the physics of *HTS* Josephson junctions has been recently given on a study on universal conductance fluctuations in magnetic field in *YBCO* biepitaxial Josephson junctions. This is of relevance to investigate coherent quantum behavior in *HTS* [175, 176]. The experimental conductance fluctuations as a function of the magnetic field Fig. 4.11a have been studied using autocorrelation function protocols typical of traditional mesoscopic systems.

At low temperatures, quantum coherence can be monitored in the conductance G of a normal metallic sample of length L_x attached to two reservoirs [177–183]. The electron wave packets that carry current in a diffusive wire have minimum size of the order of $L_T > L_x \gg l$. Here, l is the electron mean free path in the wire and L_T is the thermal diffusion length (D is the diffusion constant). The first inequality is satisfied at relatively low temperatures as far as $k_B T \ll \varepsilon_C \approx \hbar D/L_x^2$, where ε_C is the Tholuess energy. The critical current I_C oscillations in magnetic field point to a flux periodicity which is roughly consistent with the typical size (10–100 nm) expected for the microbridge from structural investigations. Studies have been carried out at different voltages and nonequilibrium conditions. Conductance fluctuations become appreciable at low temperatures in the whole magnetic field range Fig. 4.11a. At low voltages ($eV \ll \varepsilon_C$), the system is in the regime of universal conductance fluctuations: the variance $< g^2 >$ of the dimensionless conductance $g = G/(2e^2/\hbar)$ is of the order of unity. The fluctuations are nonperiodic and have all the typical characteristics of mesoscopic fluctuations [177–183].

By performing the ensemble average of G_N over H, additional insights can be gained. The autocorrelation function is defined as:

$$\Delta G(\delta V) = = \frac{1}{N_V} \sum_V \sqrt{\left\langle (G_{V+\delta V}(H) - \overline{G}_N)(G_V(H) - \overline{G}_N) \right\rangle_H} \quad (4.11)$$

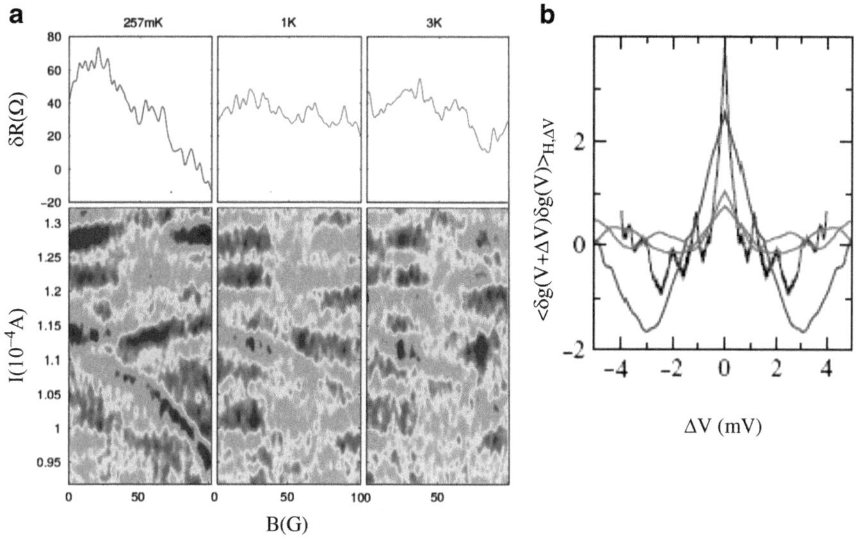

Fig. 4.11 (**a**) Color plot of the resistance fluctuations $\delta R(H, I)$ as a function of the applied current and of the magnetic field for three temperatures: 257 mK, 1 K, and 3 K. The three top panels show single magnetoconductance trace for each temperature at fixed bias current ($I = 128\,\mu$A). (**b**) The measured autocorrelation of the conductance vs. the voltage difference ΔV, for average voltages $V \approx 7.5, 12.5, 17.5, 25.0$ mV. Adapted from Tagliacozzo et al. [176]

Here, N_V is the number of V intervals in which the voltage range has been divided. $\Delta G(\delta V)$ is reported in Fig. 4.11b at about $T = 250$ mK. The fit is obtained by calculating the contribution to the autocorrelation due to diffusions in the limit $k_B T \gg \hbar/\tau_\varphi$ [175, 176]. An energy scale of the order of 1 meV arises naturally in correspondence to the half width at its half maximum [175, 176], which has been identified as the Thouless energy [183]. This is proportional to the inverse time an electron spends in moving coherently across the mesoscopic sample. Quasi-particles seem to travel coherently across the junction even if $V \gg \varepsilon_C$. Hence, microscopic features of the weak link appear as less relevant, in favor of mesoscopic, nonlocal properties. In this case, the quasi-particle phase coherence time τ_φ does not seem to be limited by energy relaxation due to voltage-induced non-equilibrium. The remarkably long lifetime of the carriers, found in these experiments, appears to be a generic property in high-T_C *YBCO* junctions as proved by optical measurements [184] and by macroscopic quantum tunneling [16, 17].

4.7 Conclusions

The great interest of the results reviewed in this chapter lies in the combination of the stimulating subject of macroscopic quantum phenomena (*MQT*, *ELQ*) with the new important clues that such phenomena provide on underlying aspects of the physics of *HTS*. We have discussed some significant aspects of the Josephson effect in *HTS* structures. The phenomenology of the Josephson effect in *HTS* is enriched by a series of effects, directly related to a predominant d-wave order parameter symmetry, to the unusual scaling energies and lengths of *HTS*, to an effective, different granularity and so on. *HTS* junctions have promoted novel insights also in the relation between the vortex matter and the Josephson effect, which is one of the main themes of the NES Project. We have focused on macroscopic quantum decay phenomena as one of the most exciting expressions of the Josephson effect. A system that displays macroscopic quantum effects, despite the presence of nodes in the order parameter symmetry and therefore of low energy quasi-particles, raises several challenging issues on dissipation mechanisms and on the peculiar coherence phenomena occurring in Josephson systems and in *HTS*. We believe that the progress in quantum engineering and in nanotechnologies will represent an invaluable additional drive force to further address advanced topics also on macroscopic quantum phenomena.

Acknowledgements Special thanks to John Kirtley for sharing so many common interests in several experiments reviewed in this chapter. We would also like to thank Thilo Bauch, Franco Carillo, Luigi Longobardi, Giampaolo Papari, Giacomo Rotoli, Daniela Stornaiuolo, and Arturo Tagliacozzo for valuable discussions on the topics of this chapter.

References

1. B.D. Josephson, Phys. Lett. 1, 251 (1962)
2. A. Barone, G. Paternò, Physics and Applications of the Josephson Effect (Wiley, New York, 1982)
3. I.O. Kulik, K. Yanson, The Josephson Effect in Superconductive Tunneling Structures (Israel Program of Scientific Translations, Jerusalem, 1972)
4. K.K. Likharev, Dynamics of Josephson Junctions and Circuits (Gordon and Breach, New York, 1986)
5. K.K. Likharev, Rev. Mod. Phys. 51, 101 (1979)
6. T. Van Duzer, C.W. Turner, Principles of Superconductive Devices and Circuits (Elsevier, New York, 1981)
7. M. Tinkham, Introduction to Superconductivity, 2nd edn. (Mc-Graw-Hill Book Co, New York, 1996)
8. A.O. Caldeira, A.J. Leggett, Phys. Rev. Lett. 46, 211 (1981)
9. A.J. Leggett J. Phys. (Paris) Colloq. 39 C6-1264 (1980)
10. A.O. Caldeira, A.J. Leggett, Ann. Phys. (N.Y.) 149, 374 (1983)
11. M.H. Devoret, J.M. Martinis, J. Clarke, Phys. Rev. Lett. 55, 1908 (1985)
12. J.M. Martinis, M.H. Devoret, J. Clarke, Phys. Rev. B 35, 4682 (1987)
13. J.M. Martinis, M.H. Devoret, J. Clarke, Phys. Rev. Lett. 55, 1543 (1985)
14. J.G. Bednorz, K.A. Muller, Z. Phys. B 64, 189 (1986)
15. J.G. Bednorz, K.A. Muller, Rev. Mod. Phys. 60, 585 (1988)
16. T. Bauch, F. Lombardi, F. Tafuri, A. Barone, G. Rotoli, P. Delsing, T. Claeson, Phys. Rev. Lett. 94, 87003 (2005)
17. T. Bauch, T. Lindstrom, F. Tafuri, G. Rotoli, P. Delsing, T. Claeson, F. Lombardi, Science 311, 57 (2006)
18. D.J. Van Harlingen, Rev. Mod. Phys. 67, 515 (1995)

19. C.C. Tsuei, J.R. Kirtley, Rev. Mod. Phys. 72, 969 (2000)
20. H. Hilgenkamp, J. Mannhart, Rev. Mod. Phys. 74, 485 (2002)
21. F. Tafuri, J.R. Kirtley, Rep. Prog. Phys. 68, 2573(2005)
22. J. Mannhart, P. Chaudhari, D. Dimos, C.C. Tsuei, T.M. McGuire, Phys. Rev. Lett. 61, 2476 (1988)
23. P. Chaudari, J. Mannhart, D. Dimos, C.C. Tsuei, C.C. Chi, M.M.Oprysko, M. Scheuermann, Phys. Rev. Lett. 60, 1653 (1988)
24. D. Dimos, P. Chaudhari, J.Mannhart, Phys. Rev. B 41 4038 (1990)
25. H. Hilgenkamp, J. Mannhart, Appl. Phys. Lett. 73, 265 (1998)
26. Z.G. Ivanov, P.A. Nilsson, D. Winkler, J.A. Alarco, T. Claeson, E.A. Stepansov, A.Ya. Tsalenchuk, Appl. Phys. Lett. 59, 3030 (1991)
27. R. Gross, P. Chaudari, M. Kawasaki, A. Gupta, Phys. Rev. B 42, 10735 (1992)
28. N.F. Heinig, R.D. Redwing, J.E. Nordman, D.C. Larbalestier, Phys. Rev. B 60, 1409 (1999)
29. N.F. Heinig, R.D. Redwing, I.F. Tsu, A. Gurevich, J.E. Nordman, S.E. Babcock, D.C. Larbalestier, Appl.Phys. Lett. 69 577 (1996)
30. U. Schoop, S. Kleefisch, S. Meyer, A. Marx, L. Alff, R. Gross, M. Naito, H. Sato, IEEE Trans. Appl. Supercond. 9, 3409 (1999)
31. R.W. Simon, J.F. Burch, K.P. Daly, W.D. Dozier, R. Hu, A.E. Lee, J.A. Luine, H.M. Manasevit, C.E. Platt, S.M. Scharzberg, D. St. John, M.S. Wire, M.J. Zani, in Science and Technology of Thin Film Superconductors, vol. 2, ed. by R.D. McConnel, R. Noufi (Plenum, New York, 1990) p. 549
32. C.L. Jia, B. Kabius, K. Urban, G.J. Cui, J. Schubert, W. Zander, A.I. Braginsky, C. Heiden, Physica C 196, 211 (1991)
33. H.R. Yi, Z.G. Ivanov, D. Winkler, Y.M. Zhang, H. Olin, P. Larsson, Appl. Phys. Lett. 65, 1177 (1994)
34. F. Lombardi, Z.G. Ivanov, G.M. Fischer, E. Olsson, T. Claeson, Appl. Phys. Lett. 72, 249 (1998)
35. K. Char, M.S. Colclough, S.M. Garrison, N. Newman, G. Zaharchuk, Appl. Phys. Lett. 59, 733 (1991)
36. F. Tafuri, F.M. Granozio, F. Carillo, A. Di Chiara, K. Verbist, G. Van Tendeloo, Phys. Rev. B 59, 11523 (1999)
37. F. Lombardi, F. Tafuri, F. Ricci, F.M. Granozio, A. Barone, G. Testa, E. Sarnelli, J.R. Kirtley, C.C. Tsuei, Phys. Rev. Lett. 89, 207001 (2002)
38. F.M. Granozio, U. Scotti di Uccio, F. Lombardi, F. Ricci, F. Bevilacqua, G. Ausanio, F. Carillo, F. Tafuri, Phys. Rev. B 67, 184506 (2003)
39. C.C. Tsuei, J.R. Kirtley, C.C. Chi, L.S. Yu-Jahnes, A. Gupta, T. Shaw, J.Z. Sun, M.B. Ketchen, Phys. Rev. Lett. 73, 593 (1994)
40. D.A. Wollman, D.J. Van Harlingen, J. Giapintzakis, D.M. Ginsberg, Phys. Rev. Lett. 74, 796 (1995)
41. D.A. Wollman, D.J. Van Harlingen, W.C. Lee, D.M. Ginsberg, A.J. Leggett, Phys. Rev. Lett. 71, 2134 (1993)
42. T. Lofwander, V. Shumeiko, G. Wendin, Supercond. Sci. Technol. 14, R53 (2001)
43. J.R. Kirtley, F. Tafuri, in Treatise of Superconductivity, ed. by R. Schrieffer (Springer Verlag, New York, 2007) pp. 19–85
44. L.N. Bulaevskii, V.V. Kuzii, A.A. Sobyanin, JETP Lett. 25, 290 (1977)
45. V.B. Geshkenbein, A.I. Larkin, JETP Lett. 43, 395 (1986)
46. V.B. Geshkenbein, A.I. Larkin, A. Barone, Phys. Rev. B 36, 235 (1987)
47. A.V. Andreev, A.I. Buzdin, R.M. Osgood, Phys. Rev. B 43, 10124 (1991)
48. V.V. Ryazanov, V.A. Oboznov, A.V. Veretennikov, A.Yu. Rusanov, Phys. Rev. B 65, 020501 (2001)
49. V.V. Ryazanov et al. , Phys. Rev. Lett. 86, 2427 (2001)
50. A. Bauer et al., Phys. Rev. Lett. 92, 217001 (2004)
51. T. Kontos et al., Phys. Rev. Lett. 89, 137007 (2002)
52. E. Goldobin, D. Koelle, R. Kleiner, Phys. Rev. B 70, 174519 (2004)

53. J.J.A. Baselmans, A.F. Morpurgo, B.J. van Wees, T.M. Klapwijk, Nature 397, 43 (1999)
54. J.J.A. Baselmans, T.T. Heikkila, B.J. van Wees, T.M. Klapwijk, Phys. Rev. Lett. 89, 207002 (2002)
55. M. Sigrist, T.M. Rice, J. Phys. Soc. Jpn. 61 4283 (1992)
56. M. Sigrist, T.M. Rice, Rev. Mod. Phys. 67, 503 (1995)
57. M. Sigrist, T.M. Rice, Phys. Rev. B 55, 14647 (1997)
58. P.K. Svendlindh, P. Niskanen, P. Nordling, P. Nordblad, L. Lundgren, L.B.T. Lundstrom, Physica C 162–164, 1365 (1989)
59. W. Braunisch, N. Knauf, V. Kateev, S. Neuhausen, A. Grutz, A. Kock, B. Roden, D. Khomskii, D. Wohlleben, Phys. Rev. Lett. 68, 1908 (1992)
60. W. Braunisch, N. Knauf, G. Bauer, A. Kock, A. Becker, B. Freitag, A. Grutz, V. Kataev, S. Neuhausen, B. Roden, D. Khomskii, D. Wohlleben, Phys. Rev. B 48, 4030 (1993)
61. Ch. Heinzel, Th. Theileg, P. Ziemann, Phys. Rev B 48, 3445 (1993)
62. H.J.H. Smilde, Ariando, D.H.A. Blank, H. Hilgenkamp, H. Rogalla, Phys. Rev. B 70 024519 (2004)
63. R.R. Schulz, B. Chesca, B. Goetz, C.W. Schneider, A. Schmehl, H. Bielefeldt, H. Hilgenkamp, J. Mannhart, Appl. Phys. Lett. 76, 912 (2002)
64. H. Hilgenkamp, J. Mannhart, B. Mayer, Phys. Rev. B 53, 14586 (1996)
65. J. Mannhart, H. Hilgenkamp, B. Mayer B, Ch. Gerber, J.R. Kirtley, K.A. Moler, M. Sigrist, Phys. Rev. Lett. 77, 2782 (1996)
66. C.A. Copetti, F. Ruders, B. Oelze, Ch. Buchal, B. Kabius, J.W. Seo, Physica C 253, 63 (1995)
67. J.R. Kirtley, P. Chaudhari, M.B. Ketchen, N. Khare, S.-Y. Lin, T. Shaw, Phys. Rev. B 51 R12057 (1995)
68. F. Tafuri, J.R. Kirtley, F. Lombardi, F.M. Granozio, Phys. Rev. B 67, 174516 (2003)
69. J.R. Kirtley, K.A. Moler, D.J. Scalapino, Phys. Rev. B 56, 886 (1997)
70. M. Sigrist, D.B. Bailey and R.B. Laughlin, Phys. Rev. Lett. 74, 3249 (1995)
71. D.B. Bailey, M. Sigrist, R.B. Laughlin, Phys. Rev. B 55, 15239 (1997)
72. M.H.S. Amin, A.N. Omelyanchouk, A.M. Zagoskin, Phys. Rev. B 63, 212502 (2001)
73. M.H.S. Amin, S.N. Rashkeev, M. Coury, A.N. Omelyanchouk, A.M. Zagoskin, Phys. Rev. B 66, 174515 (2002)
74. T. Lofwander, V.S. Shumeiko, G. Wendin, Phys. Rev. B 62, R14653 (2003)
75. R. Carmi, E. Polturak, G. Koren, A. Auerbach, Nature 404, 853 (2000)
76. F. Tafuri, J.R. Kirtley, Phys. Rev. B 62, 13934 (2000)
77. H.J.H. Smilde, Ariando, D.H.A. Blank, G.J. Gerritsma, H. Hilgenkamp, H. Rogalla, Phys. Rev. Lett. 88, 057004 (2002)
78. H. Hilgenkamp, Ariando, H.J.H. Smilde, D.H.A. Blank, G. Rijnders, H. Rogalla, J.R. Kirtley, C.C. Tsuei, Nature 422, 50 (2003)
79. J.R. Kirtley, C.C. Tsuei, Ariando, H.J.H. Smilde, H. Hilgenkamp, Phys. Rev. B 72, 214521 (2005)
80. D. Stornaiuolo, G. Rotoli, K. Cedergen, D. Born, T. Bauch, F. Lombardi, F. Tafuri, J. Appl. Phys. 107, 113901 (2010)
81. H.J.H. Smilde, A.A. Golubov, Ariando, G. Rijnders, J.M. Dekkers, S. Harkema, D.H.A. Blank, H. Rogalla, H. Hilgenkamp, Phys. Rev. Lett. 95, 257001 (2005)
82. J.R. Kirtley, C.C. Tsuei, Ariando, C.J.M. Verwijs, S. Harkema, H. Hilgenkamp, Nat. Phys. 2, 190 (2006)
83. E. Ilichev, V. Zakosarenko, R.P.J. Ijsselsteijn, H.E. Hoenig, V. Schultze, H.G. Meyer, M. Grajcar, R. Hlubina, Phys. Rev. B 60, 3096 (1999)
84. E. Il'ichev, M. Grajcar, R. Hlubina, R.P.J. IJsselsteijn, H.E. Hoenig, H.-G. Meyer, A. Golubov, M.H.S. Amin, A.M. Zagoskin, A.N. Omelyanchouk, M.Yu. Kupriyanov, Phys. Rev. Lett. 86, 5369 (2001) Phys. Rev. Lett. 86, 5369 (2001)
85. E. Il'ichev, M. Grajcar, R. Hlubina, R.P.J. IJsselsteijn, H.E. Hoenig, H.-G. Meyer, A. Golubov, M.H.S. Amin, A.M. Zagoskin, A.N. Omelyanchouk, M.Yu. Kupriyanov, Phys. Rev. Lett. 86, 5369 (2001)
86. T. Lindstrom, S.A. Charlebois, A.Ya. Tzalenchuk, Z. Ivanov, M.H.S. Amin, A.M. Zagoskin, Phys. Rev. Lett. 90 117002 (2003)

87. C.H. Gardiner, R.A.M. Lee, J.C. Gallop, A.Ya. Tzalenchuk, J.C. Macfarlane, L. Hao, Supercond. Sci. Technol. 17, S234 (2004)
88. G. Rotoli, T. Bauch, T. Lindstrm, D. Stornaiuolo, F. Tafuri, F. Lombardi, Phys. Rev. B 75, 144501 (2007)
89. S. Kawabata, T. Bauch, T. Kato, Phys. Rev. B 80, 174513 (2009)
90. H.A. Kramers, Physica (Utrecht) 7, 284 (1940)
91. Yu.M. Ivanchenko, L.A. Zilberman, Zh. Eksp. Teor.Fiz. 55, 2396 (1968) (Sov.Phys. JETP 28, 1272 (1969)
92. A. Barone, Yu.N. Ovchinnikov, J. Low Temp. Phys. 55, 297 (1984)
93. Yu.N. Ovchinnikov, A. Barone, J. Low Temp. Phys. 67, 323 (1987)
94. A. Barone, J. Supercond.: Incorp. Novel Mag. 17, 585 (2004)
95. P. Hanggi, P.Talkner, M. Borkovec, Rev. Mod. Phys. 62, 251 (1990)
96. E. Ben-Jacob, D.J. Bergman, B.J. Matkowsky, Z. Schuss, Phys. Rev. A 26, 2805 (1982)
97. H. Grabert, P. Olschoswki, U. Weiss, Phys. Rev. B 36, 1931 (1987)
98. J.M. Martinis, H. Grabert, Phys. Rev. B 38, 2371 (1988)
99. R.L. Kautz, J.M. Martinis, Phys. Rev. B 42, 9903 (1990)
100. R.F. Voss, R.A. Webb, Phys. Rev. Lett. 47, 265 (1981)
101. L.D. Jackel, J.P. Gordon, E.L. Hu, R.E. Howard, L.A. Fetter, D.M. Tennant, R.W. Epworth, J. Kurkijarvi, Phys. Rev.Lett. 47, 697 (1981)
102. W. der Boer, R. de Bruyn Ouboter, Physica 98B, 185 (1980)
103. D.W. Bol, R. van Weelderen, R. de Bruyn Ouboter, Physica 122B, 2 (1983)
104. D.W. Bol, J.J.F. Scheffer, W. Giele, R. de Bruyn Ouboter, Physica 113B, 196 (1985)
105. R.J. Prance, A.P. Long, T.D. Clarke, A. Widom, J.E. Mutton, J. Sacco, M.W. Potts, G. Negaloudis, F. Goodall, Nature 289, 543 (1981)
106. I.M. Dimitrenko, V.A. Khlus, G.M. Tsoi, V.I. Shnyrkov, Fiz. Nizk. Temp. 11, 146 (1985) (Sov. J. Low Temp. Phys. 11, 77 (1985))
107. S. Washburn, R.A. Webb, R.F. Voss, S.M. Farris, Phys. Rev. Lett. 54, 2712 (1985)
108. D.B. Schartz, B. Sen, C.N. Archie, J.E. Lukens, Phys. Rev.Lett. 55, 1547 (1985)
109. T.A. Fulton, L.N. Dunkleberger, Phys. Rev. B 9, 4760 (1974)
110. A. Wallraff, T. Duty, A. Lukashenko, A.V. Ustinov, Phys. Rev. Lett. 90, 370031 (2003)
111. J.E. Marchese, M. Cirillo, N. Gronbech-Jensen, Europ. Phys. J. On Special Topics 147, 333 (2007)
112. N. Gronbech-Jensen, M. Cirillo, Phys. Rev. Lett. 95, 067001 (2005)
113. J.E. Marchese, M. Cirillo, N. Gronbech-Jensen, Phys. Rev. B 73, 174507 (2006)
114. Yu. N. Ovchinnikov, A. Barone, A.A. Varlamov, Phys. Rev. B 78, 054521 (2008)
115. Y.N. Ovchinnikov, A. Barone, A.A. Varlamov, Phys. Rev. Lett. 99, 037004 (2007)
116. J. Clarke, F. Wilhelm, Nature 453, 1031 (2008)
117. A.J. Leggett, in Chance and Matter: Les Houches Session XLVI, eds. by J. Souletie, J. Vannimenus, R. Stora (Elsevier, Amsterdam, 1987), pp. 395–506
118. Y. Nakamura, C.D. Chen, J.S. Tsai, Phys. Rev. Lett. 79, 2328–2331 (1997)
119. Y. Nakamura, Y.A. Pashkin, J.S. Tsai, Nature 398, 786–788 (1999)
120. J.R. Friedman, V. Patel, W. Chen, S.K. Tolpygo, J.E. Lukens, Nature 406, 43 (2000)
121. C.H. van der Wal, A.C.J. ter , Haar, F.K. Wilhelm, R.N. Schouten, C.J. P.M. Harmans, T.P. Orlando, S. Lloyd, J.E. Mooij, Science 290, 773 (2000)
122. D. Vion, A. Aassime, A. Cottet, P. Joyez, H. Pothier, C. Urbina, D. Esteve, M.H. Devoret, Science 296, 886 (2002)
123. G. Ithier, E. Collin, P. Joyez, P.J. Meeson, D. Vion, D. Esteve, F. Chiarello, A. Shnirman, Y. Makhlin, J. Schriefl, G. Schon, Phys. Rev. B 72, 134519 (2005)
124. J.M. Martinis, S. Nam, J. Aumentado, C. Urbina, Phys. Rev. Lett. 89, 117901 (2002)
125. P. Bertet, I. Chiorescu, G. Burkard, K. Semba, C.J. Harmans, D.P. DiVincenzo, J.E. Mooij, Phys. Rev. Lett. 95, 257002 (2005)
126. M. Steffen, M. Ansmann, R. McDermott, N. Katz, R.C. Bialczak, E. Lucero, M. Neeley, E.M. Weig, A.N. Cleland, J.M. Martinis, Phys. Rev. Lett. 97, 050502 (2006)
127. Y. Nakamura, Y.A. Pashkin, T. Yamamoto, J.S. Tsai, Phys. Rev. Lett. 88, 047901 (2002)

128. J. Koch, T.M. Yu, J. Gambetta, A.A. Houck, D.I. Schuster, J. Majer, A. Blais, M.H. Devoret, S.M. Girvin, R.J. Schoelkopf, Phys. Rev. A 76, 042319 (2007)
129. P. Dutta, P.M. Horn, Rev. Mod. Phys. 53, 497 (1981)
130. D.J. van Harlingen, T.L. Robertson, B.L.T. Plourde, P.A. Reichardt, T.A. Crane, J. Clarke, Phys. Rev. B 70, 064517 (2004)
131. Y. Makhlin, G. Schon, A. Shnirman, Rev. Mod. Phys. 73, 357 (2001)
132. L.B. Ioffe, V.B. Geshkenbein, M.V. Feigel'man, A.L. Fauchere, G. Blatter, Nature 398, 679 (1999)
133. G. Blatter, V.B. Geshkenbein, L.B. Ioffe, Phys. Rev. B 63, 174511 (2001)
134. A. Blais, A.M. Zagoskin, Phys. Rev. A 61, 42308 (2000)
135. A.M. Zagoskin, Preprint cond-mat 9903170 (1999)
136. C.C. Tsuei, J.R. Kirtley, Physica C 367 1 (2002)
137. D.M. Newns, C.C. Tsuei, US Patent No US6,495,854 B1 (2002)
138. A. Zagoskin, Preprint cond-mat 0506039 (2005)
139. F. Tafuri, F. Carillo, F. Lombardi, F.M. Granozio, F. Ricci, U. Scotti di Uccio, A. Barone, G. Testa, E. Sarnelli, J.R. Kirtley, Phys. Rev. B 62, 14431 (2000)
140. Y.V. Fominov, A.A. Golubov, M.Y. Kupriyanov, JETP Lett. 77, 587 (2003)
141. M.H.S. Amin, A.Y. Smirnov, Phys. Rev. Lett. 92, 017001 (2004)
142. S. Kawabata, S. Kashiwaya, Y. Asano, Y. Tanaka, Phys. Rev. B 70, 132505 (2004)
143. S. Kawabata, S. Kashiwaya, Y. Asano, Y. Tanaka, Phys. Rev. B 72, 052506 (2005)
144. T. Yokoyama, S. Kawabata, T. Kato, Y. Tanaka, Phys. Rev. B 76, 134501 (2007)
145. K. Inomata, S. Sato, K. Nakajima, A. Tanaka, Y. Takano, H.B. Wang, M. Nagao, H. Hatano, S. Kawabata, Phys. Rev. Lett. 95, 107005 (2005)
146. X.Y. Jin, J. Lisenfeld, Y. Koval, A. Lukashenko, A.V. Ustinov, P. Muller, Phys. Rev. Lett. 96, 177003 (2006)
147. G. Rotoli, T. Bauch, T. Lindstrom, D. Stornaiuolo, F. Tafuri, F. Lombardi, Phys. Rev. B 75, 144501 (2007)
148. X.G. Wen, P.A. Lee, Phys. Rev. Lett. 80, 2193 (1998)
149. P.W. Anderson, Science 288, 480 (2000)
150. R. Kleiner, F. Stenmeyer, G. Kunkel, P. Muller, Phys. Rev. Lett. 68, 2394 (1992)
151. R. Kleiner, P. Muller, Phys. Rev. B 49, 1327 (1994)
152. H.B. Wang, P.H. Wu, T. Yamashita, Appl. Phys. Lett. 78, 4010 (2001)
153. V.M. Krasnov, T. Golod, T. Bauch, P. Delsing, Phys. Rev. B 76, 224517 (2007)
154. V.M. Krasnov, T. Bauch, S. Intiso, E. Hurfeld, T. Akazaki, H. Takayanagi, P. Delsing, PRL 95, 157002 (2005)
155. J. Mannik, S. Li, W. Qiu, W. Chen, V. Patel, S. Han, J.E. Lukens, Phys. Rev. B 71, 220509 (2005)
156. J. Kivioja, T. Nieminen, J. Claudon, O. Buisson, F. Hekking, J. Pekola, Phys. Rev. Lett. 94, 247002 (2005)
157. R.L. Kautz, J.M. Martinis, Phys. Rev. B 42, 9903 (1990)
158. D. Vion, M. Gotz, P. Joyez, D. Esteve, M.H. Devoret, Phys. Rev. Lett. 77, 3435 (1996)
159. J.C. Fenton, P.A. Warburton, Phys. Rev. B 78, 054526 (2008)
160. A. Franz, Y. Koval, D. Vasyukov, P. Muller, H. Schneidewind, D.A. Ryndyk, J. Keller, C. Helm, Phys. Rev. B 69, 14506 (2004)
161. Y. Koval, M.V. Fistul, A.V. Ustinov, Phys. Rev. Lett. 93, 087004 (2004)
162. M.-H. Bae, M. Sahu, H.-J. Lee, A. Bezryadin, Phys. Rev. B 79, 104509 (2009)
163. I. Petkovic, M. Aprili, Phys. Rev. Lett. 102 157003 (2009)
164. J. Pfeiffer, T. Gaber, D. Koelle, R. Kleiner, E. Goldobin, M. Weides, H. Kohlstedt, J. Lisenfeld, A.K. Feofanov, A.V. Ustinov, Pre-print Cond Mat. 0903.1046 (2008)
165. J. Pfeiffer, M. Kemmler, D. Koelle, R. Kleiner, E. Goldobin, M. Weides, A.K. Feofanov, J. Lisenfeld, A.V. Ustinov, Phys. Rev. B 77, 214506 (2008)
166. B.I. Ivlev, V.I. Melnikov, Phys. Rev. Lett. 55, 1614 (1985)
167. B. Ivlev, G. Pepe, R. Latempa, A. Barone, F. Barkov, J. Lisenfeld, A. V. Ustinov, Phys. Rev. B 72, 094507 (2005)
168. A. Barone, G. Kurizki, G. Kofman, Phys. Rev. Lett. 92, 200403-1 (2004)

169. G. Kurizki, G. Kofman, Nature 405, 546 (2000)
170. A. Ya. Tzalenchuk, T. Lindstrom, S.A. Charlebois, E.A. Stepantsov, Z. Ivanov, A.M. Zagoskin, Phys. Rev. B 68, 100501(2003)
171. F. Herbstritt, T. Kemen, L. Alff, A. Marx, R. Gross, Appl. Phys. Lett. 78, 955 (2001)
172. G. Testa, A. Monaco, E. Esposito, E. Sarnelli, D.J. Kang, S.H. Mennema, E.J. Tarte, M.G. Blamire, Appl. Phys. Lett. 85, 1202 (2004)
173. G. Testa, E. Sarnelli, A. Monaco, E. Esposito, M. Ejrnaes, D.J. Kang , S.H. Mennema, E.J. Tarte, M.G. Blamire, Phys. Rev. B 71, 134520 (2005)
174. D. Stornaiuolo, E. Gambale, T.Bauch, D. Born, K. Cedergren, D. Dalena, A. Barone, A. Tagliacozzo, F. Lombardi, F. Tafuri, Physica C 468, 310 (2008)
175. A. Tagliacozzo, D. Born, D. Stornaiuolo, E. Gambale, D. Dalena, F. Lombardi, A. Barone, B.L. Altshuler, F. Tafuri, Phys. Rev. B 75, 012507 (2007)
176. A. Tagliacozzo, F. Tafuri, E. Gambale, B. Jouault, D. Born, P. Lucignano, D. Stornaiuolo, F. Lombardi, A. Barone, B.L. Altshuler, Phys. Rev. B 79, 024501 (2009)
177. B.L. Altshuler, A.G. Aronov, in Electron-Electron Interaction in Disordered Systems, ed. by A.L. Efros, M. Pollak (Elsevier, Amsterdam, 1985)
178. P.A. Lee, T.V. Ramakrishnan, Rev. Mod. Phys. 57, 287 (1985)
179. P.A. Lee, A.D. Stone, Phys. Rev. Lett. 55, 1622 (1985)
180. P.A. Lee, A.D. Stone, H. Fukuyama, Phys. Rev. B 35, 1039 (1987)
181. B. Al'tshuler, P. Lee, Phys. Today 41,12, 36 (1988)
182. R.A. Webb, S. Washburn, Phys. Today 41, 46 (1988); For a review see Mesoscopic Phenomena in Solids, edited by B. L. Altshuler, P. A. Lee, and R. A. Webb (North-Holland, New York, 1991)
183. Y. Imry Introduction to Mesoscopic Physics (Oxford University Press, Oxford (London/Melbourne), 1997)
184. N. Gedik, J. Orenstein, R. Liang, D.A. Bonn, W.N. Hardy, Science 300, 1410 (2003)

Chapter 5
Intrinsic Josephson Tunneling in High-Temperature Superconductors

A. Yurgens and D. Winkler

Abstract Intrinsic Josephson tunneling takes place between copper-oxide (CuO) planes of several anisotropic high-temperature superconductors, such as $Bi_2Sr_2Ca Cu_2O_{8+\delta}$ (Bi2212). To study the intrinsic tunneling effects, mesas are usually patterned into the surfaces of Bi2212 single crystals. The mesa structure contains an array of typically 5–30 intrinsic Josephson junctions formed between adjacent superconducting CuO planes. The number of defects that could affect measurements is limited in the small volume of the mesa. Variations in the carrier concentration and introduction of columnar defects by heavy-ion irradiation can be used to control the correlations and pinning of magnetic vortex pancakes. The intrinsic tunneling provides information on the magnetic phase diagram of Abrikosov vortices, allows for direct measurements of the superconducting critical current of the individual CuO planes, permits for the observation of macroscopic quantum tunneling, allows for tunneling spectroscopy of phonons, the superconducting- and pseudo gaps, and shows coherent THz radiation of significant power.

5.1 Introduction

The layered crystal structure is characteristic for high-temperature superconductors (HTS), with conducting layers of copper-oxide separated by other layers which form charge reservoirs and internal tunneling barriers (Fig. 5.1). Intrinsic Josephson tunneling can be seen as multivalued and hysteretic current–voltage (I-V) characteristics in the c-axis direction. For any finite voltage to appear, the current should be higher than the superconducting critical one, I_c. Given the relatively high critical current density $j_c = 10^3\,\mathrm{A/cm^2}$ in the c-axis direction of typical HTSs, the cross-sectional area of the samples should be less than 10^3 to $10^4\,\mu\mathrm{m}^2$, to have $I_c < 0.1\,\mathrm{A}$, thereby avoiding too much of Joule heating in the contacts to the samples. Using very small $Bi_2Sr_2CaCu_2O_{8+\delta}$ (Bi2212) single crystals, Kleiner et al. [1] experimentally showed for the first time that such an intrinsic tunneling of Cooper pairs indeed takes place in this material. Later on, the intrinsic Josephson effect was demonstrated also in other layered superconductors such as $Tl_2Sr_2Ca_2Cu_3O_{10+\delta}$ (Tl2223), or $(Pb_xBi_{1-x})_2Sr_2CaCu_2O_{8+\delta}$ [(Bi,Pb)-2212] [2].

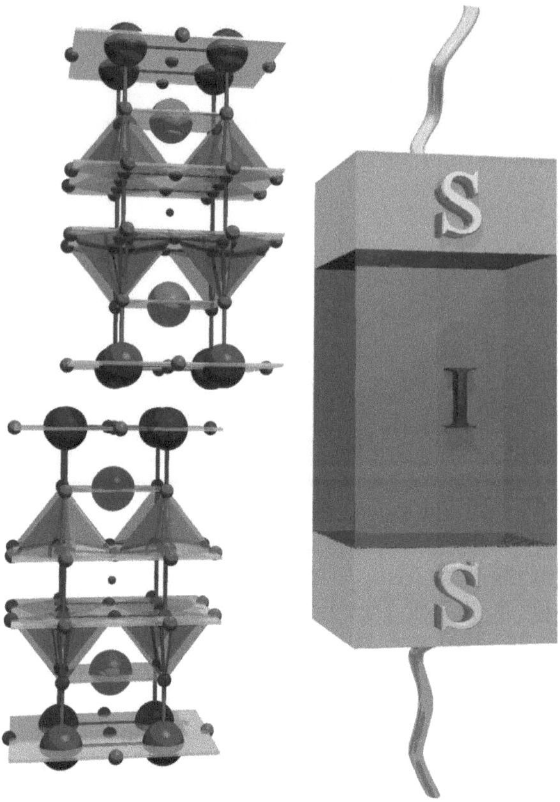

Fig. 5.1 Schematic of the unit-cell crystal structure of Bi2212 (*left*) and the equivalent intrinsic Josephson junction corresponding to the crystal lattice (*right*). BiO and SrO layers are considered as insulating while double CuO layers with Sr atoms in between them as superconducting. The Bi2212 unit cell contains two IJJs in the c-axis direction

To further decrease the sample area, standard microfabrication techniques can be used to pattern small towers (mesas) at the surfaces of the HTS single crystals or make zigzag structures with a thin slice of the single crystal in the middle where the current largely flows in the c-axis direction (for reviews, see [3] and [4]). Here, e-beam or photolithography together with argon ion milling or wet etching is a common technique to obtain a few intrinsic Josephson junctions of small area. By controlling time of etching, just a few layers of high mesas can be patterned, in contrast to the first experiments that used thin crystals enclosing some 10^4 layers. The mesas thus contain an array of typically $N = 5$–30 intrinsic Josephson junctions (IJJs) formed between adjacent superconducting CuO planes. The CuO – CuO interlayer distance is 1.5 nm making the mesa height to be 10–50 nm only. The total mesa volume is therefore $\sim 10^6$ times smaller than the volume of a HTS single crystal. It is statistically much less probable to have defects in the mesas (Fig. 5.2). For the small lateral dimensions of the mesa (below 1 μm), maybe only a few major defects, if any, will be present within the tunnel-junction area. As a result, all the

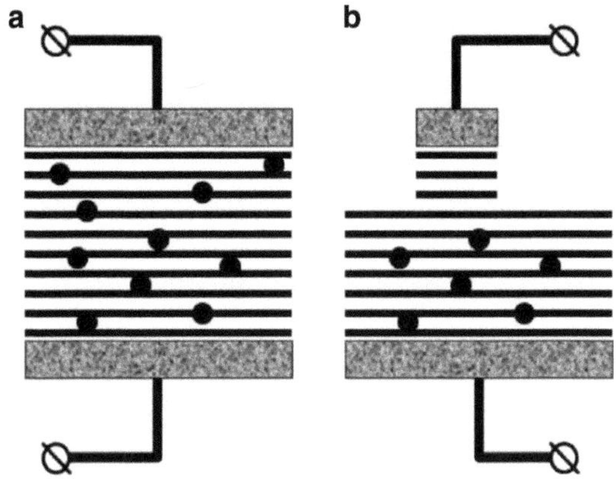

Fig. 5.2 Schematic illustration demonstrating advantages of small mesas on the surface of Bi2212 single crystals. The defects present in the single crystal all contribute to the c-axis transport measurements (**a**) while they are largely excluded from measurements on mesas (**b**)

observed effects are much clearer and stronger in the mesas as compared to single crystals.

Since intrinsic junctions are formed inside a single crystal, they allow for tunneling spectroscopy studies without problems of deteriorated surfaces and interfaces. Characteristic properties of HTS can be measured with a view from "inside" the single crystal complementing the surface methods of angle-resolved photoemission spectroscopy (ARPES) or scanning-tunneling microscopy (STM) providing information from "outside" the single crystal.

Experimental studies of IJJs give valuable information regarding mechanism and applications of HTS. This includes phonon- [5, 6], pseudogap- and superconducting gap spectroscopy [7–9], vortex matter [10–12], high-frequency applications such as detectors [13–15] and sources of tetrahertz (THz) radiation [16–18], genuine superconducting characteristics of individual CuO planes [19], verification of some models of HTS [20], macroscopic quantum effects [21, 22], and several other material-specific effects such as intercalation [7], engineering of defects by heavy-ion irradiation [23], and thin-film growth [24, 25].

However, intrinsic tunneling is not free from characteristic difficulties. The main problem is that all HTS materials have a relatively low thermal conductivity. Joule dissipation in an array of IJJs is set by j_c and the gap voltage $V_g = 2\Delta \times N \approx$ 0.2–2 V, where $\Delta \approx 50$ meV is the superconducting energy gap parameter characteristic for Bi2212 and several other HTSs. This gives the power density $P = j_c V_g \approx 0.2$–$2\,\mathrm{kW/cm^2}$, which should be effectively removed from the mesa. This is difficult at times and leads to a mesa internal temperature that can significantly exceed the bath temperature and even T_c at a high bias current $I \sim (10$–$20) I_c$.

This chapter will review some of key results from research done at Chalmers University of Technology during the last 5 years emphasizing advantageous characteristics of IJJs as compared with similar studies on bulk materials or with surface-bound techniques.

5.2 Sample Fabrication

5.2.1 Simple Mesa

There exist several techniques for isolating- and studying a limited number of IJJs in microstructures. Figure 5.3 schematically shows the most popular ones including the simplest mesa with one electrode on top. The fabrication process usually starts with finding a good quality Bi2212 single crystal that is big enough ($\sim 1 \times 1\,\mathrm{mm}^2$) and has flat and smooth surfaces. The single crystal is then glued to a substrate (silicon or sapphire) using epoxy (Stycast), polyimide, or PMMA. Next step is cleavage of the crystal which is usually done by first attaching a piece of Scotch tape to the surface of the single crystal and then tearing it off. The single crystal breaks/cleaves in two pieces, one follows with the tape and is removed leaving behind another half at the substrate with a fresh surface revealed. It is now very important to deposit a layer of gold or silver as soon as possible to protect the surface from deterioration, mainly due to the contact with water during patterning.

The patterning can then safely be done using common photo- or e-beam lithography, and dry- or wet chemical etching. The height of the resulting mesa is controlled mainly by the time of the etching. There were also experiments that used the in situ monitoring of the resulting I-V characteristics *during* the etching [26]. A layer-by-layer increase in the mesa height could be seen.

A top electrode to the mesa is patterned from a 100- to 200-nm-thick gold or silver thin film. To avoid electrical contact of the electrode to the base of the single crystal outside the mesa area, an insulating layer of different materials can be used,

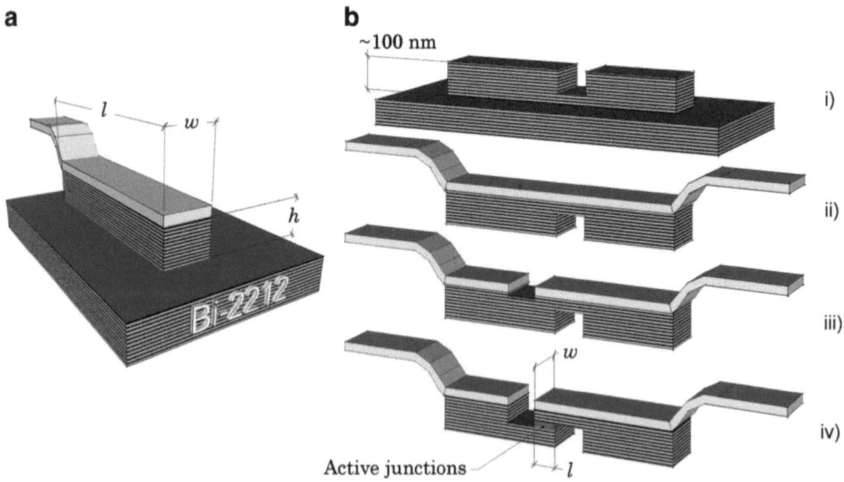

Fig. 5.3 Schematics of a simple mesa (**a**) and flip-chip zigzag structure (**b**). The length l, width w, and height h of simple mesas can be from $l \times w \times h \sim 1 \times 1 \times 0.01\,\mu\mathrm{m}^3$ and up to $\sim 300 \times 100 \times 1\,\mu\mathrm{m}^3$. The latter sizes are common for experiments on THz radiation (see Sect. 5.3.5)

like CaF_2. CaF_2 is chemically inert to YBCO [27], and much likely to Bi2212. Moreover, the material can conveniently be deposited by evaporation, making it suitable for lift-off patterning technique. The insulating layer can also be made of hard-baked photoresist, although the layer can be rather thick, of the order of 1 μm. However, for simple structures, the insulation layer can be neglected entirely while using the very surface of Bi2212 that becomes insulating after photolithography steps and contact with water and photoresist developer.

Two electrodes to the mesa can be formed with some spacing between the contacts on the top of the mesa. Some Bi2212 layers could also be removed during the formation of the electrodes resulting in several "extra" IJJs underneath each electrode. Only common to the both electrodes IJJs can be seen in the four-probe measurements. The extra IJJs under the current lead can result in additional Joule dissipation and increased heating. This is one of the reasons why back-bending of the I-V curves is more pronounced in the four-probe measurements. The back-bending of the I-V curves is a characteristic feature of self-heating in stacks of Josephson junctions [28].

5.2.2 Flip-Chip Zigzag Bridges

This technique was first successfully tested by Wang et al. [29] and is schematically shown in Fig. 5.3b.

First, a single crystal is glued onto a sapphire substrate using polyimide. Then, the single crystal is cleaved using Scotch tape and a gold thin film is deposited by evaporation. A mesa with a microbridge in the center is formed on the crystal using the conventional photolithography and Ar-ion etching. The height of the mesa d is $\sim$100 nm which is controlled by etching time and rate. A slit is etched across the bridge, to approximately half the mesa height.

Next, we glue a second substrate to the sample so that the single crystal with the mesa turns out to be sandwiched in between the two substrates. Separating the substrates breaks the single crystal into two pieces. The piece with the mesa is now flipped upside down at the second substrate. By iteratively using Scotch tape, we remove unnecessary parts of that piece only keeping the mesa.

A new gold thin film is deposited and patterned to make four electrodes attached to the mesa. Finally, an etch-protection layer is made by lift-off from CaF_2 thin film, having an open window across the bridge. The Ar-ion etching through the window incises a second slit being shifted from the first one by some distance along the bridge.

At some moment during the etching of the second slit, the two slits overlap in the out-of-plane direction. Electrical current applied to the bridge is then pushed to flow in the c-axis direction of the central zigzag part between the slits, across a newly formed stack of IJJs. By adjusting the etching rate to about 1.5 nm/min and by monitoring the resulting low-temperature ($T < T_c$) I-V curves of the bridge after consecutive sessions of Ar-ion etching each 1–2 min long, a specified number of IJJs can easily be tailored in the stack.

To obtain a sample with only one IJJ in the stack, it is important to know exactly when the two slits are about to overlap. The total thickness of the bridge (equal to the height d of the first mesa in the ideal case only) becomes somewhat uncertain after flipping the mesa and removing the unnecessary parts of the single crystal. The time of overlap can be determined from the value of the superconducting critical current of the bridge after the second slit becomes sufficiently deep.

The superconducting critical current of the bridge is limited by its geometry and a combination of the in-plane- and out-of-plane critical current densities, j_{ab} and j_{c}, respectively:

$$I_c \sim ws\,(n+1)\,j_{\text{ab}} + wlj_{\text{c}}, \tag{5.1}$$

where $n = \ldots 3, 2, 1, 0, -1$ is the number of layers between the slits to be etched before the overlap occurs; s is the interlayer distance (i.e., the thickness of IJJ), w and l are the width and the length of the stack, respectively (Fig. 5.3b).

Due to the fact that $j_{\text{ab}} \gg j_{\text{c}}$, I_{c} changes in large steps with etching of every layer before the overlap. The moment of overlap is characterized by a sudden decrease in the superconducting critical current and the appearance of a distinctive hysteresis in the I-V curves typical for the SIS-type IJJ.

5.2.3 *Other Methods*

Other methods for isolation of IJJs for different experiments largely rely on thin films on different substrates [30]. A Bi2212 thin film can be epitaxially grown on a substrate with its surface oriented slightly away from the major crystallographic directions. On such a "miscut" substrate, the thin film will have CuO layers glide-shifted along the substrate, resembling a deck of cards that is spread out over desk while having overlap between the individual cards. For a microbridge patterned in the direction of the miscut tilt, the electrical current will pass from layer to layer and cross internal barriers thereby revealing intrinsic junctions. The number of active IJJs is set by the bridge length, film thickness, and miscut angle; this number is normally very large.

Focused-ion-beam (FIB) systems are being conveniently used to mill zigzag structures from the whole single crystal, usually whiskers [31], and also relatively thick films [32]. This method allows for great flexibility in obtaining different sample shapes. Some doping with Ga atoms should, however, be anticipated. Rather, high etching rate and non-flat beam profile does not allow for layer-by-layer tailoring of IJJ stacks, as in the case of flip-chip samples and Ar-ion etching.

Nakajima et al. used Si-ion implantation to destroy superconductivity at specific areas of Bi2212 single crystals that were unprotected by photoresist [33]. Superconducting stacks wholly embedded in the non-superconducting material can thereby be obtained.

As it will be clear from the following sections, it is possible to see intrinsic tunneling even without patterning stacks and mesas. A sufficiently high current through a small contact formed at the surface of Bi2212 single crystal will destroy superconductivity in the surface layer along the perimeter of the contact. The current will then be redistributed and have an increasing component in the c-axis direction across IJJs below the contact. The same branch-like structure of the I-V curves can be observed although shifted along the current axis by the critical current of the surface layer.

5.3 Electrical Characterization

5.3.1 I-V Curves of Intrinsic Josephson Junctions in Bi2212

Typical three-probe I-V characteristics are shown in Fig. 5.4 for a mesa $\sim 5 \times 5\,\mu m^2$ large. The I-V characteristics are multivalued and hysteretic, which is typical for Josephson tunnel junctions with large capacitances. There are about 60 branches, each corresponding to one, two, three, and so on, IJJ leaving the zero-voltage state as the current exceeds the critical currents for the first (I_{c1}), second (I_{c2}), and so

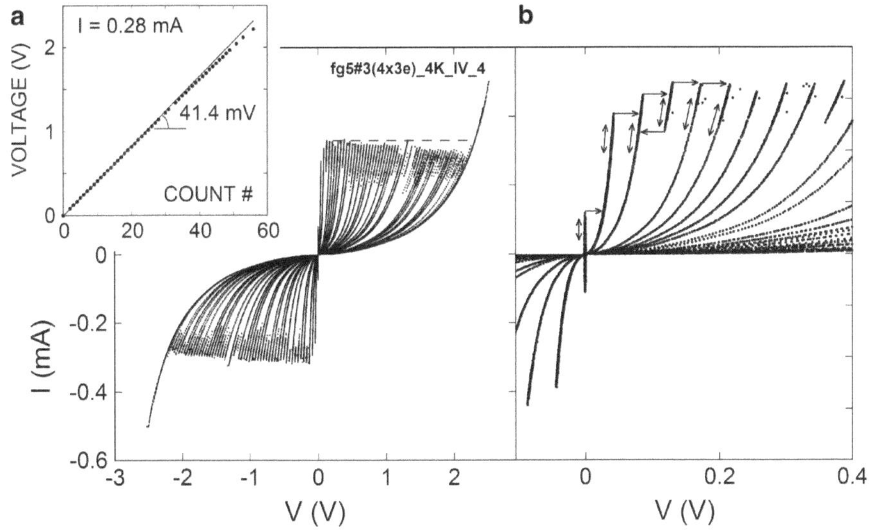

Fig. 5.4 (**a**) Typical current-voltage characteristics of an array of stacked IJJs $\sim 5 \times 5\,\mu m^2$ large at $T = 4.2$ K. (**b**) Zoomed-in central part of the characteristics. Each quasiparticle branch corresponds to 1, 2, 3... N IJJs in series that have switched to the quasiparticle state. The critical current of the subsequent branches I_{ci} decreases indicating the presence of Joule heating. The internal temperature can then be determined from the difference $I_{c1} - I_{ci}$, where $I_{c1}(T)$ is the temperature dependence of the critical current for one of the first branches which is least affected by the self-heating. The inset shows V at $I = 0.28$ mA as a function of the branch count number i. Nearly linear dependence indicates uniformity of the junction properties. Some downward deviation from the *straight solid line* indicates Joule heating for $i > 20$–30

on, IJJ. The branches were traced in multiple bias sweeps where the amplitude of the ac current ($f \sim 17$ Hz) varies at some frequency $f_a \sim 1$ Hz, while the current and voltage are measured simultaneously at some higher frequency ~1 kHz. The switching from one branch to the next one may not necessarily occur, but there can be a jump to a higher-count branch if the critical currents of all the junctions have similar values. This is often called a phase locking of tunneling junctions in the stack.

The I-V curves shown in Fig. 5.4 belong to a stack that was intentionally made high to investigate self-heating and effective ways of its removal. The single crystal was soldered rather than glued to a 30-μm-thick copper foil. The heat removal then became more effective; it may be seen from the moderate decrease of I_{ci} with the branch number i (Fig. 5.4a) and decrease of branch spacing in voltage (see the inset of Fig. 5.4). The heating in zigzag stacks is much higher [34, 35] due to lower thermal conductivity of the glue and also due to the absence of a heat-spreading single crystal between the stack and a substrate.

5.3.2 *Critical Current Density of Individual CuO Plane*

It is important to know the ultimate performance of the HTS cables that are made of Bi2212 and related layered compounds. The critical current densities published in the literature are diverse, citing both very high values, up to a few MA/cm^2 measured in thin films and some single crystals with the help of noncontact magnetization measurements.

The variety of results can be explained not only by the quality of samples but also by the way the critical current is measured. In the measurements involving electrical contacts, the values of the critical current are usually smaller. The current distribution across the sample is highly nonuniform in the case when the contacts are placed on one and the same side of a single crystal. When the current becomes higher than the critical current of the surface layers, a finite voltage appears that can mistakenly be taken as a transition to the normal state of the whole sample despite that the lower part of the sample can still be superconducting. Likewise, defect regions of single crystals can effectively make the cross-section smaller, resulting, again, in lower estimates of the critical current density.

We measure genuine properties of the superconducting planes in Bi2212 using mesas. The current flow in the mesas can be squeezed into narrow bridges thereby easily reaching the critical current [36].

5.3.3 *Superconducting Critical Current of Individual CuO Planes in Bi2212*

In the first experiment, a mesa approximately 3–5-μm wide is etched in the single crystal. The mesas have two thin-film electrodes separated by a gap (Fig. 5.5a).

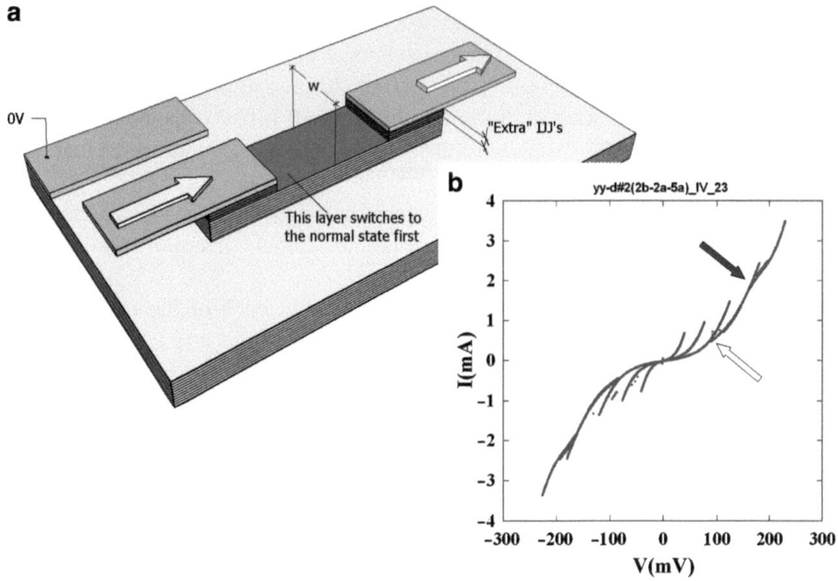

Fig. 5.5 (**a**) Measurement schematic of $I_c^{(ab)}$ of the top plane. The *arrows* indicate the current flow. The electric potential V of one of the electrodes is measured relative to the base of the mesa ("0V"). (**b**) The resulting I-V curves at $T = 4.2\,\mathrm{K}$. The three first I-V branches correspond to "extra" IJJs under each of the electrodes because of some excess etching during formation of the electrodes. The bifurcation points marked by the *arrows* indicate the current-driven transitions to the normal state of the first (*top*) and the second superconducting CuO planes

The gap should be made carefully to avoid too many "extra" junctions under the electrodes. The total height of the mesas is not important.

To see the *ab*-plane superconducting transition of a single plane in these mesas containing many junctions, we apply current between the topmost electrodes and measure the voltage on either of the contacts relative to the bulk of the single crystal, as is schematically shown in Fig. 5.5a. In the case of no extra junctions under the electrodes, no voltage is expected to appear unless the current exceeds the sum of the in-plane- $\left(I_c^{ab}\right)$ and c-axis (I_c) critical currents, $I > I_c^{ab} + I_c = j_c^{ab} w s + j_c w x$, where j_c^{ab} and j_c are the corresponding in- and out-of-plane critical current densities, and w, s, and x are [36] the width of the mesa, interlayer spacing, and length of the mesa covered by the electrode, respectively. This also means that the intrinsic Josephson junction can be revealed even without etching away layers to make a mesa. It is enough that the current through an electrode attached to a flat surface of a large single crystal exceeds the (sheet) critical current for a single plane corresponding to the whole circumference of the electrode contact area.

Indeed, when the current exceeds the critical value for the plane, the current redistributes, with one part still flowing along the plane and the rest flowing downward through an effective "hidden" IJJ under the electrode. At this moment, however, no voltage appears since that junction is still in the superconducting state. The current which is branching off downward must become larger than the critical current of that hidden junction for the finite voltage to appear at that electrode.

If the current is set to decrease after that, the hidden junction will not leave the quasiparticle-tunneling state until the current through it becomes less than the re-trapping current I_c^{retr}. All-in-all, to return to the zero-voltage state again, one has to decrease the total current below $I_c^{ab} + I_c^{retr}$. The re-trapping current I_c^{retr} is usually small and can be ignored, making the current at which the system switches back to the zero-voltage state (a return point) to be the critical current of a single plane: $I_c^{ab} + I_c^{retr} \approx I_c^{ab}$, $I_c^{retr} \ll I_c \leq I_c^{ab}$.

In the case when there are several extra IJJs under each of the electrodes, the return point is seen as a bifurcation point (a break) at the last I-V quasiparticle branch of these IJJs (Fig. 5.5b). The break is also accompanied by the appearance of an extra I-V branch having its origin at this point. The superconducting transitions of the second and even third from the top layer can also be seen at high enough current. At high current, however, the Joule heating progressively increases and makes estimations of the corresponding critical current densities somewhat understated.

From the current of the first bifurcation point and the width of the mesa $w = 7.5\,\mu\text{m}$, we find the sheet critical current density to be $\approx 0.64\,\text{A/cm}$ corresponding to the bulk in-plane current density of $4.3\,\text{MA/cm}^2$ [36]. This value is among the largest ever reported in the literature.

In the second geometry, a single mesa with one electrode is used. The electrode is isolated from the rest of the single crystal by an insulating layer (evaporated CaF_2). Then, deep cuts are made by the FIB etching (Fig. 5.6) to limit the current return path to a narrow bridge below the electrode. Note that the electrode and insulation layer protect the surface CuO layers from Ga ions. In FIB systems, Ga ions are also used for imaging the sample, meaning that not only the cuts but also the whole image area are exposed to the ions. The ions can damage unprotected surfaces even at small energies and currents.

In this geometry, the current first flows toward the mesa, than downward across a few IJJs below the contact and then return to another contact placed somewhere else on the surface of the single crystal. The deep cuts force the current to only flow within the surface layers of the narrow bridge under the electrode. The bifurcation point in the last branch will indicate when this current exceeds the critical value of the surface layer.

The return path for the current can be made narrower by additional FIB etching (Fig. 5.6b–d). The original mesa and silver thin-film electrode have ill-defined shape due to wet chemical etching that was used for that sample. FIB cutting allows correcting for this and have a nicely defined mesa in the end. Additional cuts are then made and the I-V curves are taken at low temperature after each trimming.

The resulting I-V curves are shown in Fig. 5.7a–c, corresponding to three snapshots of Fig. 5.6. Figure 5.7a shows the initial I-V curve when the mesa area has been shaped up to acquire a rectangular form. The bifurcation point denoted by an arrow is clearly seen at about 0.2 mA. After additional etching, this bifurcation point decreases in current with decreasing width w (Fig. 5.7b) and eventually becomes indiscernible from the return-current jumps of the mesas regular

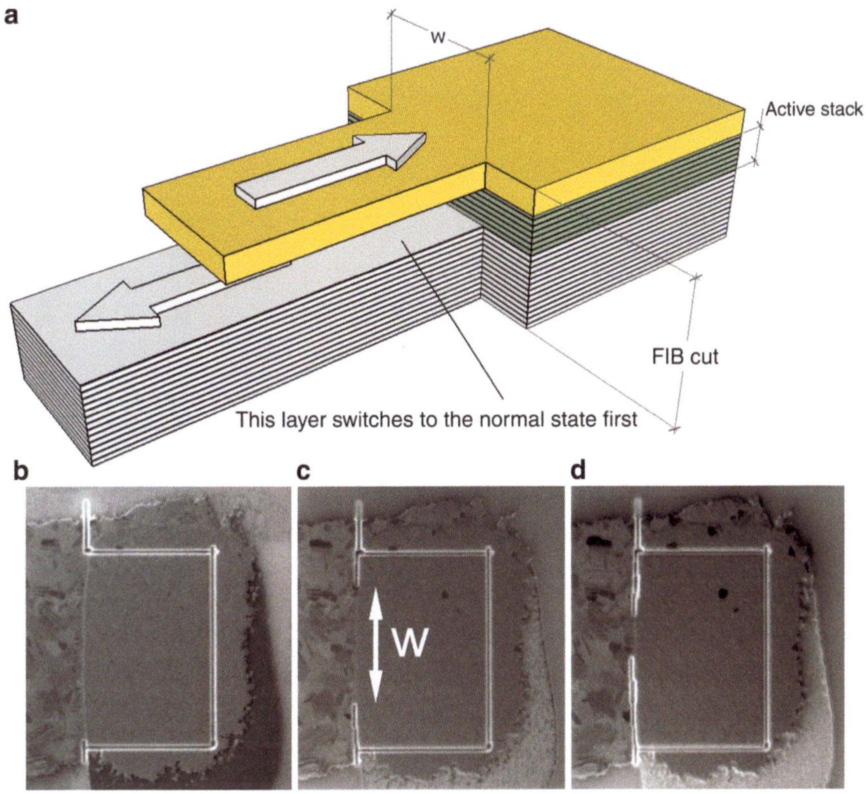

Fig. 5.6 (**a**) Schematic of alternative method of $I_c^{(ab)}$ measurements. The current flows toward the active stack, down through it, and then is forced to flow back along the surface layers of a narrow bridge under the electrode. The deep FIB cuts prevent the current to flow elsewhere. The insulation layer between the electrode and single crystal is not shown. (**b**) A top view on the mesa with a metal electrode. The original mesa shape is ill-defined due to wet etching. The sharp lines are deep cuts made by the FIB milling. (**c**)–(**d**) Additional FIB cuts to make the current return path narrower

branches (Fig. 5.7c). Instead, two more bifurcation points appear corresponding to the second- and third CuO layers that are driven to the normal state by current.

Plotting currents of the bifurcation points as functions of w, we obtain nearly linear dependencies (Fig. 5.8). Slopes of these lines give the sheet current densities for the CuO planes, $\sim$0.07 and $\sim$0.2 mA/μm, for the first and second bifurcation points, respectively. The first value is in agreement with the result mentioned above, while the second one is almost three times larger. Moreover, the best linear fit corresponding to the first bifurcation point does not go through the axis origin, i.e., the current becomes zero even for nonzero width. This suggests that the surface layer became damaged after contact with different chemicals during the mesa fabrication and is not homogeneous anymore.

It is interesting to note that the second value of the sheet current density corresponds to a very high bulk current density, $\sim$13 MA/cm^2, if using the whole

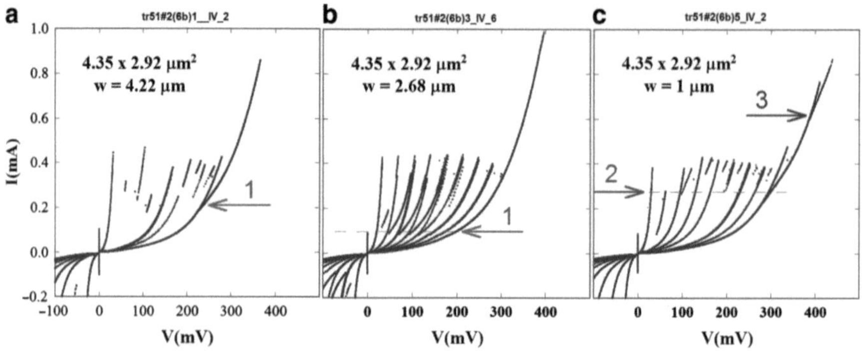

Fig. 5.7 The I-V curves at $T = 4.2\,\mathrm{K}$ of the stack shown in Fig. 5.6 after subsequent etching steps to decrease the width w of the bridge. Note an approximately constant critical current of all the branches. It is determined by the constant area of the initial FIB cuts. The current, corresponding to bifurcation point 1 in (**a**), decreases once w is made smaller. It decreases even more for smaller w and cannot be detected anymore for $w = 1\,\mu\mathrm{m}$ (**c**). However, new bifurcation points appear (2 and 3) that correspond to the current-induced transitions to the normal state of the second and third CuO planes of the narrow bridge under the electrode

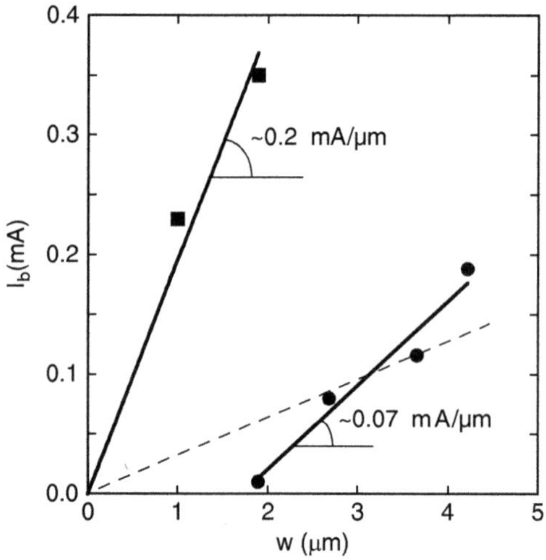

Fig. 5.8 The current of I-V bifurcation point versus width of the bridge where current flows. *Circles* and *squares* refer to the first and second bifurcation points, respectively. The *solid lines* are the best linear fits with one of which is forced to go through the axis origin. The *dashed line* is the guide to the eye drawn through the middle two circles and the axis origin. The fact that the current of the first bifurcation point becomes zero even for nonzero width suggests that the surface layer is damaged and is not homogeneous

interlayer distance (1.5 nm) as the sheet thickness. However, if only taking the thickness of CuO double layer (0.3 nm), the critical current density becomes five times larger, $65\,\mathrm{MA/cm^2}$, which is comparable with both the pinning-limited critical current density $j_{\mathrm{pin}} = c\Phi_0/(64\pi^2\lambda^2\xi) \sim 40\,\mathrm{MA/cm^2}$ and the Ginsburg–Landau depairing current density $j_{\mathrm{GL}} = c\Phi_0/(12\sqrt{3}\pi^2\lambda^2\xi) \sim 130\,\mathrm{MA/cm^2}$, where c, Φ_0, $\lambda \sim 200\,\mathrm{nm}$, and $\xi \sim 2\,\mathrm{nm}$ are the speed of light, flux quanta, London penetration depth, and coherence length, respectively [37]. It is unreasonable to demand higher accuracy of comparison from these order-of-magnitude estimates, although it clear that the critical current density measured in experiment is very close to theoretical limit. Moreover, some part of the total current can also flow in the deeper-lying layers because of their nonzero kinetic inductance and Josephson coupling between them. The exact current distribution can be intricate although the major part of the current is still expected to flow in the surface layers only.

5.3.4 Tunneling Spectroscopy

Tunneling spectroscopy is a proven experimental method giving information about excitations and quasiparticles in a conducting material [38,39]. STM allows even for spatially resolved spectroscopic studies at different locations of a specimen [40,41]. STS was used in studies of HTS earlier [42] and continues to be successful for, i.e., mapping superconducting gap and pseudogap [43–45], and, possibly, the van Hove singularity of electronic spectra in some HTSs [46]. It has been observed that the most common layered HTS like Bi2212 have large gap variations on the scale of few nanometers [43, 44]. The non-uniformity of the superconducting gap distribution has been taken as genuine property of the material. However, some of the interstitial oxygen atoms that promote doping in the CuO planes can easily diffuse out from the surface layers of Bi2212 after cleaving single crystals. Some random distribution of the remaining oxygen within surface layers will then give rise to a disordered pattern of local doping and the corresponding pattern of local gaps traced out by STM. Indeed, it is sometimes possible to see the spectra that are very identical and uniform over macroscopic distances [47]. It is difficult to surely rule out such a possibility and be confident in that the material is disordered everywhere from the beginning.

Intrinsic tunneling allows for complementary spectroscopy information, providing a view from inside the single crystal, away from the deteriorated surfaces [7–9, 48].

First, experiments to study the pseudogap signatures were done on mesas fabricated on the surfaces of Bi2212 single crystals. Unfortunately, intrinsic tunneling experiments have a problem of self heating (see Sect. 5.1 above). To alleviate it, a pulsed I-V-curve measurement technique was applied by Suzuki et al. [49]. Simulations, however, show that much shorter pulses would be needed to completely get around the problem with this technique.

Decreasing the number of junctions and also decreasing their area [50] are straightforward methods to deal with the adverse effect of self-heating. Still, it is rather difficult to fabricate stacks with sizes below 1 μm using common microlithography methods because of the nonflat surface of the substrate with a single crystal glued to it. Also, with all its flexibility and accuracy, FIB milling does not allow for isolation of small enough number of IJJs in the stacks.

The flip-chip technique described above allows for both relatively small lateral sizes and ultimately small number of junctions involved in obtaining intrinsic tunneling spectra. Samples with only few IJJs can be made. The typical intrinsic tunneling spectra obtained with such samples are shown in Fig. 5.9. $I(V)$- (main panel) and $dI/dV(V/N)$-curves (insets) of zigzag stacks of different sizes at 4.2 K demonstrate somewhat unexpected results [51].

The I-V curves show several wiggles at high bias current $\gg I_c$. These wiggles correspond to peaks in the $dI/dV(V)$-plots. The wiggles persist even when the number of junctions increases after additional etching. Samples with different sizes and samples made from different Bi2212 single crystals have also been tested. According to the common practice, the tunneling-conductance peaks should correspond to the maxima of the electronic density of states. In superconductors, the maxima occur at the energy equal to the superconducting energy gap. Now, which of the peaks should be assigned to the superconducting gap and what is the origin of the others?

A way to understand this is to follow how the peak positions change with number of junctions, their area, and also temperature (Fig. 5.10). The lower inset of Fig. 5.9 shows scaling of the peak positions with the number of IJJs in one and the same stack of increasing height. It is seen that only the peaks at the lowest voltage preserve their position in the normalized voltage (V/N). It may therefore be tempting to assign them to the superconducting gap.

The most probable scenarios for the observed peaks are phonon-related effects or resonant tunneling. Indeed, phonons effects are known to be very strong in IJJs [5,6]

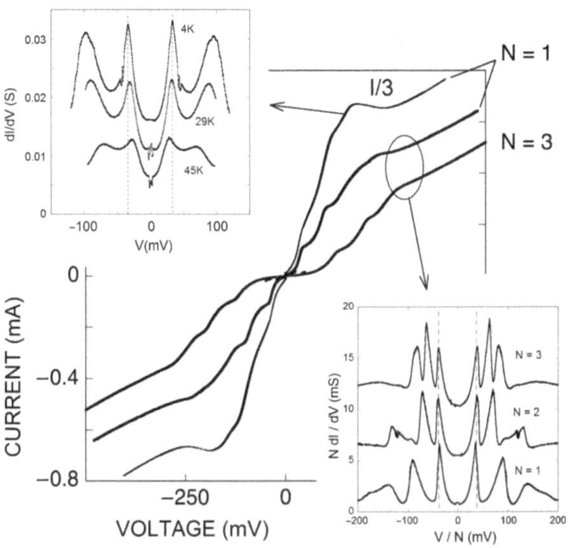

Fig. 5.9 $I(V)$- (*main panel*) and $dI/dV(V/N)$ curves (*insets*) of zigzag stacks of different sizes at 4.2 K. The vertical *dashed lines* mark the voltage where scaling of the assumed small gap is observed. The voltage is nearly the same for these samples coming from different sources

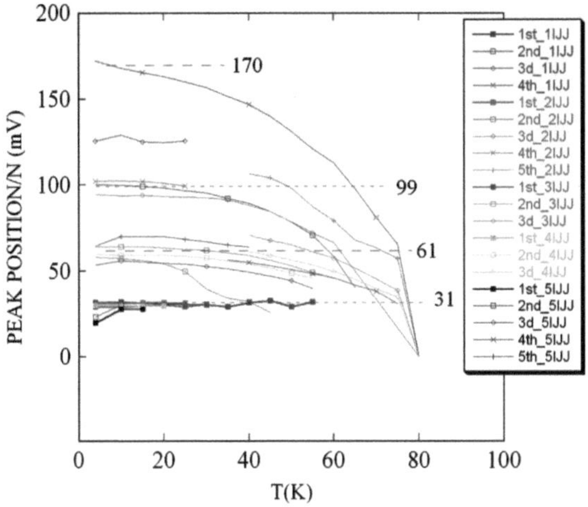

Fig. 5.10 The temperature dependence of the peak positions. Different symbols (■, □, ◇, ×, +) and different colors (*blue, red, green, cyan*, and *black*) correspond to the count numbers of the peaks and the total number of IJJs in the stacks (1–5), respectively. The *solid squares* and *thick lines* correspond to the low-energy peaks for which scaling with number of junctions is observed

and are seen directly in the I-V curves. In the ordinary superconductors, phonon-related features can only be detected when using the second derivative of the I-V curves [38, 39]. In the case of IJJs, two models have been put forward to explain the observed effects. In the first model, the Josephson radiation couples to the optical LO phonons of the dielectric layers between the superconducting electrodes [6]. In the second model, a phonon-assisted tunneling of Cooper pairs leads to the appearance of I-V peaks at voltages corresponding to the Raman-active phonons [52]. Both models explain the appearance of I-V features at some voltages below 2Δ and their independence on temperature.

In the present experiments, phonons at 269 or 287 cm^{-1} corresponding to vibrations of oxygen atoms in the CuO planes [53] would have to be involved in the out-of-plane tunneling to explain the low-energy peaks with the help of these models. Simultaneous Raman and intrinsic tunneling experiments could possibly reveal the exact mechanism.

Another likely explanation of the low-lying peaks is a double-barrier resonant-tunneling scenario [54, 55]. In the simplified picture, an intrinsic tunneling junction consists of two electrodes and one barrier. Bi-O and Sr-O layers are assumed to be totally insulating. However, a Bi-O layer inside IJJ can very likely represent an intermediate charge reservoir separated from the CuO layers by two barriers, resulting in a classical double-barrier structure. It is then possible to have a noticeable current enhancement at voltages corresponding to approximately one-half of the superconducting gap voltage [56, 57], i.e., corresponding to the low-energy peaks observed in our experiments. To test this hypothesis, systematic doping dependencies of the peak positions are desirable. Also, stacks with four superconducting electrodes should be used to exclude possible contribution from the in-plane transitions described above (see Fig. 5.3 in [34]). I-V branch bifurcation associated with such transitions could be smeared at $I > I_c$ and give rise to wiggles in the I-V curves and peaks in the tunneling conductance.

5.3.5 *THz Radiation*

Electromagnetic waves in the frequency range from about 0.1 to 10 THz, i.e., between the microwave and infrared (IR) bands, are known as THz radiation. THz technology is a relatively new and extremely rapidly expanding area of research and development. This technology is expected to provide solutions to a number of important applications that are highly requested by the society. Terahertz waves propagate through various materials such as plastic, wood, ceramics, and cloths, while they undergo strong absorption by water, organic molecules, and biological matter. This makes THz technology a unique tool for such applications as screening for concealed weapons and explosives, detection of cancer tissues, imaging and nondestructive testing, and evaluation in industrial processes. At the same time, availability of powerful, efficient, and at the same time inexpensive THz emitters or receivers on the market is currently limited [58].

Brian Josephson predicted the appearance of an oscillating tunnel current in *superconducting junctions* under applied constant voltage in 1962. The frequency of these oscillations ω is related to the applied voltage V : $\hbar\omega = 2\,eV$, where $\hbar$ and e are the Planck's constant and electron charge, respectively. This makes Josephson junction a fundamental voltage-to-frequency converter across a wide frequency band. While 1 mV corresponds to a frequency of 483 GHz, very high frequencies sources can be produced. The upper frequency limit is set by the superconducting energy gap in the electronic spectrum of the junction material. The upper frequency is about 3 THz in the case of low-temperature superconductors (e.g., NbN), while it is theoretically as high as 10–20 THz for HTSs. However, the electromagnetic radiation generated by a single Josephson junction was found to be very weak [59]. To get higher radiation power for use in practical applications, cooperative action of many Josephson junctions is required. However, attempts to generate powerful radiation from large arrays of Josephson junction repeatedly failed, basically due to insufficient coupling, small junction impedance, and large spread of the individual junction characteristics [60].

The situation is different in Bi2212. IJJs in Bi2122 are atomically identical and are very tightly packed in the single crystal. This significantly enhances the inter-junction coupling and synchronization of radiation, thereby allowing generation of coherent radiation with enhanced power. Although this was qualitatively understood long time ago, soon after the discovery of the intrinsic Josephson effect in 1992, the crucial step has been made only recently. It has theoretically been pointed out that the efficiency of the Josephson emission can significantly be enhanced using resonant accumulation of radiation in the volume of IJJs stack, and that this enhancement is sufficient to trigger the transition to a super-radiation state resulting in a coherent THz emission of high power [61]. These theoretical predictions have found their confirmation in recent experiments, where a relatively strong coherent radiation ($\sim$0.5 μW at 0.5 THz) has been detected from large mesas ($\sim 300 \times 100 \times 1\ \mu m^3$) patterned on Bi2212 single crystals when a voltage is applied in the c-axis direction [17, 18].

It should be emphasized that these stacks of IJJs represent a conceptually new, all-solid-state *primary* THz-radiation source with estimated output power of up to $\sim$1 mW at frequencies 0.3–2 THz and maximum theoretical efficiency of up to 30%. The nearest competitors are optical methods of THz generation that include optical mixing (down-conversion), parametric generation, and quantum cascade lasers (QCL). All these techniques cover the high-end part of the THz band and have some difficulties at the lower end of the THz spectra. In particular, the lowest frequency that has been achieved with QCLs is 1.6 THz. The low-end part of THz spectra is covered by other sources such as Gunn oscillators (combined with a chain of frequency multipliers) or the uni-travelling-carrier photodiode (photomixing), backward wave oscillators (BWOs). These sources, however, do not offer sufficiently high power at frequencies approaching 1 THz [58]. A millimeter-large Bi2212 single-crystal chip is a very viable alternative for an efficient source at the low-end part of THz spectra. It is pertinent to mention also a few earlier works on microwave emission from IJJs, which, however, have not demonstrated high output power [16, 62–64].

We use small Bi2212 mesas for in situ detection of THz radiation from large mesas at high bias at which Joule heating is important. Experiments are performed with simultaneous bias of the emitter- and detector mesas. A dc bias is applied to the emitter mesa while monitoring the swept I-V curves of the detector mesa with an oscilloscope. Corresponding I-V curves of both mesas are shown in Fig. 5.11. The detector I-V curves are shown in the insets (a)–(d) of Fig. 5.11, each corresponding to a particular bias point of the emitter mesa (main panel).

First, when we increase the emitter bias, the mesa switches from superconducting to a normal-tunneling state. The switching occurs via multiple step, first through some intermediate quasiparticle branches and then to the wholly resistive voltage state that corresponds to voltages $\sim 2\Delta N$, with $N \sim 1000$. This voltage is very high and Joule heating quickly sets in leading to smaller $\Delta(T)$ and decreasing voltage. The I-V curve in Fig. 5.11 shows the final state of the system when the mesa temperature becomes stable.

The last quasiparticle branch of the I-V curve of the emitter mesa is highly nonlinear and has parts with a negative differential conductance (NDC). There is a narrow range of currents corresponding to a detector response indicating the presence of THz radiation. The detector is a small IJJ stack placed either on the same single crystal in close vicinity to the emitter one or on another single crystal placed above the emitter stack. There is no electrical connection between the two in the latter case.

The insets of Fig. 5.11 show the detector I-V curves at different points of the emitter I-V curve, in agreement with [18].

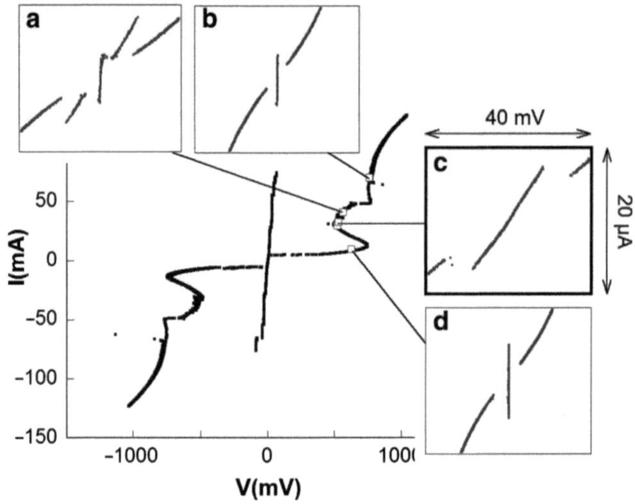

Fig. 5.11 The I-V characteristics of the emitter (*main panel*) and detector (*insets*) at 4.2 K. The detector I-V curves are measured at certain emitter I-V points indicated by the *empty squares*. The detector is on another substrate. Note the very narrow range of currents where the suppression of the detector critical current is observed. Both below and above that current range there is no detector response. This excludes black-body radiation as the origin for the detector response

One of the central problems with IJJs is Joule heating by the bias current which becomes especially acute in THz-emission experiments where a synchronous action of large number of IJJs is necessary to enhance the output power of the radiation. It is a challenge to effectively remove the generated heat, especially at low temperatures, given a low thermal conductivity of Bi2212. Preliminary tryouts have shown that *soldering* of Bi2212 single crystals instead of gluing them at a substrate is very promising (Fig. 5.4).

When Joule heating is strong, the mesa temperature can be much higher than the bath temperature T_0. To calculate a realistic temperature increase and distribution in the mesa, the nonlinear diffusion equation

$$\nabla[-\kappa(\mathrm{T})\nabla T] = \rho(T)j^2 \tag{5.2}$$

should be solved while taking into account the temperature dependencies of thermal conductivities and a particular geometry of sample. We will show results of such calculations below.

Equation 5.2 assumes also that the resistivity does not depend on the current density j. Fortunately, this simplification is well justified for large- and high mesas where the self-heating becomes essential already at currents much smaller than the superconducting critical current I_c. I-V curves can be regarded as linear *in the absence of self-heating* at such currents.

5.3.6 Joule Heating in Mesas

The removal of the Joule heating depends on material characteristics and geometry of the mesa. Placing the sample in liquid helium does not help much in removing the heat from the mesa. Helium starts boiling and losing its cooling capability already at quite small power densities, of the order of $1\,\mathrm{W/cm^2}$ [28] to be compared to several hundred $\mathrm{W/cm^2}$ which can be dissipated in mesas. Below, we numerically calculate a realistic temperature increase due to Joule heating using (5.2) and typical thermal- and electrical parameters and their temperature dependencies of all the materials involved in the heat generation and its removal.

For calculations, we use a model mesa depicted in the bottom of Fig. 5.3a with $l = 300\,\mu\mathrm{m}$, $h = 1\,\mu\mathrm{m}$ and w either 50 or $100\,\mu\mathrm{m}$. The single crystal under the mesa is $1 \times 1 \times 0.02\,\mathrm{mm^3}$ large. It is glued to a sapphire substrate by a 20-μm-thick layer of polymethylmethacrylate (PMMA). The bath temperature $T_0 = 10$–$80\,\mathrm{K}$. There is no helium cooling gas around the sample.

The anisotropic thermal conductivity $k(T)$ of Bi2212 single crystals is adopted from a number of works [65–68]. The ratio of thermal conductivities along and perpendicular to the CuO layers, $k_{\mathrm{ab}}/k_{\mathrm{c}}$, is ~6–8 in a wide range of temperatures [69].

To model the transition to the normal-metal state, the Heaviside step function smeared over some temperature interval $\delta T = 0.5\,\mathrm{K}$ is used for the resistivity: $\rho_{\mathrm{ab}} = 0.013\,\theta(T - T_{\mathrm{c}},\, \delta T)\,T$ (in μohm m). The c-axis resistivity ρ_{c} is taken from [70,71]. ρ_{c} shows a tendency to saturate at low temperatures.

A 100-nm gold thin film makes an electrode on top of the mesa. The thermal conductivity of such a film is calculated from its experimentally measured resistivity using the Wiedemann–Franz law.

The thermal conductivity of the PMMA layer is taken from [72]. It is not much different from the thermal conductivity of other possible epoxies and glues [73]. We note also that any "silver glue" does not provide any better thermal contact between the single crystal and substrate. Such a glue has a high electrical resistivity, ~0.5–1.5 Ω cm in most cases [74], thus accounting for barely 10% of the increased thermal conductivity k of the glue. This small gain in k is, however, wasted because of a thicker glue layer. Viscosity of glue increases a factor of 10–100 when filler is added (e.g., Stycast 1265).

Contour plots of the lateral temperature distribution in a 100-μm mesa wide are shown in Fig. 5.12 (left) for several current densities. The contour plot of the vertical temperature distribution in a 50-μm-wide mesa is shown in Fig. 5.12 (right). It is seen that the temperature is spatially uneven and much higher than T_0, even exceeding T_{c} in the middle part of the mesa at high bias. The mesa therefore breaks up into two smaller ones connected by the common electrode and the normal-state region.

Recently, THz radiation with a relatively high output power has been observed at high currents corresponding to regions of back-bending in these I-V curves [75] (see also Fig. 5.11). The fact that the mesa temperature can exceed T_{c} makes it reasonable to suggest that the heating might play an active role in the synchronization of IJJs in large mesas.

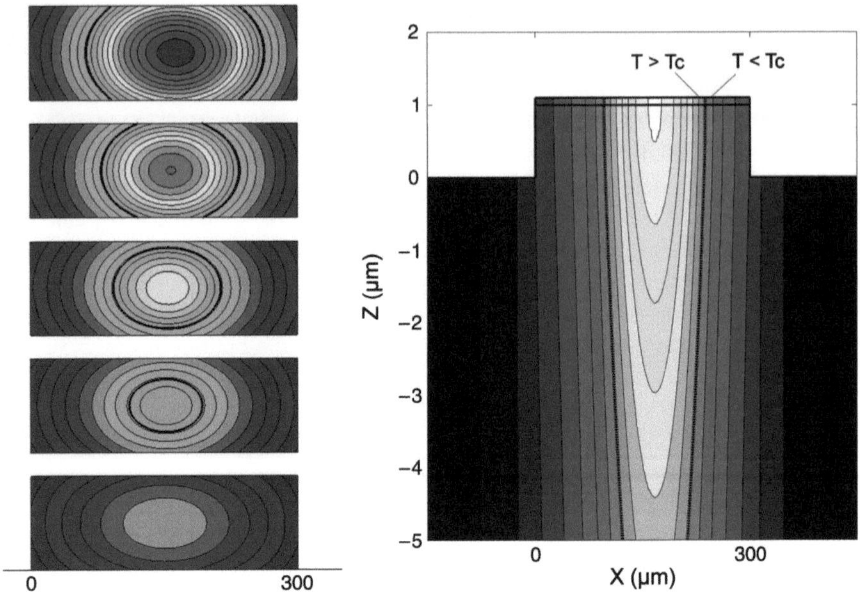

Fig. 5.12 *Left*: The temperature distributions in a 100-μm-wide mesa at the different current densities $j = 1.33$, 1.00, 0.80, 0.67, and $0.50\,\mathrm{MA/m^2}$, from *top* to *bottom*. The distributions are shown for a lateral plane at half the mesa height. The temperature difference between neighboring contour lines is 5 K. *Right*: The vertical cross-section of the 3D temperature distribution in a 50-μm-wide mesa for $j = 1.33\,\mathrm{MA/m^2}$. The bath temperature is $T_0 = 10\,\mathrm{K}$. The thick contour lines indicate the boundary between the superconducting and normal-metal regions

Indeed, the self-heating-induced normal-state region can effectively create a shunting resistor R_{shunt} for the rest of the mesa. Moreover, the temperature is nonuniform having large temperature gradients (Fig. 5.12). The shunting resistance and uneven $j_c(T(x, y))$ can help synchronization of IJJs in the mesa. It is known, for instance, that a common load impedance can stimulate synchronization of Josephson junctions in series [76], although it is unclear whether a distributed shunting resistance when every IJJ is connected to all others would provide any stronger synchronization. It requires further theoretical analysis.

It is possible to estimate some average value of R_{shunt} due to the presence of the normal-state parts in the mesa, by dividing an average voltage drop across the mesa height by the current integrated over the normal-state area. Determined in such way, $R_{\mathrm{shunt}} = 100$–$200\ \Omega$ is equivalent to 0.15–0.3 Ω per junction and is of the order of the intrinsic normal-state resistance at $T < T_c$.

5.3.7 The C-Axis Positive and Negative Magneto-Resistance in a Perpendicular Magnetic Field

The temperature dependence of the c-axis resistance $R_c(T)$ shows a peak just above T_c. The peak in $R_c(T)$ increases in magnitude with an applied perpendicular magnetic field, and the transition to the zero resistance state moves to lower temperature [16–22,77].

The conductance G of a Josephson junction can be divided in pair- (G_p), quasiparticle- (G_{qp}), and shunt (G_n) contributions, $G = G_p + G_{qp} + G_n$. G_n is assumed here to be due to the presence of defects that electrically short-circuit the junction. In the case of Bi2212 mesas, $R_c(T) = N\,G^{-1}(T)$. G_p and G_{qp} have opposite temperature dependences, and the competition between the two determines the position in temperature and the magnitude of the peak in $R_c(T)$ [78].

In a perpendicular magnetic field, no significant effect on $R_c(T)$ is expected at not too large fields because there is no Lorentz force acting upon Abrikosov vortices penetrating a superconductor when the current and the magnetic field are parallel. However, a discrepancy between the expectations and experiments is found in the layered nature of the HTS. The vortex lines in Bi2212 consist of pancake vortices which tend to stack on top of each other. Irregular pinning potential from oxygen vacancies or crystal defects results in breaking the straightness of a vortex line when a pancake jump to a nearby pinning center. The straight vortex line turns into a zigzag one with a nonzero in-plane component of the magnetic field, producing a local phase variation and a depression of the c-axis critical current and G_p [79,80]. At low temperatures, G_{qp} vanishes as well, and G_n takes over the electrical transport across IJJs. Thus, $R_c(T)$ should saturate at low temperature and high magnetic field. Moreover, the $R_c(T)$-peak magnitude should reflect the quality of the crystal. The fewer the defects in the crystal, the stronger the magnetoresistance peak. Therefore, it is always stronger in small-volume mesas than in single crystals, compare [23,81] with [77,82–84] (Fig. 5.2).

At a given temperature below T_c, resistance first increases in the perpendicular magnetic field above the irreversibility H-T line. However, the magnetoresistance becomes negative in very strong magnetic fields $\mu_B H > 10$ T [10,85].

Assuming d-wave superconductivity, the negative magnetoresistance can be explained by the increased density of electronic states (DOS) due to a shifted quasiparticle energy caused by the in-plane supercurrents around vortices (Doppler shift) [86]. The theory predicts a linear increase in the quasiparticle conductivity at high fields (>0.2–0.6 T) and low temperature: $\sigma_{qp}(H,T)/\sigma_{qp}(0,T) = \mathrm{C}_1(T) + H/H_\Delta$, where $\mu_0 H_\Delta = (\Delta_0/\hbar v_F)^2 \Phi_0$, Δ_0 is the maximum superconducting energy gap parameter, v_F is the Fermi velocity, Φ_0 is the magnetic flux quantum, $\hbar$ is the Planck constant, and $\mathrm{C}_1(T) \sim 1$; $\mu_0 H_\Delta \sim 40$–80 T [87].

Another reason for the negative magnetoresistance was suggested in [88]. It was assumed that the part of resistivity due to the PG, $\Delta\rho_c$, defined as excess resistivity above linear extrapolation of metallic $\rho_c(T)$ from high temperatures could be totally suppressed by the pseudogap quenching field H_{PG}. Using a simple Zeeman energy scaling, the value of H_{PG} could be expressed in Kelvin and thereby compared with T_c [88]. Using single crystals of different doping p, a pseudogap phase diagram could be obtained (Fig. 5.13b). It showed that the $H_{PG}(p)$ line is above the $T_c(p)$ one and gets closer to it at high p. These findings were concluded to "indicate a predominant role of spins over the orbital effects in the formation of the pseudogap" [88].

The negative magnetoresistance effect seen in Bi2212 *mesas* is by a factor of 2–8 larger than in single crystals (Fig. 5.13a). Each of the straight lines represents the maximal effect seen in the single crystals of the corresponding work. The lines have slopes $s = \mathrm{d}\ \ln(\Delta\rho_c)/\mathrm{d}B$ which are smaller than 0.010–0.014 T^{-1}, see also [89]. On the contrary, the mesa data presented in Fig. 5.13a have on average $s \approx 0.032\ \mathrm{T}^{-1}$, see for example, curve 7.

This can have crucial consequences for the PG phase diagram shown in Fig. 5.13b. Using curve 7 in Fig. 5.13a, we estimate the PG closing magnetic field H_{PG} to be about 31 T for this sample. This field corresponds to $T_{PG} = (\mu_B g/k_B)H_{PG} = 43\,\mathrm{K}$ which is lower than $T_c \approx 86\,\mathrm{K}$.

First, this means that the $H_{PG}(p)$ line for mesas crosses the $T_c(p)$ curve, in contrast to [88]. Second, there is no correspondence between H_{PG} and the temperature corresponding to the minimum in $R_c(T)$, commonly taken as the characteristic PG temperature [88]. It is more likely that the negative magnetoresistance effect is due to the Doppler-shift- and d-wave-superconductivity-related mechanism [87], rather than to a suppression of the PG by the magnetic field.

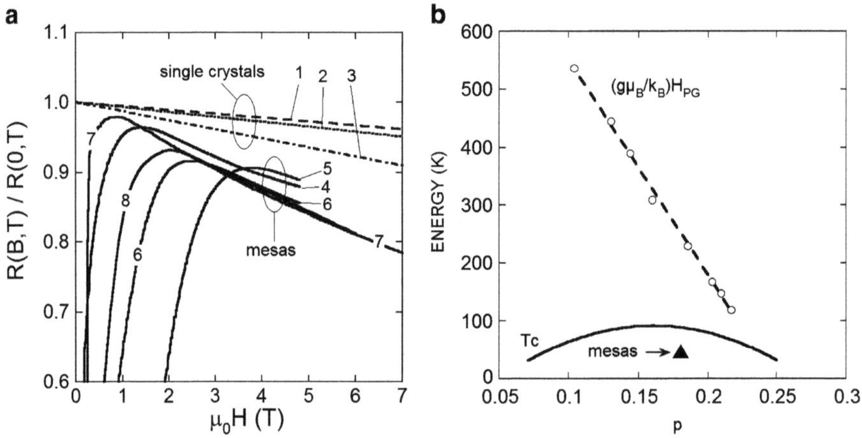

Fig. 5.13 (**a**) Negative-magnetoresistance in Bi2212 single crystals (*lines* 1–3) and mesas (*curves* 4–8). 1, 2, and 3: Bi2212 single crystals at ~50 K from [10], [85], and [88], respectively; 4, 5, and 6 – Bi2212 mesas after heavy-ion irradiation, at $T = 80$, 84, and 75 K, respectively [23]; 7– a $HgBr_2$-Bi2212 mesa at 10 K; 8 – overdoped-Bi2212 mesa from [12]. (**b**) Doping dependence of T_c (the *solid line*) and the pseudogap-quenching field H_{PG} (the *dashed line* and *solid triangle*). $H_{PG}(p)$ for single crystals is adapted from [88]. Note much smaller H_{PG} for mesas. μ_0, μ_B, and k_B are the permeability of free space, Bohr magneton, and Boltzmann constant, respectively; $g = 2$ is the g-factor

5.4 Summary

Intrinsic Josephson junctions are naturally formed inside single crystals of high-temperature superconductors. They represent one of the best types of tunneling junctions regarding uniformity and smoothness of interfaces. Small-volume mesas are favorable for studies of c-axis transport because of much fewer defects than in experiments on larger crystals.

In this work, we have shown that they allow measurements of the genuine in-plane critical current density of Bi2212 and give cardinally different result regarding the suggestive quenching of the pseudogap by the magnetic field. The strong negative c-axis magnetoresistance in mesas gives a value of the PG-quenching magnetic field which, when expressed on a Kelvin temperature scale, is smaller than T_c. This fact favors the existence of the quantum critical point in Bi2212 and the coexistence of PG and superconductivity.

Intrinsic Josephson junctions can be used for generation and detection of coherent THz microwaves. Self-heating at high bias current results in highly nonuniform temperature distribution in the mesas; this can be decisive in synchronizing the radiating junctions in a mesa stack.

The stacks with only a few active intrinsic Josephson junctions reveal the existence of reproducible low-energy peaks of tunneling conductance that can be signatures of either a phonon-assisted or resonant intrinsic tunneling.

Acknowledgements The authors are grateful for support from the Swedish Research Council through the Linnaeus Centrum "Engineering quantum systems" and The Knut and Alice Wallenberg Foundation and acknowledge valuable contributions and discussions from L.-X. You, M. Torstensson, T. Claeson, I. Kakeya, K. Kadowaki, and V. Krasnov.

References

1. R. Kleiner, F. Steinmeyer, G. Kunkel, P. Müller, Phys. Rev. Lett. **68**, 2394 (1992)
2. R. Kleiner, P. Müller, Phys. Rev. B **49**, 1327 (1994)
3. A. Yurgens, Supercond. Sci. Technol. **13**, R85 (2000)
4. A. Yurgens, V.M. Krasnov, D. Winkler, T. Claeson, in *Studies of Josephson Junction Arrays*, ed. by A. Narlikar (Nova Science Publ., Inc., Huntington, NY, 2001), p. 270
5. K. Schlenga, G. Hechtfischer, R. Kleiner, W. Walkenhorst, P. Müller, H.L. Johnson, M. Veith, W. Brodkorb, E. Steinbeiß, Phys. Rev. Lett. **76**, 4943 (1996)
6. K. Schlenga, R. Kleiner, G. Hechtfischer, M. Mößle, S. Schmitt, P. Müller, C. Helm, C. Preis, F. Forsthofer, J. Keller, H.L. Johnson, M. Veith, E. Steinbeiß, Phys. Rev. B**57**, 14518 (1998)
7. A. Yurgens, D. Winkler, T. Claeson, S.-J. Hwang, J.-H. Choy, Int. J. Mod. Phys. B **13**, 3758 (1999)
8. V.M. Krasnov, A. Yurgens, D. Winkler, P. Delsing, T. Claeson, Phys. Rev. Lett. **84**, 5860 (2000)
9. V.M. Krasnov, A.E. Kovalev, A. Yurgens, D. Winkler, Phys. Rev. Lett. **86**, 2657 (2001)
10. N. Morozov, L. Krusin-Elbaum, T. Shibauchi, L.N. Bulaevskii, M.P. Maley, Y.I. Latyshev, T. Yamashita, Phys. Rev. Lett. **84**, 1784 (2000)
11. N. Morozov, L.N. Bulaevskii, M.P. Maley, Y.I. Latyshev, T. Yamashita, Phys. Rev. B **62**, R14681 (2000)

12. A. Yurgens, D. Winkler, T. Claeson, G. Yang, I.F.G. Parker, C.E. Gough, Phys. Rev. B **59**, 7196 (1999)
13. H.B. Wang, Y. Aruga, T. Tachiki, Y. Mizugaki, J. Chen, K. Nakajima, T. Yamashita, P.H. Wu, Appl. Phys. Lett. **74**, 3693 (1999)
14. H.B. Wang, Y. Aruga, J. Chen, K. Nakajima, T. Yamashita, P.H. Wu, Appl. Phys. Lett. **77**, 1017 (2000)
15. H.B. Wang, P.H. Wu, T. Yamashita, Phys. Rev. Lett. **87**, 107002 (2001)
16. I.E. Batov, X.Y. Jin, S.V. Shitov, Y. Koval, P. Muller, A.V. Ustinov, Appl. Phys. Lett. **88**, 262504 (2006)
17. L. Ozyuzer, A.E. Koshelev, C. Kurter, N. Gopalsami, Q. Li, M. Tachiki, K. Kadowaki, T. Yamamoto, H. Minami, H. Yamaguchi, T. Tachiki, K.E. Gray, W.-K. Kwok, U. Welp, Science **318**, 1291 (2007)
18. K. Kadowaki, H. Yamaguchi, K. Kawamata, T. Yamamoto, H. Minami, I. Kakeya, U. Welp, L. Ozyuzer, A. Koshelev, C. Kurter, K.E. Gray, W.K. Kwok, Physica C **468**, 634 (2008)
19. L.X. You, A. Yurgens, D. Winkler, Phys. Rev. B **71**, 224501 (2005)
20. A. Yurgens, D. Winkler, T. Claeson, T. Murayama, Y. Ando, Phys. Rev. Lett. **82**, 3148 (1999)
21. K. Inomata, S. Sato, K. Nakajima, A. Tanaka, Y. Takano, H.B. Wang, M. Nagao, H. Hatano, S. Kawabata, Phys. Rev. Lett. **95**, 107005 (2005)
22. X.Y. Jin, J. Lisenfeld, Y. Koval, A. Lukashenko, A.V. Ustinov, P. Muller, Phys. Rev. Lett. **96**, 177003 (2006)
23. A. Yurgens, M. Konczykowski, N. Mros, D. Winkler, T. Claeson, Phys. Rev. B **60**, 12480 (1999)
24. M. Sakai, A. Odagawa, H. Adachi, K. Setsune, Physica C **293**, 55 (1997)
25. K. Endo, P. Badica, H. Sato, H. Akoh, Supercond. Sci. Technol. **19**, S221 (2006)
26. A. Yurgens, D. Winkler, N.V. Zavaritsky, T. Claeson, Appl. Phys. Lett. **70**, 1760 (1997)
27. H.M. Meyer, J.H. Weaver, in *Physical Properties of High Temperature Superconductors*, vol. II, ed. by D.M. Ginsberg (World Scientific Publishing Co., Singapore, 1990), p. 369
28. A.V. Gurevich, R.G. Mints, Rev. Mod. Phys. **59**, 941 (1987)
29. H.B. Wang, P.H. Wu, T. Yamashita, Appl. Phys. Lett. **78**, 4010 (2001)
30. A. Yurgens, T. Claeson, Physica C **341–348**, 2277 (2000)
31. Y.I. Latyshev, T. Yamashita, L.N. Bulaevskii, M.J. Graf, A.V. Balatsky, M.P. Maley, Phys. Rev. Lett. **82**, 5345 (1999)
32. P.A. Warburton, J.C. Fenton, M. Korsah, C.R.M. Grovenor, Supercond. Sci. Technol. **19**, S187 (2006)
33. K. Nakajima, S. Sudo, T. Tachiki, T. Yamashita, IOP Conf. Series **167**, 273 (2000)
34. B. Verreet, N. Sergeant, D.M. Negrete, M. Torstensson, D. Winkler, A. Yurgens, Supercond. Sci. Technol. **20**, S48 (2007)
35. H.B. Wang, T. Hatano, T. Yamashita, P.H. Wu, P. Müller, Appl. Phys. Lett. **86**, 023504 (2005)
36. L.X. You, A. Yurgens, D. Winkler, Phys. Rev. B **71**, 224501 (2005)
37. V.V. Schmidt, *The Physics of Superconductors* (Springer, Berlin Heidelberg New York, 1997)
38. C.B. Duke, in *Solid State Physics. Advances in Research and Applications*, vol. 10, ed. by F. Seitz, D. Turnbull, H. Ehrenreich (Academic, New York and London, 1969)
39. E.L. Wolf, *Principles of Electron Tunneling Spectroscopy* (Oxford University Press, New York, 1985)
40. G. Binnig, H. Rohrer, C. Gerber, E. Weibel, Phys. Rev. Lett. **50**, 120 (1983)
41. R. Wiesendanger, *Scanning Probe Microscopy and Spectroscopy: Methods and Applications* (Cambridge University Press, Cambridge, 1994)
42. T. Hasegawa, H. Ikuta, K. Kitazawa, in *Physical Properties of High Temperature Superconductors III*, ed. by D.M. Ginsberg (World Scientific Publishing Co., Singapore, 1992), pp. 525
43. K.M. Lang, V. Madhavan, J.E. Hoffman, E.W. Hudson, H. Eisaki, S. Uchida, J.C. Davis, Nature **415**, 412 (2002)
44. K. McElroy, J. Lee, J.A. Slezak, D.-H. Lee, H. Eisaki, S. Uchida, J.C. Davis, Science **309**, 1048 (2005)

45. C. Renner, B. Revaz, J.-Y. Genoud, K. Kadowaki, Ø. Fisher, Phys. Rev. Lett. **80**, 149 (1998)
46. G.L. deCastro, C. Berthod, A. Piriou, E. Giannini, Ø. Fischer, Phys. Rev. Lett. **101**, 267004 (2008)
47. C. Renner, Ø. Fischer, Phys. Rev. B **51**, 9208 (1995)
48. M. Suzuki, T. Watanabe, A. Matsuda, Phys. Rev. Lett. **82**, 5361 (1999)
49. M. Suzuki, T. Watanabe, Phys. Rev. Lett. **85**, 4787 (2000)
50. V.M. Krasnov, A. Yurgens, D. Winkler, P. Delsing, J. Appl. Phys. **89**, 5578 (2001)
51. L.X. You, M. Torstensson, A. Yurgens, D. Winkler, C.T. Lin, B. Liang, Appl. Phys. Lett. **88**, 222501 (2006)
52. E.G. Maximov, P.I. Arseyev, N.S. Maslova, Solid State Commun. **111**, 391 (1999)
53. O.V. Misochko, E.Y. Sherman, N. Umesaki, K. Sakai, S. Nakashima, Phys. Rev. B **59**, 11495 (1999)
54. M.G. Blamire, E.C.G. Kirk, J.E. Evetts, T.M. Klapwijk, Phys. Rev. Lett. **66**, 220 (1991)
55. D.R. Heslinga, T.M. Klapwijk, Phys. Rev. B **47**, 5157 (1993)
56. G. Johansson, V.S. Shumeiko, E.N. Bratus, G. Wendin, Physica C **293**, 77 (1997)
57. G. Johansson, E.N. Bratus, V.S. Shumeiko, G. Wendin, Phys. Rev. B **60**, 1382 (1999)
58. M. Tonouchi, Nat. Photonics **1**, 97 (2007)
59. R.P. Robertazzi, R.A. Buhrman, Appl. Phys. Lett. **53**, 2441 (1988)
60. K. Wan, A.K. Jain, J.E. Lukens, Appl. Phys. Lett. **54**, 1805 (1989)
61. L.N. Bulaevskii, A.E. Koshelev, Phys. Rev. Lett. **99**, 057002 (2007)
62. E. Kume, I. Iguchi, H. Takahashi, Appl. Phys. Lett. **75**, 2809 (1999)
63. K. Lee, I. Iguchi, Physica C **367**, 376 (2002)
64. K. Lee, W. Wang, I. Iguchi, M. Tachiki, K. Hirata, T. Mochiku, Phys. Rev. B **61**, 3616 (2000)
65. Y. Ando, J. Takeya, Y. Abe, K. Nakamura, A. Kapitulnik, Phys. Rev. B **62**, 626 (2000)
66. K. Krishana, N.P. Ong, Q. Li, G.D. Gu, N. Koshizuka, Science **277**, 83 (1997)
67. C. Uher, in *Physical Properties of High Temperature Superconductors*, vol. III, ed. by D.M. Ginzberg (World Scientific Publishing Co., Singapore 1992)
68. N.V. Zavaritsky, A.V. Samoilov, A. Yurgens, Physica C **169**, 174 (1990)
69. M.F. Crommie, A. Zettl, Phys. Rev. B **43**, 408 (1991)
70. Y.I. Latyshev, V.N. Pavlenko, S.-J. Kim, T. Yamashita, L.N. Bulaevskii, M.J. Graf, A.V. Balatsky, N. Morozov, M.P. Maley, Physica C **341**, 1499 (2000)
71. A. Yurgens, D. Winkler, N.V. Zavaritsky, T. Claeson, Phys. Rev. Lett. **79**, 5122 (1997)
72. D.G. Cahill, R.O. Pohl, Phys. Rev. B **35**, 4067 (1987)
73. G. Hartwig, in *Handbook of Cryogenic Engineering*, ed. by J.G. Weisend (Taylor & Francis, Philadelphia, 1998)
74. A. Yurgens, D. Winkler, T. Claeson, N. Zavaritsky, in *European Conference on Applied Superconductivity, EUCAS'95*, ed. by D. Dew-Huges (IOP Publishing Ltd., Edinburgh, 3–6 July, 1995), p. 1423
75. K. Kadowaki, H. Yamaguchi, K. Kawamata, T. Yamamoto, H. Minami, I. Kakeya, U. Welp, L. Ozyuzer, A. Koshelev, C. Kurter, K.E. Gray, W.-K. Kwok, Physica C **468**, 634 (2008)
76. P. Hadley, M.R. Beasley, K. Wiesenfeld, Phys. Rev. B **38**, 8712 (1988)
77. G. Briceno, M.F. Crommie, A. Zettl, Phys. Rev. Lett. **66**, 2164 (1991)
78. K.E. Gray, D.H. Kim, Phys. Rev. Lett. **70**, 1693 (1993)
79. L.N. Bulaevskii, V.L. Pokrovsky, M.P. Maley, Phys. Rev. Lett. **76**, 1719 (1996)
80. A.E. Koshelev, L.I. Glasman, A.I. Larkin, Phys. Rev. B **53**, 2786 (1996)
81. A. Yurgens, D. Winkler, N.V. Zavaritsky, T. Claeson, Phys. Rev. Lett. **79**, 5122 (1997)
82. M.C. Hellerqvist, S. Ryu, L.W. Lombardo, A. Kapitulnik, Physica C **230**, 170 (1994)
83. J.H. Cho, M.P. Maley, S. Fleshler, A. Lacerda, L.N. Bulaevskii, Phys. Rev. B **50**, 6493 (1994)
84. A.S. Alexandrov, V.N. Zavaritsky, W.Y. Liang, P.L. Nevsky, Phys. Rev. Lett. **76**, 983 (1996)
85. V.N. Zavaritsky, M. Springford, A.S. Alexandrov, Europhys. Lett. **51**, 334 (2000)
86. G.E. Volovik, JETP Lett. **58**, 469 (1993)
87. I. Vekhter, L.N. Bulaevskii, A.E. Koshelev, M.P. Maley, Phys. Rev. Lett. **84**, 1296 (2000)
88. T. Shibauchi, L. Krusin-Elbaum, M. Li, M.P. Maley, P.H. Kes, Phys. Rev. Lett. **86**, 5763 (2001)
89. Y.F. Yan, P. Matl, J.M. Harris, N.P. Ong, Phys. Rev. B **52**, R751 (1995)

Chapter 6
Stacked Josephson Junctions

S. Madsen, N.F. Pedersen, and P.L. Christiansen

Abstract Long Josephson junctions have for some time been considered as a source of THz radiation. Solitons moving coherently in the junctions is a possible source for this radiation. Analytical computations of the bunched state and bunching-inducing methods are reviewed. Experiments showing THz radiation have recently been reported and are discussed at the end of this chapter.

6.1 Introduction

The fluxon dynamics in long Josephson junctions is a topic that is of strong interest because it is a very special nonlinear system and because it may have applications in fast electronics. An extension of that system is to have several junctions stacked on top of each other giving rise to a system of N-coupled partial differential equations. Such a system is often used as a model for high temperature superconductors [1, 2]. In this chapter, we go even one step further in complexity and include results on a stacked junction coupled to a cavity. Such a model is needed to understand and interpret the experimental measurements. For frequencies in the GHz, or even THz range, a cavity is often needed to enhance the radiated power to useful levels.

This chapter, is structured in the following way. In Sect. 6.2 the model is described. Numerical methods and analytical solutions are discussed. Section 6.3 discusses some methods to obtain bunched fluxons which is needed obtain useful radiation levels. In Sect. 6.4, we relate some experimental results to our model. Section 6.5 gives our conclusions.

6.2 Model

Our starting point is the model by Sakai, Bodin, and Pedersen [1] describing the interactions between a general system of N junctions. In the following, this model is reviewed, following [1] for the case of identical junctions. A similar type of

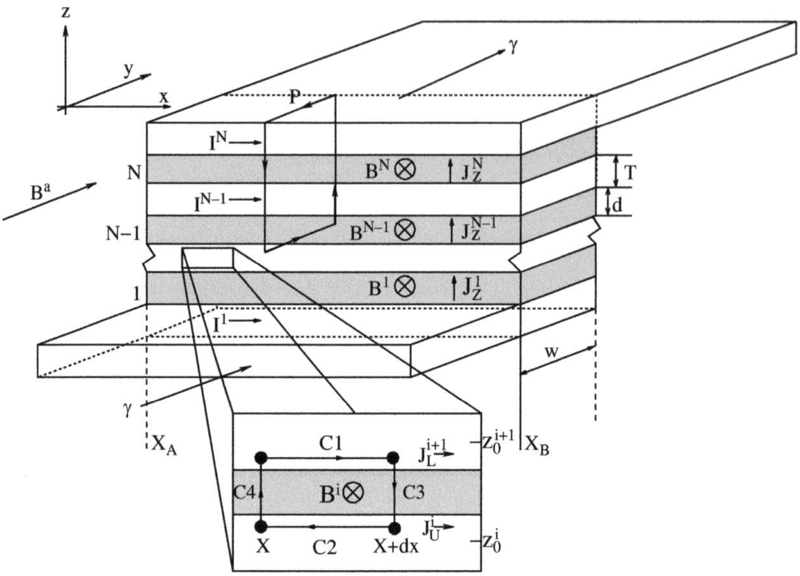

Fig. 6.1 Schematic view of the stacked Josephson junctions. *White layers* are superconductors and *gray insulators*. B^a is an applied magnetic field in the y-direction and η is an applied bias current

derivation can be found in [2], although Kleiner and Müller only consider the static case. The model geometry is sketched in Fig. 6.1. There are $N + 1$ superconductors in the stack resulting in N Josephson junctions. The superconductors have thickness T, the insulators have thickness d, the width is w, and the length is $x_B - x_A = L$. Each junction is characterized by a superconductor phase, θ, satisfying

$$\nabla\theta = \frac{\lambda^2 \mu_0 q}{\hbar}\mathbf{J} + \frac{q}{\hbar}\mathbf{A}, \tag{6.1}$$

where λ is the London penetration depth, μ_0 is the vacuum permeability, $\hbar$ is Planck's constant divided by 2π, q is the Cooper-pair charge, $\mathbf{A}$ the vector potential, and $\mathbf{J}$ the current density. Integrating the phase along C_1 and C_2 in Fig. 6.1 and adding the integral of $\mathbf{A}$ along C_3 and C_4 result in

$$\frac{\hbar}{\mu_0 q}\frac{\partial \phi^i}{\partial x} = \lambda^2\left(J_L^{i+1} - J_U^i\right) + \frac{1}{\mu_0}B^i d, \tag{6.2}$$

where J_L^{i+1} is the current density at the lower edge of superconductor $i+1$, J_U^i is the current density along the upper edge of superconductor i, and

$$\phi^i(x) \equiv \theta^{i+1}(x) - \theta^{i-1}(x) - \frac{q}{\hbar}\int_{z_i}^{z_i+d} A_z(x,z)\mathrm{d}z \tag{6.3}$$

is the gauge invariant phase difference across junction number i. The screening current densities, J_L^{i+1} and J_U^i, may be related to the magnetic field between the two superconductors by solving the second London Equation to obtain the magnetic fields inside the superconductors. For the geometry in Fig. 6.1, the solution becomes

$$B^i = \frac{B^{i-1}\sinh\left(\frac{T-2z-2z_0^i}{2\lambda}\right) + B^i\sinh\left(\frac{T+2z+2z_0^i}{2\lambda}\right)}{\sinh(T/\lambda)} \tag{6.4}$$

leading to the current densities

$$J_L^{i+1} = \frac{-1}{\mu_0}\frac{\partial B^{i+1}}{\partial z}\bigg|_{z=z_0^{i+1}-T/2} = \frac{B^i\cosh(T/\lambda) - B^{i+1}}{\mu_0\lambda\sinh(T/\lambda)} \tag{6.5}$$

and

$$J_U^i = \frac{-1}{\mu_0}\frac{\partial B^i}{\partial z}\bigg|_{z=z_0^i+T/2} = \frac{B^{i-1} - B^i\cosh(T/\lambda)}{\mu_0\lambda\sinh(T/\lambda)}. \tag{6.6}$$

Equation (6.2) can now be written as

$$\frac{\hbar}{q}\frac{\partial\phi^i}{\partial x} = d'B^i + s\left(B^{i-1} + B^{i+1}\right) \tag{6.7}$$

with the definitions

$$d' \equiv d + \frac{2\lambda\cosh(T/\lambda)}{\sinh(T/\lambda)} \quad \text{and} \quad s \equiv -\frac{\lambda}{\sinh(T/\lambda)}. \tag{6.8}$$

These equations are valid for all junctions, $i = 1,\ldots,N$, and we define $B^0 = B^{N+1} \equiv B_a$. The corresponding equation for the electric field is obtained using $\nabla\times\mathbf{E} = -\partial\mathbf{B}/\partial t$, resulting in

$$\frac{\hbar}{q}\frac{\partial\phi^i}{\partial t} = d'E^i + s\left(E^{i-1} + E^{i+1}\right), \tag{6.9}$$

where E^i is the electric field in the z-direction in the ith junction and $E^0 = E^{N+1} \equiv 0$ are defined.

Integrating the magnetic induction along the path P gives

$$\oint_P \mathbf{B}\cdot\mathbf{dl} = w\left(B^i - B^a\right) = \mu_0\sum_{n=i+1}^{N+1} I^n, \tag{6.10}$$

where I^n is the current along the x-direction in the n'th superconducting layer. This current may be related to the current in the z-direction using $\nabla\times\mathbf{B} = \mu_0\mathbf{J}$ with $\mathbf{B} = B_y\hat{y}$, resulting in

$$\frac{\partial J_x}{\partial x} = -\frac{\partial J_z}{\partial z}. \tag{6.11}$$

Using $\partial J_z^i/\partial z|_{z=z_i} \approx (J_z^i - J_z^{i-1})/T$ and $I^i \approx wTJ_x^i$, the relation

$$\frac{1}{w}\sum_{n=k}^{N+1}\frac{\partial I^n}{\partial x} = J_z^{k-1}, \tag{6.12}$$

valid for $1 \leq k \leq N$ is obtained, with z_i being the middle of the i'th superconducting layer. By differentiating (6.7) with respect to x, the system of equations can now be written as [1]

$$\frac{\hbar}{\mu_0 q}\frac{\partial^2 \boldsymbol{\phi}}{\partial x^2} = \mathbf{S}\,\mathbf{J}_z\,, \tag{6.13}$$

where the i'th element of $\boldsymbol{\phi}$ is ϕ^i, and the $N \times N$ coupling matrix, $\mathbf{S}$, has the form [1]

$$\mathbf{S} = \begin{pmatrix} d' & s & & & & \\ s & d' & s & & & \\ & s & d' & s & & \\ & & \ddots & \ddots & \ddots & \\ & & & s & d' & s \\ & & & & s & d' \end{pmatrix}, \tag{6.14}$$

and the i'th element of $\mathbf{J}_z$ is the current across junction i in the z-direction, J_z^i. Comparing with the perturbed sine-Gordon equation of a single junction, the total current density in the z-direction can be written as [1]

$$J_z^i = \frac{\hbar C}{q}\frac{\partial^2 \phi^i}{\partial t^2} + \frac{\hbar}{qR}\frac{\partial \phi^i}{\partial t} + J_\mathrm{J}\sin\phi^i - J_\mathrm{B}, \tag{6.15}$$

where $C \approx \varepsilon_0\varepsilon_r/d'$ is the unit area capacitance, R is the unit area resistance, J_J is the maximum Josephson current density, and J_B is an added bias current density.

Boundary conditions are obtained by evaluating (6.7) at $x = 0$ and L, giving [1]

$$\frac{\hbar}{q}\frac{\partial \phi^i}{\partial x}\bigg|_{x=x_A,x_B} = (d' + 2s)B^a \tag{6.16}$$

showing that a uniform applied magnetic field enters through the boundaries.

It is often more convenient to work with normalized equations. The equations for the long Josephson stack can be normalized by introducing the following dimensionless variables and quantities

$$x \to \lambda_J x = \sqrt{\frac{\hbar}{q\mu_0 J_J d'}}\, x, \qquad t \to \omega_0^{-1} t = \sqrt{\frac{\hbar C}{q J_J}}\, t, \tag{6.17}$$

$$B \to \sqrt{\frac{\hbar \mu_0 J_J}{q d'}}\, B, \qquad J_z^i \to J_J J_z^i, \tag{6.18}$$

where λ_J is the Josephson penetration depth and ω_0 is the Josephson plasma frequency. Defining

$$S \equiv \frac{s}{d'} \quad , \quad \alpha \equiv \sqrt{\frac{\hbar}{q J_J C R^2}} \quad , \quad \eta \equiv \frac{J_B}{J_J}, \tag{6.19}$$

the set of normalized equations for the inductively coupled stacked long Josephson junctions become

$$\frac{\partial^2 \boldsymbol{\phi}}{\partial x^2} = \mathbf{S}\, \mathbf{J}_z\,, \tag{6.20}$$

$$J_z^i = \frac{\partial^2 \phi^i}{\partial t^2} + \alpha \frac{\partial \phi^i}{\partial t} + \sin \phi^i - \eta, \tag{6.21}$$

$$\mathbf{S} = \begin{pmatrix} 1 & S & & & & \\ S & 1 & S & & & \\ & S & 1 & S & & \\ & & \ddots & \ddots & \ddots & \\ & & & S & 1 & S \\ & & & & S & 1 \end{pmatrix}, \tag{6.22}$$

with boundary conditions

$$\left.\frac{\partial \phi^i}{\partial x}\right|_{x=x_A, x_B} = (1 + 2S) B^a. \tag{6.23}$$

Frequently in analytical calculations, the length of the system is assumed to be infinite and the boundary conditions in (6.23) can of course not be used. To model an infinite system, periodic boundary conditions on a finite system of length L are often used in numerical computations. In the case of fluxon-solutions, these read

$$\phi^i(0, t) = 2\pi + \phi^i(L, t), \tag{6.24}$$

for $i = 1, \ldots, N$, since the fluxon will provide a phase-shift of 2π along the junction.

For the simplest low T_c stack and a typical high T_c stack, the parameters may be [3]

Material	N	λ	λ_J	c_0	J_J	
$NbAlO_x$	2	90 nm	25 μm	7×10^6 m/s*	200 A/cm^2	
$Bi_2Sr_2CaCu_2O_8$	120	170 nm	0.34 μm	4×10^5 m/s*	1015A/cm^2	
Material	L	w	d	T	S	ω_0
$NbAlO_x$	200 μm	40 μm	5 nm*	30 nm	−0.46*	100GHz*
$Bi_2Sr_2CaCu_2O_8$	10 μm	1.8 μm	12Å	3Å	−0.5*	500GHz*

The values marked with * have been estimated from the values given in [3, 4].

It is easily seen that the width of both the insulators, d, and the superconductors, T, is somewhat different in the two examples shown here. Also, the London penetration depth is about two times larger in the intrinsic stack than in the artificial stack. The smaller width of the insulator and the larger London penetration depth of the intrinsic stack lead to the larger absolute value of S because the magnetic field from surrounding junctions penetrates the superconductors more easily. It should be noted that the values provided above are temperature dependent, and the strength of the inductive coupling in the low T_c stack may easily be made much smaller, e.g., by increasing T.

6.2.1 Numerical Method

Equations (6.20)–(6.23) are straightforward to discretize in space using symmetric second-order finite differences with spacing Δx. The boundary conditions (6.23) are also discretized using second-order finite differences. This gives, for example, the following

$$\frac{\phi_1^i(t) - \phi_{-1}^i(t)}{2\Delta x} = \Gamma \tag{6.25}$$

condition at $x = 0$, where the subscript refers to the x-coordinate by $\phi^i(x_n, t) \equiv \phi_n^i(t)$ with $x_n = n\Delta x$. The above condition enables one to determine the value of ϕ at the virtual point x_{-1} in such a way that the desired boundary condition is satisfied. In this case, it gives

$$\phi_{-1}^i(t) = \phi_1^i(t) - 2\Delta x\Gamma. \tag{6.26}$$

The $NL/\Delta x$ ordinary differential equations (ODEs) in t, resulting from the discretization of the spatial variable, can be effectively solved using a fifth-order Runge–Kutta method with adaptive step-size control [5]. Sometimes it is desired to have the solution at evenly spaced times, and a second-order Verlet integrator works efficiently [6].

6.2.2 Analytic Solutions

The uncoupled and unperturbed system, $S = \alpha = \eta = 0$ in (6.20)–(6.22), reduces to N identical equations which allow simple single fluxon solution given by the expression

$$\phi(x,t) = 4\arctan\left[\exp\sigma\,(x - vt - x_0)\,\gamma(v)\right], \tag{6.27}$$

where v is the velocity of the wave and $\gamma(v) = 1/\sqrt{1-v^2}$ is the Lorentz factor. It is called a fluxon ($\sigma = 1$) or an antifluxon ($\sigma = -1$), depending on the polarity. The maximum speed for one single equation is the velocity of the plasma waves, $|v| = 1$ [7]. A fluxon–antifluxon collission at $t = \tau$ and $x = x_0$ is described by the phase

$$\phi = 4\sigma\arctan\left[\frac{1}{v}\frac{\sinh\left(v(t-\tau)\gamma(v)\right)}{\cosh\left((x-x_0)\gamma(v)\right)}\right], \tag{6.28}$$

which also models the collission of a fluxon with a boundary at $x = x_0$ for the condition $\phi_x(t, x_0) = 0$. Several perturbation methods have been developed to study the model in detail [8,9].

The so-called Swihart velocities [10], c_m, can be found by a small-signal analysis on the system, as done in [11]. The result is

$$c_m = \frac{1}{\sqrt{1 - 2S\cos\left(\frac{m\pi}{N+1}\right)}}, \quad m = 1, \ldots, N. \tag{6.29}$$

The Swihart velocities enters the profiles in (6.27) and (6.28) as v/c_m in the case of coupled junctions. $c_\pm$ is usually used to refer to the lowest and highest velocities, i.e. $c_- = c_1$ and $c_+ = c_N$.

When a constant magnetic field is applied to the junctions, Fiske resonances and flux-flow solutions are found. A derivation of these solutions and comparison with experiments were done on a single junction by Cirillo et al. in [12]. Fiske-resonances and flux-flow solutions were treated for the case of two coupled junctions by Grønbech-Jensen et al. in [13]. The analytical solutions are derived using trial functions on the form

$$\phi = \phi_0 + \Gamma x + \omega t + \psi(x, \omega t) \tag{6.30}$$

where ϕ_0 is a constant, Γ is the applied field, and ω is related to the junction voltage. The solutions are found by assuming $\psi(x, \omega t)$ small and then expressing $\psi(x, \omega t)$ as a Fourier series in x. For more details, see [12] and [13] and references therein.

6.3 Bunching of Fluxons

In this section, we consider bunched fluxons in the case where the stability of the configuration is due to the coupling between the sine-Gordon equations governing the fluxon motion in Sect. 6.3.1, which is based on the work by Gorria et al. [14,15] for $N = 3$.

In Sect. 6.3.2, we then present results for the case where the bunching of the fluxons is due to high-Q cavity coupling at the boundaries of the stacked junctions. This subsection is based on the work by Madsen et al. [16,17].

6.3.1 Bunching due to Coupling Between Equations

For fluxon type solutions in each junction, the coupling effect leads to attraction between the center portion of the fluxons of opposite polarity and repulsion between the fluxons of identical polarity. This conclusion is easily obtained investigating the minimum of the potential energy of the system [18–24]. On the other hand, a unidirectional η force drives the fluxon and antifluxon in opposite directions [8].

We shall consider the case where $N = 3$ and the fluxon in the upper and lower layer has the samme polarity, given by $\sigma = \sigma_1$ in (6.27), while the polarity on the fluxon in the intermediate layer is denoted by σ_2. The corresponding traveling wave type solutions are $\phi^i(x,t) = \varphi_i(z)$, where $z = x - vt$, with $i = 1$ and 2.

In the fluxon–antifluxon–fluxon (f–a–f) case corresponding to $\sigma_1 = 1,\ \sigma_2 = -1$ in (6.27), the threshold between the driving and the coupling parameters for the bunched state is deduced in [15].

Here we shall restrict ourselves to the fluxon–fluxon–fluxon (f–f–f), corresponding to $\sigma_1 = \sigma_2 = 1$.

Inserting into (6.20)–(6.22) with $N = 3$, we find that he traveling waves satisfy

$$\begin{aligned}(\mu^2 - v^2)\varphi_1''(z) + \alpha v\varphi_1'(z) - \sin\varphi_1(z)\\ -2S\mu^2\varphi_2''(z) + \eta = 0,\\ (\mu^2 - v^2)\varphi_2''(z) + \alpha v\varphi_2'(z) - \sin\varphi_2(z)\\ -S\mu^2\varphi_1''(z) + \eta = 0,\end{aligned} \tag{6.31}$$

where $\mu^2 = 1/(1-2S^2) = c_-^2 \cdot c_+^2$.

Since (6.31) cannot be solved analytically, we shall use the piece-wise linear approximation for the nonlinear terms $\sin\varphi_i \to P(\varphi_i)$ in (6.31), where

$$P(\varphi) = \begin{cases} \varphi, & \text{for } \ 0 \le \varphi < \pi/2,\\ \pi - \varphi, & \text{for } \pi/2 \le \varphi < 3\pi/2,\\ \varphi - 2\pi, & \text{for } \ 3\pi/2 \le \varphi < 2\pi.\end{cases} \tag{6.32}$$

This was used in [25] to describe fluxon dynamics in a single junction with a surface loss term.

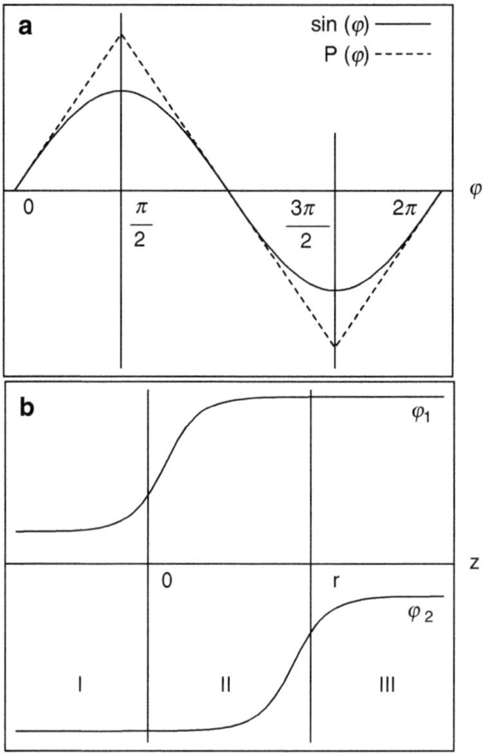

Fig. 6.2 (**a**) Piece-wise linear approximation $P(\varphi(z))$ (*dashed curve*) of the $\sin\varphi(z)$ function (*full curve*). (**b**) Fluxon, φ_2, forwarded a distance r from the fluxon, φ_1 due to the interaction from its two neighbors [15]

As seen in Fig. 6.2a, the continuous function $P(\varphi)$ provides a good approximation to the fluxon tails as $\varphi \to 0$ or 2π and also to the center portion of the fluxons at $\varphi \approx \pi$.

We now distinguish between three intervals:

$$\begin{aligned} &(\mathrm{I}):\ z \leq 0 \Longleftrightarrow 0 < \varphi_1 \leq \tfrac{\pi}{2} \\ &(\mathrm{II}):\ 0 < z \leq r \Longleftrightarrow \varphi_1 > \tfrac{\pi}{2} \text{ and } \varphi_2 \leq \tfrac{3\pi}{2} \\ &(\mathrm{III}):\ r < z \Longleftrightarrow \tfrac{3\pi}{2} < \varphi_2 \leq 2\pi. \end{aligned} \tag{6.33}$$

Inserting solutions of the type $\varphi_i(z) = D_i \mathrm{e}^{\lambda z} \mp \eta$ (+ in regions I and III and − in region II) in the linear system of ODEs obtained from (6.31) substituting $\sin\varphi_i \to P(\varphi_i)$, we get the characteristic equations

$$\left[\left(\frac{1}{1-2S^2} - v^2\right)\lambda^2 + \alpha v \lambda \pm 1\right]^2 = \frac{2S^2}{(1-2S^2)^2}\lambda^4. \tag{6.34}$$

The positive sign (+) corresponds to regions I and III and the negative one (−) to the region II. The roots of (6.34) in regions I and III are

$$\overline{\lambda}_{1,2} = -p_+ \pm \sqrt{p_+^2 + q_+},\ \overline{\lambda}_{3,4} = -p_- \pm \sqrt{p_-^2 + q_-}, \tag{6.35}$$

where $p_\pm$ and $q_\pm$ are given by

$$p_\pm = \frac{\alpha v}{2(c_\pm^2 - v^2)},\ q_\pm = \frac{1}{(c_\pm^2 - v^2)}. \tag{6.36}$$

For speeds higher than the Swihart velocity, $v > c_-$, the roots $\overline{\lambda}_{3,4}$ become complex and they are responsible for the emergence of oscillatory tails of the fluxons. For convenience, we denote $\overline{\lambda}_{3,4} = \overline{\lambda}_r \pm i\overline{\lambda}_l$, where $\overline{\lambda}_r = -p_-$ and $\overline{\lambda}_l = \sqrt{-(p_-^2 + q_-)}$ are real. In region II, the roots are

$$\overline{\lambda}_{5,6} = -p_+ \pm \sqrt{p_+^2 - q_+},\ \overline{\lambda}_{7,8} = -p_- \pm \sqrt{p_-^2 - q_-}. \tag{6.37}$$

The roots $\overline{\lambda}_{7,8}$ are real while $\overline{\lambda}_{5,6}$ are complex. For convenience, we denote as $\overline{\lambda}_{5,6} = \overline{\lambda}_m \pm i\overline{\lambda}_h$, where $\overline{\lambda}_m = -p_-$ and $\overline{\lambda}_h = \sqrt{-(p_-^2 + q_-)}$ are real. In the three regions, the eigenvalues corresponding to the solution of the characteristic equations are $(D_1, D_2) = (1, \sqrt{2})$ and $(D_1, D_2) = (1, -\sqrt{2})$ for the in-phase and the antiphase modes, respectively. The bounded solutions of the linearized version of (6.31) in regions I, II, and III become

$$(\mathrm{I}) \begin{cases} \hat{\varphi}_1(z) = \overline{H}_1 e^{\overline{\lambda}_1 z} + e^{\overline{\lambda}_r z}(\overline{H}_2 \cos\overline{\lambda}_l z \\ \qquad + \overline{H}_3 \sin\overline{\lambda}_l z) + \eta, \\ \hat{\varphi}_2(z) = \sqrt{2}(\overline{H}_1 e^{\overline{\lambda}_1 z} - e^{\overline{\lambda}_r z}(\overline{H}_2 \cos\overline{\lambda}_l z \\ \qquad - \overline{H}_3 \sin\overline{\lambda}_l z)) + \eta, \end{cases} \tag{6.38}$$

$$(\mathrm{II}) \begin{cases} \hat{\varphi}_1(z) = e^{\overline{\lambda}_m z}(\overline{G}_1 \cos\overline{\lambda}_h z + \overline{G}_2 \sin\overline{\lambda}_h z) \\ \qquad + \overline{G}_3 e^{\overline{\lambda}_7 z} + \overline{G}_4 e^{\overline{\lambda}_8 z} + \pi - \eta, \\ \hat{\varphi}_2(z) = \sqrt{2}(e^{\overline{\lambda}_m z}(\overline{G}_1 \cos\overline{\lambda}_h z + \overline{G}_2 \sin\overline{\lambda}_h z) \\ \qquad - \overline{G}_3 e^{\overline{\lambda}_7 z} - \overline{G}_4 e^{\overline{\lambda}_8 z}) + \pi - \eta, \end{cases} \tag{6.39}$$

$$(\mathrm{III}) \begin{cases} \hat{\varphi}_1(z) = \overline{K}_1 e^{\overline{\lambda}_2 z} + 2\pi + \eta, \\ \hat{\varphi}_2(z) = \sqrt{2}(\overline{K}_1 e^{\overline{\lambda}_2 z}) + 2\pi + \eta. \end{cases} \tag{6.40}$$

The eight constants $(\overline{H}_1, \overline{H}_2, \overline{H}_3, \overline{G}_1, \overline{G}_2, \overline{G}_3, \overline{G}_4, \overline{K}_1)$ are determined by the ten matching conditions at the points, $z = 0$ and $z = r$, and the corresponding equations are

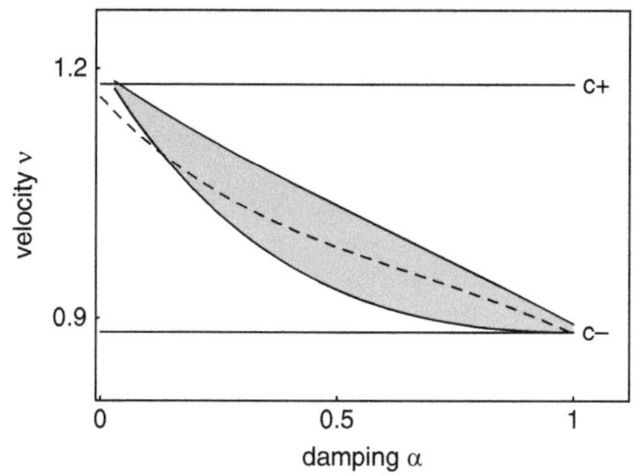

Fig. 6.3 The *gray*-shaded area indicates the region in the corresponding (α, v) space for $S = -0.2$, where bunching of (f–f–f) occurs in the numerical simulations. The bottom contour of this region is approximated by piece-wise linear approximation (*dashed line*) but top contour cannot be found by this approach [15]

$$\begin{gathered}\hat{\varphi}_1^I(0) = \hat{\varphi}_1^{II}(0) = \frac{\pi}{2},\ \hat{\varphi}_2^{II}(r) = \hat{\varphi}_2^{III}(r) = \frac{3\pi}{2},\\ \hat{\varphi}_2^I(0) = \hat{\varphi}_2^{II}(0),\ \hat{\varphi}_1^{II}(r) = \hat{\varphi}_1^{III}(r),\\ \frac{\mathrm{d}\hat{\varphi}_i^I(0)}{\mathrm{d}z} = \frac{\mathrm{d}\hat{\varphi}_i^{II}(0)}{\mathrm{d}z},\ \frac{\mathrm{d}\hat{\varphi}_i^{II}(r)}{\mathrm{d}z} = \frac{\mathrm{d}\hat{\varphi}_i^{III}(r)}{\mathrm{d}z},\ i = 1, 2.\end{gathered} \tag{6.41}$$

The remaining two conditions can be written in the following way as function of $\overline{G}_1$ and $\overline{G}_4$ obtained previously by (6.41)

$$\begin{gathered}\left(\frac{\mathrm{e}^{\overline{\lambda}_m r}}{\cos \overline{\lambda}_h r}\right)\overline{G}_1 = -\left(\frac{\overline{\lambda}_m + \dfrac{\overline{\lambda}_2}{\sqrt{2}}}{\overline{\lambda}_h} - \tan \overline{\lambda}_h r\right)\left(\eta + \frac{\pi}{2}\right),\\ \left(\frac{\overline{\lambda}_8}{\overline{\lambda}_7} - 1\right)\mathrm{e}^{\overline{\lambda}_8 r}\overline{G}_4 = -\left(1 - \frac{1}{\sqrt{2}}\right)\left(\eta + \frac{\pi}{2}\right).\end{gathered} \tag{6.42}$$

From (6.42), the values of the distance between fluxons, r, and the velocity, v, are fixed as function of the parameters $S, \alpha,$ and η. Solving (6.42) for each coupling constant, S, we obtain the region in the (α, η) parameter space where bunching in (f–f–f) mode exists. Figure 6.3 shows the corresponding region in (α, v) space, calculated numerically between the Swihart velocities, $c_- < v < c_+$, where bunching exists. The bottom contour of this region may be approximated by the piece-wise linear approximation, while the top contour of the region cannot be approximated in this manner, due to creation of new pairs of fluxons–antifluxons above this contour.

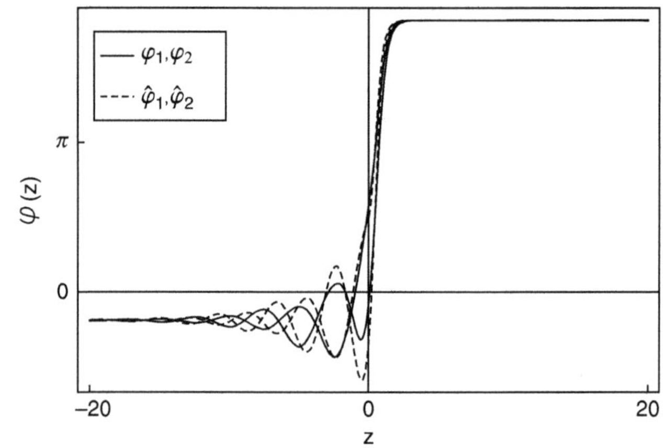

Fig. 6.4 Bunched fluxons from the numerical simulation (*full curve*) and as a result of the piece-wise linear approximation (*dashed curve*). The parameters are $S = -0.2$, $\alpha = 0.2$, and $\eta = -0.6$. The velocity is $v = 1.1 > c_-$. Good agreement between both results [15]

The piece-wise linear approximation, (6.38)–(6.40), reproduces quite accurately the shape of the fluxons for low values of the driving force, $|\eta|$, where the bunching of (f–f–f) takes place as shown in Fig. 6.4.

In [15] a symmetric central finite difference method of second order for both, space and time, has been implemented for the numerical simulations. The total length of the junctions is $L = 40$ and the spatial mesh size is $\Delta x = 0.05$. We have chosen periodic boundary conditions, $\varphi_i(L/2) = \varphi_i(-L/2) + 2\pi$, $\varphi_{i,x}(L/2) = \varphi_{i,x}(-L/2)$, $i = 1, 2, 3$, corresponding to an annular geometry to avoid ambiguities due to reflection from edges.

As seen in Fig. 6.3, in the (f–f–f) configuration the bunching is possible for speeds between the lowest and the highest Swihart velocities, $c_- < v < c_+$. Thus simulations were made for high driving force, $|\eta|$. Three types of fluxon motion were detected. When $|\eta|$ is lower than a threshold value, the bunched state does not exist. Fluxons in external junctions split from the fluxon in the internal junction (Fig. 6.5a) and they propagate with different velocities. Increasing the bias parameter, $|\eta|$, we find the range of values where bunched state exists. This bunching interval depends on the coupling S, and the dissipation, α.

When bunching takes place, the fluxons move their centers with the same velocity, v, belonging to the interval (c_-, c_+), and their centers are separated by a small distance in Fig. 6.5b. The emergence of oscillating tails in the numerical solution of (6.20)–(6.22) for this high velocity, v, induces the three fluxon bunching.

For too high bias ($|\eta| = 0.69$ in Fig. 6.5c), the equilibrium of bunching is broken by the creation of a new pair of fluxon–antifluxon [26].

It is worth mentioning that fluxon splitting is an irreversible process. For example, having initially a bunched state, one can destroy it by decreasing the bias term, $|\eta|$, and crossing the bottom contour in Fig. 6.3. As a result, the exterior fluxons and the interior one split off and start to move with different velocities. It is impossible to rebunch these fluxons by increasing the driving force, $|\eta|$.

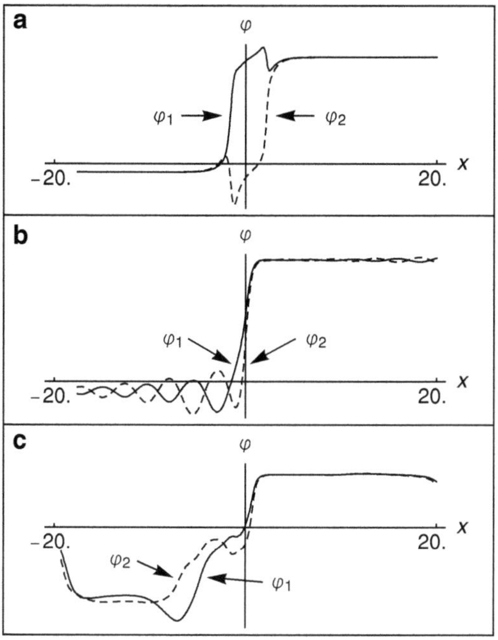

Fig. 6.5 Behavior of fluxons φ_1 (*full curve*) and φ_2 (*dashed curve*) for coupling $S = -0.2$, damping $\alpha = 0.1$, and bias current: (**a**) $\eta = -0.43$ fluxons split with velocities $v_1 = 0.868$ for φ_1 and $v_2 = 0.88$ for φ_2; (**b**) $\eta = -0.44$ bunched fluxons with velocity $v_1 = v_2 = 1.118$; (**c**) $\eta = -0.69$ creation of new fluxon–antifluxon pair due to excess energy [15]

The bias current versus the numerically found fluxon velocities (i.e., the current-voltage characteristic with voltage replaced by velocity) are plotted in Fig. 6.6. When the fluxons move slower than c_-, they split and travel with different velocities, $\varphi_1 = \varphi_3$ with v_1 and φ_2 with v_2, where $v_1 < v_2$. Bunching state branches are observed in narrow ranges of velocities between c_- and c_+.

6.3.2 Bunching due to Boundary Conditions

The repulsive nature of equal polarity fluxons in different junctions naturally leads to antiphase motion when the bias current is applied to the long Josephson junction stack. A periodic signal applied to the boundary of the system may phase-lock the fluxon motion to the signal frequency [27, 28], which in some cases leads to an attractive force between the fluxons [16, 17]. We consider one fluxon in each junction arranged in a triangular fluxon lattice with a fluxon–fluxon distance r, in the case of weakly coupled junctions, i.e., $S \ll 1$. The system is perturbed through the boundary by a periodic signal, with angular frequency ω.

The case of weakly coupled junctions allows for an expansion of the inverse of the coupling matrix (6.22) to first order in S,

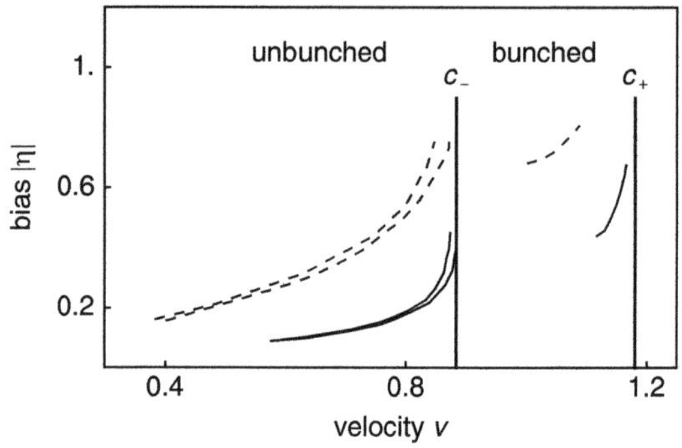

Fig. 6.6 Bias current, $|\eta|$, versus fluxon velocity, v, for coupling for $S = -0.2$. Full (*dashed*) curves represent velocity versus bias for $\alpha = 0.1$ ($\alpha = 0.3$). Below c_-, fluxons split and two different velocity branches are observed for φ_1 and φ_2. Fluxon bunching occurs in a velocity interval between c_- and c_+ [15]

$$\mathbf{S}^{-1} \approx \begin{pmatrix} 1 & -S & & & \\ -S & 1 & -S & & \\ & -S & 1 & -S & \\ & & \ddots & \ddots & \ddots \end{pmatrix}. \tag{6.43}$$

The Hamiltonian of the stack of weakly coupled Josephson junctions then becomes

$$H = \int_0^L \sum_{i=1}^{N} \left(\frac{1}{2}(\phi_t^i)^2 + 1 - \cos\phi^i + \frac{1}{2}\phi_x^i \times \right. \tag{6.44}$$
$$\left. \left(\phi_x^i - S(1-\delta_{i,N})\phi_x^{i+1} - S(1-\delta_{i,1})\phi_x^{i-1}\right) \right) \mathrm{d}x,$$

with $\delta_{i,j}$ being the Kronecker delta function. Using (6.20), (6.21), and (6.43), the rate of change in energy is

$$\frac{\mathrm{d}H}{\mathrm{d}t} = \sum_{i=1}^{N} \int_0^L \left(-\alpha \left(\phi_t^i\right)^2 + \eta\phi_t^i\right) \mathrm{d}x + \left[\sum_{i=1}^{N} \phi_x^i \times \right. \tag{6.45}$$
$$\left. \left(\phi_t^i - S(1-\delta_{i,N})\phi_t^{i+1} - S(1-\delta_{i,1})\phi_t^{i-1}\right) \right]_0^L.$$

The first condition for steady state is that the energy-change averaged over one period should be zero, thus

$$\Delta H = \int_{t_0-\frac{\pi}{\omega}}^{t_0+\frac{\pi}{\omega}} \frac{\mathrm{d}H}{\mathrm{d}t}\mathrm{d}t = 0, \tag{6.46}$$

where t_0 is some arbitrary time. To model the collisions with the boundaries, the profile in (6.28) is used for each junction with Swihart velocity c_-. The time-integral is split into two half-periods, with central collision times τ_L and $\tau_0 = \tau_L + \pi/\omega$. The fluxon collision time is taken to be $\tau^i = \tau_L + (-1)^i r/(2v)$ such that the fluxons are in a triangular lattice with central collision time τ_L at $x = L$. From condition (6.46), the current-voltage characteristic of the system is obtained to be

$$\sigma\eta = \frac{\alpha}{IN}\int\int\sum_{i=1}^{N}\left(\phi_t^i(x,t)\right)^2 \mathrm{d}x\mathrm{d}t + \frac{\Delta H_b^0}{2IN} - \frac{\Delta H_b^L}{2IN}, \tag{6.47}$$

where the double integral is over the system size and half a period of motion, I is determined from [29]

$$\sinh\left(\frac{I\gamma(v/c_-)}{4\pi c_-}\right) = \frac{c_-}{v}\sinh\left(\frac{\pi v\gamma(v/c_-)}{2\omega c_-}\right), \tag{6.48}$$

and the asymptotic velocity, v, is determined from [29]

$$\frac{c_-}{v}\sinh\left(\frac{\pi v\gamma(v/c_-)}{2\omega c_-}\right) = \cosh\left(\frac{L\gamma(v/c_-)}{2c_-}\right). \tag{6.49}$$

ΔH_b^0 and ΔH_b^L are the energy exchanges at the two boundaries at $x = 0$ and $x = L$, respectively, i.e.

$$\begin{aligned}\Delta H_\mathrm{b} &= \sum_{i=1}^{N}\int_{t_0-\frac{\pi}{\omega}}^{t_0+\frac{\pi}{\omega}}\left[\phi_x^i\left(\phi_t^i - S(1-\delta_{i,N})\phi_t^{i+1} - S(1-\delta_{i,1})\phi_t^{i-1}\right)\right]_0^L \mathrm{d}t \\ &= \Delta H_\mathrm{b}^L - \Delta H_\mathrm{b}^0,\end{aligned} \tag{6.50}$$

with

$$\begin{aligned}\Delta H_\mathrm{b}^L \approx \sum_{i=1}^{N}\int_{\tau_L-\frac{\pi}{2\omega}}^{\tau_L+\frac{\pi}{2\omega}} &\phi_x^i(L,t)\Big(\phi_t^i(L,t) - S(1-\delta_{i,N})\phi_t^{i+1}(L,t) \\ &-S(1-\delta_{i,1})\phi_t^{i-1}(L,t)\Big)\mathrm{d}t\end{aligned} \tag{6.51}$$

and similarly for ΔH_b^0. This approximation is valid when the fluxon–fluxon distance is much smaller than the junction length.

The second condition for steady state is that the total force on the fluxon configuration over one period should be zero. The force, $\partial H/\partial r$, will have three components, one from the fluxon–fluxon repulsion and one from each of the boundaries, thus

$$\int_{t_0-\frac{\pi}{\omega}}^{t_0+\frac{\pi}{\omega}}\frac{\partial H}{\partial r}\mathrm{d}t = \int_{t_0-\frac{\pi}{\omega}}^{t_0+\frac{\pi}{\omega}}\left(\frac{\partial H_\mathrm{I}}{\partial r} + \frac{\partial H_b^L}{\partial r} - \frac{\partial H_b^0}{\partial r}\right)\mathrm{d}t = 0 \tag{6.52}$$

where H_{I} is the fluxon–fluxon interaction energy

$$H_{\mathrm{I}} = -S \sum_{i=1}^{N} \int_{-\infty}^{\infty} \mathrm{d}x \; \phi_x^i \left((1-\delta_{i,1}) \phi_x^{i-1} + (1-\delta_{i,N}) \phi_x^{i+1} \right), \tag{6.53}$$

resulting in the well-known fluxon–fluxon force

$$\begin{aligned} F_I &= \int_{t_0-\frac{\pi}{\omega}}^{t_0+\frac{\pi}{\omega}} \frac{\partial H_{\mathrm{I}}}{\partial r} \mathrm{d}t \\ &\approx \frac{16\pi S(N-1)\gamma^2(v/c_-)}{\omega c_-^2 \sinh\left(r\gamma(v/c_-)/c_-\right)} \left(1 - \frac{r\gamma(v/c_-) \cosh\left(r\gamma(v/c_-)/c_-\right)}{c_- \sinh\left(r\gamma(v/c_-)/c_-\right)} \right), \end{aligned} \tag{6.54}$$

when the profile in (6.27) is used and the integration is taken to be from $-\infty$ to ∞.

6.3.3 External Microwave Signal

To model the case of an external microwave signal, we use the boundary conditions

$$\phi_x^i(0,t) = \phi_x^i(L,t) = \Gamma \sin(\omega t) \tag{6.55}$$

for $i = 1, \ldots, N$, modeling microwaves with frequency ω and amplitude Γ. The energy exchanges with the boundaries then become

$$\Delta H_{\mathrm{b}}^L = -\Delta H_{\mathrm{b}}^0 \approx \kappa\Gamma \sum_{i=1}^{N} (1 - S(2-\delta_{i,1}-\delta_{i,N})) \sin\left(\omega\tau_L + (-1)^i \frac{r\omega}{2v}\right), \tag{6.56}$$

with

$$\kappa = 4\pi\sigma \frac{\cosh\left(\frac{\omega c_-}{2v\gamma(v/c_-)} \cos^{-1}\left(2v^2/c_-^2 - 1\right)\right)}{\cosh\left(\frac{\pi\omega c_-}{2v\gamma(v/c_-)}\right)}. \tag{6.57}$$

The forces from the boundaries are

$$\begin{aligned} \int_{t_0-\frac{\pi}{\omega}}^{t_0+\frac{\pi}{\omega}} \frac{\partial H_{\mathrm{B}}^L}{\partial r} \mathrm{d}t &= -\int_{t_0-\frac{\pi}{\omega}}^{t_0+\frac{\pi}{\omega}} \frac{\partial H_{\mathrm{B}}^0}{\partial r} \mathrm{d}t \\ \approx \frac{-\kappa\Gamma}{2v} \sum_{i=1}^{N} (-1)^i &\left(1 - S(2-\delta_{i,1}-\delta_{i,N})\right) \sin\left(\omega\tau_L + (-1)^i \frac{\omega r}{2v}\right). \end{aligned} \tag{6.58}$$

To simplify things, we now look at the case of two coupled junctions, $N = 2$. Using (6.56) in (6.47), the current-voltage characteristic in the phase-locked state is

$$\sigma\eta = \eta_0 + \Delta\eta \tag{6.59}$$

with

$$\eta_0 \equiv \frac{\alpha}{2I}\int\int\left(\left(\phi_t^1\right)^2 + \left(\phi_t^2\right)^2\right)\mathrm{d}x\mathrm{d}t, \tag{6.60}$$
$$\Delta\eta \equiv \frac{\kappa\Gamma}{I}(1-S)\sin\left(\omega\tau_L\right)\cos\left(\frac{r\omega}{2v}\right).$$

For $N = 2$, the condition (6.52) becomes

$$2\frac{\kappa\Gamma}{v}(1-S)\cos\left(\omega\tau_L\right)\sin\left(\frac{\omega r}{2v}\right) = \frac{16\pi S\gamma^2(v/c_-)}{\omega c_-^2\sinh\left(r\gamma(v/c_-)/c_-\right)}\left(1-\frac{r\gamma(v/c_-)\cosh\left(r\gamma(v/c_-)/c_-\right)}{c_-\sinh\left(r\gamma(v/c_-)/c_-\right)}\right). \tag{6.61}$$

The maximum value of $\Gamma\cos(\omega\tau_L)$, necessary to ensure $r \to 0$, is found for $r = 0$. Expanding the above condition around $r = 0$ and solving for Γ, the critical value of the amplitude of the microwaves necessary to induce bunching is obtained to be

$$\Gamma_{\mathrm{c}} = \frac{16\pi S\gamma^3(v/c_-)v^2}{3\kappa\omega^2c_-^3(1-S)\cos(\omega\tau_L)}, \tag{6.62}$$

showing that the smallest amplitude is obtained for $\Delta\eta = 0$.

Figure 6.7 show the fluxon-separation r as a function of the microwave-amplitude Γ for three different system lengths. The analytical curve is obtained by solving (6.61) for r. The critical value of Γ, Γ_{c}, necessary to get bunching is also shown. Equation (6.62) gives $\Gamma_{\mathrm{c}} = 0.17,\ 0.42$, and 0.80 for lengths $L = 5$, 7.5, and 10 and parameters as in the figure. In general, good agreement between theory and numerical experiments is seen.

6.3.4 External Cavity

The previous section demonstrated that an external microwave signal can induce fluxon bunching. From a practical perspective, it might be more feasible with a cavity, such that the Josephson junctions themselves generate the signal. A model of such a system is sketched in Fig. 6.8. The cavity is described by a charge, q, satisfying

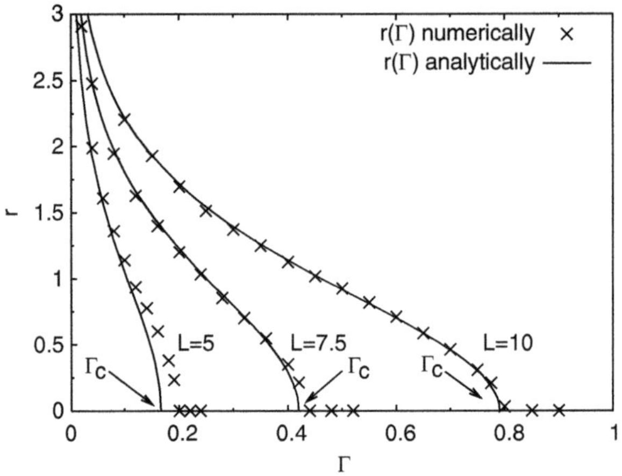

Fig. 6.7 Fluxon-separation, r, as a function of microwave amplitude, Γ. *Crosses* show results from numerical simulations and the *curves* show solutions of (6.61). The critical value of Γ, Γ_c, necessary to get bunching is also shown. Parameters are $\alpha = 0.1$, $\eta = \eta_0$, $\omega = 2.5/L$, and $S = -0.015$ [16]

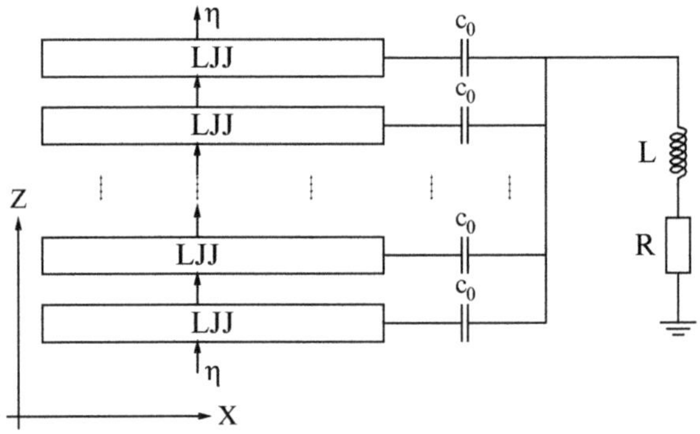

Fig. 6.8 Model of a stack of Josephson junctions coupled to an external cavity at $x = L$

$$\frac{d^2q}{dt^2} + \frac{\Omega}{Q}\frac{dq}{dt} + \Omega^2 q = \Omega^2 \frac{c}{N} \sum_{i=1}^{N} \phi_t^i(L,t), \tag{6.63}$$

where $Q = \sqrt{L/(NR^2 c_0)}$ is the quality factor, $\Omega = 1/(\omega_0 \sqrt{NLc_0})$ is the normalized cavity frequency, and $c = Nc_0/c_J$ is the normalized capacitance. The cavity is coupled to (6.20) and (6.21) through the boundary condition at $x = L$, while the boundary at $x = 0$ is kept as a reflecting boundary, i.e. we use

$$\phi_x^i(0,t) = 0 \quad , \quad \phi_x^i(L,t) = \frac{\dot{q}}{N} \quad , \quad q(0) = \dot{q}(0) = 0 \tag{6.64}$$

as the boundary conditions for (6.20), (6.21), and (6.63), where we have put the direct coupling between the junctions through the capacitors c_0 to zero.

To perform the same kind of analysis as in the previous section, a solution to (6.63) is needed. We look at the case where there is one fluxon in each junction, and $\phi_t^i(L,t)$ then becomes a voltage pulse. To simplify the integrations, these pulses are approximated by delta functions, i.e.

$$\phi_t^i(L,t) = \sum_{n=0}^{\infty} A\delta\left(t - \tau^i - 2\pi n/\omega^i\right), \tag{6.65}$$

approximating voltage pulses at $t = \tau^i + 2\pi n/\omega^i$, $n = 0, 1, \ldots$ with ω^i being the fluxon shuttling frequency in junction i, and τ^i is the phase shift of junction i. Note that since the present analysis is performed in the case of small $|S|$, all the δ-functions will have approximately the same amplitude, A. Assuming a high Q resonator and all fluxons shuttling with the same frequency, $\omega = \omega^i$, the solution to (6.63) is [17]

$$\dot{q}(t) = \varepsilon \sum_{i=1}^{N} \cos\left(\omega\left(t - \tau^i\right) + \varphi\right), \tag{6.66}$$

with

$$\varepsilon \equiv \frac{\Omega^2 A c}{N\sqrt{1 + \mathrm{e}^{-\frac{2\pi\Omega}{Q\omega}} - 2\mathrm{e}^{-\frac{\pi\Omega}{Q\omega}}\cos\left(2\pi\Omega/\omega\right)}}, \tag{6.67}$$

$$\cos\varphi = \frac{\mathrm{e}^{\frac{\pi\Omega}{Q\omega}} - \cos\left(2\pi\Omega/\omega\right)}{\sqrt{1 + \mathrm{e}^{\frac{2\pi\Omega}{Q\omega}} - 2\mathrm{e}^{\frac{\pi\Omega}{Q\omega}}\cos\left(2\pi\Omega/\omega\right)}}, \tag{6.68}$$

$$\sin\varphi = \frac{\sin\left(2\pi\Omega/\omega\right)}{\sqrt{1 + \mathrm{e}^{\frac{2\pi\Omega}{Q\omega}} - 2\mathrm{e}^{\frac{\pi\Omega}{Q\omega}}\cos\left(2\pi\Omega/\omega\right)}}. \tag{6.69}$$

Thus, the cavity current is very simple when the cavity has reached a steady state. Note that the amplitude of the cavity current for an in-phase mode ($\tau^i = \tau^j$, $i, j = 1, \ldots, N$) is $N\varepsilon$. For an antiphase mode ($\tau^i = \tau^{i+1} - (-1)^{i+1}\pi/\omega$, $i = 1, \ldots, N-1$), the amplitude is ≈ 0 for N even and $\approx \varepsilon$ for N odd.

The energy exchanges with the boundaries are given by (6.51) and (6.50), yielding

$$\Delta H_{\mathrm{b}}^0 = 0, \tag{6.70}$$

$$\Delta H_{\mathrm{b}}^L = \frac{\kappa\varepsilon}{N}\sum_{i=1}^{N}\Big[(1 - S(2 - \delta_{i,1} - \delta_{i,N}))\sum_{j=1}^{N}\cos\left(\omega(\tau^i - \tau^j) + \varphi\right)\Big],$$

with κ defined in (6.57). To determine the unknown amplitude A, it is required that (6.51) gives the same value for the δ-function approximation in (6.65) as with the fluxons in (6.28). This gives $A = \kappa$. The current-voltage characteristic of the junction-cavity system is given by (6.47), using the above expressions for $\Delta H_{\mathrm{b}}^{0}$ and $\Delta H_{\mathrm{b}}^{L}$.

The force on the fluxons from the boundary can be calculated by noting that $\phi_r^i = -(-1)^i \phi_t^i / 2v$, resulting in

$$
\begin{aligned}
\int_{t_0-\frac{\pi}{\omega}}^{t_0+\frac{\pi}{\omega}} \frac{\partial H_{\mathrm{b}}^{0}}{\partial r} \mathrm{d}t &= 0 \\
\int_{t_0-\frac{\pi}{\omega}}^{t_0+\frac{\pi}{\omega}} \frac{\partial H_{\mathrm{b}}^{L}}{\partial r} \mathrm{d}t &= -\frac{\kappa\varepsilon}{2vN} \sum_{i=1}^{N} \Big[(-1)^i \left(1 - S(2 - \delta_{i,1} - \delta_{i,N})\right) \\
&\quad \times \sum_{j=1}^{N} \cos\left(\frac{\omega r}{2v}\left((-1)^i - (-1)^j\right) + \varphi\right) \Big],
\end{aligned}
\tag{6.71}
$$

which enables the fluxon seperation, r, to be calculated by solving (6.52), as was done in the previous section for the external microwave field.

Going to the simple case of two coupled junctions, the current-voltage characteristic becomes

$$
\sigma\eta = \eta_0 + \Delta\eta \tag{6.72}
$$

with η_0 given by (6.60) and

$$
\Delta\eta \equiv \frac{2\kappa\varepsilon}{l}(1 - S)\cos(\varphi)\cos^2\left(\frac{r\omega}{2v}\right). \tag{6.73}
$$

The center of the phase locking range is seen to be for $\varphi = \pi/2$. φ is not "free" as the phase τ_L was for the microwave case, but it is determined by (6.68) and (6.69) and depends on the fluxon shuttling frequency, ω, which in turn depends on the bias current, η.

Condition (6.52) for $N = 2$ results in

$$
\begin{aligned}
&\frac{\kappa\varepsilon}{2v}(1 - S)\sin(\varphi)\sin\left(\frac{\omega r}{v}\right) = \\
&\frac{16\pi S\gamma^2(v/c_-)}{\omega c_-^2 \sinh\left(r\gamma(v/c_-)/c_-\right)}\left(1 - \frac{r\gamma(v/c_-)\cosh\left(r\gamma(v/c_-)/c_-\right)}{c_- \sinh\left(r\gamma(v/c_-)/c_-\right)}\right).
\end{aligned}
\tag{6.74}
$$

Contrary to the microwave case, the boundary force (l.h.s.) is seen to be zero when $\Delta\eta = 0$ and $\varphi = \pi/2$, which happens when $\Omega = \omega$. For fluxon shuttling frequencies slightly above the cavity frequency, Ω, the force from the boundary will counteract the repulsive fluxon–fluxon force and slightly below Ω, the force from the boundary will be repulsive. It is thus important to drive the system at a frequency

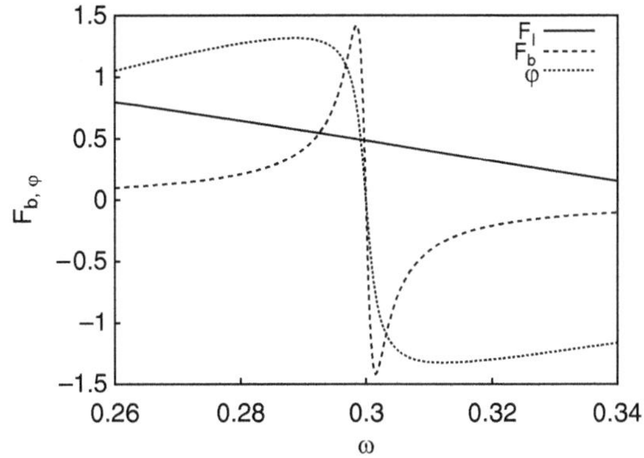

Fig. 6.9 φ is from (6.68) and (6.69), F_b is the l.h.s of (6.74), and F_I is the r.h.s. of (6.74). $N = 2,\ l = 8, \alpha = 0.1,\ Q = 100,\ c = 0.02,\ \Omega = 0.3,\ S = -0.05$, and $r = 4$ [17]

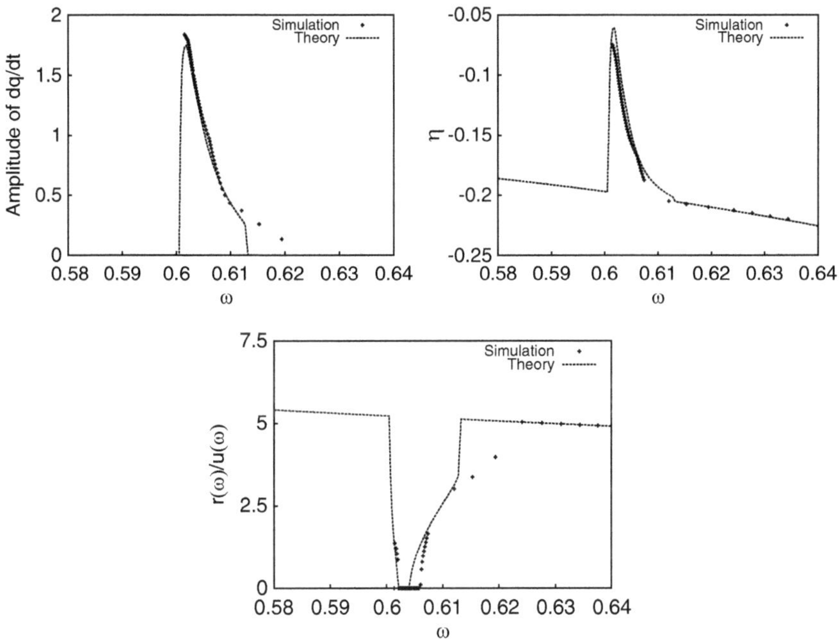

Fig. 6.10 Amplitude of $\dot{q}$ (*top left*), current-voltage (*top right*), and r/u (*bottom*). $N = 2,\ L = 4, \alpha = 0.1,\ Q = 100,\ c = 0.02,\ \Omega = 0.6$, and $S = -0.05$. ω is the fluxon shuttling frequency, controlled by the bias current (η) in the numerical experiment [17]

slightly above the cavity resonance frequency to obtain bunching from the cavity. The forces and the phase are plotted in Fig. 6.9.

Numerical simulations have been performed on the junction-cavity system and compared to the analytical calculations in Fig. 6.10. The cavity is seen to be able to induce bunching of the fluxons when the fluxons are driven above the cavity resonance frequency. Below the resonance frequency, the boundary force is repulsive and the fluxons shuttle in antiphase, resulting in almost no current in the cavity (not shown).

6.4 Experimental Work

Since many years, there has been an interest in radiation from Josephson junctions of all kinds. For long Josephsons, the spatially uniform phase is not a possible solution, and the dynamics of solitons (fluxons) as discussed for Josephson stacks in the previous section (in [30] in particular) may lead to the emission of electromagnetic radiation. At THz frequencies, it was, in [30], demonstrated that the ac Josephson effect existed in BSCCO stacks, and potentially it should be possible also to observe radiation emission in long BSCCO stacks at such high frequencies. In order to have the fluxons in the different layers move coherently, it was in most cases attempted to use a high magnetic field parallel to the layers, as $\mathbf{B}^a$ in Fig. 6.1, to give a unidirectional in-phase flux flow in all the junction. Several interesting results were reported – like oscillations in the flux flow resistance and the flux flow voltage [31], but not THz radiation emission.

Quite recently, however, THz radiation from BSCCO stacks was reported with different experimental conditions [32]. There was no applied magnetic field, and only the dc bias current was used as a variable for changing the conditions of the emitted radiation. For this latter experiment, it was assumed that an intrinsic cavity (zero field steps and Fiske steps) made the observation of THz emission possible. Thus, the fluxon dynamics of a long Josephson stack coupled to a cavity as discussed in the previous sections is very appropriate for the interpretation of the THz emissions provided by the experiments. Several groups are now adapting the methods of [32], and hopefully we will see a development toward larger amplitude of THz emissions and new applications in electronics and imaging.

We note here that the results from [32] exhibit more similarities with a much earlier and much simpler experiment in [33]. Table 1 shows a comparison between these two experiments.

Table 6.1 Comparison between experiments from [32] and [33]

	[32]	[33]
Number of layers	Large	2
Material	BSCCO	Niobium
Radiation out	Yes	Yes
Frequency	800 GHz	8 GHz
Magnetic field	No	No
Fiske resonance	Yes	Yes
Negative resistance	Yes	Yes

6.5 Summary

Emission of electromagnetic radiation from Josephson junctions is a promising technique for very high frequency applications (THz – frequencies). Typically, experimentalists use cavities as well as coupled Josephson junctions to enhance the emitted radiation. We considered such a system numerically and theoretically and found that the two methods supplement each other and provide a better understanding of the fluxon dynamics. We also considered the same system coupled to a cavity and found analytically and numerically that the cavity enhances the emitted radiation. Finally, we compared the results to experimental measurement at GHz and THz frequencies. Qualitative agreement was found.

Acknowledgements The authors wish to thank Dr. Carlos Gorria for his help in preparing this manuscript.

References

1. S. Sakai, P. Bodin, N.F. Pedersen, J. Appl. Phys. 73, 2411 (1993)
2. R. Kleiner, P. Müller, Phys. Rev. B 49, 1327 (1994)
3. A.V. Ustinov, H. Kohlsted, M. Cirillo, N.F. Pedersen, G. Hallmanns, C. Heiden, Phys. Rev. B 48, 10614 (1993)
4. S.M. Kim, H.B. Wang, T. Hatano, S. Urayama, S. Kawakami, M. Nagano, Y. Takano, T. Yamashita, K. Lee, Phys. Rev. B 72, 140504(R) (2005)
5. W.H. Press (ed), S.A. Teukolsky (ed), W.T. Vetterling, B.P. Flannery, Numerical Recipes in C++, 2nd edn. (Cambridge University Press, Cambridge, 2002)
6. L. Verlet, Phys. Rev. 159, 98 (1967)
7. J.K. Perring, T.H.R. Skyrme, Nucl. Phys. 31, 550 (1962)
8. D.W. McLaughlin, A.C. Scott, Phys. Rev. A 18, 1652 (1978)
9. Y. Tang, W. Wang, Phys. Rev. E 62, 8842 (2000)
10. J.C. Swihart, J. Appl. Phys. 32, 461 (1961)
11. N.F. Pedersen, S. Sakai, Phys. Rev. B 58, 2820 (1998)
12. M. Cirillo, N. Grønbech-Jensen, M.R. Samuelsen, M. Salerno, G.V. Rinati, Phys. Rev. B 58, 12377 (1998)
13. N. Grønbech-Jensen, J.A. Blackburn, M.R. Samuelsen, Phys. Rev. B 53, 12364 (1996)
14. C. Gorria, P.L. Christiansen, Yu. B. Gaididei, V. Muto, N.F. Pedersen, M.P. Soerensen, Phys. Rev. B 66, 172503 (2002)
15. C. Gorria, P.L. Christiansen, Yu. B. Gaididei, V. Muto, N.F. Pedersen, M.P. Soerensen, Phys. Rev. B 68, 035415 (2003)
16. S. Madsen, N. Grønbech-Jensen, Phys. Rev. B 71, 132506 (2005)
17. S. Madsen, N. Grønbech-Jensen, N.F. Pedersen, P.L. Christiansen, Phys. Rev. B 78, 174525 (2008)
18. Y.S. Kivshar, B.A. Malomed, Phys. Rev. B 37, 9325 (1988)
19. N. Grønbech-Jensen, D. Cai, M.R. Samuelsen, Phys. Rev. B 48, 16160 (1993)
20. A.V. Ustinov, H. Kohlstedt, Phys. Rev. B 54, 6111 (1996)
21. A. Wallraff, E. Goldobin, A.V. Ustinov, J. Appl. Phys. 80, 6523 (1996)
22. G. Carapella, Phys. Rev. B 59, 1407 (1999)
23. M.B. Mineev, G.S. Mkrtchyan, V.V. Schmidt, J. Low. Temp. Phys. 45, 497 (1981)
24. E. Goldobin, A. Wallraff, N. Thyssen, A.V. Ustinov, Phys. Rev. B 57, 130 (1998)
25. S. Sakai, N.F. Pedersen, Phys. Rev. B 34, 3506 (1986)

26. M. Peyrard, D. Kruskal, Phys. D 14, 88 (1984)
27. M. Salerno, M.R. Samuelsen, G. Filatrella, S. Pagano, R.D. Parmentier, Phys. Lett. A 137, 75 (1989)
28. N. Grønbech-Jensen, Phys. Rev. B 47, 5504 (1993); Phys. Lett. A 169, 31 (1992)
29. N. Grønbech-Jensen, Phys. Rev. B 47, 5504 (1993)
30. H.B. Wang, P.H. Wu, T. Yamashita, Phys. Rev. Lett. 87, 107002-1 (2001)
31. S. Ooi, T. Mochiku, K. Hirata, Phys. Rev. Lett. 89, 247002 (2002)
32. L. Ozyuzer, A.E. Koshelev, C. Kurter, N. Gopalsami, Q. Li, M. Tachiki, K. Kadowaki, T. Yamamoto, H. Minami, H. Yamaguchi, T. Tachiki, K.E. Gray, W.-K. Kwok, U. Welp, Science 23, 1291 (2007)
33. G. Carapella, G. Costabile, J.Mygind, N.F. Pedersen, IEEE Trans. Appl. Supercond. 9, no 2, 4558 (1999)

Chapter 7
Point-Contact Spectroscopy of Multigap Superconductors

P. Samuely, P. Szabó, Z. Pribulová, and J. Kačmarčík

Abstract Point-contact spectroscopy offers a unique possibility to study the fundamental superconducting properties. Namely, the superconducting energy gap, its symmetry, multiplicity, the temperature and magnetic field dependence can be addressed usually on the scale of hundreds of nanometers.

From the very beginning of the discovery of superconductivity in magnesium diboride, this technique has been applied for the investigation of the two-gap superconductivity in this compound. Very recently discovered superconducting iron pnictides are intensively studied by this technique as well. Here, we shortly review the point contact experiments leading to one of the first experimental evidences of the fact that MgB_2 represents an extraordinary example of the multigap superconductivity. We show that particularly the measurements by point contacts in magnetic fields with the sample in the vortex state provide additional important informations directly in the raw data, thus not depending on a particular model used for fitting. Namely, the direct experimental evidence of the coexistence of two well-distinct superconducting energy gaps up the common transition temperature is shown. The small gap and small superconducting coupling below the BCS value characterize the π band, while the large gap and the strong coupling are found in the hole σ band.

Also the carbon and aluminum doped MgB_2 samples have been intensively studied. It is shown that the hole band filling effect leading to a decrease in the density of states due to electron doping by carbon and aluminum is very important. It prevails over the interband scattering introduced by doping MgB_2 by these elements in the investigated doping range. This is the reason why the two gap superconductivity is preserved also in the case when T_c is significantly suppressed. The effect of applied magnetic field on the point-contact spectra is also used to study the intraband scattering processes within the two bands indicating that the carbon doping enhances significantly the scattering inside the π band. This leads to a strong increase in the upper critical magnetic field, particularly at the low temperatures, with importance for practical applications.

Recent point contact measurements performed on the iron pnictides also show a presence of multigap superconductivity underlying the multiband character of this new class of the high temperature superconductors.

7.1 Point-Contact Andreev Reflexion Spectroscopy

Microconstriction or point contact (PC) between two metals can be modeled as a hole of radius a in an insulating sheet separating them. If the electron mean free path is much longer that the orifice radius $l \gg a$, the charge transfer through the contact is ballistic and the resistance is due to the geometrical impeding this ballistic transport through the hole. In such a point contact, the excitation energy eV of charge carriers is controlled by the applied voltage V. Let one of the metals forming the point contact be a superconductor. Below T_c, a phase coherent state of Cooper pairs is formed in a superconductor with the superconducting energy gap in the quasiparticle excitation spectrum. For a quasiparticle incident on the N/S interface with an excitation energy $eV < \Delta$, a direct transfer of the charge carriers is forbidden because of the existence of the energy gap Δ in the quasiparticle spectrum of the superconductor. Then, the transport of charge carriers through the point contact with transmission probability $T = 1$ is accomplished by the unique process of Andreev reflection. The electron transfer takes place via the retroreflection of a hole back into the normal metal with the formation of a Cooper pair in the superconductor. At excitation energies above the gap, quasiparticles can be transferred directly across the interface. The Andreev reflection process leads at $V < \Delta/e$ and in the zero-temperature limit to a current as well as differential conductance twice as large as in the normal state or as what is at large bias where the coupling via the gap is inefficient. When an insulating barrier is formed at the interface in the point contact, with $T \ll 1$ the opposite limit to the ballistic transport is given by the tunneling process. For this Giaever-like tunneling between a superconductor and a normal metal, it is well known that the conductance drops to zero for $eV < \Delta$ (again in the zero-temperature limit). A more general case for arbitrary transmission T has been treated by Blonder, Tinkham and Klapwijk (BTK) [1]. In this model, the voltage(energy) dependence of the conductance of a N/S point contact is defined as

$$G = \frac{\mathrm{d}I}{\mathrm{d}V} \propto \int_{-\infty}^{\infty} \left[\frac{\mathrm{d}f(E-eV)}{eV}\right] [1 + A(E) - B(E)]\mathrm{d}E, \qquad (7.1)$$

where *A(E)* is the Andreev reflection probability and *B(E)* is the probability of the normal reflection. The experimentally measured PC conductance data can be compared with this model using as input parameters the energy gap Δ, the parameter Z (measure for the strength of the interface barrier with transmission coefficient $T = 1/(1+Z^2)$ in the normal state), and a parameter Γ for the quasi-particle lifetime broadening [2]. The evolution of the $\mathrm{d}I/\mathrm{d}V$ vs. V curves for different interfaces characterized by the barrier strength Z is schematically presented in Fig.7.1a.

In the case of a two-gap superconductor, the normalized PC conductance represents a weighted sum of two contributions g_S and g_L

$$G/G_n(V) = \alpha g_S + (1-\alpha) g_L, \qquad (7.2)$$

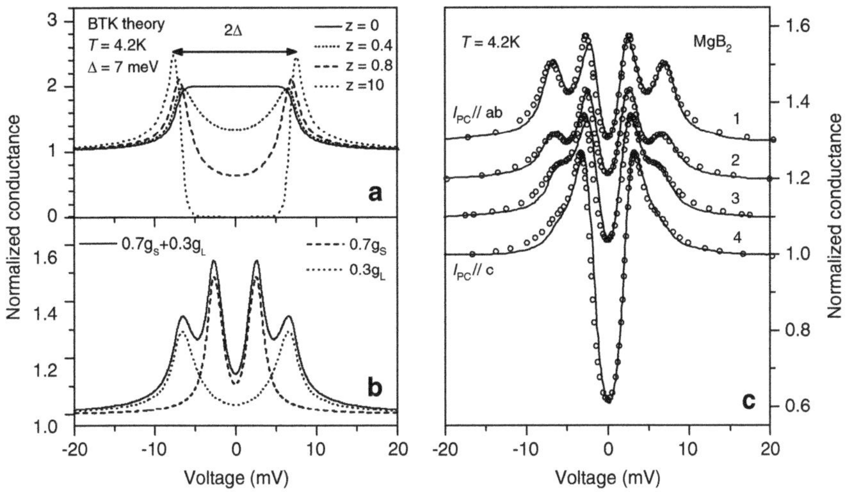

Fig. 7.1 (**a**) Numerical simulation of the BTK model for different values of the barrier strength Z, representing behavior of the point-contact spectra with $\Delta = 7$ meV between Giaever tunneling ($Z = 10$) and clean Andreev reflection ($Z = 0$) at $T = 4.2$ K. (**b**) Numerical simulation of the BTK model with two gaps involved. Full line is a sum $0.3g_S + 0.7g_L$ of two partial contributions with parameters $Z = 0.5$, $\Delta_S = 2.8$ meV, $\Delta_L = 7$ meV, $\Gamma = 0$ with a weight $\alpha = 0.7$ at $T = 4.2$ K ($0.3g_S$ and $0.7g_L$ curves are shifted to unity). (**c**) Representative Cu - MgB_2 point-contact spectra measured at $T = 4.2$ K (*full lines*). The upper curves are vertically shifted for the clarity. *Open symbols* display fitting results for the thermally smeared BTK model with $\Delta_S = 2.8 \pm 0.1$ meV, $\Delta_L = 6.8 \pm 0.3$ meV for different barrier transparencies and weight factors

where α is the weight of the band specified by the subscript S. Figure 7.1b shows the numerically simulated two gap point contact spectra with indicated parameters.

Point-contact spectroscopy measurements have been performed in a special point-contact approaching system which allows for the temperature and magnetic field measurements. The microconstrictions were prepared in situ by pressing a copper or platinum tip formed mechanically or by electrochemical etching on the freshly polished surface of the superconductor. A standard lock-in technique has been used to measure the differential resistance as a function of applied voltage on the point contacts.

7.2 Two Gaps in MgB_2 and Doped MgB_2 Systems

7.2.1 MgB_2

Discovery of superconductivity in magnesium diboride at the record temperature of 40 K in the simple binary compound was very surprising [3].

MgB_2 is a metal with a layer structure similar to graphite. As shown in Fig. 7.2, the boron atoms form honeycombed layers and the magnesium atoms are located above the centre of the hexagons in-between the boron planes. MgB_2, like graphite, has strong σ bonds in the (boron) planes and weak π bonds between them. In MgB_2, both the σ- and π-bonding boron orbitals form the electronic states at the Fermi level. Since boron atoms have fewer electrons than carbon atoms, not all the sigma bonds in the boron planes are occupied. And because not all the sigma bonds are filled, lattice vibration in the boron planes has a much stronger effect, resulting in superconducting state in the formation of strong electron pairs confined to the planes.

Very soon, the theoretical calculation appeared [4] suggesting that such a high T_c was due to an interplay between those two distinct electronic bands crossing the Fermi level. According to calculations, about half of the quasiparticles are located on quasi two-dimensional hole-like σ band and Cooper pairs there are strongly coupled via the boron vibrational mode E_{2g}. The rest of the quasiparticles resides in the three-dimensional π band with a rather weak electron–phonon interaction. In other words, electrons on different parts of the Fermi surface form pairs with two different binding energies, or two superconducting energy gaps – one large and one small. Simplified version of two MgB_2 Fermi surface sheets suggested by Dahm and Schopohl [5] is shown in Fig. 7.3. Quasi two-dimensional σ holes are located inside a warped cylinder and 3D π electrons are placed in a torus. Important theoretical

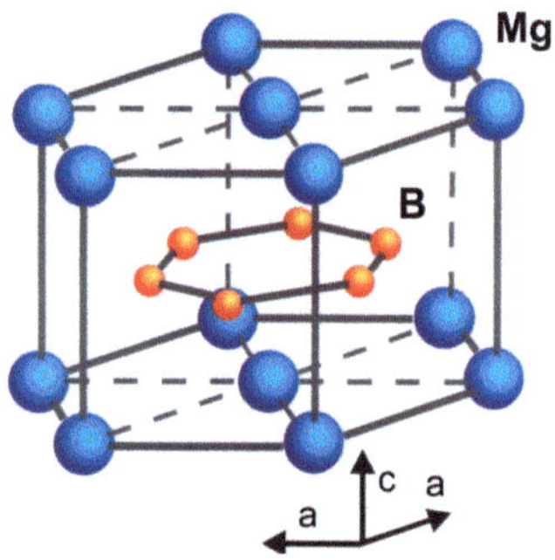

Fig. 7.2 Layered structure of MgB_2 with boron atoms forming honeycombed layers and magnesium atoms located above the centre of the hexagons in-between the boron planes

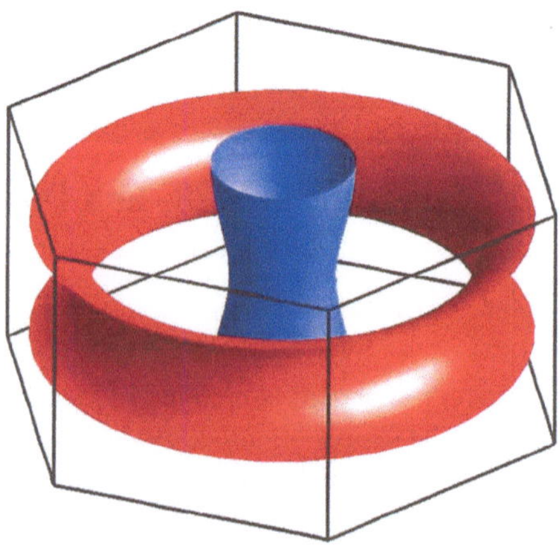

Fig. 7.3 Simplified version of the MgB_2 Fermi surface sheets with a warped cylinder of the σ band and a half-torus of the π band. Reproduced from T. Dahm and N. Schopohl, Phys. Rev. Lett. 91, 017001 (2003). Copyright 2003 by the American Physical Society

conclusions are that not all the boron electrons are needed in strong pair formation to achieve high T_c [4, 6, 7].

Although quite scattered, the first spectroscopy measurements on MgB_2 [8–11] yielded to superconducting gap values surprisingly smaller than the BCS weak coupling limit $2\Delta/k_BT_c = 3.52$. Later experiments by different techniques such as specific heat measurements [12, 13], tunneling [14–16] and Raman spectroscopy [17] revealed that beside the small energy gap there is another one with the strength $2\Delta/k_BT_c \sim 4.2$. Spectroscopy based on the Andreev reflection process gave one of the first proofs of such multigap superconductivity [18].

Our point contact experiments have been realized on polycrystalline MgB_2 samples prepared at the Institut Néel CNRS Grenoble and the Ames Laboratory. More details about the samples can be found in [18, 19]. The crystallites of the polycrystalline samples were larger than few microns. For a typical size of the metallic constriction below 10 nm, the point contacts probed therefore individual crystallites but of unknown orientation.

Figure 7.1c shows typical examples of the differential conductance dI/dV versus voltage spectra of the Cu–MgB_2 point contacts at 4.2 K (lines). The three upper curves reveal a clear twogap structure corresponding to the maxima placed symmetrically around the zero bias. However, 90% of the contacts reveal a spectrum dominantly showing the small-gap structure with only a small shoulder-like contribution around the large-gap voltage. These variations from contact to contact are due to the PC current injection along different crystal directions of locally probed crystallites. According to the above-mentioned calculations [4], the small gap is situated on the isotropic π band ($\Delta_S = \Delta_\pi$) and the large gap on the two-dimensional σ band, cylindrically shaped parallel to the c axis ($\Delta_L = \Delta_\sigma$). The Fermi-surface topology of the two bands would explain the observed differences in the spectra for different contact orientations. The small gap, being isotropic, will be observed for all directions of the current injection. However, the two-dimensional cylindrical Fermi surface will give most contribution for the current injected parallel to the ab plane. The observation of two-gap structures in a minority of the investigated contacts means that the crystallites in the polycrystalline samples are mostly textured with the ab planes parallel to the surface. Thus, the spectra in Fig. 7.1c represent an evolution from the c-axis tunneling to the π band (bottom curve) to a larger ab-plane contributions (upper curves) revealing both π and σ gap. Later on, such an interpretation was experimantally approved by the tunneling [15] as well as the point contact [20] experiments on oriented single crystals.

The conductance curves of Fig. 7.1c have been fitted to the two-band BTK model (7.2). Results of the fitting of each curve to the sum of two BTK conductances are shown by open symbols. The energy gap values obtained from fitting of different spectra are $\Delta_\pi = 2.8 \pm 0.1$ meV for the small gap and $\Delta_\sigma = 6.8 \pm 0.3$ meV for the large one. The weight factor α varied from 0.65 to 0.95 for the different contacts.

The smaller gap could a priori be caused by different reasons, too including a surface layer of reduced superconductivity or surface proximity effects. However, most of the scenarios would still require a scaling of the gap with the critical temperature. That is why it was important to show an existence of the small gap at high

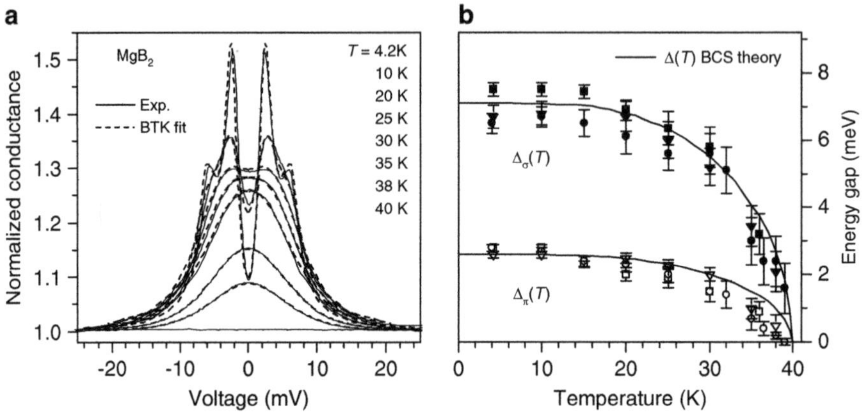

Fig. 7.4 (**a**) Differential conductances of Cu–MgB_2 point contact measured (*full lines*) and fitted (*dashed lines*) for the thermally smeared BTK model at indicated temperatures. The fitting parameters $\alpha = 0.71$, $Z = 0.52 \pm 0.02$, $\Gamma = 0.02$ meV were kept constant at all temperatures. (**b**) Temperature dependencies of both energy gaps ($\Delta_\pi(T)$ – *solid symbols*, $\Delta_\sigma(T)$ – *open symbols*) determined from the fitting on three different point contacts are displayed with three corresponding different symbols. ($\Delta_\pi(T)$ and $\Delta_\sigma(T)$ points determined from the same contact are plotted with the same (*open and solid*) symbols). *Full lines* represent BCS predictions

temperatures to establish its origin in the two-band superconductivity. The temperature dependencies of the point-contact spectra have been examined on different samples. Due to thermal smearing, the two well-resolved peaks in the spectrum merge as the temperature increases. Consequently, the presence of two gaps is not so evident in the raw data at higher temperatures (see Fig. 7.4a). For instance at 25 K, the spectrum is reduced to one smeared maximum around zero bias. Such a spectrum itself could be fitted by the BTK formula with only one gap, but the transparency coefficient Z would have to change significantly in comparison with the lower temperatures, and moreover a large smearing factor Γ would have to be introduced. However, our data could be well fitted at all temperatures by the sum of two BTK contributions with the transparency coefficient Z and the weight factor α kept constant and without any parallel leakage conductance. The point-contact conductances at different temperatures are shown in Fig. 7.4a together with the corresponding BTK fits. The resulting energy gaps Δ_σ and Δ_π with those obtained for two other point contacts with different weight values α are shown in Fig. 7.4b. In the BCS theory, an energy gap with $\Delta = 2.8$ meV could not exist for a system with T_c above $2\Delta/3.5k_B \simeq 19$ K, but as shown in [21] even a one-gap spectrum (shown as the bottom curve in Fig. 7.1c) leads to the same temperature dependence $\Delta_\pi(T)$ with the bulk T_c as those indicated in Fig. 7.4b. Our measurements thus give experimental support for the two gap model.

Even much stronger evidence for the inherent presence of two gaps in the superconductivity of MgB_2 is obtained by our magnetic field measurements. Figure 7.5 displays the effect of the magnetic field on the two gaps measured at three different temperatures. At $T = 4.2$ K, the two gaps are clearly visible for zero magnetic

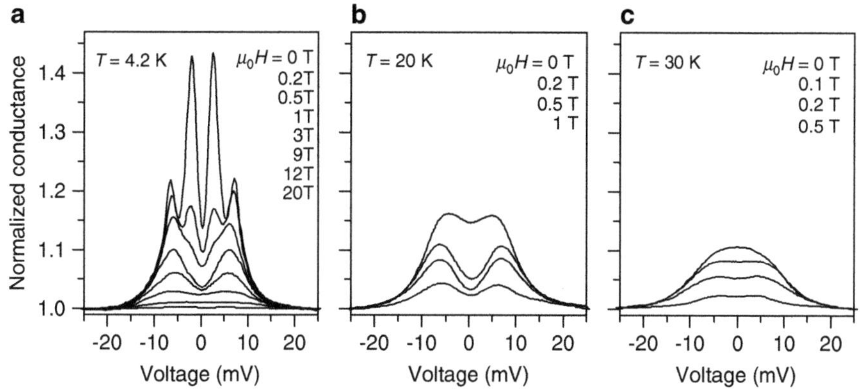

Fig. 7.5 Experimentally observed influence of the applied magnetic field on the two-gap structure of the normalized point-contact spectra at indicated temperatures. These spectra clearly reveal that both gaps exist in zero field up to T_c as shown by the rapid suppression of the small gap structure ($\Delta_\pi = 2.8$ meV) with magnetic field which leads to a deepening at zero bias

field $H = 0$, but the peaks corresponding to the small gap are rapidly affected by a magnetic field. The contribution of this small gap almost completely disappears at 1 T, whereas the large gap peaks still remain clearly visible. The other sets of spectra have been recorded at 20 and 30 K. As already shown above, the spectra are so smeared out above 20 K that the presence of the two gaps is not clearly visible and we have to rely on the fitting procedure. However, in magnetic field a peculiar effect occurs. The smaller-gap contribution to the overall point-contact conductance is rapidly suppressed by the field. As clearly visible at 4.2 K, this suppression leads to the formation of a deep broad minimum in the conductance at the zero bias voltage with increasing magnetic field with some traces of the smaller gap still surviving. Similarly, at 20 and 30 K, the peaks corresponding to the large gap become much better resolved in a magnetic field. This could not happen if there was not a contribution of the small gap in the zero-field point-contact conductance at these elevated temperatures. Indeed, a one-gap spectrum would only smear out with increasing magnetic field due to the pair-breaking effect. This effect thus unambiguously shows that the two gaps coexist near T_c, thus supporting the two band model of superconductivity in MgB_2. The system is characterized by the strongly coupled Cooper pairs in the σ band with $2\Delta/k_B T_c \sim 4.1$ and weakly coupled pairs in the π band with $2\Delta/k_B T_c \sim 1.6$.

The co-existence of two superconducting order parameters or energy gaps gives rise to very peculiar physical properties of the system, especially in the presence of applied magnetic field. At low fields, the superconducting properties are governed mainly by the π band, while at higher fields the 2D σ band dominates. The anisotropy of the upper critical field H_{c2} given as $\Gamma_H = H_{c2||ab}/H_{c2||c} = \xi_{ab}/\xi_c$, where ξ_i is the respective coherence length, is at low temperatures given just by the anisotropy of the σ band. Indeed, the point contact spectra shown in Fig. 7.5a evidence that the π gap is filled fast at low magnetic fields. The direct

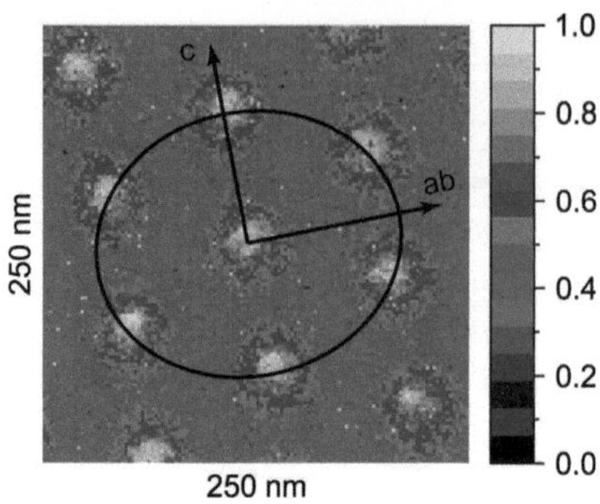

Fig. 7.6 Vortex imaging in MgB_2 single crystal by STM. The crystallographic orientation is indicated by the *arrows*. The ratio of major to minor axis of the ellipse is 1.19 giving also the superconducting anisotropy Γ_λ measured at 0.3 Tesla and 1.8 K. Reproduced from M. R. Eskildsen et al., Phys. Rev. B 68 (2003) 100508(R). Copyright 2003 by the American Physical Society

experimental measurements [22,23] of $H_{c2||c}$ and $H_{c2||ab}$ at low temperatures yield $\Gamma_H = 5 - 6$ in agreement with the anisotropy of the Fermi velocities in the σ band. The superconducting anisotropy can also be defined via the penetration depths λ_i as $\Gamma_\lambda = \lambda_c/\lambda_{ab}$. Usually, the two anisotropies Γ_λ and Γ_H are identical. It is not the case of MgB_2. Vortex lattice imaging in fields perpendicular to the uniaxial c direction can directly visualize the superconducting anisotropy Γ_λ. Eskildsen et al. [24] obtained vortex images in fields parallel to the ab planes of MgB_2 single crystals by the scanning tunneling microscope (STM). The vortex lattice measured at 0.3 T and 1.8 K is shown in Fig. 7.6. The crystallographic orientation is indicated by the arrows. In the case of isotropic vortex lattice, the vortices would lie on a circle. Here, they lie on an ellipse. The vortex lattice anisotropy proportional to Γ_λ is given by the ratio of the major to the minor axis, yielding $\Gamma_\lambda = 1.19$. This is in strong disagreement with the upper critical field anisotropy $\Gamma_H = 5 - 6$ found at this temperature, indicating that the superconducting screening currents in the essentially isotropic π band dominate the vortex–vortex interactions at low fields. Eskildsen et al. have also analysed the size of the vortex core in MgB_2 at low temperatures and very small magnetic fields (50 mT) parallel to the c axis [25]. The obtained $\xi \approx 50$ nm would correspond to the upper critical field $H'_{c2} = 0.13$ T much below the real H_{c2} value. These facts can be understood by taking into account that while at low fields the superconductivity of the π band is important and the vortex core is given as ξ_π determined by the π gap; at higher fields, the σ band prevails and the superconductivity in the π band is just induced by the σ band. Kogan and Bud'ko [26] have predicted a very pronounced difference between the Γ_λ and Γ_H anisotropies in case of the two-gap superconductor of MgB_2. They have predicted variances not only in the absolute values of the anisotropies, but also both of them should be temperature dependent with opposite tendencies. Our data published in [27], obtained on single crystals of MgB_2 by magnetization and transport measurements have fully validated the Kogan's scenario.

Many other new physical phenomena can occur in a system with two superconducting order parameters. We mention just two more examples. Blumberg et al. [28] have observed in MgB_2 signatures of the intrinsic Josephson effect between the σ and the π bands. Recently, Moshchalkov et al. [29] have detected the so-called type-1.5 superconductivity in MgB_2, where the π band creates virtually a type-1 superconductor with $\kappa_\pi = \lambda_\pi/\xi_\pi < 1/\sqrt{2}$ and the σ quasiparticles have virtually type-2 character with $\kappa_\sigma = \lambda_\sigma/\xi_\sigma > 1/\sqrt{2}$.

7.2.2 Aluminum and Carbon-Doped MgB_2

Due to coexistence of charge carriers from two almost completely separated bands, two *intraband* and two *interband* scattering channels have to be distinguished. The *interband* scattering by nonmagnetic scatterers is supposed to have a particularly strong effect in a two-band superconductor: it will blend the strongly and weakly coupled quasiparticles, merge two gaps and consequently decrease T_c. Thus, the Anderson's theorem on the negligible effect of non magnetic impurities is violated in the multigap superconductors. In the case of MgB_2, T_c can drop to about 20 K upon significant increase of the interband scattering [6, 7]. Fortunately, a different symmetry of the bands ensures that the interband scattering remains small also in very dirty MgB_2 samples, which show about the same T_c as the purest material [30]. A systematic decrease of T_c is achieved in substituted MgB_2 samples. The only on-site substitutions by nonmagnetic elements that are known so far in the case of MgB_2 are carbon for boron and aluminium for magnesium. The carbon and aluminium atoms in MgB_2 take indeed a role of scatterers, but they also dope the system with one extra electron which inevitably leads to a decrease in the density of states in the σ band due to electron filling effect. Kortus *et al.* [31] have introduced a model incorporating both effects: interband scattering and band filling in MgB_2. The former effect leads to an increase in the small gap Δ_π and a decrease in the large Δ_σ, while the latter suppresses both Δ_π and Δ_σ. One of the goals of our point contact experiments on the doped MgB_2 was to resolve the problem whether the decrease of T_c was due to a merging of two gaps upon increased interband scattering, or whether it was a consequence of the decreased density of states in the hole σ band. As a result, the experiments have shown that two distinct superconducting energy gaps remain in all samples with the highest doping suggesting the dominance of the band filling effects in our samples. For the increased Al doping, the enhanced interband scattering approaching two gaps must also be considered.

In contrast to the interband scattering, the increase of the scattering within the bands is not supposed to have any effect on the two gaps. A tuning of the intraband scattering can lead to significant changes of the upper critical magnetic field $H_{c2}(0)$ and its anisotropy, as shown, for example, by Angst et al. [32]. There, upon increased C doping, $H_{c2}(0)$ is enhanced significantly in both principal crystallographic directions (parallel to the ab hexagonal boron layers and in the c-axis direction). On the other hand, the Al substitution suppresses $H_{c2}(0)$ in the ab-plane direction and has almost no effect on H_{c2} along the c axis. This different behavior is apparently due to a different influence of C and Al doping on intraband scatterings. While the strong increase of $H_{c2}(0)$ with the C doping is due to graded dirty limit conditions, Al-doped samples stay probably in the clean limit and the decrease of H_{c2} is just a consequence of the lower T_c [32]. A weight of scatterings in separated bands was one of the items addressed by our point-contact experiments on the doped samples. A relative weight of the intraband scattering was estimated from the fitting of the field dependence of the PC spectra to a simple empirical model. Significant changes were found in the case of C doping with increased scattering particularly in the π band. The Al doping does not change the relative weight of the scatterings within the bands and the samples seem to stay in the clean limit.

A systematic study of carbon- and aluminum- doped samples with different substitutions: $Mg_{1-x}Al_xB_2$ and $Mg(B_{(1-y)}C_y)_2$ with x up 0.2 and y up to 0.1[1] has been performed [19, 33–35]. A large number of the point-contact measurements have been performed on every substition. The point-contact spectra had different barrier transparencies changing from more metallic interfaces to intermediate barriers. By trial and error, we looked for the spectra showing both gaps of polycrystalline specimens. For more detailed studies, we chose those junctions revealing the spectra with a low spectral broadening Γ, i.e., with an intensive signal in the normalized PC conductance. Figure 7.7a displays the representative spectra of the

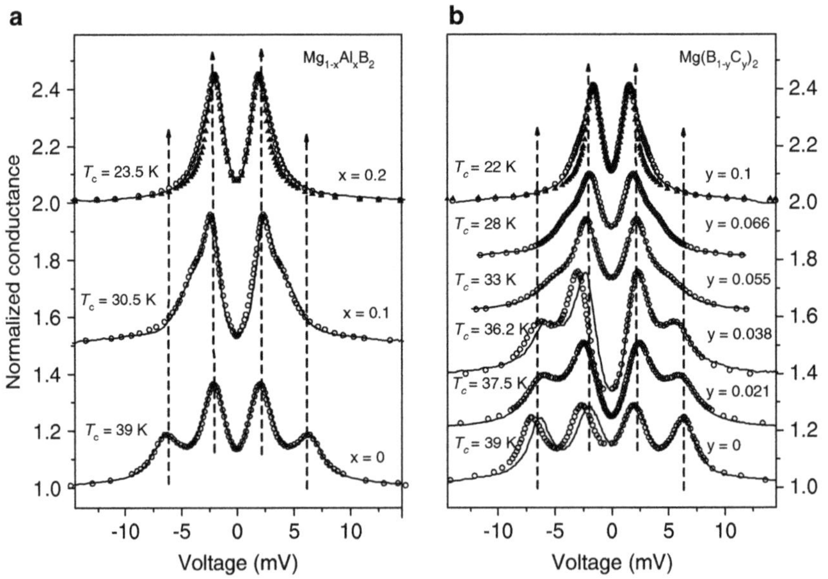

Fig. 7.7 (**a**) Normalized point-contact spectra of $Mg_{1-x}Al_xB_2$ with $x = 0, 0.1, 0.2$ and $Mg(B_{(1-y)}C_y)_2$ with $y = 0, 0.021, 0.038, 0.055, 0.065$, and 0.1 – *solid lines*. Symbols show the fitting curves: *open circles* for the two-gap BTK model and *solid triangles* for the one-gap BTK model. The spectra are vertically shifted for the clarity. The *dashed vertical arrows* emphasizes the evolution tendency of the gaps

[1] Within this notation, the equivalent doping is achieved for $x = 2y$.

aluminum-doped series and Fig. 7.7b resumes the results for the carbon-doped MgB_2. The experimental data are presented by the full lines, while the fits to the two-gap formula are indicated by the open circles. All shown point-contact spectra have been normalized to the spectra measured in the normal state at a temperature above T_c. The upper curves are shifted for the clarity. The retention of two gaps for all dopings apart from the highest substitution is evident. For the highest dopings (always with a larger spectral broadening, $\Gamma \simeq 0.2\Delta$), the spectra reveal only one pair of peaks without an apparent shoulder at the expected position of the second gap. For those data measured at 1.6 K (the rest of the data were measured at 4.2 K) also the one-gap fit is presented. The one-gap fit (solid symbols) can reproduce neither the height of the peaks nor the width of them, while the two-gap fit (open symbols) can reproduce both. The size of the apparent gap is well indicated by the peak position, which is about 2 meV for $Mg_{0.8}Al_{0.2}B_2$ and 1.6 meV for $Mg(B_{0.9}C_{0.1})_2$. However, this size is too small to explain the superconductivity with the respective T_c's equal to 23.5 K and 22 K within the single-gap BCS scenario. This suggests that we are dealing with a smeared two-gap structure.

The existence of both gaps is strongly evidenced in the magnetic field dependence of the PC spectra measured on 10% C-doped MgB_2, shown in the top panel of Fig. 7.9b. Here, the spectrum at $\mu_0 H = 0.5$ T shows that the contribution from the π band was partially suppressed and the large-gap shoulder clearly appeared near 3 mV besides the gap peak at 1.6 mV. The magnetic field has a similar effect in the case of 20 % Al-doped MgB_2: As shown in the top panel of Fig. 7.9a, the peak position which is about 2 mV in zero field is shifted toward 3 mV at 1 Tesla. In the case of a single-gap superconductor, the application of magnetic field can only lead to a shrinkage of the distance between the peak positions in the point-contact spectrum. It is an obvious consequence of the introduction of vortices and magnetic pair breaking [36]. The shift in the peak of the PC spectra to higher voltages demonstrated in the top panel of Fig. 7.9a is then due to an interplay between the two gaps: the dominance of the Δ_π peak at zero field is suppressed in increasing magnetic field and Δ_σ contributes more. Thus, we can draw the conclusion that the retention of two gaps is experimentally verified in all samples studied with T_c's ranking from 39 down to 22 K.

Figure 7.8 displays the values of the energy gaps as a function of T_c resuming all our PC data, supplemented also by the gaps inferred from the specific heat measurements performed on the Al-doped MgB_2 single crystals [35]. From this data collection, the following conclusion can be found: the large gap Δ_σ is essentially decreased linearly with the respective T_c. The behavior of Δ_π is more complicated. For both kinds of substitution, the small gap is almost unchanged at smaller doping. In the case of the C-doped MgB_2, it holds clearly down to $T_c = 33$ K. In (Mg, Al)B_2, Δ_π of the 10% Al-doped sample seems to even slightly increase. But for the highest doping, Δ_π decreases in both cases. The solid lines in Fig. 7.8 are the calculations of Kortus et al. [31] for the case of a pure band filling effect and no interband scattering. The dashed lines show the calculations including also the interband scattering with the rate $\Gamma_{IB} = 1{,}000 \times \text{cm}^{-1}$ (Al-doping) or 2,000 y cm^{-1} (C doping). As can be seen, the gaps in the Al-doped samples reveal stronger tendency to

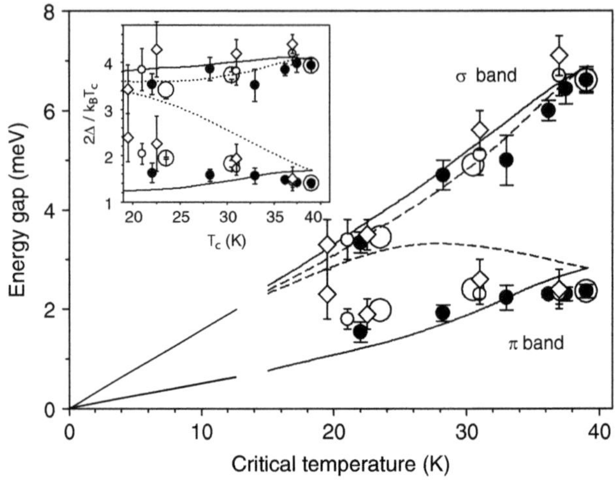

Fig. 7.8 Averaged values of the superconducting energy gaps as a function of the critical temperatures obtained from the point-contact spectroscopy on C-doped polycrystals (*solid circles*), Al-doped polycrystals (*large open circles*), and Al-doped single crystals (*small open circles*). The *open diamonds* are the gaps from the specific heat data on Al-doped single crystals [35]. The lines are explained in the text

approach each other compared to C doping, and their evolution cannot be explained without an increased interband scattering but it is smaller than in the presented calculations. The inset of Fig. 7.8 presents the reduced gaps or the coupling strengths $2\Delta/k_BT_c$ as a function of T_c. All the presented data show a modest tendency to approach two gaps with decreased T_c without any merging in the investigated range of T_c's. It is worth noting that the tendency shown here is very general and has been found in the ARPES measurements [37] not only on the C-doped MgB_2 but also on the neutron-irradiated MgB_2 samples with T_c above 21 K [38,39]. We can conclude that even if the interband scattering can vary depending on the substitution (bigger in the Al-doped material than in the C-substituted MgB_2) or the sample treatment (neutron irradiation), it is difficult to increase it in MgB_2 to the level necessary to blend both gaps. Extrapolating the data in Fig. 7.8, the merging of two gaps can be expected on the samples with T_c below 10–15 K. Putti et al. [39] claimed from their specific heat measurements an observation of the crossover from the two-gap to single-gap superconductivity in the neutron-irradiated MgB_2 polycrystals with T_c of 11 K.

The theoretical calculations of Erwin and Mazin [40] on merging of the gaps due to a substitution also support the picture presented here. According to this work, the carbon substitution on the boron site should have no effect on the merging of the gaps. This is due to the fact that replacing boron by carbon does not change the local point symmetry in the π and σ orbitals which are both centered at the boron sites. On the other hand, a much bigger effect is expected for out of plane substitutions (Al instead of Mg) or defects which would indeed change the local point symmetry. The substitution of Al instead of Mg leads also to a significant decrease in the c-lattice parameter [39] (it is basically absent in case of carbon doping), which helps the interlayer hopping from a p_z orbital (π band) in one atomic layer to a σ bond orbital in the next atomic layer. Anyway the retention of the two gaps superconductivity in the carbon- and the aluminum- doped MgB_2 samples helps to keep a relatively high T_c. Of course, a particular strength of the interband scattering can differ among samples due to some defects, and thus the merging of two gaps could be sample dependent [41].

In the following, we present the effect of magnetic fields on the point-contact spectra of the aluminum- and carbon- doped MgB_2 samples. The field effect has been measured on a number of samples of each substitution. The field was always oriented perpendicularly to the sample surface and parallel with the tip having the point-contact area in the vortex state. As the C doping leads the MgB_2 samples deeply into the dirty limit (with an increase of H_{c2} despite a decreased T_c), we have chosen the undoped MgB_2 sample being already in a dirty limit for a comparison with the 6.5% and 10% C-doped samples in the following discussion. On the other hand, the aluminum doping seems to keep the clean limit conditions, so we have chosen the pure MgB_2 wire as a reference in the studies of the magnetic field effects.

Figure 7.9a, b displays typical magnetic field dependencies of the PC spectra (solid lines) obtained on the samples (a) $Mg_{(1-x)}Al_xB_2$ with x = 0, 0.1, 0.2 and (b) $Mg(B_{(1-y)}C_y)_2$ with $y = 0, 0.065$, and 0.1. All the magnetic field dependencies of the PC spectra shown here reveal two-gap behavior typical for MgB_2: i.e., the peaks of the small gap are suppressed in a much lower field than those of the large gap.

From the in-field spectra, the excess current I_{exc} has been determined. It was calculated by integrating the PC spectra (normalized conductance) after subtracting an area below unity. When a single gap superconductor is in the vortex state, each vortex core represents a normal state region and this region/vortex density increases linearly up to the upper critical magnetic field. Thus, the PC excess current ($I_{exc} \propto \Delta$) is reduced linearly with increasing magnetic field as $I_{exc} \propto (1-n)\Delta$, where $n = N^A/A$ is a factor characterizing the normal state region N^A of the PC junction area A. In the case of MgB_2, I_{exc} will be represented as a weighted sum of the contributions of both bands [42]

$$I_{exc} \approx \alpha(1-n_\pi)\Delta_\pi + (1-\alpha)(1-n_\sigma)\Delta_\sigma, \tag{7.3}$$

where n_π and n_σ represents the normal state (vortex core) contributions of the π and σ bands, respectively.

The field dependencies of the excess currents I_{exc} obtained on the carbon- and aluminum- doped samples are shown in Fig. 7.10a, b, respectively. The magnetic field coordinates have been normalized to an appropriate $H_{c2||c}$ referred to as $h = H/H_{c2||c}$ [32]. In none of the cases, the linear decrease of I_{exc} was observed. On the contrary, a strong nonlinearity observed in all curves approves the presence of two gaps in the quasiparticle spectrum of the material. A steeper decrease of $I_{exc}(h)$ at low magnetic fields is ascribed to a strong filling of the π-band gap states up to a virtual upper critical field $H_{c2,\pi}$ of the π band. Then, at higher fields, the low temperature superconductivity is maintained mainly by the σ band [26], with

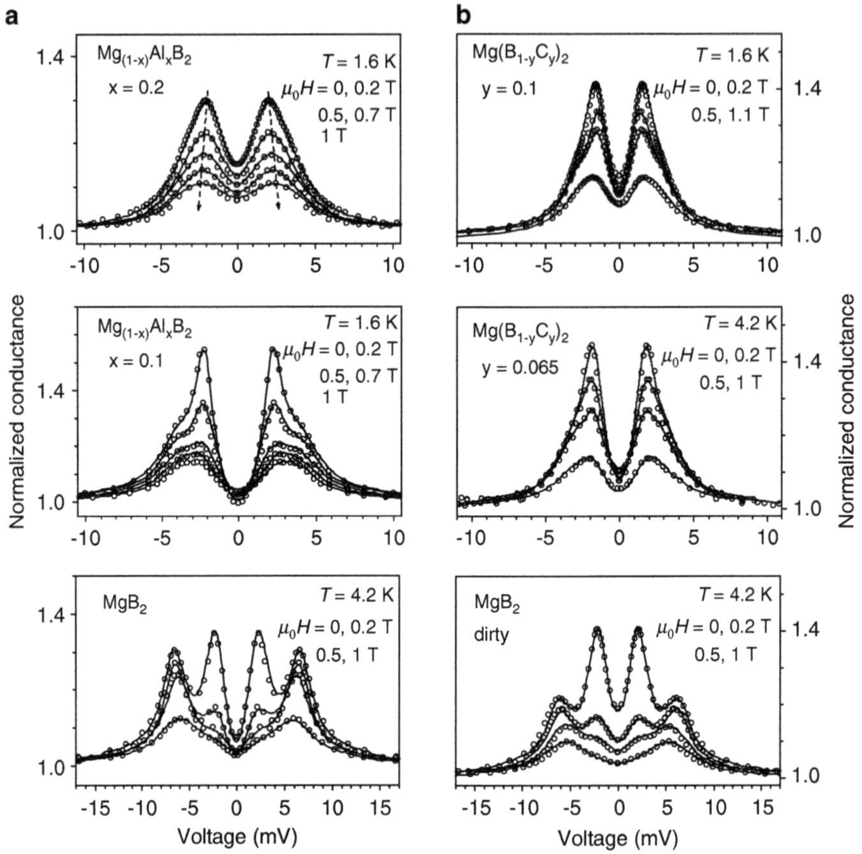

Fig. 7.9 Influence of the applied magnetic field on the point-contact spectra of (**a**) $Mg_{1-x}Al_xB_2$ with $x = 0, 0.1, 0.2$ and (b) $Mg(B_{(1-y)}C_y)_2$ with $y = 0, 0.065$, and 0.1 – *solid lines. Open circles* show fitting results for the two-band mixed-state BTK formula

the σ gap getting filled smoothly up to the real upper critical field of the material $H_{c2} = H_{c2,\sigma}$. Due to a small but finite coupling of the two bands, the superconductivity is maintained also in the π band above $H_{c2,\pi}$, but, as shown, e.g., in the tunneling data of Eskildsen et al. [25], the π-gap states are filled much more slowly here than in the low field region. A magnitude of the upper critical fields of the two bands can be estimated in the clean limit [12, 13] as $H_{c2,\pi} \approx \Delta_\pi^2/v_{F,\pi}^2$ and $H_{c2,\sigma} \approx \Delta_\sigma^2/v_{F,\sigma}^2$, with $v_{F,\pi}$ or $v_{F,\sigma}$ representing the Fermi velocity of the respective band (since only $H_{c2||c}$ is considered the anisotropy due to an effective mass tensor is neglected). A similar estimate can be done in the dirty limit when taking into account the respective diffusion coefficients $D_{\pi(\sigma)}$. Then, $H_{c2,\pi(\sigma)}$ is proportional to $\Delta_{\pi(\sigma)}/D_{\pi(\sigma)}$.

Figure 7.10b shows that the carbon doping leads to changes in the field dependence of the excess current. The undoped MgB_2 reveals a very steep fall of I_{exc}

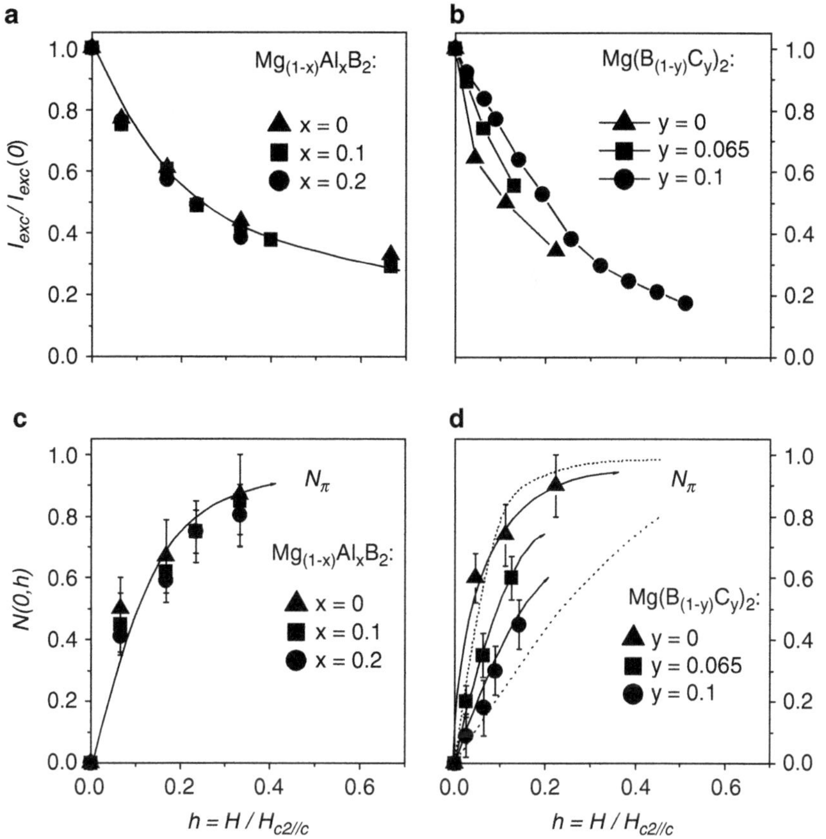

Fig. 7.10 (**a**) Excess currents $I_{exc}(h)$ of $Mg(B_{(1-y)}C_y)_2$. (**b**) $I_{exc}(h)$ of $Mg_{1-x}Al_xB_2$. (**c**) Zero-bias density of states $N(0,h)$ of $Mg(B_{(1-y)}C_y)_2$. (**d**) Zero-bias DOS of $Mg_{1-x}Al_xB_2$ determined from the fitting for the two-band mixed-state BTK formula. The *solid lines* are guides for the eyes. All field dependencies presented here have been determined from the point-contact spectra with similar weight of the π-band contribution $\alpha \simeq 0.75 \pm 0.05$. The *dashed* and *dotted lines* in Fig. 7.9c represent the theoretical predictions of $N_\pi(0,h)$ [43] for the dirty MgB_2 samples with the ratio of in-plane diffusivities $D_\sigma/D_\pi = 0.2$ and 1, respectively

already at small fields indicating also proportionally small crossover field $H_{c2,\pi}$. By doping, the π-band crossover field apparently shifts to higher values. From the previous subsection, we know that the carbon doping does not influence significantly the interband scattering in MgB_2. The ratio between the two gaps, Δ_σ/Δ_π, remains almost unchanged. Then, following from the dirty limit theory, the shift of the normalized crossover field $H_{c2,\pi}/H_{c2||c}$ is due to an increase in the ratio of the diffusion coefficients D_σ/D_π. This is a first indication that the carbon doping leads to a significant enhancement of the π-band scattering.

In Fig. 7.10a, the excess currents on the aluminum- doped samples show a different picture: there is no change in the field dependence of I_{exc} upon the aluminum doping up to 20% indicating a constant value of the crossover field $H_{c2,\pi}$. Qualitatively, it can be understood within the framework of the clean limit persisting in the Al-doped MgB_2 up to 20% Al doping region. Indeed, the constant value of the upper critical field $H_{c2||c}$, which is equal to approximately 3 Tesla for all three dopings, indicates the constant ratio between Δ_σ and $v_{F,\sigma}$. It means that the electron filling of the hole σ band decreases both quantities proportionally. Then, $H_{c2,\pi}$ must be only weakly doping dependent. As $H_{c2,\pi} \approx \Delta_\pi^2/v_{F,\pi}^2$ and Δ_π remains almost constant for $x = 0.1$ and 0.2, this would suggest that $v_{F,\pi}$ also remains unchanged.

The experimental study of the density of states at the Fermi level $N(0)$ can give further information about the relative weights of intraband scatterings in MgB_2. The calculations of Koshelev and Golubov [43] have shown that in a two-band dirty limit, the gap filling in the presence of magnetic field is a function of the intraband diffusivities. The field-induced increase in the π-band zero bias DOS $N_\pi(0, H)$ is controlled by the ratio of the diffusion coefficients D_σ/D_π. Bugoslavsky *at al.* proposed an empiric model for the point-contact conductance of a two-band superconductor in the mixed state [44]. This model, similar to the (7.3), describes the effect of magnetic field by taking into account the contribution of the normal cores of vortices. There again, n_π and n_σ are the area of the respective vortex cores normalized to the junction area. Then, the normalized PC conductance is

$$G/G_n(V) \approx \alpha[n_\pi + (1-n_\pi)g_\pi] + (1-\alpha)[n_\sigma + (1-n_\sigma)g_\sigma]. \qquad (7.4)$$

Considering that n_π and n_σ represent the number of normal-state vortex-core excitations, these parameters can be identified with the zero-energy density of states $N_\pi(0, H)$ and $N_\sigma(0, H)$ averaged over the vortex lattice. In this way, the field-dependent energy gaps and zero-energy DOS $N(0, H)$ can be determined. The magnetic field dependencies of the PC spectra measured on different Al- and C-doped MgB_2 samples shown in Fig. 7.9a, b have been fitted for this model (fitting curves ploted with open symbols). The fitting of the PC conductance for the two-band BTK model in the mixed state is not trivial because of too many (nine) parameters. Details about the fitting procedure can be found in [19]. The resulting zero-energy DOS $N_\pi(0, H)$ and $N_\sigma(0, H)$ are plotted in Fig. 7.10d for the carbon-doped samples and in Fig. 7.10c for the Al-substituted MgB_2. Again, the carbon and aluminum dopings behave very differently. The steep increase of $N_\pi(0, h)$ in the undoped MgB_2 can be within the framework of the dirty superconductivity model [43] ascribed to a small value of the ratio of the in-plane diffusivities $D_\sigma/D_\pi < 1$ (see dotted line, calculated for $D_\sigma/D_\pi = 0.2$). It means that the σ band is dirtier than the π band. The subsequent decrease in the low-field slope of $N_\pi(0, h)$ upon the C doping can be explained by an increase in the D_σ/D_π ratio, i.e., by a more rapid enhancement of the π-band intraband scattering as compared with the scattering in the σ band. The dashed line plots the theoretical $N_\pi(0, h)$ dependence at $D_\sigma/D_\pi = 1$. The situation is different for the Al doping. Here, the field dependencies of $N_\pi(0, h)$ reveal no change upon doping. The same explanation as used above for the excess current versus field can be applied, underlying the conservation of the superconducting clean limit for Al doping.

7.3 Multiband Superconductivity in the 122-type Iron Pnictides

Discovery of superconductivity in compounds containing iron was very surprising since magnetism and superconductivity are contradicting physical phenomena. Iron, the archetypal ferromagnet, with locally polarized spins pointing to one direction, was supposed to break apart any Cooper pairs, the basic constituent of superconductors. Studies of superconducting iron pnictides, where iron is always placed together with one of two pnictogens (phosphorus or arsenic) are just at the beginning. In many aspects, the new systems are similar but in others very different from the cuprates. Similar to the high-T_c cuprates, the superconductivity in iron pnictides is enabled by chemical doping of the antiferromagnetic parental compounds. As a consequence, a complicated phase diagram *Transition temperature versus doping* arises with underdoped, optimally doped (with the highest T_c), and overdoped superconducting samples revealing different properties. But in contrast to the cuprates where the parent, undoped compounds are insulators, in the case of iron pnictides, these are (semi)metallic. The highest transition temperature (up to 56 K) among different iron pnicitides has been achieved in the optimally doped REFeAsO(F), or the 1111 group with Gd, Nd, or Sm [45] standing for a rare element RE where doping is performed by fluorin partly replacing oxygen. Considerable interest has also been attracted by another class of iron pnictide superconductors based on AFe_2As_2 with A = Ba, Sr, and Ca, referred to as the 122-type group. The 122-type compounds are chemically and structurally simpler and less anisotropic than the 1111 ones. The maximum T_c of 38 K is obtained in the optimally hole-doped $Ba_{0.6}K_{0.4}Fe_2As_2$ system [46] and also the electron-doped $Ba(Fe_{1-x}Co_x)_2As_2$ crystals with T_c about of 25 K are available [47]. In contrast to the 1111 systems, the 122-type parent compounds show magnetic (from paramagnetic to SDW antiferromagnetic phase) and structural transition (from tetragonal to orthorhombic phase) at the same temperature of about 140 K. This transition is gradually suppressed by chemical doping, but the phase diagram *Temperature versus doping* shows an overlap between the SDW/orthorhombic and superconducting phases for $x = 0.2 - 0.4$ in $Ba_{1-x}K_xFe_2As_2$ [48]. Figure 7.11 shows the crystal structure of $CaFe_2As_2$ below antiferromagnetic transition [49]. Iron moments are oriented along a axis of the orthorhombic unit cell.

Both the cuprates and iron pnictides are layered and anisotropic systems. The cuprate superconductors are basically single-band compounds, where the conduction band is formed from the Cu $3d$ orbitals. There is a single superconducting energy gap which has a nontrivial $d_{x^2-y^2}$-wave symmetry. Band structure calculations on iron pnictides [50] have shown that the multiple Fe $3d$ bands located near the Fermi energy are responsible for the appearance of multiple Fermi surface (FS) sheets. The multiband/multigap superconductivity with interband interactions leading to an exotic s-wave pairing with a sign reversal of the order parameter between different FS sheets [51] (Fig. 7.12) stands among the hot candidates to explain the high-T_c superconductivity in iron pnictides. Since the superconductivity appears when a doping suppresses the magnetic order, the magnetic fluctuations are considered as a clue for the attractive superconducting interaction.

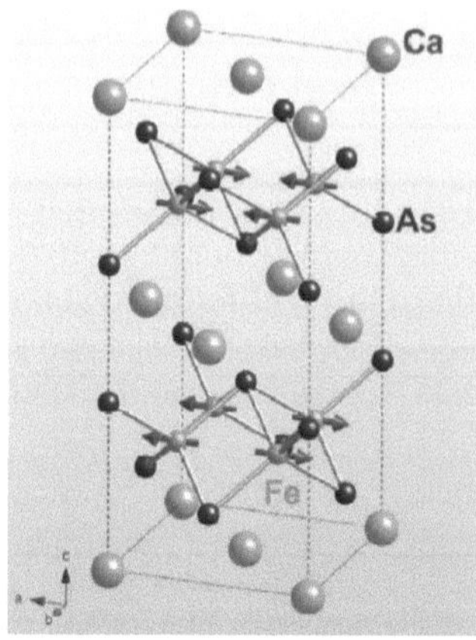

Fig. 7.11 Structure of 122 $CaFe_2As_2$ compound. *Arrows* depict the local magnetic moments of Fe ions. Reproduced from A.I. Goldman et al., Phys. Rev. B 78 (2008) 100506(R). Copyright 2008 by the American Physical Society

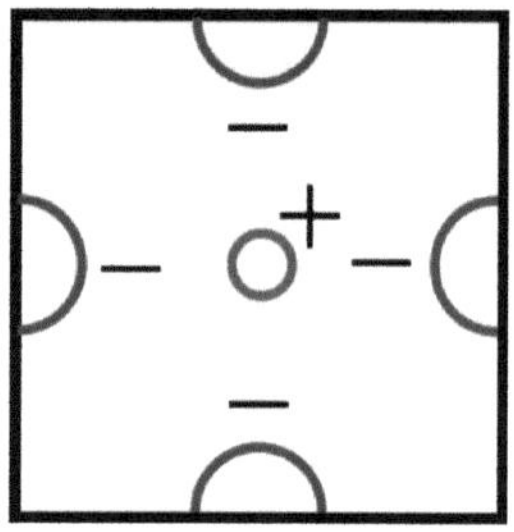

Fig. 7.12 Sketch of two Fermi surface sheets of pnictides with a sign reversal of the superconducting order parameter between them

Several authors have addressed the problem of the superconducting energy gap in pnictides by the point-contact spectroscopy. While in the early studies some indications for the nodal order parameter have been claimed [52, 53], later most of the papers point to the nodeless superconductivity in the optimally doped 1111 and 122 systems. T.Y. Chen et al. [54] found in $SmFeAs(O_{1-x}F_x)$ ($x = 0.15$ and 0.30) that one s-wave superconducting gap is independent of contacts, while many other 'gap-like' features vary appreciably for different contacts. The determined gap value gives $2\Delta_S/k_BT_c = 3.68$, close to the BCS prediction of 3.52. The superconducting gap decreases with temperature and vanishes at T_c, in a manner similar to the BCS behavior but dramatically different from that of the nodal pseudogap behavior in cuprate superconductors. R.S. Gonnelli et al. [55] in their point-contact Andreev-reflection spectroscopy experiments performed on LaFeAsO(F) polycrystals with $T_c = 27$ K and SmFeAsO(F) polycrystals with $T_c = 53$ K obtained evidences for two nodeless gaps in the superconducting state characterized by the coupling ratios below and much above the BCS value, respectively.

In the following, we report our point-contact spectroscopy study on the $Ba_{1-x}K_xFe_2As_2$, x = 0.45 single crystals recently published in [56]. The crystals were grown from a Sn flux. The resistive measurements showed the onset of the superconducting transitions below 30 K and the zero resistance at 27 K. A local transition temperature measured by the point-contact technique showed superconducting T_c's between 23 and 27 K. The specific heat and the resistivity measurements on these crystals showed features at about 85 K [57]. They are found at the same temperature as on the undoped $BaFe_2As_2$ samples where the tetragonal-to-orthorhombic structural phase transition takes place. This transition is revealed at a lower temperature in our samples as compared with 140 K found in [46] on the samples made by a different method. The decreased structural/magnetic transition temperature is due to the amounts of Sn up to 1% incorporated into the bulk. A presence of tin does not significantly effect the high-temperature superconducting phase transition in $Ba_{0.55}K_{0.45}Fe_2As_2$.

For the point contact measurements, the crystals were cleaved to reveal fresh surface. For the measurements in the c direction, the fresh shiny surface was obtained by detaching the degraded surface layers by a scotch tape. The microconstrictions were prepared in situ by pressing a metallic Pt tip on a fresh surface of the superconductor. For the measurements with the point-contact current in the ab plane, a reversed tip-sample configuration was used. The freshly cleaved edge of the single crystal jetting out in ab direction was pressed on a piece of chemically etched copper.

$Ba_{0.55}K_{0.45}Fe_2As_2$ single crystals are highly resistive materials where the reaching of the ballistic regime is quite challenging. Point contacts made on such a highly resistive material have resistances between tens and hundreds of ohms which corresponds to the contact diameter of tens of nanometers. Indeed, the electron mean free path can also be of the same order, here. Then, precautions should be made to avoid junctions heating effects. In our work on iron pnictides, only the junctions without the conductance dips and irreversibilities in voltage dependencies are presented. Moreover, to preserve the spectroscopic conditions, we focus on PC junctions having a finite barrier strength parameter Z with a tunneling component in the spectrum.

Figure 7.13a shows typical point contact spectra obtained on $Ba_{0.55}K_{0.45}Fe_2As_2$ with the PC current preferably within the ab plane. The spectra present double enhanced conductances, the typical features of the Andreev reflection of quasiparticles coupled via two superconducting energy gaps. The first enhancement starts below 20 mV with gap-like humps at about 10 mV, while the second one is located below $\simeq$ 5–7 mV. On the spectra 2 and 3 also two symmetrical maxima at 2–3 mV are visible. Majority of the spectra measured in the ab direction revealed a heavily broadened enhanced conductance near zero bias as indicated by the spectrum 4. This is most probably caused by the sample inhomogeneities on the nanoscale. The presented spectra are normalized to their respective normal state and fitted to the two gap BTK model (7.2) (symbols). The resulting values of energy gaps are spread in the range of 2–5 meV and 9–10 meV for the small and large gap, respectively. The values of smearing parameters were 10%, 60%, 30%, and 100% of each energy gap value for curves 1, 2, 3, and 4, respectively. For each presented fit, different values of Z for the two bands were also necessary. Typically, Z_L for the band with a large gap was about 0.4–0.8, while Z_S was twice smaller. Although the s-wave two gap BTK formula has been successfully used to fit our point contact data, a possibility of unconventional pairing symmetry cannot be completely ruled out. Obviously, rather strongly broadened spectra as presented here could be in principle fitted also

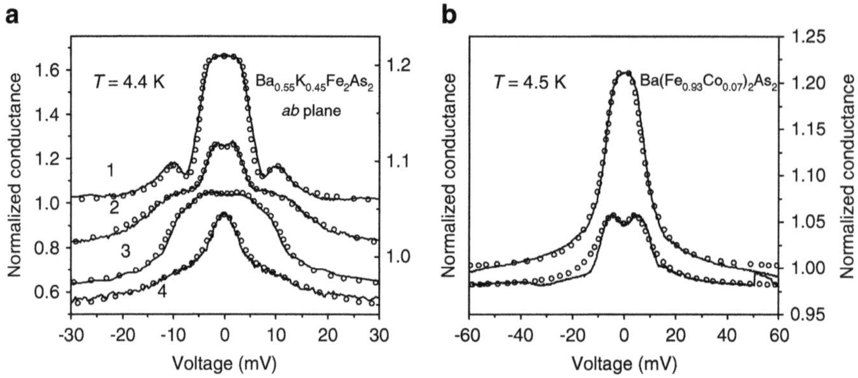

Fig. 7.13 (**a**) Typical point-contact spectra on $Ba_{0.55}K_{0.45}Fe_2As_2$ (*lines*) measured at 4.4 K with the PC current mostly within the *ab* plane of the sample. Spectra are normalized to their respective normal state and fitted to the two gap BTK model (*symbols*). For curves 2 and 4 right *y* axis applies. Spectra 2, 3 and 4 are vertically shifted. (**b**) Spectra of Pt–$Ba(Fe_{0.93}Co_{0.07})_2As_2$ junctions measured at 4.5 K (*lines*). Fits to the single gap BTK model are shown by *circles*

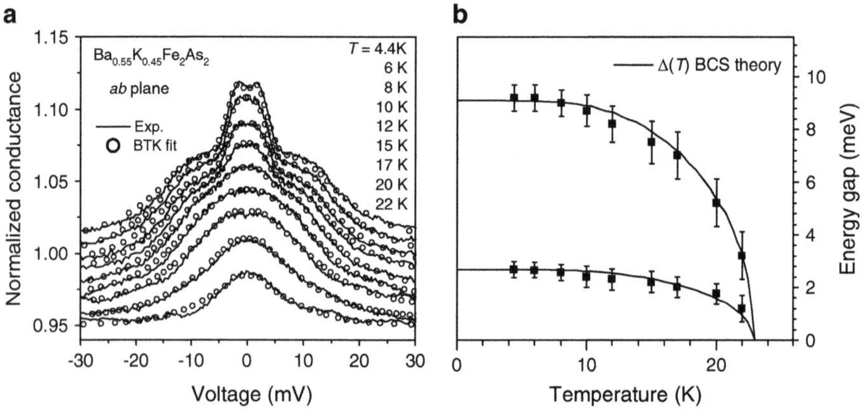

Fig. 7.14 (**a**) Temperature evolution of spectrum 2 from Fig. 7.13 (*lines*). Fits to the BTK model are indicated by symbols. (**b**) Values of the energy gaps obtained from fitting procedure for distinct temperatures (*symbols*). *Lines* represent the BCS type temperature dependencies of energy gaps

by a model taking into account anisotropic or nodal gaps, if an appropriate current injecting angle was selected [58].

In Fig. 7.14a, the temperature evolution of the second spectrum from the previous figure is presented. All the spectra (lines) were normalized to the conductance measured at 27 K and fitted to the two-band BTK model (7.2) with a proper temperature smearing involved. Obviously, the spectrum at the lowest temperature reminds the two gap spectrum of MgB_2 for a highly transparent junction with conductance enhancements due to Andreev reflection of quasiparticles. As the temperature is increased, the double enhanced point-contact conductance corresponding to two energy gaps is gradually smeared out and spectrum intensity decreases. The best fit for each temperature is shown by open circles. The extracted values of the gaps at different temperatures are shown in Fig. 7.14b (symbols) following nicely a BCS prediction (lines) rescaled to the size of the respective gap. The values of the energy gaps at lowest temperature are for the small gap $\Delta_S \simeq 2.7$ meV and the large gap $\Delta_L \simeq 9.2$ meV, which corresponds to the coupling strengths $2\Delta_S/kT_c \simeq 2.7$ and $2\Delta_L/k_BT_c \simeq 9$ for $T_c = 23$ K. The smearing parameters (about 60% of the respective gap values), the barrier strengths $Z \simeq 0.3$ and 0.6, and the weight factor $\alpha \simeq 0.5$ obtained at $T = 4.4$ K were kept constant at higher temperatures. From the data obtained on more junctions, we have observed that the gaps are scattered as $2\Delta_L/k_BT_c \simeq 2.5$–4 and $2\Delta_L/k_BT_c \simeq 9$–10. These results are in a good agreement with ARPES results of Ding et al. [59]. They observed three FS sheets in $Ba_{0.6}K_{0.4}Fe_2As_2$ crystals with T_c of 37 K: an inner hole-like FS pocket, an outer hole-like Fermi surface sheet, both centered at the Brillouin zone center and a small electron-like FS at the M point. A large superconducting energy gap ($\Delta_L = 12$ meV) was detected on the two small hole-like and electron-like FS sheets, while a small gap ($\Delta_S = 6$ meV) was found on the large hole-like FS. Both gaps have been found as isotropic closing at the same T_c. Two small FS sheets with a very strong coupling strength $2\Delta_L/k_BT_c \simeq 8$ are connected by the $(\pi, 0)$ SDW vector in the parent compound, indicating the importance of the interband interaction between these two nested FS sheets also for superconductivity.

Recently, we have extended our point-contact measurements to the electron- doped system of the 122-type family, namely to the optimally doped $Ba(Fe_{0.93}Co_{0.07})_2As_2$ samples with T_c of approximately 23 K [60]. The single crystals were grown from FeAs/CoAs flux from a starting load of Ba, FeAs, and CoAs precursors. Details of the preparation resulting to large crystals of a few millimeter size can be found elsewhere [61]. In Fig. 7.13b, the typical spectra of the Pt-$Ba(Fe_{0.93}Co_{0.07})_2As_2$ junctions measured at 4.5 K are presented. The spectra show enhanced differential conductances and the bottom curve reveal also a single pair of peaks at about 5 mV. In no case, a double enhanced conductance spectrum indicative for two gap superconductivity was found. Indeed, the spectra on $Ba(Fe_{0.93}Co_{0.07})As_2$ can be fitted just by a single s-wave gap BTK formula. This is in strong contrast to the point-contact spectra of the hole-doped $Ba_{0.55}K_{0.45}Fe_2As_2$. The fits on the electron-doped system have yielded the superconducting energy gap of around 5–6 meV. In all spectra, a significant broadening was observed which is witnessed by a large value of the Γ parameter spreading between 50% and 100% of the respective gap value. The superconducting transition temperature at the junctions was 22 K very close to the bulk T_c determined from transport measurements. With this T_c, the coupling strength $2\Delta/k_BT_c$ between 5.3 and 6.3 can be determined. Our results are in a reasonable agreement with the ARPES measurements of Terashima et al. [62, 63]. They have shown that in the sample of $Ba(Fe_{0.85}Co_{0.15})_2As_2$ with $T_c = 25$ K due to the electron doping, the inner hole-like FS sheets is absent and the nesting conditions are switched from the inner hole FS sheet to the outer one which is connected to the electron FS sheets by the $(\pi, 0)$ SDW vector. Strong coupling strengths $2\Delta_S/k_BT_c \simeq 4.5$ and $2\Delta_S/k_BT_c \simeq 6$ have been found on the hole and electron FS's, respectively. Although our results are only preliminary and the statistics is not yet sufficient, the data seem to prove that if there are multiple gaps in the optimally electron doped $Ba(Fe_{0.93}Co_{0.07})_2As_2$ superconductors they are quite close to each other.

7.4 Conclusions

We have demonstrated that point-contact spectroscopy has been an effective tool for studies of the superconducting energy gap, its multiplicity, temperature, and magnetic field dependence. In our measurements, the multigap s-wave superconductivity has been detected in the magnesium diboride and related doped materials. Similarly, the nodeless multiple superconducting energy gaps have been observed in the multiband 122-type iron pnictides.

The point-contact spectroscopy allowed also for the studies of the interband and the intraband scattering effects. Strong reluctance to an increase in the interband scattering in the MgB_2-related compounds-by doping but also by irradiation or technology treatment is caused by the different local point symmetry of the π and σ bands, respectively. It helps to protect the two gap superconductivity even in the samples with high level of carbon and aluminum doping, thus keeping a relatively high T_c. The studies on intraband scatterings show that the carbon doping enhances significantly the scattering inside the π band. This leads to a strong increase in the upper critical magnetic field, particularly at the low temperatures with importance for practical applications. The partial aluminum substitution of magnesium in MgB_2 bringing an excess of electrons does not affect significantly the scattering in the sample and mostly leads to a decrease of T_c due to the decrease of the hole density of states in the σ band.

In the hole-doped $Ba_{0.55}K_{0.45}Fe_2As_2$, the point-contact data point to an existence of two very distinct superconducting energy gaps with the strength $2\Delta_S/k_BT_c \sim 2.5$–4 and $2\Delta_L/k_BT_c \sim 9$–10, respectively. In the optimally electron-doped $Ba(Fe_{0.93}Co_{0.07})_2As_2$, if there are two gaps present they are very close to each other having a strong coupling $2\Delta_S/k_BT_c$ between 5 and 6. Our experiments could not resolve the problem of a potential sign reversal between different superconducting energy gaps which rests among many open issues in the high-T_c superconducting pnictides.

References

1. G.E. Blonder, M. Tinkham, T.M. Klapwijk, Phys. Rev. B 25, 4515 (1982)
2. A. Plecenik et al., Phys. Rev. B 49, 10016 (1996)
3. J. Nagamatsu, N. Nakagawa, T. Muranaka, Y. Zenitani, J. Akimitsu, Nature (Lond.) 410, 63 (2001)
4. A.Y. Liu, I.I. Mazin, J. Kortus, Phys. Rev. Lett. 87, 087005 (2001)

5. T. Dahm, N.Schopohl, Phys. Rev. Lett. 91, 017001 (2003)
6. H.J. Choi, D. Roundy, H. Sun, M.L. Cohen, S.G. Louie, Nature 418, 758 (2002)
7. H.J. Choi, D. Roundy, H. Sun, M.L. Cohen, S.G. Louie, Phys. Rev. B 66, 020513 (2002)
8. G. Rubio-Bollinger, H. Suderow, S. Vieira, Phys. Rev. Lett. 86, 5582 (2001)
9. G. Karapetrov, M. Iavarone, W.K. Kwok, G.W. Crabtree, D.G. Hinks, Phys. Rev. Lett. 86, 4374 (2001)
10. H. Schmidt, J.F. Zasadzinski, K.E. Gray, D.G. Hinks, Phys. Rev. B 63, 220504 (2001)
11. T. Takahashi, T. Sato, S. Souma, T. Muranaka, J. Akimitsu, Phys. Rev. Lett. 86, 4915 (2001)
12. F. Bouquet, Y. Wang, I. Sheikin, T. Plackowski, A. Junod, S. Lee, S. Tajima, Phys. Rev. Lett. 89, 257001 (2002)
13. Y. Wang et al., Physica C 355, 179 (2001)
14. F. Giubileo, D. Roditchev, W. Sacks, R. Lamy, D.X. Thanh, J. Klein, S. Miraglia, D. Fruchart, J. Marcus, Ph. Monod, Phys. Rev. Lett. 87, 177008 (2001)
15. M. Iavarone, G. Karapetrov, A.E. Koshelev, W.K. Kwok, G.W. Crabtree, D.G. Hinks, W.N. Kang, E.-M. Choi, H.J. Kim, H.-J. Kim, S.-I. Lee, Phys. Rev. Lett. 89, 187002 (2002)
16. P. Martinez-Samper, J.G. Rodrigo, G. Rubio-Bollinger, H. Suderow, S. Vieira, S. Lee, S. Tajima, Physica C 385, 233 (2003)
17. X.K. Chen et al., Phys. Rev. Lett. 87, 157002 (2001)
18. P. Szabó, P. Samuely, J. Kačmarčík, T. Klein, J. Marcus, D. Fruchart, S. Miraglia, C. Marcenat, A.G.M. Jansen, Phys. Rev. Lett. 87, 137005 (2001)
19. P. Szabó, P. Samuely, Z. Pribulová, M. Angst, S.L. Bud'ko, P.C. Canfield, J. Marcus, Phys. Rev. B 75, 144507 (2007)
20. R.S. Gonnelli, D. Daghero, G.A. Ummarino, V.A. Stepanov, J. Jun, S.M. Kazakov, J. Karpinski, Phys. Rev. Lett. 89, 247004 (2002)
21. P. Szabó, P. Samuely, J. Kačmarčik, T. Klein, J. Marcus, A.G.M. Jansen, Physica B 328, 10 (2003)
22. S.L. Bud'ko, V.G. Kogan, P.C. Canfield, Phys. Rev. B 64, 180506 (2001)
23. L. Lyard, P. Samuely, P. Szab, C. Marcenat, T. Klein, K.H.P. Kim, C.U. Jung, H.-S. Lee, B. Kang, S. Choi, S.-I. Lee, L. Paulius, J. Marcus, S. Blanchard, A.G.M. Jansen, U. Welp, G. Karapetrov, W.K. Kwok, Phys. Rev. B 66, 180502(R) (2002)
24. M.R. Eskildsen, N. Jenkins, G. Levy, M. Kugler, O. Fischer, J. Jun, S.M. Kazakov, J. Karpinski, Phys. Rev. B 68, 100508(R) (2003)
25. M.R. Eskildsen, M. Kugler, S. Tanaka, J. Jun, S.M. Kazakov, J. Karpinski, O. Fischer, Phys. Rev. Lett. 89, 187003 (2002)
26. V.G. Kogan, S.L. Bud'ko, Physica C 385, 131 (2003)
27. L. Lyard, P. Szabó, T. Klein, J. Marcus, C. Marcenat, K.H.P. Kim, B.W. Kang, H.-S. Lee, S.-I. Lee, Phys. Rev. Lett. 92, 057001 (2004)
28. G. Blumberg, A. Mialitsin, B.S. Dennis, M.V. Klein, N.D. Zhigadlo, J. Karpinski, Phys. Rev. Lett. 99, 227002 (2007)
29. V. Moshchalkov, M. Menghini, T. Nishio, Q.H. Chen, A.V. Silhanek, V.H. Dao, L.F. Chibotaru, N.D. Zhigadlo, J. Karpinski, Phys. Rev. Lett. 102, 117001 (2009)
30. I.I. Mazin, O.K. Andersen, O. Jepsen, A.A. Golubov, O.V. Dolgov, J. Kortus, Phys. Rev. B 69, 056501 (2004)
31. J. Kortus, O.V. Dolgov, R.K. Kremer, A.A. Golubov, Phys. Rev. Lett. 94, 027002 (2005)
32. M. Angst, S.L. Bud'ko, R.H.T. Wilke, P.C. Canfield, Phys. Rev. B 71, 144512 (2005)
33. P. Samuely, Z. Hol̆anová, P. Szabó, J. Kačmarčík, P.C. Canfield, S.L. Bud'ko, Phys. Rev. B 68, 020505(R) (2003)
34. Z. Hol̆anová, P. Szabó, P. Samuely, P.C. Canfield, S.L. Bud'ko, Phys. Rev. B 70, 064520 (2004)
35. T. Klein, L. Lyard, J. Marcus, C. Marcenat, P. Szabó, Z. Hol̆anová, P. Samuely, B. Kang, H.-J. Kim, H.-S.Lee, S.I. Lee, Phys. Rev. B 73, 224528 (2006)
36. P. Samuely, P. Szabó, J. Kačmarčík, T. Klein, Physica C 385, 244 (2003)
37. S. Tsuda, T. Yokoya, T. Kiss, T. Shimojima, S. Shin, T. Togashi, S. Watanabe, C. Zhang, C.T. Chen, S. Lee, H. Uchiyama, S. Tajima, N. Nakai, K. Machida, Phys. Rev. B 72, 064527 (2005)

38. M. Iavaronne, R. Di Capua, A.E. Koshelev, W.K. Kwok, F. Chiarella, R. Vaglio, W.N. Kang, E.M. Choi, H.J. Kim, S.I. Lee, A.V. Pogrebnyakov, J.M. Redwing, X.X. Xi, Phys. Rev. B 71, 214502 (2005)
39. M. Putti, M. Affronte, C. Ferdeghini, P. Manfrinetti, C. Tarantini, E. Lehmann, Phys. Rev. Lett. 96, 077003 (2006)
40. S.C. Erwin, I.I. Mazin, Phys. Rev. B 68, 132505 (2003)
41. R.S. Gonnelli, D. Daghero, A. Calzolari, G.A. Ummarino, V. Dellarocca, V.A. Stepanov, S.M. Kazakov, N. Zhigadlo, J. Karpinski, Phys. Rev. B 71, 060503 (2005)
42. Yu.G. Naidyuk, O.E. Kvitnitskaya, I.K. Yanson, S. Lee, S. Tajima, Solid State Commun. 133, 363 (2005)
43. A.E. Koshelev, A.A. Golubov, Phys. Rev. Lett. 90, 177002 (2004)
44. Y. Bugoslavsky, Y. Miyoshi, G.K. Perkins, A.D. Caplin, L.F. Cohen, A.V. Pogrebnyakov, X.X. Xi, Phys. Rev. B 72, 224506 (2005)
45. Z.-A. Ren, J. Yang, W. Lu, W. Yi, X.-L. Shen, Z.-C. Li, G.-C. Che, X.-L. Dong, L.-L. Sun, F. Zhou, Z.-X. Zhao, Europhys. Lett. 82, 7002 (2008)
46. M. Rotter, M. Tegel, D. Johrendt, Phys. Rev. Lett. 101, 107006 (2008)
47. A.S. Sefat, R. Jin, M.A. McGuire, B.C. Sales, D.J. Singh, D. Mandrus, Phys. Rev. Lett. 101, 117004 (2008)
48. H. Chen, Y. Ren, Y. Qiu, W. Bao, R.H. Liu, G. Wu, T. Wu, Y.L. Xie, X.F. Wang, Q. Huang, X.H. Chen, Europhys. Lett. 85, 17006 (2009)
49. A.I. Goldman, D.N. Argyriou, B. Ouladdiaf, T. Chatterji, A. Kreyssig, S. Nandi, N. Ni, S.L. Bud'ko, P.C. Canfield, R.J. McQueeney, Phys. Rev. B 78, 100506(R) (2008)
50. D.J. Singh, M.H. Du, Phys. Rev. Lett. 100, 237003 (2008)
51. I.I. Mazin, D.J. Singh, M.D. Johannes, M.H. Du, Phys. Rev. Lett. 101, 057003 (2008)
52. L. Shan, Y. Wang, X. Zhu, G. Mu, L. Fang, C. Ren, H.-H. Wen, Europhys. Lett. 83, 57004 (2008)
53. K.A. Yates, L.F. Cohen, Z.-A. Ren, J. Yang, W. Lu, X.-L. Dong, Z.-X. Zhao, Supercond. Sci. Technol. 21, 002003 (2008)
54. T.Y. Chen, S.X. Huang, Z. Tesanovic, R.H. Liu, X.H. Chen, C.L. Chien, Physica C 469, 521 (2009)
55. R.S. Gonnelli, D. Daghero, M. Tortello, G.A. Ummarino, V.A. Stepanov, R.K. Kremer, J.S. Kim, N.D. Zhigadlo, J. Karpinski, Physica C 469, 521 (2009)
56. P. Szabó, Z. Pribulová, G. Pristaš, S.L. Bud'ko, P.C. Canfield, P. Samuely, Phys. Rev. B 79, 012503 (2009)
57. N. Ni, S.L. Bud'ko, A. Kreyssig, S. Nandi, G.E. Rustan, A.I. Goldman, S. Gupta, J.D. Corbett, A. Kracher, P.C. Canfield, Phys. Rev. B 78, 014507 (2008)
58. S. Kashiwaya, Y. Tanaka, Rep. Prog. Phys. 63, 1641 (2000)
59. H. Ding, P. Richard, K. Nakayama, L. Sugawara, T. Arakane, Y. Sekiba, A. Takayama, S. Souma, T. Sato, T. Takahashi, Z. Wang, X. Dai, Z. Fang, G.F. Chen, J.L. Luo, Europhys. Lett. 83, 47001 (2008)
60. P. Samuely, Z. Pribulová, P. Szabó, G. Pristaš, S.L. Bud'ko, P.C. Canfield, Physica C 469, 507 (2009)
61. N. Ni, M.E. Tillman, J.-Q. Yan, A. Kracher, S.T. Hannahs, S.L. Bud'ko, P.C. Canfield, arXiv:0811.1767 (2008)
62. K. Terashima, Y. Sekiba, J.H. Bowen, K. Nakayama, T. Kawahara, T. Sato, P. Richard, Y.-M. Xu, L.J. Li, G.H. Cao, Z.-A. Xu, H. Ding, T. Takahashi, arXiv:0812.3704 (2008)
63. Y. Sekiba, T. Sato, K. Nakayama, K. Terashima, P. Richard, J.H. Bowen, H. Ding, Y.-M. Xu, L.J. Li, G.H. Cao, Z.-A. Xu, T. Takahashi, arXiv:0812.4111 (2008)

Chapter 8
Nanoscale Structures and Pseudogap in Under-doped High-Tc Superconductors

M. Saarela and F.V. Kusmartsev

Abstract We show that superconductor–insulator transitions in oxides and FeAs-based high Tc superconducting multilayers may arise due to a charge density wave instability induced by charged impurities and the over-screening of the long-ranged part of the Coulomb interaction, which is enhanced due to decreasing carrier density [1]. When the carrier density is low enough, impurities begin to trap particles and form bound states of clusters of charge carriers, which we call Coulomb bubbles. These bubbles are embedded inside the superconductor and form nuclei of the new insulating state. The growth of a bubble is terminated by the Coulomb force and each of them has a quantized charge and a fluctuating phase. When clusters first appear, they are covered by superfluid liquid due to the proximity effect and invisible. However, when the carrier density decreases the size of bubbles increases and the superconducting proximity inside them vanishes. The insulating state arises via a percolation of these insulating islands, which form a giant percolating cluster that prevents the flow of the electrical supercurrent through the system. We also show the formation of two groups of charge carriers in these compounds associated with free and localized states. The localized component arises due to the Coulomb bubbles. Our results are consistent with the two-component picture for cuprates deducted earlier by Gorkov and Teitelbaum [2] from the analysis of the Hall effect data and ARPES spectra. The Coulomb clusters induce nanoscale superstructures observed in scanning tunneling microscope (STM) experiments [3] and are responsible for the pseudogap [4].

8.1 Introduction

The discovery of high transition temperature superconductors in cuprates and more recently, for instance, in pnictides has created a long standing excitement in the study of superconductors, because new electronic devices have become feasible, and also because these materials show unconventional behavior as superconductors. In the conventional BCS theory of superconductivity [5], electrons are paired in momentum space, forming Cooper pairs. Cooper pairs are bosons and can occupy a coherent, macroscopic Bose-condensate state. At sufficiently low temperatures, the system becomes superfluid and the superconductivity may then be described as the superfluid flow of the charged condensed liquid. Yet, the conventional Cooper pairing corresponds to momentum space correlations between the motion of two electrons, and in this sense, they are not point-like bosons at all. In the system of charged boson particles, the superconducting state arises in a similar way when a macroscopic number of bosons is condensed on the lowest possible energy level and becomes superfluid below a critical temperature. According to the Landau theory [6], the flow will be superfluid when its velocity is lower than the critical velocity associated with the lowest energy elementary excitation.

The size of Cooper pairs is assumed to be inversely proportional to the square root of the superconducting gap. At optimal doping, the coherence length and therewith the size of Cooper pairs in high T_c cuprates is of the order of interparticle distances. This is in contrast to low temperature superconductors where the size of Cooper pairs is usually of hundreds or even thousands of interatomic distances. In over-doped cuprates, the value of superconducting gap decreases, and therefore, the "size" of Cooper pairs increases and becomes of the order of dozens of interparticle spacings, but the under-doped case is much more complicated because of the large pseudogap. The superconducting phase is very inhomogeneous and the coherence length is small, i.e., of the order of the Bohr radius. There are many experimental indications that pairs of electrons (or, more precisely, holes) are bound in the coordinate space, although the nature of the pairing in high T_c superconductors at any doping is not yet established and many debates are still going on. However, a very appealing possibility is that in the under-doped region the size of pairs decreases with decreasing doping and therefore the type of pairing has crossed from the BCS regime to the BEC regime, where molecular-like pairs form the Bose-Einstein condensed state. In such a case, the charge carriers could be treated as charged bosons in the homogeneous fluid unless the temperature is so high that pairs break, and then we need to take into account the Fermi kinetic energy E_F, which in that strongly correlated regime is significantly smaller than the characteristic energy of the Coulomb interaction E_c. This ratio determines the parameter $r_s = E_c/E_F$. Our studies here concentrate to the physics in the under-doped region of cuprates where $r_s \gg 1$.

Increasing temperature or disorder destroys the superconducting state, but we will show that the superconducting state vanishes even at zero temperature when the charge carrier density *decreases*. The effect arises due to the presence of charged impurities and strong over-screening of the Coulomb interaction at low densities. As a result, superconducting charge carriers spontaneously form self-trapped clusters around each impurity, which we will call *Coulomb bubbles* (CB). The mechanism of formation of Coulomb bubbles and electronic phase separation described here for charged bosons is equally well applicable to charged fermions, because in both cases the dominating role is played by the Coulomb interaction and its nonlinear screening. However, it is also important to note that for fermionic Coulomb bubbles the internal, energetical, and spatial structure will be slightly different as well as the critical density when CBs first appear. For fermions bound inside the Coulomb bubble, the Pauli principle must be obeyed and thus lower density is required for the bubble formation. The degeneracy of the band structure, however, reduces differences between fermionic and bosonic clusters by lowering the Fermi energy and increasing the number of particles allowed in one quantum state. That property leads, for instance, to the electron-hole liquid in highly degenerate semiconductors such as silicon and germanium.

With these arguments in mind, we present here a fully microscopic many-body description and numerical simulations of the quantum charged boson fluid that includes disorder in the form of charged point impurities and unconstrained charge fluctuations around them. The Coulomb fluids studied correspond to mobile holes located on Cu sites and charged impurities located on oxygen sites. The mobile and localized states are co-existing, and their balance is controlled by doping. At high doping, a large Fermi surface is open. There the density of real charge carriers (holes) is significantly larger than the density of the doped ones. The charged impurities are very heavy, they are localized on oxygen sites and therefore they do not contribute into transport. When doping decreases, more and more carriers are localized to these impurities. Finally in the limit of low doping, the bubbles form a regular lattice. We demonstrate explicitly how the over-screening leads to the spontaneous formation of Coulomb bubbles in layered or quasi-two-dimensional superconductors, when the density of current carriers decreases. The charge neutrality of the whole system is maintained by a charged, structureless jellium background, which is mostly associated with polarons localized on oxygen sites in Cu–O planes of cuprates.

In a closer look, we arrive at the following scenario of the quantum superconductor – insulator phase transition. In the attempt to screen the individual charges of impurities, the local superconducting density develops a giant charge density oscillation (GCDO) around each impurity [1]. The GCDO amplitude increases when the density of superfluid boson decreases. An electronic phase separation into localized and superconducting states takes place when the amplitude at the first minimum around the absolute maximum of the GCDO is comparable with the average boson density. As a result, a droplet of charged bosons is self-trapped around the impurity and decoupled from the rest of the superconducting condensate. The number of particles inside each droplet is fixed and therefore the phase is fluctuating, while in the rest of the condensate the phase is fixed and charge is fluctuating, giving the possibility to form a superconducting current. In order to destroy the global superconducting state, the CBs form an infinite cluster, which prevents percolation of the supercurrent from one side of the sample to the other side.

The Coulomb bubble is like a microscopic molecule, which is formed around a doped impurity as holes are trapped due to many-body over-screening. It behaves like a quantum dot under Coulomb blockade, but because of its small size the superconducting order parameter penetrates inside the bubble due to the proximity effect [1,7–10]. Such superconducting state with embedded heavy CBs behaves like a bulk superconductor and is a new kind of a ground state where superfluid density is small, and there are hidden nanoscale superstructures formed by these molecules. At very low temperatures, these molecules are localized and therefore they can be revealed by STM.

The rest of this article is organized in four chapters. In the next chapter, we define our microscopic model, which gives the theoretical foundation to the idea of two types of charge carriers. The loalized component of them forms Coulomb bubbles around impurities. Results on the simulations are presented here. In Chap. 3 we discuss connections to experiments based on the extensive analysis done by Gorkov and Teitelbaum [2, 11]. We will show that the doping-dependent activation energy in their model can be understood as the binding energy of holes into the Coulomb bubble. Furthermore, that energy is responsible of the pseudogap temperature. In Chap. 4 we show that nanoscale structures seen in STM measurements in underdoped region can be described by appearance of Coulomb bubbles, and finally in Chap. 5, we conclude our work and discuss future possibilities with pnictide superconductors.

8.2 Microscopic Origin of Two Types of Charge Carriers

The original model for cuprate compounds, the t-J model, has been derived by Zhang and Rice (ZR) in 1988 from a two-band model [12]. Since that time, the majority of the theoretical research has been focused around this single band t-J model. In their derivation, Zhang and Rice have started with the Hamiltonian describing a single layer of square planar coordinated Cu and O atoms [12] having the form:

$$H = \sum_{i,\sigma} \varepsilon_d d^\dagger_{i\sigma} d_{i\sigma} + \sum_{i,\sigma} \varepsilon_p p^\dagger_{i\sigma} p_{i\sigma} + U \sum_i d^\dagger_{i\uparrow} d_{i\uparrow} d^\dagger_{i\downarrow} d_{i\downarrow} + H', \tag{8.1}$$

where H' is the hybridization between d holes located on Cu sites and p holes located on oxygen sites.

$$H' = \sum_{i,\sigma} \sum_{l\in\{i\}} V_{il} d^\dagger_{i\sigma} p_{l\sigma} + H.C. \tag{8.2}$$

with $V_{il} = \pm t$, i.e., it is proportional to the overlapping integral t and the sign depends on the sign of the integral [12]. When we neglect the on-site Coulomb interaction on Cu- sites, i.e., put $U = 0$, we obtain the following energy spectrum consisting of three branches:

$$\varepsilon_\pm(k_x, k_y) = \pm\sqrt{\varepsilon_p^2/4 + 4t - 2t\cos(k_x) - 2t\cos(k_y)}, \text{ and } \varepsilon_0(k_x, k_y) = \varepsilon_p/2, \tag{8.3}$$

where we put for simplicity $\varepsilon_d = 0$. The spectrum indicates that there are two types of states, localized (the flat band) and mobile, which may be associated with any of the other two empty or partially filled bands. ZR have considered the hole localized on a single Cu-site which is hybridized with the hole located on four joint oxygens. They have also shown that such a state known now as the ZR-singlet may be mobile and its behavior including the doping can be described with the aid of the t-J model [12]. Our starting point is the same as Zhang and Rice had, in a sense that charge carriers in cuprates are holes and they move in the Cu–O planes. The second band is a homogeneous, inert, flat band responsible of neutralizing the system. However, to address the issue of localized states observed recently in numerous experiments, we extend the ZR-model to include the long-range Coulomb interaction. This made a dramatic difference to the appearance of the localized states [1, 13, 14]. Here, we show that besides the mobile ZR singlets or free spinless holes on a square lattice, there are localized states with energy levels located below the empty hole band, which is filled with doping. These states arise due to a nonlinear many-body screening. We assume that the localized states are induced by randomly distributed heavy impurities by doping. They may trap holes due to attractive Coulomb interaction at low enough doping.

In the quantitative description of such a system, we start with the microscopic Hamiltonian where charged, random impurities are embedded into the two-dimensional charged gas of current carriers,

$$H_e + H_I = -\sum_{i=1}^{N} \frac{\hbar^2}{2m} \nabla_i^2 + \frac{1}{2} \sum_{i,j=1}^{N}{}' \frac{e^2}{\varepsilon |\mathbf{r}_i - \mathbf{r}_j|} - \frac{\hbar^2}{2M_I} \nabla_0^2 - \sum_{i}^{N} \frac{e^2}{\varepsilon |\mathbf{r}_0 - \mathbf{r}_i|} . \quad (8.4)$$

We have separated the purely electronic (or bosonic) part H_e from terms containing the impurity with the mass M_I, and described by the Hamiltonian, H_I. For simplicity, we consider only one single impurity. The number of mobile holes (or bosons) is N, and they have the mass m, charge e, and position $\mathbf{r}_i$. The impurity is placed at $\mathbf{r}_0$ and its kinetic energy is controlled by the mass M_I. For a localized impurity, we let the mass grow to infinity. We also add into the Hamiltonian a structureless, charged jellium, which neutralizes the system. The strength of the Coulomb interaction depends on the dielectric constant ε. Both the hole (or boson) mass and dielectric constant depend on the band structure of the material, and it is convenient to use the atomic units where all distances are given in units of $r_0 = r_s r_B$ with the Bohr radius r_B and energies in Rydbergs.

The ground-state wave function which contains correlations between bosons and impurities is chosen in the form of the Jastrow-type variational ansatz [15]

$$\Psi(\mathbf{r}_1, \mathbf{r}_2, \ldots, \mathbf{r}_N) = \mathrm{e}^{\frac{1}{2}\sum_{i,j=1}^{N} u^{\mathrm{bb}}(|\mathbf{r}_i - \mathbf{r}_j|)} \mathrm{e}^{\frac{1}{2}\sum_{i=1}^{M} u^{\mathrm{It}}(|\mathbf{r}_i - \mathbf{r}_0|)} \mathrm{e}^{\frac{1}{2}\sum_{i=M+1}^{N} u^{\mathrm{In}}(|\mathbf{r}_i - \mathbf{r}_0|)} \quad (8.5)$$

We have extended the conventional many-body variational theory to the case where M bosons are trapped by the impurity. The wave function includes now a product of three components, boson–boson (bb) correlations, impurity-trapped boson (It) correlations, and impurity-nontrapped boson (In) correlations. Since particles are indistinguishable bosons the wave function should be symmetrized with respect to trapped and nontrapped particles, but as we will show later the overlap of their distributions becomes vanishingly small and in such a case we can ignore additional terms coming from the symmetrization.

The key ingredients of the theory are, along with the correlation functions, the two-particle distribution functions,

$$
\begin{aligned}
\rho^{\mathrm{bb}}(|\mathbf{r}_1-\mathbf{r}_2|) &= N(N-1)\frac{\int \mathrm{d}^2 r_3 \ldots \mathrm{d}^2 r_N \, |\Psi(\mathbf{r}_1,\ldots,\mathbf{r}_N)|^2}{\langle\Psi|\Psi\rangle} \\
&= n_0^2 g^{\mathrm{bb}}(|\mathbf{r}_1-\mathbf{r}_2|) \\
\rho^{I}(|\mathbf{r}-\mathbf{r}'|) &= \sum_{i=1}^{N} \frac{\langle\Psi^{\mathrm{IM}}|\delta(\mathbf{r}_0-\mathbf{r})\delta(\mathbf{r}_i-\mathbf{r}')|\Psi^{\mathrm{IM}}\rangle}{\langle\Psi^{\mathrm{IM}}|\Psi^{\mathrm{IM}}\rangle}, \\
&= \frac{n_0}{\Omega}\left[g^{\mathrm{It}}(|\mathbf{r}-\mathbf{r}'|)+g^{\mathrm{In}}(|\mathbf{r}-\mathbf{r}'|)\right]
\end{aligned}
\tag{8.6}
$$

where $g^{\mathrm{bb}}(|\mathbf{r}_1-\mathbf{r}_2|)$, $g^{\mathrm{It}}(|\mathbf{r}_0-\mathbf{r}_1|)$ and $g^{\mathrm{In}}(|\mathbf{r}_0-\mathbf{r}_1|)$ are the boson–boson, impurity-trapped boson, and impurity-nontrapped boson radial distribution functions, respectively, Ω is the total volume occupied by the system, and n_0 is the boson density.

The perfect screening condition requires the following normalizations, when we assume that the charge of the impurity, $-e$, has opposite sign to the boson charge.

$$
\begin{aligned}
\int \mathrm{d}\mathbf{r}\rho^{I}(r) &= \frac{N}{\Omega} \\
n_0 \int \mathrm{d}\mathbf{r} g^{\mathrm{It}}(r) &= M \\
n_0 \int \mathrm{d}\mathbf{r}[g^{\mathrm{In}}(r)-1] &= 1-M
\end{aligned}
\tag{8.7}
$$

The Fourier transforms of the distribution functions define the static structure functions,

$$
\begin{aligned}
S^{\mathrm{bb}}(k) &= 1 + n_0 \int \mathrm{d}^2 r \mathrm{e}^{\mathrm{i}\mathbf{k}\cdot\mathbf{r}}[g^{\mathrm{bb}}(r)-1] \\
S^{\mathrm{IM}}(k) &= n_0 \int \mathrm{d}^2 r \mathrm{e}^{\mathrm{i}\mathbf{k}\cdot\mathbf{r}}[g^{\mathrm{It}}(r)+g^{\mathrm{In}}(r)-1] \,.
\end{aligned}
\tag{8.8}
$$

The additional information needed to solve the variational many-body problem is the connection between the correlation functions and the physically observable distribution functions. This is provided by the infinite order hypernetted-chain (HNC) equations with the sum of nodal diagrams [15].

$$
\begin{aligned}
g^{\mathrm{bb}}(r) &= \mathrm{e}^{u^{\mathrm{bb}}(r)+N^{\mathrm{bb}}(r)} \\
g^{\mathrm{It}}(r) &= \mathrm{e}^{u^{\mathrm{It}}(r)+N^{I}(r)} \\
g^{\mathrm{In}}(r) &= \mathrm{e}^{u^{\mathrm{In}}(r)+N^{I}(r)}
\end{aligned}
\tag{8.9}
$$

Here, we have made the essential assumption that in the impurity radial distribution functions the dependences on $\exp(u^{\mathrm{It}}(r))$ and $\exp(u^{\mathrm{In}}(r))$ can be factored out and the sum of nodal diagrams $N^I(r)$ is the same in both quantities. This is strictly valid only when the spatial overlap of those correlation factors vanishes, which happens when the fluid clusterizes.

Within these assumptions, we can calculate the radial distribution function of the homogeneous Bose gas by minimizing the total energy without the impurity ([16]). The impurity distribution functions, $g^{\mathrm{It}}(r)$ and $g^{\mathrm{In}}(r)$, can then be solved by minimizing the chemical potential of one charged impurity [13, 17, 18],

$$\mu^{\mathrm{IM}} = n_0 \int \mathrm{d}^2 r \left[-\frac{e^2}{\varepsilon}\frac{1}{r} \left(g^{\mathrm{It}}(r) + g^{\mathrm{In}}(r) - 1 \right) + \frac{\hbar^2}{2m} |(\nabla \sqrt{g^{\mathrm{It}}(r)}|^2 + |\nabla \sqrt{g^{\mathrm{In}}(r)}|^2) \right] \\ + \frac{1}{2} \int \frac{\mathrm{d}^2 k}{(2\pi)^2 n_0} S^{\mathrm{IM}}(k)\, \tilde{w}^{\mathrm{IM}}_{\mathrm{ind}}(k) - M E_{\mathrm{bin}}\,, \qquad (8.10)$$

with the constraints of (8.7). In the last line, we introduce the induced potential which collects the many-body correlation effects between the impurity and bosons,

$$\tilde{w}^{\mathrm{IM}}_{\mathrm{ind}}(k) = \frac{\hbar^2 k^2}{4m} S^{\mathrm{IM}}(k) \left[\frac{1}{(S^{\mathrm{bb}}(k))^2} - 1 \right]. \qquad (8.11)$$

The set of Euler equations which minimize μ^{IM} can be written as,

$$-\frac{\hbar^2}{2m}\nabla^2 \sqrt{g^{\mathrm{It}}(r)} + \left[-\frac{e^2}{\varepsilon r} + w^{\mathrm{IM}}_{\mathrm{ind}}(r) \right] \sqrt{g^{\mathrm{It}}(r)} = E_{\mathrm{bin}} \sqrt{g^{\mathrm{It}}(r)} \\ -\frac{\hbar^2}{2m}\nabla^2 \sqrt{g^{\mathrm{In}}(r)} + \left[-\frac{e^2}{\varepsilon r} + w^{\mathrm{IM}}_{\mathrm{ind}}(r) \right] \sqrt{g^{\mathrm{In}}(r)} = 0 \qquad (8.12)$$

with the Fourier transform $w^{\mathrm{IM}}_{\mathrm{ind}}(r)$ of the induced interaction (8.11). Equation (8.12) defines the effective, coordinate space interaction

$$V_{\mathrm{eff}}(r) = -\frac{e^2}{\varepsilon r} + w^{\mathrm{IM}}_{\mathrm{ind}}(r) \qquad (8.13)$$

experienced by bosons around the bubble. If we think eqs. (8.12) as a set of Schrödinger-like equations, then the first equation gives the bound state solution in that effective potential and the second one is the zero energy solution with the boundary condition $\lim_{r\to\infty} g^{\mathrm{In}}(r) = 1$. However, this set of equations is not that simple, because the induced potential depends on the radial distribution functions. A consistent solution is found by iteration and it converges within a few iteration cycles.

By studying the boson-impurity pair distribution function, $g^{\mathrm{In}}(r)$, which describes the spatial charge-distribution of bosons around a single impurity, we demonstrate the formation of CBs when the density decreases (Figs. 8.1 and 8.2). One can clearly resolve the giant oscillation, the GCDO, closest to the impurity sitting at the origin, where the high density area is surrounded by lower charged

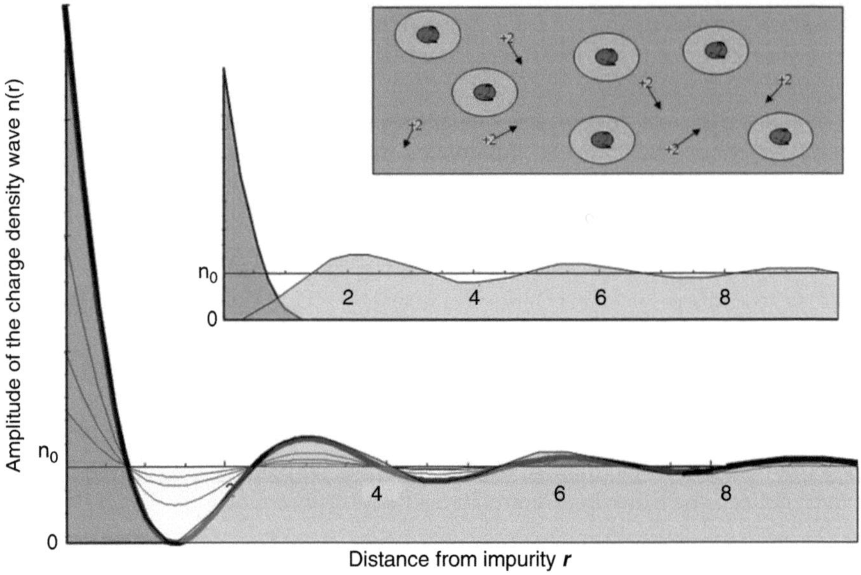

Fig. 8.1 A schematic distribution of dense bose fluid around an impurity at the origin is shown at different r_s values for the range $r_s = 1.5, 2, 2.5$, and 4. The first minimum decreases with increasing r_s. The density distribution is normalized to unity far away from the impurity. At the value $r_s = 4$, six holes are trapped into the bubble shown in *red*. They are separated from the superconducting holes shown in *green*. In the lower insert, the trapped and superconducting density distributions are overlapping, and in the upper insert, we show the top view of the density distribution. The *light blue* rings around trapped holes represent the revealed background charge

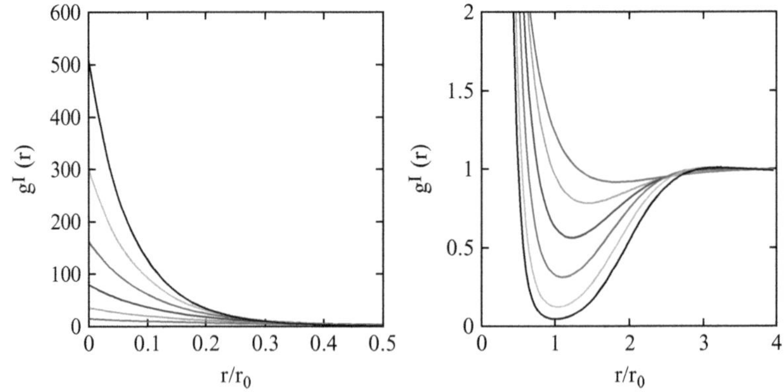

Fig. 8.2 The *left figure* shows the radial distribution function $g^{\mathrm{In}}(\mathbf{r})$ describing the distribution of bosons around the impurity located at $r = 0$ for r_s values 1., 1.5, 2., 2.5, 3., 3.5, the higher the peak the larger the r_s -value. In the *right figure*, the same radial distribution function $g^{\mathrm{In}}(\mathbf{r})$ are shown, but now the focus is on the region of the first minimum. One notices that when $r_s \to 3.5$ the boson density at the minimum vanishes. This indicates the separation of boson liquid into two fragments: bosons localized near impurity forming the Coulomb bubble and the rest, superfluid liquid

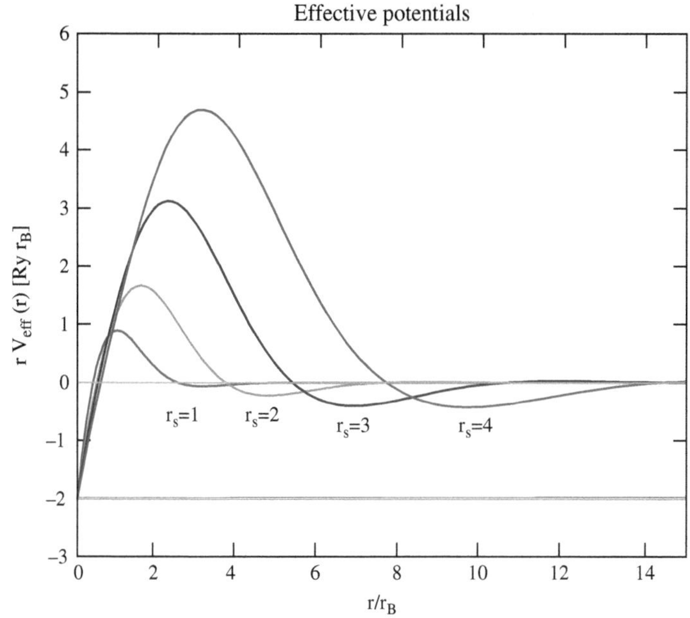

Fig. 8.3 The effective potentials multiplied by r are shown for r_s values indicated in the figure. The constant line at −2 is the bare Coulomb potential in atomic units. The behavior of the effective potential shows a huge over-screening which increases with r_s

density. The amplitude of this oscillation increases with r_s. At the value $r_s = 3.5$, this amplitude is so large that the charged fluid separates into two orthogonal phases. At that density, five bosons or holes are trapped around the impurity and decoupled from the rest of the superfluid forming CB, because the strong repulsive over-screening of the Coulomb interaction prevents tunneling from the trapped states to continuum states. The effective potential multiplied with r is shown in Fig. 8.3. In the atomic units used here, the pure Coulomb potential multiplied by r is a constant -2, and as shown the induced many-body effects strongly over-screen at intermediate ranges the bare potential.

The results plotted in Figs. 8.2 and 8.3 are for the case where the impurity is in the two-dimensional plane, and its charge is equal but opposite to the boson charge. They are very typical, and similar ones are obtained for cases where the impurity is outside the superconducting layer or its charge is a multiple or a fraction of the boson charge. The critical r_s value depends on these conditions. For instance, if the impurity charge is minus half of the boson charge, then the critical $r_s \sim 6.2$, or if the distance from an impurity to the superconducting layer is one Bohr radius, then the critical $r_s \sim 12$.

The behavior of the condensate phase around the CB for larger $r_s = 5, 10$, and 15 with six trapped bosons is shown in Fig. 8.4. The radius of the bubble grows linearly with r_s and it is $r \approx 28 r_B$ at $r_s = 15$. The almost vertical lines near the origin depict

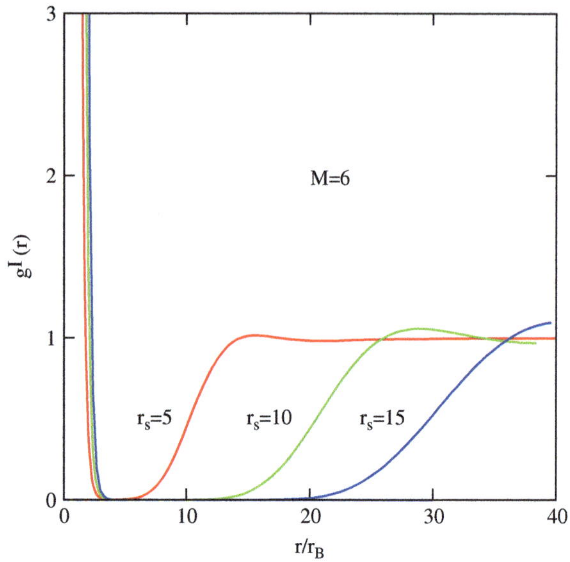

Fig. 8.4 The distribution of charge around the impurity for $r_s = 5, 10, 15$ when six bosons are trapped into the Coulomb bubble. The peaks near the origin show the distribution of the six trapped bosons for each of the r_s-values and the smooth part gives the distribution of the rest of the bosons. The distance is here in units of Bohr radius r_{B}

the lower part of the trapped boson distribution. The difference between the bound state distributions for different r_s values comes from the normalization (8.9), where we require that six particles are trapped within the area of one background particle. Within a very good approximation, the bound state distribution functions behave at this large r_s-value exponentially $\phi^2(r) = 8N_B r_s^2 \exp(-4r/r_{\mathrm{B}})$.

The dependence of the chemical potential of Coulomb bubbles having different number of particles trapped on r_s is presented in Fig. 8.5. The larger number of particles is trapped into the bubble, the lower the chemical potential it has. From the Euler equations (8.12), we get also the binding energy per trapped particle into the Coulomb bubble. It increases with r_s as shown in Fig. 8.6. and has the limiting value -4 Rydbergs of the two-dimensional Hydrogen-like atom. The mutual repulsion reduces the binding energy when more particles are bound.

8.3 Pseudogap and Two Types of Charge Carriers

Gorkov and Teitelbaum (GT) [2, 11] have made a detailed analysis of experimental data, which include the temperature dependence both for resistivity and the Hall coefficient in LSCO [19, 20], the ARPES experiments [21], and STM [22]. The Hall data were available for the whole range of concentrations and temperatures up to 900 K [19, 20]. By the analyses of these data, they have shown that there are two

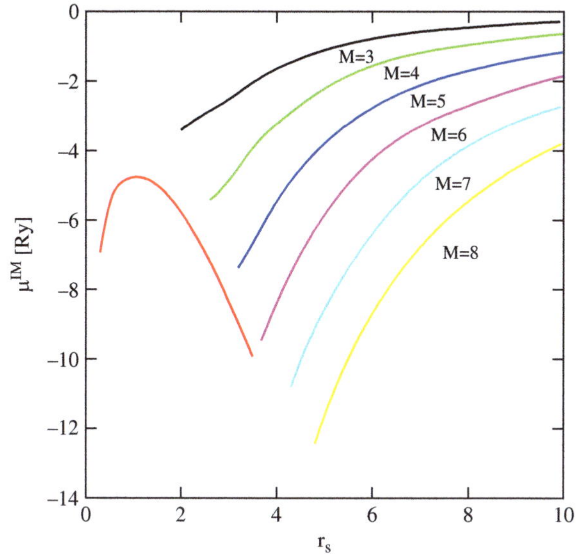

Fig. 8.5 The chemical potentials as a function of r_s are shown for Coulomb bubbles having different number of particles trapped as indicated in the figure. The larger number of particles is trapped into a bubble, the lower the chemical potential becomes

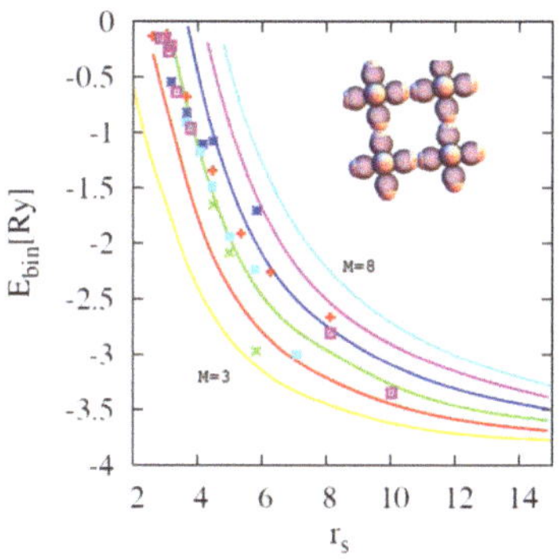

Fig. 8.6 The binding energy/particle into the Coulomb bubble is shown as a function of r_s for different number of particles $M = 3, 4, 5, 6, 7$, and 8. The bottom *yellow* curve corresponds to $M = 3$ particles trapped, while the top *cyan* curve corresponds to $M = 8$ trapped particles. We plot also here the experimental data used by Gorkov and Teitelbaum (GT) [2,11] to fit the resistivity and the Hall coefficient in LSCO [19, 20], ARPES [21] and STM experiments [22]. The ARPES data are presented by *red crosses*. The GT fitting data taken from the *ab* resistivity and the Hall coefficient in LSCO [19, 20] are presented by *squares (cyan and magenta colors)*. The *crossed points (green and blue)* correspond to STM data [3,22,23]. On insert, we present a possible shape of the eight particle cluster, which can comprise up to four CuO plaquettes

types of charge carriers in cuprates. In a broad range of doping x and temperatures T, the number of current carriers satisfies the expression:

$$n(x) = n_0(x) + n_1 \exp(-\Delta(x)/T), \tag{8.14}$$

where $n_0(x) = x$ for small doping x and low temperatures. The second component leads to a strong temperature dependence. GT argued that the value of $\Delta(x)$ must be associated with the activation energy of some localized states, which may introduce inhomogeneity into these systems. It was also found there that $\Delta(x)$ is a universal function of doping x and independent of temperature. The factor in front, $n_1 \sim 2.8$, is independent of temperature and doping up to $x \sim 0.2$ and then drops rapidly to zero. This expression fits well numerous experimental data in LSCO, taken from Hall effect studies [20], from ARPES [21], and from STM measurements presented by Davis' group [3, 22, 23].

In our approach, the GT function $\Delta(x)$ corresponds to the binding energy of M particles self-trapped into the Coulomb bubble, that is, $\Delta(x) = -E_{\text{bin}}(x, M)$. With that identification, the number of the current carriers must satisfy the relation:

$$n(x) = n_0(x) + n_1 \exp(E_{\text{bin}}(x, M)/T). \tag{8.15}$$

Exactly in the same way as Gorkov and Teitelbaum, we have taken instead of the GT's function $\Delta(x)$ the binding energy $-E_{\text{bin}}(x, M)$ and plotted this function as a function of r_s in Fig. 8.6 for different number of particles M trapped into the bubble. The lowest yellow curve corresponds to the bubble with three particles, while the top curve corresponds to the Coulomb bubble with $M = 8$ particles trapped. For intermediate curves (red, green, blue, and purple), the number of particles increases from $M = 4$ to $M = 7$, respectively.

The saturation in $E_{\text{bin}}(r_s \to \infty)$, or in $\Delta(x \to 0)$, is about 0.25 eV in LSCO, which, according to our microscopic calculations, corresponds to $-4Ry = -2me^4/(\hbar^2\varepsilon_0^2)$. From this expression for the saturation of the binding energy at small doping or large r_s, we can extract the static dielectric constant and compare this with the value obtained from other experiments. For example if $-4Ry = -2me^4/\varepsilon_0^2/\hbar^2 = 0.25$ eV, and the hole charge $e = 1$, we can estimate that $\varepsilon_0 = 20$, when $m = 2m_e$; however, when we use for effective mass its optical value for LSCO $m = 4m_e$ taken from the reference [24], we obtain that $\varepsilon_0 = 29.5$. These values for ε_0 are in a perfect agreement with independent experiments in BSCO and LSCO where this value of dielectric constant has been measured directly [25].

For a comparison, we also plot in Fig. 8.6 the experimental data which were used by Gorkov and Teitelbaum to fit the resistivity and the Hall coefficient in LSCO [19, 20], ARPES [21], and STM experiments [22]. These data coincide with theoretical curves (see, green and blue curves in Fig. 8.6) associated with the Coulomb bubbles having $M = 5$ and $M = 6$ particles. For small values of r_s the bubbles have $M = 5$ particles, while with increasing r_s further, i.e., for $r_s > 5$, the bubbles will have $M = 6$ particles. This is exactly following the prediction of our theory. The GT fitting data taken from the *ab* resistivity and from the measurements of the

Hall coefficient in LSCO [19, 20], and presented by squares on the Fig. 8.6. (cyan and magenta colors) are consistent with the Coulomb bubbles having five to six particles self-trapped. In fact at relatively small values of r_s, $3 < r_s < 5$, these data correspond to bubbles having $M = 5$ particles, while at larger r_s there are more fluctuations in comparison with the calculated curves. The rest of the crossed points correspond to the STM data [22], which also fit nicely to the Coulomb bubbles picture and numerical calculations shown in Fig. 8.6. Thus, the binding energy of the Coulomb bubbles agrees well with the experimental data for the pseudogap in LSCO, extracted by GT [2, 11]. Indeed, we have shown here that the GT function $\Delta(x(r_s))$ perfectly coincides with the function $-E_{\mathrm{bin}}(r_s, M)$, which is the binding energy of M particles forming the Coulomb bubble in (8.12). This comparison with the experimental data extracted by GT [2, 11] indicates that with increasing r_s, the average number of particles in the Coulomb bubble increases from $M = 5$ to $M = 6$. It is important to note that in under-doped cuprates, Coulomb bubbles will always arise with different number of particles trapped. This is consistent with STM data [3, 22, 23], where various nanoscale superstructures in HTSC have been observed as well as with transport data [26, 27].

This comparison with experimental data reinforces the conclusion deducted by Gorkov and Teitelbaum that there are two types of current carriers in cuprates: doped, associated with the function $n_0(x) \sim x$ and activated associated with the activation energy $\Delta(x)$, which is nothing but an activation energy of M holes, $-E_{\mathrm{bin}}(r_s, M)$, bound into the Coulomb bubbles. GT argue then that the pseudogap temperature T^* is not a critical temperature of a sharp phase transition, but rather a smooth transition at which the number of doped and activated charge carriers is roughly equal. With this assumption, they got from (8.14) the definition of the pseudogap temperature,

$$T^*(x) = -\Delta(x)/\ln(x). \tag{8.16}$$

Following our finding and identification of $E_{\mathrm{bin}}(x, M) = \Delta(x)$, we can perform a microscopic estimation of the T^* temperature, i.e.,

$$T^*(x) = E_{\mathrm{bin}}(x, M)/\ln(x). \tag{8.17}$$

Our schematic phase diagram in the plane defined by the temperature and concentration of dopants is shown in Fig. 8.7. Below each rainbow color lines bubbles exist, and this is valid both inside the superconducting dome indicated by the orange color and in the insulating and metallic phases. The coexistence region of the superconducting fluctuating phase consisting of insulating and metallic islands is located between red thin curves in the figure. Below the superconducting transition inside the orange dome in Fig. 8.7, the density of CBs is nearly the same as above the transition. Of course due to the activation process, the number of bubbles with smaller number of particles increases with temperature until all bubbles will evaporate. In the under-doped region, we have below the superconductor–insulator transition bubbles and the superconducting condensate, while above the transition there are the CBs and free bosons (Fig. 8.7). We assume here that the condensate fraction of

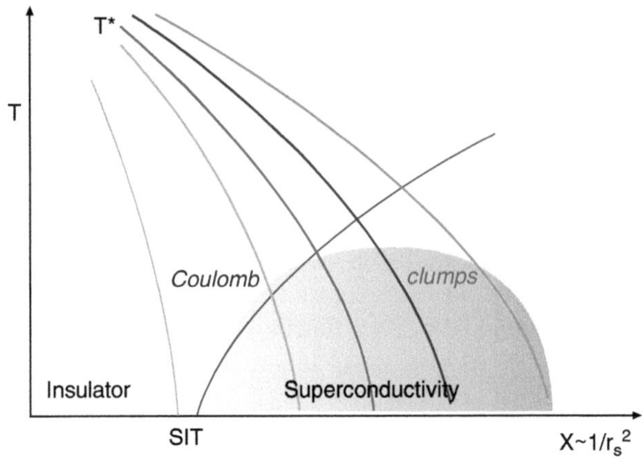

Fig. 8.7 A schematic phase diagram, which emerges from the first principle microscopic many-body studies of the charged fermionic and bosonic liquid with embedded charged impurities. The superconducting dome is noted by the *orange color*; the *stronger orange* color corresponds to larger concentration of the superconducting current carriers. The *bold rainbow color* lines corresponds to the pseudogap critical temperatures. The pseudogap is determined as the binding energy of particles forming the Coulomb bubble. The different colors of the rainbow corresponds to 4, 5, 6, and 7 holes trapped into the Coulomb bubbles; the curve in the left has four and the curve on the right has seven trapped holes. The whole superconducting dome represents a coexistence region of Coulomb bubbles with superconductivity. Inside the coexistence region, there are superconducting microscopic bubbles of different size randomly distributed. In the region separated with *thin red lines* below the pseudogap lines, superconducting droplets and insulating islands can coexist. This agrees with an analysis of the transport experimental data [26, 27]

bosons vanishes at T_c. With increasing temperature more and more free bosons evaporate from CBs. We associate the temperature, at which all bosons have evaporated from the CBs (the energy of their complete ionization), with the critical temperature, at which the pseudogap state disappears.

In Fig. 8.8, we show the doping dependence of $T^*(x)$ both in our theoretical estimations (the solid colored curves) and in various experimental data collected by Gorkov and Teitelbaum [2, 11]. Following again arguments given by GT, such values of T^* calculated in our model for 4, 5, 6, and 7 trapped holes are in agreement with existing experimental data on T^* [2, 11]. This shows clearly that the pseudogap is not a sharp phase transition, but rather a smooth crossover from the free charge carrier dominated region to the region dominated by the activated ones.

To summarize this section on a comparison with the GT fitting data [2, 11], we have obtained that with decreasing doping from the over-doped state toward the optimal doping the broad band splits into two sub-bands. One of them remains always a broad one, and the density of the current carriers there depends weakly on the temperature and doping $n_0(x) = x$. The second band is very narrow or just flat and arises due to a Coulomb instability and formation of the Coulomb bubbles and possibly electron-hole strings [28]. The gap separating these bands is associated with the binding energy of electrons and holes forming the Coulomb bubbles,

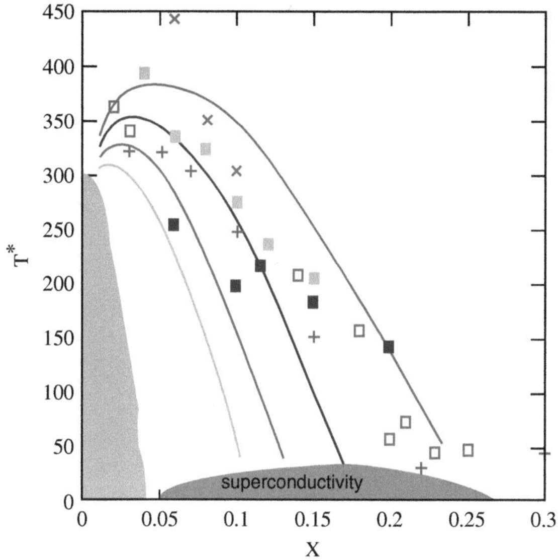

Fig. 8.8 A comparison of the experimental data for the pseudogap temperature with the critical temperature of dissociation of the Coulomb bubbles calculated with the variational many-body formalism. The resistivity and the Hall coefficient in LSCO data are taken from [19, 20], ARPES from [21] and STM experiments from [22]. The ARPES data are presented by *red crosses*. The transport data are presented by *squares (cyan and magenta colors)*. The *crossed points (green and blue)* correspond to STM data [3, 22, 23]. This comparison indicates that these r-space states of the Coulomb bubbles may indeed be associated with the copper oxide pseudogap states identified in numerous experiments

which also induce nanoscale inhomogeneities in these systems. Thus, our results consistent with GT ones [2, 11] state that there are two types of sources for current carriers: (1) from localized "bubbles" and (2) just free carriers arising due to a direct doping effect. We have also found that similar situation arises in YBCO samples, where analogous gap, $\Delta(x)$, has been identified in optical experiments [29,30]. There in heavily under-doped samples, the resistivity and optical conductivity has similar activation character associated with the quantity $\Delta(x)$ which is in our picture associated with $E_{\text{bin}}(x, M)$. It is even more striking that this function is very similar for all cuprates, not only qualitatively but also quantitatively defined as the binding energy of the Coulomb bubbles.

8.4 Nanostructures in STM Measurements

In numerous scanning tunneling microscope (STM) experiments [7, 8, 31, 32], various nanoscale structures have been observed in the superconducting state both in under-doped and in optimally doped cuprates. The appearance of these structures depends on the value of the bias voltage used in STM experiments. At small values, no such structures are visible. However, when the bias voltage is larger than some critical value, these structures are unveiled. This fact has been intensively discussed in the literature and considered as the main puzzle of HTSC, which cannot be explained by any of the existing theories [9].

This puzzling fact is very naturally described in the framework of our results. Each CB may be viewed as a quantum dot, a mountain located on the bottom of the deep sea of the superfluid charged liquid. When a bias voltage is applied, the depth of the superfluid sea (a level of chemical potential) or the thickness of the superfluid density decreases. The effect is dramatic for areas where the CBs are densely packed. Then at some critical bias voltage, the superfluid density will be locally broken and the CBs begin to take part in the STM current and therewith become visible as nanoscale structures.

McElroy et al. [3, 9] have identified populations of atomic-scale impurity states whose spatial distribution follows closely that of the oxygen dopant atoms (Fig. 8.9). They also found a close connection between the nanoscale structure of the superconducting order parameter (electronic disorder) and locations of dopant atoms.

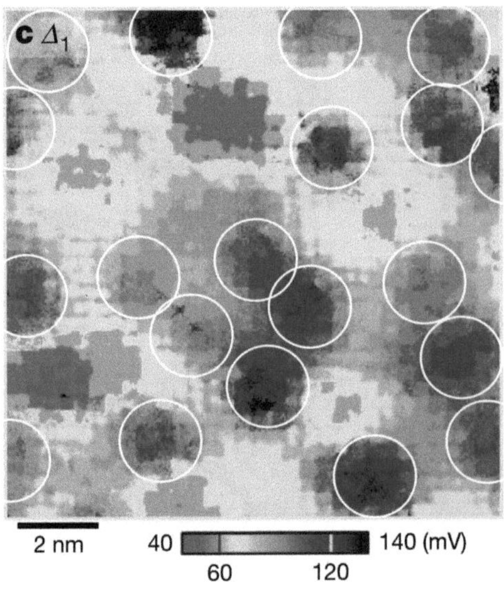

Fig. 8.9 The comparison of experimental STM data on the spatial map of the pseudogap magnitude over the sample surface in $Bi_2Sr_2Ca\ Cu_2O_{8+\delta}$ with theoretical estimations of the Coulomb bubbles distribution made in a framework of the model of the charged Bose fluid. The data are taken from [3]. The picture indicates the color-coded nanoscale superstructures associated with the pseudogap excitations. Here the higher pseudogap value corresponds to the *darker color*. The distribution of Coulomb bubbles is schematically indicated by *white circles*. Each *circle* shows the position of the Coulomb bubble. There is a wide distribution of heterogeneous pseudogap values associated with bubbles of different sizes and shapes that are associated with a distribution of charged impurities over the whole 3D sample. This comparison indicates that these r-space excited states of the Coulomb bubbles are indeed the copper oxide pseudogap excitations

This is obviously possible only if dopant atoms pin CBs. As stated before, CBs are like quantum dots in the Coulomb blockade regime. Then due to the large charging energy, the low-energy spectral weight shifts to higher energies, and the superconducting coherence peak becomes strongly suppressed near impurities. At the same time, the low-energy, condensate bosons scatter very weakly from the CBs. Probably, the mechanism of the superconductivity is related to an interplay between Coulomb bubbles and mobile liquid of holes. Such speculation is consistent, in particular with ARPES [21] and anomalous criticality in electrical resistivity data [27].

In the first type of experiments each, emitted photoelectron leaves behind a hole surrounded by local lattice distortions associated with the electronic strings on oxygens. The energy of these lattice distortions is about the deformation energy per electron per string. In this case, ARPES experiments show that lower energy is needed to break the bound state of electrons and holes forming the Coulomb bubble than for ionization of the electrons or holes in the thermal activation process. There the ARPES data [21] correspond to the lower binding energy, or lower pseudogap value. The difference is about 20 m eV, which is about the typical energy of the strings formation per electron (see for details [28, 33–36]). The pedagogical review about the string formation is presented in [37]. In our approach, the Coulomb bubbles are naturally and originally formed due to over-screening of the Coulomb interaction when the density of the current carriers changes. Additional factors such as the existence of the staggered field on the Cu–O plane and strong electron–phonon interaction reinforce our conclusion. As GT have also noted that the binding energy does not go to zero (see recent ARPES data [2, 11, 21]). It indicates that when Coulomb bubbles vanish (slightly above the doping $x \sim 0.27$ in LSCO), strings still remain since the energy gap still persists and that is about the energy of local lattice distortions existing in the oxygen electronic strings, i.e., it is about 20 m eV.

In the second group of experiments [27], they found that, when the carrier number falls, the effective interaction responsible for the (anisotropic) T-linear scattering term in electrical resistivity becomes progressively stronger. This is consistent with our result that when the carrier number falls, the over-screening of Coulomb interaction increases leading to the Coulomb bubble formation. Moreover, they found that this term closely correlates with superconducting critical temperature T_c and the condensation energy. The Coulomb bubbles described here may play a role of scatterers responsible for a linear T-resistivity analyzed in [27]. The important conclusion made there is that the region of the linear T-resistivity increases when temperature decreases. Moreover, they claim that such scatterers, which we ascribe to the Coulomb bubbles, do exist even in strongly over-doped regime where there is no superconductivity. So in the framework of our theory, it is possible that bubbles arise before (i.e., at higher doping) the superconducting state emerges.

8.5 Conclusions

We have presented here an emerging theoretical scenario of microscopic phase separation driven by the long-range Coulomb forces. This scenario grabs the key physical aspects of the phase diagram of cuprates depicted in Figs. 8.7 and 8.8. It was shown that the model of strongly correlated charged Bose fluid is appropriate for the superconducting phase of all cuprates. The highest critical temperature of the superconducting transition in cuprates is attained within the phase-separated state, where the first microscopic, quantum droplets of electronic inhomogeneities (Coulomb bubbles coexisting with electronic strings) arise.

The recent discovery of high temperature superconductivity in FeAs multilayered materials [38–47] provides another class of systems where the Coulomb interaction and the two component physics of the electron-hole quantum liquids discussed above do exist. As a matter of fact, some of these undoped FeAs compounds may correspond to the state known as the excitonic insulator. However, the difference to the case of cuprates discussed above is the ratio of the electron and hole masses, which is not as large as in cuprates. The fact that in cuprates electrons are localized means that their masses tend to infinity. In contrast to that the fact that in FeAs-based materials the effective hole and electrons masses may be comparable leads to a large variety of very interesting possibilities which arise with the electron or hole doping of these excitonic insulators.

Acknowledgements This work was supported by the ESF network-programs NES and AQDJJ and European project CoMePhS.

References

1. F. Kusmartsev, M. Saarela, Supercond. Sci. Technol. 22, 014008 (2009)
2. L.P. Gorkov, G.B. Teitelbaum, Phys. Rev. Lett. 97, 247003 (2006)
3. Y. Kohsaka, C. Taylor, P. Wahl, A. Schmidt, J. Lee, K. Fujita, J.W. Alldredge, K. McElroy, J. Lee, H. Eisaki, S. Uchida, D.H. Lee, J.C. Davis, Nature, 454, 1072 (2008)
4. T. Kondo, R. Khasanov, T. Takeuchi, J. Schmalian, A. Kaminski, Nature, 457, 376 (2009)
5. J. Bardeen, L.N. Cooper, J.R. Schrieffer, Phys. Rev. 108, 1175 (1957)
6. E.M. Lifshitz, L.P. Pitaevskii, Statistical Physics, Part II (Nauka, Moscow, 1978)
7. K.K. Gomes et al., Nature, 447, 569 (2007)
8. J. Lee et al., Nature, 442, 546 (2006)
9. K. McElroy, J. Lee, J.A. Slezak, D.H. Lee, H. Eisaki, S. Uchida, J.C. Davis, Science, 309, 1048 (2005)
10. J.X. Zhu, K. McElroy, J. Lee, T.P. Devereaux, Q. Si, J.C. Davis, A.V. Balatsky, Phys. Rev. Lett. 97, 177001 (2006)
11. L.P. Gorkov, G.B. Teitelbaum, J. Phys.: Conf. Ser. 108, 012009 (2008)
12. F. Zhang, T.M. Rice, Phys. Rev. B 37, 3759 (1988)
13. M. Saarela, T. Taipaleenmäki, F.V. Kusmartsev, J. Phys. A 36, 9223 (2003)
14. F. Kusmartsev, M. Saarela, J. Phys.: Conf. Ser. 108, 012029 (2008)
15. E. Feenberg, Theory of Quantum Fluids (Academic, New York, 1969)
16. V. Apaja, J. Halinen, V. Halonen, E. Krotscheck, M. Saarela, Phys. Rev. B 55, 12925 (1997)
17. M. Saarela, E. Krotscheck, J. Low Temp. Phys. 90, 415 (1993)

18. E. Krotscheck, M. Saarela, Phys. Rep. 232, 1 (1993)
19. S. Ono, et al., Phys. Rev. B 75, 024515 (2007)
20. T. Nishikawa et al., J. Phys. Soc. Jpn. 63, 1441 (1994)
21. T. Yoshida et al., Phys. Rev. Lett. 91, 027001 (2003)
22. Y. Kohsaka et al., Phys. Rev. Lett. 93, 097004 (2004)
23. Y. Kohsaka et al., Science, 315, 13801385 (2007)
24. W.J. Padilla, Y.S. Lee, M. Dumm, G. Blumberg, S. Ono, K. Segawa, S. Komiya, Y. Ando, D.N. Basov, Phys. Rev. B 72, 060511 (2005)
25. C.Y. Chen et al., Phys. Rev. B 43, 392 (1991)
26. B. Sipos, A.F. Kusmartseva et al., Nat. Mater. 7, 960 (2008)
27. R.A. Cooper et al., Science, 323, 603 (2009)
28. F.V. Kusmartsev, D. Di Castro, G. Bianconi, A. Bianconi, Phys. Lett. A 275, 118 (2000)
29. G. Yu et al., Phys. Rev. B 48, 7545 (1993)
30. D. Mihailovic, T. Mertelj, K.A. Mueller, Phys. Rev. B 57, 6116 (1998)
31. S.H. Pan et al., Nature, 413, 282 (2001)
32. Y. Dubi, Y. Meir, Y. Avishai, Nature, 449, 876 (2007)
33. F.V. Kusmartsev, J. de Physique IV, 9, 321 (1999)
34. F.V. Kusmartsev, Phys. Rev. Lett. 84, 530 (2000)
35. F.V. Kusmartsev, Int. J. Mod. Phys. B 14, 3530 (2000)
36. A. Bianconi et al., Solid State Commun. 63, 1009 (1987)
37. F.V. Kusmartsev, Contemp. Phys. 453, 237 (2004)
38. G.F. Chen, Z. Li, D. Wu, G. Li, W.Z. Hu, J. Dong, P. Zheng, J.L. Luo, N.L. Wang, Phys. Rev. Lett. 100, 247002 (2008)
39. Y. Kamihara, T. Watanabe, M. Hirano, H. Hosono, J. Am. Chem. Soc. 130, 3296 (2008)
40. H. Takahashi, K. Igawa, K. Arii, Y. Kamihara, M. Hirano, H. Hosono, Nature, 453, 376 (2008)
41. H. Wen, G. Mu, L. Fang, H. Yang, X. Zhu, Europhys. Lett. 82, 17009 (2008)
42. Z.A. Ren, J. Yang, W. Lu, W. Yi, X.L. Shen, Z.C. Li, G.C. Che, X.L. Dong, L.L. Sun, F. Zhou, Z.X. Zhao, Europhys. Lett. 82, 57002 (2008)
43. Z.A. Ren, G.C. Che, X.L. Dong, J. Yang, W. Lu, W. Yi, X.L. Shen, Z.C. Li, L.L. Sun, F. Zhou, Z.X. Zhao, Europhys. Lett. 83, 17002 (2008)
44. Z.A. Ren, J. Yang, W. Lu, W. Yi, G.C. Che, X.L. Dong, L.L. Sun, , Z.X. Zhao, Mater. Sci. Innov. 12, 1 (2008)
45. X.H. Chen, T. Wu, G. Wu, R.H. Liu, H. Chen, D.F. Fang, Nature, 453, 761 (2008)
46. C.H. Lee, T. Ito, A. Iyo, H. Eisaki, H. Kito, M.T. Fernandez-Diaz, K. Kihou, H. Matsuhata, M. Braden, K. Yamada, J. Phys. Soc. Jpn. 77, 083704 (2008)
47. M. Rotter, M. Tegel, D. Johrendt, Phys. Rev. Lett. 101, 107006 (2008)

Chapter 9
Scanning Tunneling Spectroscopy of High T_c Cuprates

Ivan Maggio-Aprile, Christophe Berthod, Nathan Jenkins, Yanina Fasano, Alexandre Piriou, and Øystein Fischer

Abstract Tunneling spectroscopy played a central role in the experimental verification of the microscopic theory of superconductivity in the classical superconductors. In the case of high-temperature superconductors (HTS), the initial attempts to adopt the same approach were hampered by various problems related to the complexity of these materials. The progresses made in synthesizing high quality samples, and the use of scanning tunneling microscopy/spectroscopy (STM/STS) on these compounds allowed to overcome the main difficulties. In this review, we present some of the experimental highlights obtained with STM/STS techniques over the last decade. Most of the results confirm the fact that this new class of materials differ noticeably from the conventional BCS superconductors, and provide convincing arguments toward the understanding of the microscopic mechanisms at the origin of high-temperature superconductivity.

9.1 Introduction

In the quest for an understanding of high-temperature superconductors (HTS), tunneling spectroscopy was rapidly considered as a key experimental technique. In the past, electron tunneling played a central role to test the microscopic mechanism of conventional superconductivity, providing the first quantitative confirmation of the BCS and strong coupling theories [1, 2]. The invention of the scanning tunneling microscope (STM) [3] opened a new world of possibilities for tunneling spectroscopy. A beautiful demonstration was realized with the observation of the vortex lattice in $NbSe_2$, showing how the electronic structure of the vortex core can be explored in detail [4]. On the cuprate HTS, the difficulty in obtaining reproducible data with planar tunneling spectroscopy was partly due to a bad control of the tunnel barrier and material inhomogeneities. Using the STM, it was possible to overcome these difficulties, to demonstrate reproducible spectra, and to identify the essential intrinsic features of tunneling spectra on HTS [5]. The remarkable progresses made in both the STM/STS technique and the synthesis of high quality samples greatly contributed to the world-wide effort toward an understanding of the underlying mechanism of high-temperature superconductivity in the cuprates. The recent discovery of high-temperature superconductivity in the iron-based pnictides [6] provided a fantastic boost in the field of superconductivity. Sharing many resemblances with the HTS cuprates – like their bidimensional layered structure, the strong magnetic correlations found in the normal state, or the existence of a (π,π) spin resonance in the superconducting state – these materials have focused much of the attention of the community, and rapidly raised the question of whether the microscopic mechanisms responsible for high-temperature superconductivity are the same for both families of compounds. By now, still awaiting large and well-prepared surfaces, STS has not yet been able to fully exploit its capabilities on these materials. There is no doubt that reproducible and convincing electron tunneling data in iron-based pnictides will be obtained in a near future.

This article highlights some of the more striking STM/STS results obtained on HTS cuprates during the last two decades [7]. After a brief introduction to the STM/STS technique, we will describe the characteristic spectral features revealed by these compounds. We will then report measurements obtained in the mixed state, allowing a direct visualization of vortices and the study of the electronic signature of their cores. We will finally focus on the observation of spatial periodic modulations of the tunneling conductance.

9.2 Basic Principles of the STM/STS Technique

9.2.1 *Operating Principles*

Quantum tunneling of electrons between two electrodes separated by a thin potential barrier was known since the early days of quantum mechanics (Fig. 9.1a). The application of this phenomenon to study the superconducting gap was first demonstrated in superconductor/insulator/normal metal (SIN) planar junctions [1] and point-contact junctions [8]. The invention of the scanning tunneling microscope [3] allowed to overcome these rigid electrodes configurations by mounting a sharp metallic tip (local probe) on a three-dimensional piezoelectric drive (Fig. 9.1b). Applying a bias voltage between the metallic tip and the conducting sample, and approaching the tip within a few Ångströms of the sample surface, results in a measurable tunneling current. The tip is scanned in the xy-plane above the sample using the X and Y actuators, while its height is controlled using the Z actuator.

The remarkable spatial resolution that STM can achieve (a few hundredths of an Ångström) comes from the exponential dependence of the tunneling current, I, on the tip-to-sample spacing, d:

$$I \sim \mathrm{e}^{-2\kappa d}, \quad \kappa = \sqrt{\frac{2m\phi}{\hbar^2}} \approx 0.5\sqrt{\phi\ [\mathrm{eV}]}\ \text{Å}^{-1}. \tag{9.1}$$

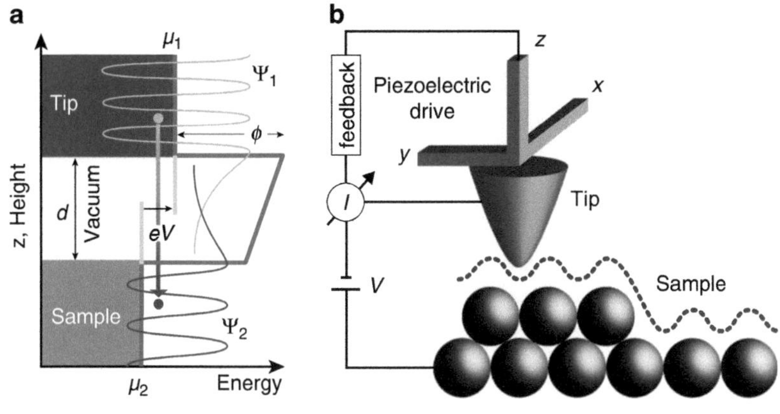

Fig. 9.1 (**a**) Tunneling process between the tip and the sample across a vacuum barrier of width d and height ϕ (for simplicity, the tip and the sample are assumed to have the same work function ϕ). The electron wave functions Ψ decay exponentially into vacuum with a small overlap, allowing the electrons to tunnel from one electrode to the other. With a positive bias voltage V applied to the sample the electrons tunnel preferentially from the tip into unoccupied sample states. (**b**) Schematic view of the scanning tunneling microscope

For a typical metal ($\phi \sim 5$ eV), the current will decrease by about one order of magnitude for every Ångström increase in the electrode spacing. The lateral resolution mainly depends on the apex geometry and electronic orbitals of the scanning tip, which confine the tunneling electrons into a narrow channel, offering the unique opportunity to perform real-space imaging down to atomic length scales. The tunneling regime is defined by a set of three interdependent parameters: the electrode spacing d (typically 5–10 Å), the tunneling current I (typically 0.01–10 nA), and the bias voltage V (typically 0.01–2 V). The parameters I and V are generally chosen to set the tunneling resistance $R_t = V/I$ in the GΩ range.

Controlling the sample surface quality is crucial. Contamination in ambient atmosphere may rapidly degrade the sample top layer, often preventing stable and reproducible tunnel junctions and the investigation of intrinsic properties. This issue is nontrivial and different for each compound, depending on its crystallographic structure and surface nature. The most suitable surfaces for STM/STS are those prepared in situ. Ideally, the top layers are mechanically removed by cleaving the sample in ultra-high vacuum.

9.2.2 *Topography*

In the topographic mode, the surface is mapped via the dependence of the tunneling current upon the tip-to-sample distance. In the standard *constant-current imaging* mode, the tunneling current I is kept constant by continuously feedback-adjusting the tip vertical position during the scan. Since the tunneling current integrates over all states above or below E_F up to an energy equal to the tunnel voltage, the constant-current mapping corresponds to a profile of constant integrated electron density of states (DOS). If the local DOS (LDOS) is homogeneous over the mapped area, this profile corresponds to constant tip-to-sample spacing, and recording the height of the tip as a function of position gives a three-dimensional image of the surface $z = z(x, y)$. In the *constant-height imaging* mode, the tip is maintained at a constant absolute height (feedback loop turned off) during the scan. For ideal tip and sample, modulations of the tunneling current $I(x, y)$ are only due to variations in the tip-to-sample spacing, and recording the current as a function of position will reflect the surface topography. In this mode, the image corrugation also depends on the local work function ϕ as $d(x, y) \sim \ln I(x, y)/\sqrt{\phi}$. Thus, unless the actual local value of ϕ is known, quantitative characterizations of topographic features are difficult to achieve.

9.2.3 Local Tunneling Spectroscopy

In spectroscopy, the DOS of the material can be accessed by recording the tunneling current $I(V)$, while the bias voltage is swept with the tip held at a fixed vertical position. If a positive bias voltage V is applied to the sample, electrons will tunnel into *unoccupied* sample states, whereas at negative bias they will tunnel out of *occupied* sample states (Fig. 9.1a). For planar junction configurations, at zero temperature, the tunneling conductance can be expressed as :

$$\sigma(V) = \frac{\mathrm{d}I}{\mathrm{d}V} = \frac{2\pi e^2}{\hbar}|T|^2 N_{\mathrm{tip}}(0) N_{\mathrm{sample}}(eV), \tag{9.2}$$

where T, the tunneling matrix element, is assumed to be constant, and $N(\omega)$ is the DOS measured from the Fermi energy. This simple formula shows the essence of tunneling spectroscopy: the bias dependence of the tunneling conductance directly probes the DOS of the sample. For understanding *local* probes like the STM, the use of real-space spectral function provides an explicit relation between the tunneling current and the sample LDOS $N_{\mathrm{sample}}(\mathbf{x}, \omega)$:

$$I \propto \int \mathrm{d}\omega\, [f(\omega - eV) - f(\omega)] \times N_{\mathrm{tip}}(\omega - eV) N_{\mathrm{sample}}(\mathbf{x}, \omega), \tag{9.3}$$

where $\mathbf{x}$ denotes the tip center of curvature[9]. Assuming a structureless tip DOS, $N_{\mathrm{tip}}(\omega) =$ constant, one finds

$$\sigma(\mathbf{x}, V) \propto \int \mathrm{d}\omega\, [-f'(\omega - eV)] N_{\mathrm{sample}}(\mathbf{x}, \omega), \tag{9.4}$$

where f' is the derivative of the Fermi function. Thus, the interpretation of scanning tunneling spectroscopy becomes remarkably simple: the voltage dependence of the tunneling conductance measures the thermally smeared LDOS of the sample at the position of the tip. Because the local tunneling matrix element is properly taken into account, (9.4) describes the STM much better than (9.2) describes planar junctions.

Practically, dI/dV spectra can be obtained either by numerical differentiation of $I(V)$ curves or by a lock-in amplifier technique. In the latter case, a small ac-voltage modulation $V_{ac}\cos(\omega t)$ is superimposed to the sample bias V, and the corresponding modulation in the tunneling current is measured, with the component at frequency ω proportional to $dI/dV(V)$. This statement is only valid if $V_{ac} \ll V$ and if $I(V)$ is sufficiently smooth. For optimal energy resolution, V_{ac} should not exceed $k_B T$, and typical values are in the few hundred μV range. The advantage offered by the lock-in technique is that the sampling frequency ω can be selected outside the typical frequency domains of mechanical vibrations or electronic noise, considerably enhancing the measurement sensitivity.

Spatially resolved tunneling spectroscopy fully exploits the capabilities of the STM. In 'current-imaging tunneling spectroscopy' (CITS) [10], the tip is scanned over the sample surface with a fixed tunneling resistance $R_t = V_t/I$, recording the topographic information. At each point of the image, based on a regular matrix of points distributed over the surface, the scan and the feedback are interrupted to freeze the tip position (x, y, and z). This allows the voltage to be swept to measure $I(V)$ and/or dI/dV, either at a single bias value or over an extended voltage range. The bias voltage is then set back to V_t, the feedback is turned on, and the scanning resumed. The result is a topographic image measured at V_t, and simultaneous spectroscopic images reconstructed from the $I(V)$ and/or dI/dV data. This technique provides a very rich set of information.

9.2.4 STS of Superconductors

The measurement of the gap is the most direct application of tunneling spectroscopy in superconductors. Since for standard isotropic BCS superconductors this gap is independent on the position in real space and on the momentum along the Fermi surface in k-space (s-wave gap symmetry), the tunneling spectra of a homogeneous sample depend neither on the tunneling direction nor on the position along the surface. The situation is very different for layered HTS cuprates, owing to their very anisotropic structural, electronic, and superconducting properties. As a consequence, the tunneling spectra measured on these compounds may differ noticeably from the spectra obtained on conventional superconductors like lead (Fig. 9.2).

In the spectroscopic images, the best contrast is obtained by selecting the energy where maximal variations in the tunneling conductance occur. In the case of vortex imaging, the mapping energy is usually selected at the position of the coherence peaks (superconducting gap), or close to zero energy where the amplitude of the localized core states is the largest (Sect. 9.4). Spatially resolved tunneling spectroscopy enables to reconstruct spectroscopic maps acquired simultaneously at different energies. One of the first example of such an analysis was provided by the measurement of the energy dependence of the star-shaped vortex core structure in $NbSe_2$ [14]. More recently, the Fourier transforms of such maps revealed periodic modulations of the LDOS in real space in Bi2212 [15] (Sect. 9.5). From the energy dependence of these modulations, energy dispersion curves could be extracted, opening to STS the door of reciprocal-space spectroscopies.

9.3 Spectral Characteristics of HTS Cuprates

9.3.1 General Spectral Features of HTS Cuprates

We discuss here the typical spectral features of HTS cuprates by describing the characteristics observed for two reference HTS compounds: the Bi-based and Y-123 cuprates.

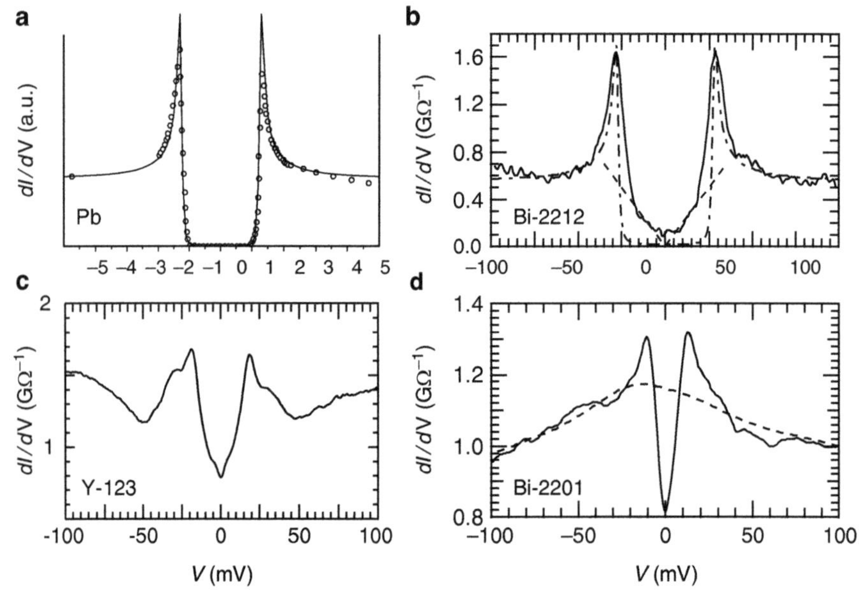

Fig. 9.2 STS spectra measured on various superconductors: (**a**) Pb at 300 mK (*circles*); *s*-wave BCS fit with $\Delta = 1.12$ meV (*solid line*); from [11]. (**b**) Optimally doped Bi-2212 ($T_c = 92$ K) at 4.8 K (*solid line*); *s*-wave BCS fit with $\Delta = 27.5$ meV and pair-breaking parameter $\Gamma = 0.7$ meV (*dash-dotted line*), and low-energy V-shaped *d*-wave BCS conductance (*dashed line*); from [5]. (**c**) Y-123 at 4.2 K; from [12]. (**d**) Overdoped Bi-2201 ($T_c = 10$ K) at 2.5 K (*solid line*) and 82 K (*dashed line*); adapted from [13]

$Bi_2Sr_2CaCu_2O_{8+\delta}$ (Bi-2212) is the most widely studied HTS using STM since it is relatively easy to prepare atomically flat and clean surfaces by cleaving. It is commonly believed that at sufficiently low energies, the measured DOS originates from the CuO_2 bilayer lying approximately 4.5 Å beneath the nonconducting surface BiO layer. This compound was the first of any HTS whose surface yielded atomic topographic resolution using STM. In addition to the atomic lattice, a nearly commensurate structural superstructure along $[\bar{1}10]$ can be seen. The prominent low-energy features of the Bi-2212 spectra are two large conductance peaks defining a clear gap, as shown in Fig. 9.2b for an optimally doped single crystal [5, 16]. The gap obtained by measuring half of the energy separating the two conductance peaks, Δ_p, – usually slightly larger than the value Δ calculated from a proper fit of the spectra [17] – yields ~35 – 40 meV for optimally doped Bi-2212, much larger than expected for a BCS superconductor with a similar T_c of 91 K (~14 meV for s-wave, ~17 meV for d-wave). The V-shaped conductance around zero bias (Fig. 9.2b) is indicative of the d-wave symmetry and the presence of nodes in the gap. The spectra display a systematic electron-hole asymmetry, the conductance at negative bias (occupied states) being slightly larger than the one at positive bias (empty states). Heavily underdoped samples reveal a more pronounced asymmetry (ascribed to strong-correlations effects [18]) and smeared coherence peaks at the gap edges, which may be a consequence of the short quasiparticle lifetime due to proximity of a Mott insulator phase. A remarkable conductance depletion (dip) develops at an energy slightly higher than the coherence peak energy. In early STM experiments, this feature was mostly obscured by an increasing (linear or parabolic) background conductance and became clearly apparent in spectra obtained with the best STM junctions. In Bi-2212 and in the three-layer compound $Bi_2Sr_2Ca_2Cu_3O_{10+\delta}$ (Bi-2223), the dip is systematically seen at negative sample bias [5], and in some exceptional cases also at positive bias [19]. The dip structure, often accompanied by a broad conductance hump at higher energy, is at the core of a heated debate. One key question is whether it is a band structure or a strong coupling effect. The tunneling spectra of Bi-2223 look very similar to those of Bi-2212, except that all features are shifted to higher energies. On the contrary, the single-layer compound $Bi_2Sr_2CuO_6$ (Bi-2201) yields significantly different spectra (Fig. 9.2d) with a finite zero-bias conductance, a bell-shaped background (reminiscent of the normal state DOS), and a much larger reduced gap ($2\Delta_p/k_BT_c \sim 28$ in overdoped Bi-2201 [13]).

$YBa_2Cu_3O_7$ (Y-123) is the second most studied HTS by STM. Y-123 is much less anisotropic than Bi-2212 and lacks any natural cleaving plane. Thus, atomic resolution on as-grown Y-123 surfaces appears very difficult to achieve and has been reported only on thin films [20]. A characteristic dI/dV spectrum measured on an as grown optimally doped Y-123 single crystal is shown in Fig. 9.2c. The spectrum reveals a number of remarkable differences compared to Bi-2212 with similar T_c. The zero-bias conductance is always high and the conductance at high bias is strongly energy-dependent. The main coherence peaks define a gap of ~20 meV, much closer to the value expected for a BCS d-wave superconductor than the gap observed in the Bi-based cuprates. Additional structures of interest are weak shoulders flanking the main peaks at higher energy, and two weak features often developing below the gap at about ± 6 meV. Similar to Bi-2212, the dip-hump feature is also measured in the spectra of Y-123, but symmetrically at positive and negative bias. The multiple peak structure does not correspond to the simple d-wave expectation. The features inside the gap have been ascribed to weak proximity-induced superconductivity in the BaO- and CuO-chain planes [21, 22], while shoulders outside the main coherence peaks have been modelled in terms of the band structure van-Hove singularity [17,21]. Although a completely satisfactory description of the tunneling spectra in YBCO has still to be devised, STS measurements underline the very high homogeneity of the electronic properties in this material, illustrated by the first successful imaging of the Abrikosov vortex lattice in a HTS on Y-123 as-grown surfaces [12] .

Attempts to fit the tunneling spectra of HTS compounds using a conventional d-wave BCS DOS model usually fail to reproduce most of the measured spectral features. This is particularly true for the global electron-hole asymmetry of the spectra, the large spectral weight in the conductance peaks at the gap edges, and the dip-hump feature. Both the asymmetry and the excessive spectral weight can be simulated simultaneously by taking into account the band structure, and especially the van-Hove singularity near $(\pi, 0)$ and $(0, \pi)$ in the Brillouin zone [17, 23]. To explain the presence of the dip-hump feature in the spectra, a coupling of quasiparticles with a collective electronic mode was considered. Specific models [24] were developed to describe the coupling of the electrons to the (π, π) spin resonance observed by neutron scattering near 41 meV in Y-123 [25], and subsequently in Bi-2212 [26]. These models were used to calculate the tunneling spectra of Bi-2212 using doping-dependent band structure parameters from ARPES [27]. Refinements in the fitting procedure allowed to reproduce the features of the tunneling spectra measured in both Bi-2212 and Bi-2223 with an extreme accuracy, and showed that the energy of the resonant mode can be accessed by measuring the energy difference between the dip and the coherence peak locations [28]. The analysis of a large number of spectra corresponding to different gap magnitudes reveals that the spin resonance energy extracted from the tunneling spectra and the gap magnitude is anticorrelated [28, 29], in agreement with neutron scattering measurements in underdoped samples [26, 30].

9.3.2 Superconducting Gap and Pseudogap

While most conventional BCS superconductors reveal a gap Δ proportional to the superconducting transition temperature T_c, as predicted by the theory, many HTS compounds do not show this behavior. In overdoped Bi-2212, the gap decreases with decreasing T_c, as expected. But in underdoped Bi-2212, the gap unexpectedly increases with decreasing T_c [31, 32]. As a consequence, the reduced gap defined as $2\Delta_p/k_B T_c$ is not anymore constant: for Bi-based cuprates, it ranges from 4.3 to values as high as 28 [13], suggesting that T_c (the temperature at which the superconducting state with macroscopic phase coherence is established) is not the appropriate energy scale to describe the gap. This striking experimental fact has to be confronted with an unconventional normal state common to all HTS, characterized by the opening of a gap in the electronic excitation spectrum at a temperature T^* above the critical temperature T_c. This so-called pseudogap (PG) has been evidenced by many experimental techniques [33], and the answer to the question of its origin may turn out to be essential for the understanding of high-T_c superconductivity. In some models, the PG is the manifestation of some order, static or fluctuating, generally of magnetic origin, but unrelated to and/or in competition with the superconducting order. In this picture, the PG progressively disappears upon doping as superconductivity sets in, possibly vanishing at a quantum critical point. In other models, the PG is the precursor of the superconducting gap and reflects pair fluctuations above T_c. At present, a definitive answer is still lacking, although the STS results favor the latter hypothesis. The doping dependence of the gap is much less documented for Y-123 than for Bi-2212. Some studies report that unlike in Bi-2212, the gap in Y-123 does scale with T_c as a function of doping, and the reduced gap does not reach values as high as in Bi-2212 [34].

The first tunneling results revealing the PG in Bi-2212 were reported on planar junctions [35] and using an STM junction [31]. These experiments showed unambiguously that some HTS display a superconducting gap that does not close at T_c. By rising the temperature, the coherence-peak intensity is rapidly reduced at the bulk T_c. The coherence peak at negative bias is suppressed and a reduced peak shifting to a slightly higher energy remains at positive bias. Crossing T_c, the spectra exhibit a pseudogap of quasi temperature-independent magnitude. Both the superconducting gap and the PG increase with underdoping in Bi-2212 (Fig. 9.3b). Overdoped samples also reveal a PG, however, with smaller magnitudes, which vanishes at lower temperatures [31,36,37]. A scaling between the amplitude of the gap and the PG has been found to hold at all oxygen doping levels (Fig. 9.3b). STS measurements on overdoped Bi-2201 revealed well-developed coherence peaks at $\Delta_p \approx \pm 12$ meV, yielding a reduced gap $2\Delta_p/k_B T_c \approx 28$, a value seven times larger than the BCS ratio. Similar to the case of Bi-2212, this extremely high value is concomitant with a PG state extending over an extremely wide temperature range above T_c [13].

The absence of scaling between T_c and Δ_p for Bi-based cuprates may be reconsidered by replacing T_c with T^*, the PG opening temperature. Doing so, as shown in Fig. 9.3c, a constant reduced gap is recovered not only for Bi-2212 at various doping levels but also for Bi-2201 [13] and optimally doped Y-123 [38]. This universality underlines the robustness of this scaling. The graph shows that $2\Delta_p/k_B T^* \approx 4.3$, which corresponds to the BCS d-wave relation if T^* is considered instead of T_c. This finding supports the idea that T^* is the mean-field BCS temperature, scaled with the interaction energy that leads to superconductivity. STS studies demonstrated that the characteristic PG signature is not only observed above T_c, but also occurs at low temperatures inside vortex cores [39], as will be discussed further in Sect. 9.4.3.2, and on strongly disordered surfaces [40,41].

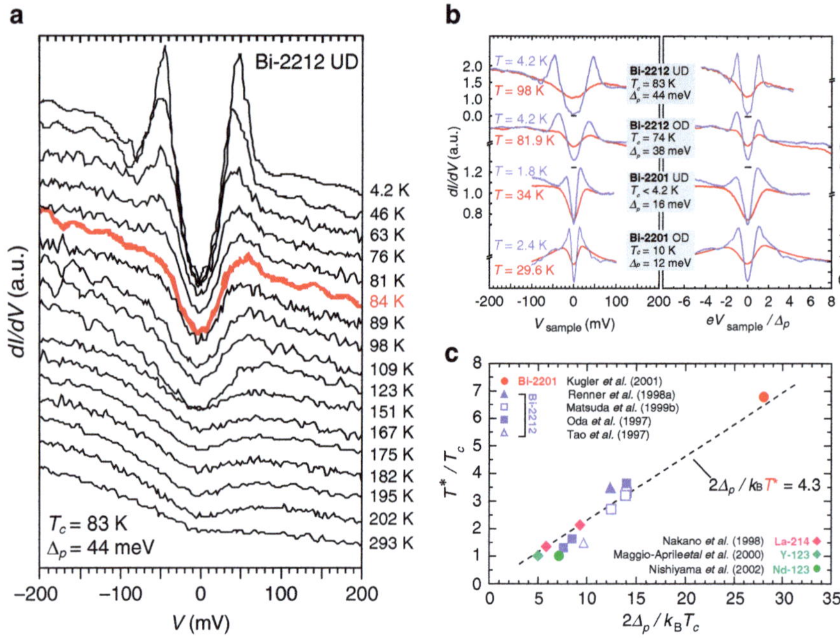

Fig. 9.3 (**a**) Temperature evolution of STS spectra in underdoped Bi-2212 with $T_c = 83$ K, $\Delta_p =$ 44 meV, and T^* near room temperature; adapted from [31]. (**b**) Comparison of the pseudogap and the superconducting gap for Bi-2212 and Bi-2201 in an absolute energy scale (*left*) and in energy rescaled by Δ_p (*right*). (**c**) T^*/T_c as a function of $2\Delta_p/k_BT_c$ for various cuprates investigated by tunneling spectroscopy. The mean-field d-wave relation $2\Delta_p/k_BT^* = 4.3$ is indicated by the *dashed line*

9.4 Revealing Vortices and the Structure of their Cores by STS

The fundamental properties of superconductors can be accessed by studying the vortex matter. Indeed, the electronic structure of the vortex cores and the interaction between the flux lines are intimately connected with the nature and the behavior of the charge carriers. STS offers the unique possibility to detect vortices with nanometer-scale resolution, and the application of this technique to HTS compounds was a considerable success. Vortex imaging finds an obvious interest in the study of the spatial distribution of individual flux lines, allowing a direct measure of the degree of vortex disorder. In addition, the probe can access electronic excitations within the cores, revealing important and often unexpected properties of the superconducting pairing state. This section will start with a brief overview of vortex matter studies by STS, before focusing on the specific electronic signature of the vortex cores.

9.4.1 Vortex Matter in Conventional Superconductors

STS imaging of the vortices relies on the fact that their cores affect the quasiparticle excitation spectrum. Images are usually obtained by mapping the tunneling conductance in real space at a particular bias voltage, usually at the energy of the coherence peaks. Using this criterion, the first imaging of flux lines with STS was performed on the conventional BCS compound $NbSe_2$ [4]. Following this remarkable experimental breakthrough, many other studies were reported on the same compound [42–46].

In Fig. 9.4a, the STS image reveals vortices arranged into a perfect hexagonal network, with a lattice parameter of ~90 nm, close to the 89 nm inter-vortex spacing $a_\triangle = (2\Phi_0/\sqrt{3}H)^{\frac{1}{2}} = 48.9/\sqrt{H\,[\mathrm{T}]}$ nm, expected at a field of 0.3 T. Vortex cores were found to be isotropic with an apparent radius of 15–20 nm – larger than the coherence length of $NbSe_2$ estimated to be 7.7 nm [49]. However, detailed measurements performed later below 1 K allowed to resolve fine structures in the imaged cores, revealing star-shaped patterns with sixfold symmetry [45, 50]. Theoretical studies based on the Bogoliubov-de Gennes formalism and using a sixfold perturbation term were able to explain the observed shape of the cores [51].

Other superconductors have been investigated by STS in the mixed state. A hexagonal vortex lattice was detected in the pyrochlore compound KOs_2O_6 (Fig. 9.4b) [47], in MgB_2 single crystals (Fig. 9.4c) [48], or in the Chevrel phases [52]. More surprisingly, a square vortex lattice was found in the borocarbide $LuNi_2B_2C$ [53], explained by considering a fourfold perturbation term in the Ginzburg–Landau free energy, arising from the underlying tetragonal structure of the crystal.

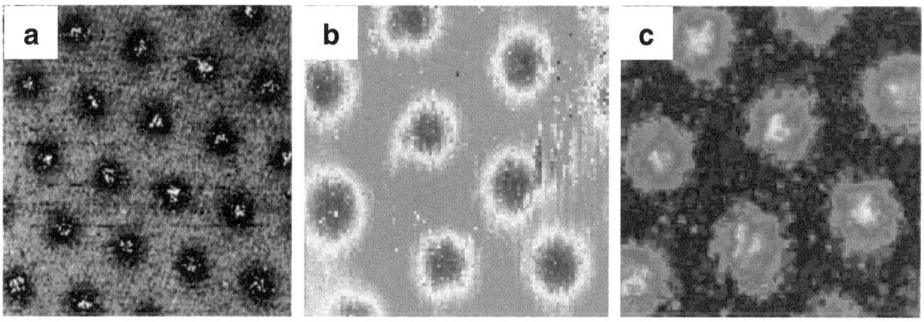

Fig. 9.4 STS images of vortices in low T_c superconductors: (**a**) 400×400 nm^2 hexagonal lattice in $NbSe_2$ ($T = 1.3$ K, $H = 0.3$ T) [42], (**b**) 60×60 nm^2 conductance map ($T = 1.3$ K, $H = 6$ T) in the β-pyrochlore KOs_2O_6 [47], and (**c**) 250×250 nm^2 hexagonal lattice in MgB_2 ($T = 2$ K, $H = 0.2$ T) [48]. The mapping energy was Δ_p for (**a**), and 0 for (**b**) and (**c**)

9.4.2 Vortex Matter in HTS

The study of vortices in HTS by scanning tunneling spectroscopy remained for a long time a challenging task, mostly because of the difficulties in obtaining surfaces with sufficient quality and homogeneity. Another reason is the very small coherence length characterizing the HTS materials. This has the direct implication that the vortex cores are extremely tiny in comparison to conventional superconductors (and thus difficult to localize) and also that the vortices get easily pinned by defects. As a consequence, flux lines lattices show no perfect order in HTS at the relatively high fields which are suitable for STS. Although a large number of STS studies have been focused on the spectral characteristics of HTS in the superconducting state, vortices have been imaged in only two compounds: Y-123 and the Bi-based cuprates.

9.4.2.1 Y-123

Y-123 was the first HTS whose vortex lattice could be investigated by STS on as-grown surfaces [12]. Since then, the STS observation of flux lines in Y-123 was reported either on as-grown surfaces [55] or on chemically etched surfaces [56]. Figure 9.5a shows an image acquired at a field of 6 T in optimally doped Y-123 ($T_c = 92$ K). The slightly disordered lattice shows a local oblique symmetry, with an opening angle of about 77°. Vortex cores reveal an apparent radius of about 5 nm, elliptical in shape with an axis ratio of about 1.5. This asymmetry was attributed to the ab-plane anisotropy of Y-123, the anisotropy factor being in agreement with the one derived from penetration depth measurements in Y-123 [57] or neutron scattering experiments [58]. Interestingly, it was found that the observed lattice can be interpreted as a square lattice distorted by an anisotropy factor of $\sim$1.3, close to the value found for the intrinsic distortion of the cores [59]. All these results are in

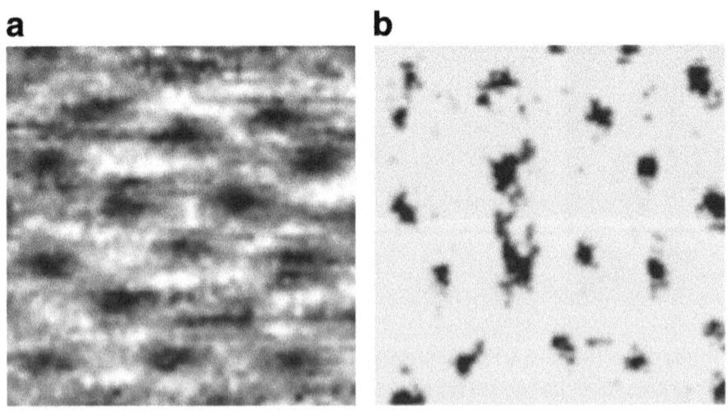

Fig. 9.5 75×75 nm^2 STS images of vortex lattices in two HTS cuprates. (**a**) A slightly disordered oblique lattice in Y-123 [12]. (**b**) A disordered vortex distribution in Bi-2212 [54]. Both images were acquired at $T = 4.2$ K and $H = 6$ T. The contrast is defined by the conductance measured at the gap energy (20 meV for Y-123 and 30 meV for Bi-2212) normalized by the zero-bias conductance

agreement with the theoretical expectations made for superconductors with d-wave pairing symmetry [60,61]. The disorder observed in the vortex lattice underlines the influence of strong pinning in compounds with small coherence lengths. The influence of twin boundaries on pinning in Y-123 single crystals has been investigated by STS, revealing that they act as very efficient barriers that block the movements of the flux lines perpendicular to the twin plane [62]. The gradients of the magnetic field could be directly measured from the flux line distribution, and allowed the authors to estimate a current density close to the depairing current limit for Y-123 ($\sim 3 \times 10^8$ A/cm^2 at 4.2 K).

9.4.2.2 Bi-2212

In Bi-2212, at fields at which STS measurements are performed, vortices imaged by STS are distributed in a totally disordered manner (Fig. 9.5b). This can be understood since Bi-2212, unlike Y-123, has a nearly two-dimensional electronic structure and vortices are therefore constituted by a stack of 2D elements (pancakes), weakly coupled between adjacent layers, and easily pinned by any kind of inhomogeneity. The observed disordered lattice fits into the commonly accepted vortex phase diagram for Bi-2212, where the low-temperature high-field region is associated with a disordered vortex solid. In the STS image of Fig. 9.5b, the average density matches the applied magnetic field of 6 T. At very low fields, an ordered Bragg glass phase is present, which is difficult to probe with STS because of the large inter-vortex distances. At 8 T, however, a short-range ordered phase was observed in Bi-2212 [63], presenting a nearly square symmetry almost aligned with the crystallographic ab directions. The cores in Bi-2212 are tiny, consistent with a coherence length of the order of a few atomic unit cells. The vortices often present very irregular shapes and are in some cases even split into several subcomponents [64].

In calculations made for d-wave BCS superconductors, the anisotropic order parameter leads to a fourfold anisotropy of the low-energy LDOS around vortices [65]. Such a characteristic fourfold signature should be visible in the conductance maps, but has up to now never been firmly reported in the STS studies of HTS, remaining a considerable challenge for future experiments.

9.4.3 Electronic Structure of the Cores

9.4.3.1 BCS Superconductors

The existence of electronic states bound to the vortex at energies below the superconducting gap, as predicted by Caroli, de Gennes, and Matricon in 1964 [66], was beautifully confirmed in STS experiments in pure $NbSe_2$ single crystals at $T = 1.85$ K [4]. In theory, these bound states form a discrete spectrum with a typical inter-level spacing Δ^2/E_F. For most superconductors, this spacing is so small

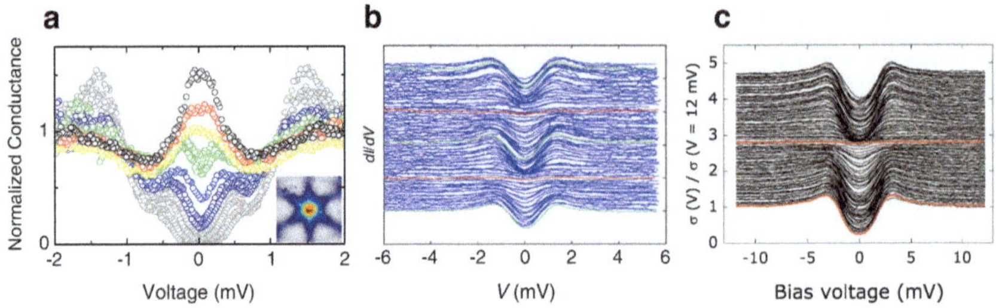

Fig. 9.6 Core spectroscopy in conventional superconductors. (**a**) Conductance spectra acquired in $NbSe_2$ at $T = 0.1$ K, $H = 0.05$ T along a 60 nm path from the center to the outside of a vortex core (shown in the insert) [45]. Examples of vortices revealing a flat conductance in their cores: (**b**) in the β-pyrochlore KOs_2O_6 ($T = 2$ K and $H = 6$ T) along a 45 nm path crossing two vortices [47], and (**c**) in MgB_2 ($T = 2$ K and $H = 0.2$ T) along a 250 nm trace crossing a single vortex [48]

that the detection of individual bound states is hampered by the finite temperature and by impurity scattering. In low temperature measurements of $NbSe_2$, the center of the flux line reveals the expected broad conductance peak located around zero energy [4, 45], as shown in Fig. 9.6a. The progressive splitting of the zero-bias peak (ZBP) at a distance r from the center, corresponding to bound states with increasing angular momenta, provided another verification of the BCS predictions for vortex cores in s-wave superconductors [67]. In dirty superconductors, when the quasiparticle mean free path becomes shorter than the coherence length, quasiparticle scattering mixes states with different angular momenta, and the ZBP transforms into a quasi-normal (flat) DOS spectrum. This effect was observed by STS in Ta-doped $NbSe_2$ [42]. In $LuNi_2B_2C$, a nearly constant tunneling conductance was initially reported in the vortex-core spectra [53], and the expected ZBP could be observed more recently only in very pure samples of YNi_2B_2C [68]. Examples of cores revealing a flat conductance are shown in Fig. 9.6b for the β-pyrochlore KOs_2O_6 and in Fig. 9.6c for the two-band superconductor MgB_2. The absence of a ZBP in MgB_2 is surprising, since the mean free path in this compound is believed to be much larger than the superconducting coherence length. This observation was attributed to the two-band superconductivity behavior, and the fact that the π-band is predominantly probed in c-axis tunneling. Since this band becomes superconducting through coupling with the σ-band, the presence of a flux line is suppressing the superconducting character of the σ-band, pushing the π-band into the metallic regime, with an energy-independent DOS [48].

9.4.3.2 High-Temperature Superconductors

Y-123 and Bi-based Cuprates

Measurements on optimally doped Y-123 single crystals [12] revealed for the first time the electronic signature of the vortex cores of a HTS. In contrast to the case of conventional superconductors, the core spectra in Y-123 show neither a broad zero-bias feature nor a flat conductance. Instead, when entering the core, the coherence peaks located at ±20 meV progressively disappear (Fig. 9.7a), and a pair of low-energy peaks emerges above the conductance background (Fig. 9.7b).

The observation of isolated core states at finite energy in Y-123 appeared contradictory in various respects. On one hand, attempts to interpret the observed bound states like classical BCS s-wave localized states failed for two reasons: first, it would imply an unusually small Fermi energy ($E_F \sim 36$ meV); second, there is a complete absence of energy dispersion at different positions within the core (Fig. 9.7a). On the other hand, this observation does not fit with the d-wave character of the cuprate compounds, where the presence of nodes in the order parameter changes the properties of the low-energy excitations. It was shown that the vortex-core energy spectrum is continuous rather than discrete, and that the LDOS at the center of the core should show a broad peak at zero energy [65, 69].

Vortex-core spectroscopy in Bi-2212 provided new insights into the debate concerning the origin of the pseudogap, starting with the STS measurements on both underdoped and overdoped Bi-2212 single crystals [39]. The evolution of the spectra across a vortex core in overdoped Bi-2212 is shown in Fig. 9.7c. Upon entering the core, the spectra evolve in the same way as if the temperature is raised above T_c, i.e., the coherence peak at negative energy vanishes over a very small distance, the coherence peak at positive energy is reduced and shifts to slightly higher energies, and the dip-hump feature disappears.

While the first investigations of the vortex cores in Bi-2212 focused on pseudogap aspects, detailed analysis performed subsequently showed that weak low-energy structures were also present in the vortex core spectra [15, 72]. Contrary to Y-123, where the peaks clearly emerge in the core center (Fig. 9.7b), the core states appear in Bi-2212 as weak peaks in an overall pseudogap-like spectrum (Fig. 9.7d). For slightly overdoped Bi-2212, the energy of these states is of the order of ±6 meV. The low-energy core states observed in both Y-123 and Bi-2212 raise important questions. A correlation was found between the core-state energy and Δ_p for both Y-123 (optimal doping) and Bi-2212 (various dopings), strongly suggesting that the core states in the two compounds have a common origin. Moreover, since this correlation is linear (with a slope of about 0.3), the interpretation of these bound states as Caroli–de Gennes–Matricon states of an s-wave superconductor with a large gap to Fermi energy ratio can be ruled out, as one would expect a Δ_p^2 dependence.

The qualitative difference between the vortex-core spectra measured by STM in HTS and the spectra expected for a d-wave BCS superconductor [65, 69, 73] is thought to reflect an intrinsic property of the superconducting ground state, which distinguishes between this state and a pure BCS ground state, and is presumably related to the anomalous normal (pseudogap) phase. It was shown [65] that the admixture of a small magnetic-field-induced complex component with d_{xy} symmetry [74] leads to a splitting of the zero-bias conductance peak due to the suppression of the d-wave gap nodes. A good qualitative agreement with the experimental spectra was also obtained using a model where short-range incoherent pair correlations coexist with long-range superconductivity in the vortex state [75]. This model

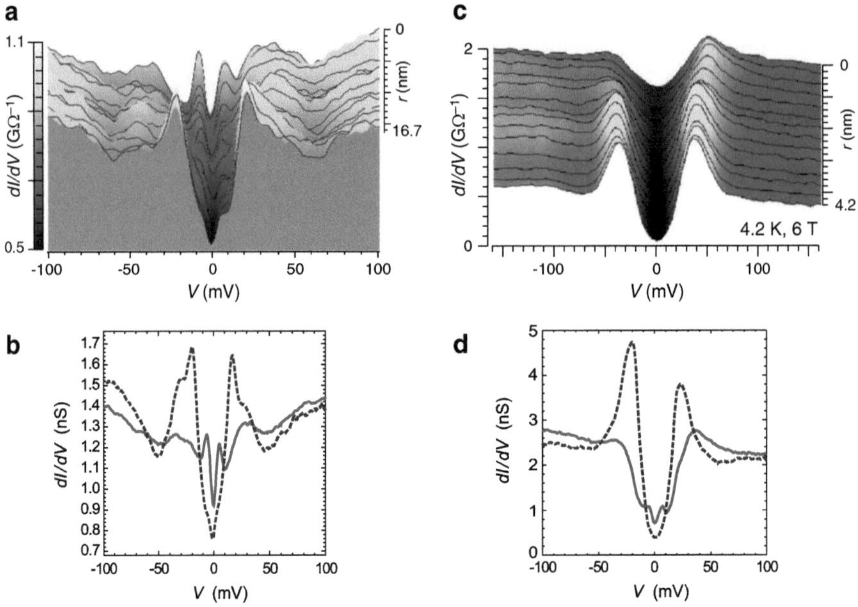

Fig. 9.7 Vortex-core spectroscopy in Y-123 and Bi-2212. (**a**) STS spectra taken along a 16.7 nm path into a vortex core in Y-123 ($T = 4.2$ K, $H = 6$ T). From [39]. (**b**) Conductance spectra acquired at the center (*red*) and at about 20 nm from the center of the vortex core (*dashed blue*). The low-energy core states are located at about ± 5.5 meV. Adapted from [12]. (**c**) Conductance spectra taken along a 12 nm path across a vortex core in overdoped Bi-2212 ($T_c = 77$ K) at $T = 2.6$ K and $H = 6$ T [70]. (**d**) Spectra taken close to the core center with two peaks at ± 6 meV corresponding to a pair of core states (*red*) and outside the vortex core (*blue*). Adapted from [71]

correctly reproduced the measured exponential decay of the LDOS at the energy of the core states [15], as well as the increase of the core-state energy with increasing gap Δ_p. However, there is still no consensus about the microscopic origin of the finite energy core states detected with STS in HTS cuprates.

9.5 Local Electronic Modulations seen by STM

The role of inhomogeneities in HTS has stirred up considerable interest during the last decade. The key question is to what extent these inhomogeneities are an *intrinsic* phenomenon at the core of high-temperature superconductivity, as suggested by models considering electronic phase separation [76]. This is in contrast to an *extrinsic* origin of the spectral inhomogeneities, a phenomenon unrelated to HTS, resulting from stoichiometric inhomogeneities such as the distribution of atoms and dopants. Recent STS studies have been focusing on the local DOS variations in these materials, exploiting both the spatial and energy resolution of the technique.

Although the question is not definitively settled, there is more and more experimental evidence for the extrinsic scenario. Two types of spatial DOS inhomogeneities emerge from these spectroscopic studies: (1) the spatial inhomogeneity of the superconducting gap amplitude, and (2) the spatial inhomogeneity of the low-energy DOS. All studies presented here have been performed in single crystals of Bi-based compounds.

9.5.1 Local Modulations of the Superconducting Gap

Inhomogeneous superconductivity in Bi-2212 was reported in a number of early STM measurements [16, 40, 77–79]. The first hints suggesting that inhomogeneous superconductivity is not essential for a high T_c came from the observation that the spread in gap magnitude (see Fig. 9.8b as an example) measured on different surface locations and the average gap value are not correlated[80]. It was then demonstrated that spatial gap inhomogeneities were intimately associated with a broad superconducting transition[27], as measured by ac-susceptibility on the same samples. Conversely, single crystals with narrow superconducting transitions ($\Delta T_c < 0.5$ K) yield very homogeneous tunneling spectroscopy. This correlation between a wide gap distribution and the ac-susceptibility transition width was also observed in Bi-2223 [81]. These measurements strongly suggest that inhomogeneities are most likely due to inhomogeneous oxygen distribution. This assumption was confirmed in a study combining high-resolution topographic and spectroscopic STM maps where a direct correlation between the gap and oxygen impurity distributions in real space was established [82].

The relationship between inhomogeneities and structural [83] or dopant atom- [82, 84] disorder was another point of interest. It has been shown that Pb substitution and disorder in the Sr and Ca layers have no effect on the gap inhomogeneity [79]. The naturally present bulk modulation of atomic positions in Bi-based cuprates

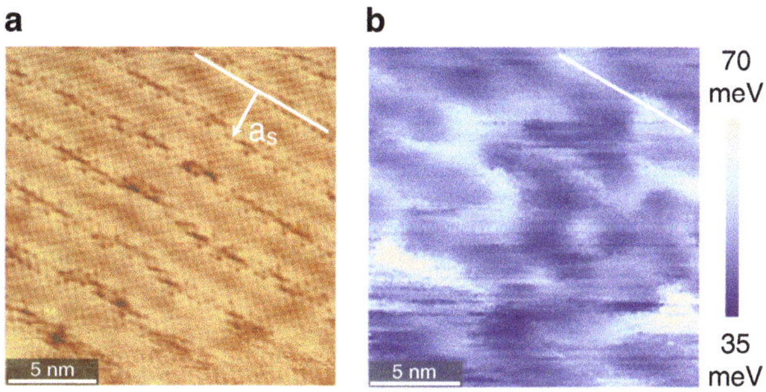

Fig. 9.8 Modulation of the superconducting gap in nearly optimally doped Bi-2223. (**a**) Topographic image of a 17×17 nm^2 region and (**b**) simultaneously measured local gap map. The *white line* is a spatial reference, and a_s is the superstructure unit vector

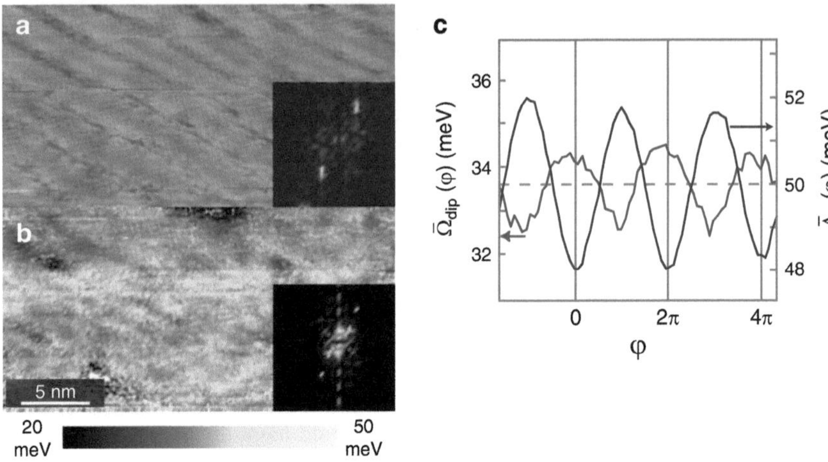

Fig. 9.9 Modulation of the collective-mode energy (CME) in Bi-2223. Spatial modulation of the CME in nearly optimally doped Bi-2223. (**a**) Topographic image of a $30 \times 15\,\mathrm{nm}^2$ region. Inset: Fourier transform of the topography depicting the two peaks at the wave-vectors of the superstructure $\pm 2\pi/a_s$. (**b**) Simultaneously acquired CME map. Inset: Fourier transform of the CME map displaying the same region in reciprocal space as in (a). (**c**) Dependence of the CME (*green line*) and gap (*blue line*) on the distance to the minima of the superstructure ($\varphi = 2n\pi$). The gap and the CME are locally anticorrelated. From [29]

[85, 86], the so-called superstructure seen in Fig. 9.8a as a one-dimensional surface corrugation, offers another interesting way to study this problem by scanning tunneling microscopy. In a recent study on Bi-2223 single crystals, the influence of the superstructure on the local DOS was analyzed [29]. As shown in Fig. 9.8b, it was observed that the gap magnitude is periodically modulated on a lengthscale of about 5 crystal unit-cells, following the superstructure modulation. As seen in the figure, the gap is the largest at the maxima of the surface corrugation. A similar observation had been reported in Bi-2212 [83].

Based on the analysis of the dip-hump structure described in Sect. 9.3, the local collective mode energy (CME) was extracted at each position of measurement and mapped (Fig. 9.9b). As seen in the map, the CME displays a modulation that also follows the superstructure shown in Fig. 9.9a. The CME and the gap are locally anticorrelated (Fig. 9.9c), in agreement with previous STS investigations performed on samples with different doping levels, and with neutron scattering experiments in underdoped cuprates [26, 30]. These findings support that the collective mode is related to superconductivity, and is most likely the anti-ferromagnetic spin resonance detected by neutron scattering. These results are in agreement with the spin-fluctuation scenario [87] that predicts the existence of a (π, π) resonance as a consequence of the feedback of pairing on the spin fluctuations [88]. Moreover, these findings challenge the theories that do not account for the very short lengthscale modulation (4–5 crystal unit cells) and the local anticorrelation of the gap and the CME, like the ones based on phonon-mediated pairing for high-temperature superconductivity.

9.5.2 Local Modulations of the DOS

9.5.2.1 Modulations in the Superconducting and Pseudogapped Regimes

The detection of electronic modulations in the superconducting state is complicated due to the strong tendency of these materials to show inhomogeneities masking the weak periodic modulations. In order to overcome this difficulty, large maps of the conductance at a given energy must be acquired. Such modulations were first observed in a magnetic field around the center of vortices, where the LDOS showed a periodicity of about $4a_0$ [89]. Subsequently, a similar structure also appeared in the absence of field, at an energy close to 25 meV, and did not disperse with energy [90]. In parallel, zero-field electronic modulations with a period varying with energy were also found in Fourier-transformed maps [91]. These structures and their characteristic dispersion trends were convincingly interpreted as modulations due to quasiparticle interference of a d-wave superconductor, resulting from scattering on impurities and other imperfections [91–94]. The fact that the intensity of such charge modulations must depend on the amount of scattering centers in the sample explain why in some cases the modulations are not observed [71]. A more detailed analysis of these results allowed to calculate the locus (k_x, k_y) of the Fermi surface [94]. The striking correspondence with ARPES measurements reinforces the interpretation that these LDOS oscillations result from quasiparticle interference. Thus, although STM on homogeneous samples does not have k-space resolution, this can be obtained when scattering centers are present, producing quasiparticle interference.

Electronic modulations are also detected in the pseudogap state. In an experiment on a slightly underdoped sample, conductance maps acquired at temperatures above T_c revealed a square superstructure with q-values equal to $2\pi/4.7a_0$ [95]. The intensity of these peaks is energy dependent and they are best seen at low energy around and below 20 meV, but the positions of these spots in q-space do not vary with energy. The most striking observation was made by studying underdoped samples at low temperature [96]. These samples showed strong inhomogeneity in the gap distribution, with large regions exhibiting pseudogap-like spectra. At low energy, the dispersing quasiparticle interference patterns were found. However, at higher energy (above 65 meV), the maps revealed a nondispersing square pattern similar to the one seen in the pseudogap state [95]. This observation suggests that the superconducting coherence is lost at high energy, i.e., for states around $(\pi, 0)$, but is still present at low energy, i.e., around $(\frac{\pi}{2}, \frac{\pi}{2})$. In this sense, the sample is partly in the superconducting state and partly in a pseudogap state, the nondispersive $4a_0 \times 4a_0$ pattern being seen in addition to the dispersive quantum interference pattern.

The study of the pseudogap state in strongly underdoped $Ca_{1-x}Na_xCuO_2Cl_2$ crystals revealed a dominant $4a_0 \times 4a_0$ superstructure similar to the one seen at high temperature in the pseudogap state in Bi-2212 [97]. The characteristic periods are energy independent and do not vary with doping. Additional periodicities at $\frac{4}{3}a_0$ were also found, revealing a noncommensurate electronic modulation and thus a more complex structure than observed in Bi-2212 [95]. More recently, detailed analysis revealed stripe-like patterns in the conductance maps of Bi-based cuprates [98].

These tunneling spectroscopy observations agree well with the picture provided by ARPES experiments, which revealed that the antinodal points near $(\pi, 0)$ are associated with an incoherent pairing (pseudogap), while the nodal regions near $(\frac{\pi}{2}, \frac{\pi}{2})$ are associated with coherent pairing and a true superconducting gap [99].

9.5.2.2 Modulations in the Vortex Cores

Since it has been found that the vortex cores possess pseudogap-like spectra, one might expect that the spatially ordered structure found around the vortex cores [89] and the one observed in the pseudogap state [95] have the same origin. To check this assumption, a detailed analysis of the behavior of the DOS modulation inside a vortex was performed [71]. Two questions motivated these investigations: (1) How does the square pattern observed in the vortex cores relate to the square pattern observed in the pseudogap state above T_c? (2) How does this square pattern in the vortex core relate to the localized state seen in the cores of HTS cuprates? Imaging an isolated vortex at 25 meV reveals the irregular core typical for Bi-2212 (Fig. 9.10a). If the same area is imaged at 6 meV, the energy of the core state in slightly overdoped Bi-2212, a DOS modulation following a square pattern oriented in the Cu–O bond direction is seen (Fig. 9.10b). The Fourier transform of this pattern shows, in addition to the atomic lattice, four spots with a period close to $4a_0$ (inset: q_1 spots). The position of the q_1 spots is found to be independent of energy, exactly as observed in the pseudogap phase. The period in this overdoped sample was $(4.3 \pm 0.3)a_0$, somewhat less than observed in the underdoped sample [95]. This difference could be due either to a difference in doping or to the difference in temperature. Note that in the $(\pi, 0)$ direction an additional $\frac{4}{3}a_0$ modulation was found, similar to the one reported in $Ca_{1-x}Na_xCuO_2Cl_2$ [97].

The relation to the localized states was also studied. It was demonstrated that the intensity of the peaks corresponding to these states is maximal at the four spots reflecting the local order, whereas in between and outside these four spots the peaks at $\pm 0.3\Delta_p$ are absent (Fig. 9.10c). Since the spatial dependence of the peaks (no dispersion with position) suggests that these two peaks reflect a single pair of localized states (see Sect. 9.4.3), the four spot pattern in the center of the vortex may be understood as a plot of the wave function of the localized state. Note that this pattern is very different from what one would obtain for a classical s-wave localized state [66]. The link established between the localized state structure observed in the vortex core spectra [12,15,72] and the local square modulation in the vortices raises the question about the precise relation between this structure and the pseudogap. While the modulations in the superconducting state can be understood in terms of quasiparticle interference, the simple square pattern observed in the pseudogapped regime may turn out to be a characteristic signature of order in this phase. Moreover, the question whether the localized state itself is a characteristics of the pseudogap remains open. So far no signature of such a state at $\sim 0.3\Delta_p$ has been observed

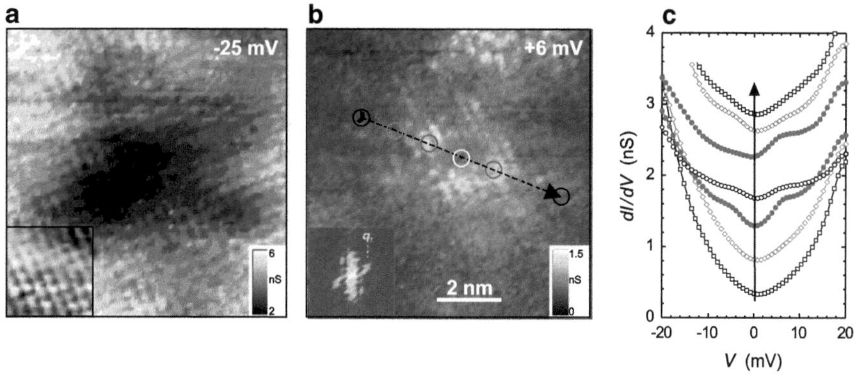

Fig. 9.10 Modulation of the LDOS inside a vortex core of Bi-2212. (**a**) $8.7 \times 8.7\ \text{nm}^2$ conductance map at $V = -25$ mV. The inset shows the simultaneously acquired topography at the same scale as the underlying conductance map. (**b**) Conductance map at $+6$ mV in the same area as (a) revealing a *square pattern* around the vortex center. Inset: central part of the Fourier transform of the conductance map. (**c**) Spectra averaged in the seven *circles* shown in (b). The curves are offset for clarity. Adapted from [71]

above T_c, but such a structure would be certainly difficult to probe due to thermal smearing. The challenge remains to determine whether such a state exists in the zero-field pseudogap regime.

9.5.3 Summary

The amount of valuable information provided by scanning tunneling spectroscopy has undoubtedly greatly improved our knowledge of high-temperature superconductivity. Although some of the questions remain unsolved, like the issue of the fundamental microscopic pairing mechanism, STS has shed light on many fundamental aspects of this complex phenomenon. Growing evidence provided by STS is that in cuprates, inhomogeneities stem from *extrinsic* stoichiometric disorder rather than *intrinsic* electronic phase separation. More importantly, the experiments reveal that an inhomogeneous gap distribution is not a prerequisite to the occurrence of high-temperature superconductivity, since there is no correlation between such a distribution and the value of T_c.

The link between the pseudogap and the superconducting state is provided by both the scaling between the superconducting gap and the pseudogap magnitudes, and by the mean-field relation established between Δ and T^*. According to these observations, the STS measurements in Bi-based cuprates suggest that the pseudogap is the precursor of the superconducting gap, and hence that the mechanisms leading to the formation of both gaps are the same. A step toward the identification of this pairing mechanism came from the possibility to de-convolve the coupling with the (π, π) resonance mode from the tunneling spectra, and to uncover an intimate relation between spin fluctuations and superconductivity. Both the dependence of the local mode energy with the superconducting gap and the ultra-short length-scales over which the coupling with these bosonic modes takes place favor the spin-mediated pairing scenario.

Finally, the complex interplay between the antinodal incoherent pairing (pseudogap) and the nodal coherent pairing (superconducting gap) is accessed through the local modulation of the quasiparticle excitations, providing k-space resolution to the STM. In addition, tunneling spectroscopy was able to firmly establish the direct link between the pseudogap regime and the vortex state, revealing that both regimes yield similar tunneling characteristics and exhibit the same $\sim 4a_0 \times 4a_0$ pattern in the conductance maps.

With its unrivaled spatial and energy resolution, STM is complementary to other techniques such as optical spectroscopy and ARPES. Forthcoming years will see this tool continuing to shine light on the key questions concerning high-temperature superconductivity in cuprates, and probably very soon in iron-based pnictides or other promising materials still to be discovered.

Acknowledgements This work was supported by the Swiss National Science Foundation through the National Centre of Competence in Research "Materials with Novel Electronic Properties MaNEP". We acknowledge D. Roditchev and H. Suderow for providing us with their tunneling data for the figures.

References

1. I. Giaever, Phys. Rev. Lett. 5, 147 (1960)
2. G.M. Eliashberg, Sov. Phys. JETP 11, 696 (1960)
3. G. Binnig, H. Rohrer, Helv. Phys. Acta 55, 128 (1982)
4. H.F. Hess, R.B. Robinson, R.C. Dynes, J.M. Valles, J.V. Waszczak, Phys. Rev. Lett. 62, 214 (1989)
5. C. Renner, Ø. Fischer, Phys. Rev. B 51, 9208 (1995)
6. Y. Kamihara, T. Watanabe, M. Hirano, H. Hosono, J. Am. Chem. Soc. 130, 3296 (2008)
7. Ø. Fischer, M. Kugler, I. Maggio-Aprile, C. Berthod, C. Renner, Rev. Mod. Phys. 79, 353 (2007)
8. H.J. Levinstein, J.E. Kunzler, Phys. Lett. 20, 581 (1966)
9. J. Tersoff, D.R. Hamann, Phys. Rev. Lett. 50, 1998 (1983)
10. R.J. Hamers, R.M. Tromp, J.E. Demuth, Phys. Rev. Lett. 56, 1972 (1986)
11. T. Cren, D. Fokin, D. Roditchev, F. Debontridder, Private communication
12. I. Maggio-Aprile, C. Renner, A. Erb, E. Walker, Ø. Fischer, Phys. Rev. Lett. 75, 2754 (1995)
13. M. Kugler, C. Renner, S. Ono, Y. Ando, Ø. Fischer, Phys. Rev. Lett. 86, 4911 (2001)
14. H.F. Hess, R.B. Robinson, J.V. Waszczak, Physica B 169, 422 (1991)
15. S.H. Pan, E.W. Hudson, A.K. Gupta, K.W. Ng, H. Eisaki, S. Uchida, J.C. Davis, Phys. Rev. Lett. 85, 1536 (2000)
16. S.H. Pan, J.P. O'Neal, R.L. Badzey, C. Chamon, H. Ding, J.R. Engelbrecht, Z. Wang, H. Eisaki, S. Uchida, A.K. Gupta, K.-W. Ng, E.W. Hudson, K.M. Lang, J.C. Davis, Nature 413, 282 (2001)
17. B.W. Hoogenboom, C. Berthod, M. Peter, Ø. Fischer, A.A. Kordyuk, Phys. Rev. B 67, 224502 (2003)
18. P.W. Anderson, N.P. Ong, J. Phys. Chem. Solids 67, 1 (2006)

19. Y. DeWilde, N. Miyakawa, P. Guptasarma, M. Iavarone, L. Ozyuzer, J.F. Zasadzinski, P. Romano, D.G. Hinks, C. Kendziora, G.W. Crabtree, K.E. Gray, Phys. Rev. Lett. 80, 153 (1998)
20. M. Nantoh, M. Kawasaki, T. Hasegawa, K. Fujito, W. Yamaguchi, H. Koinuma, K. Kitazawa, Physica C 242, 277 (1995)
21. M. Tachiki, S. Takahashi, F. Steglich, H. Adrian, Z. Phys. B 80, 161 (1990)
22. T.G. Miller, M. McElfresh, R. Reifenberger, Phys. Rev. B 48, 7499 (1993)
23. J.Y.T. Wei, C.C. Tsuei, P.J.M. van Bentum, Q. Xiong, C.W. Chu, M.K. Wu, Phys. Rev. B 57, 3650 (1998)
24. M. Eschrig, M.R. Norman, Phys. Rev. Lett. 85, 3261 (2000)
25. H.F. Fong, B. Keimer, P.W. Anderson, D. Reznik, F. Dogan, I.A. Aksay, Phys. Rev. Lett. 75, 316 (1995)
26. H.F. Fong, P. Bourges, Y. Sidis, L.P. Regnault, A. Ivanov, G.D. Gu, N. Koshizuka, B. Keimer, Nature, 398, 588 (1999)
27. B.W. Hoogenboom, K. Kadowaki, B. Revaz, Ø. Fischer, Physica C, 391, 376 (2003)
28. G. Levy de Castro, C. Berthod, A. Piriou, E. Giannini, Ø. Fischer, Phys. Rev. Lett. 101, 267004 (2008)
29. N. Jenkins, Y. Fasano, C. Berthod, I. Maggio-Aprile, A. Piriou, E. Giannini, B.W. Hoogenboom, C. Hess, T. Cren, Ø. Fischer, Phys. Rev. Lett. 103, 227001 (2009)
30. H.F. Fong, P. Bourges, Y. Sidis, L.P. Regnault, J. Bossy, A. Ivanov, D. Milius, I. Aksay, B. Keimer, Phys. Rev. B 61, 14773 (2000)
31. C. Renner, B. Revaz, J.Y. Genoud, K. Kadowaki, Ø. Fischer, Phys. Rev. Lett. 80, 149 (1998)
32. N. Miyakawa, J.F. Zasadzinski, L. Ozyuzer, P. Guptsarama, D.G. Hinks, C. Kendziora, K.E. Gray, Phys. Rev. Lett. 83, 1018 (1999)
33. T. Timusk, B. Statt, Rep. Prog. Phys. 62, 61 (1999)
34. G. Deutscher, Nature, 397, 410 (1999)
35. H.J. Tao, F. Lu, E.L. Wolf, Physica C 282–287, 1507 (1997)
36. A. Matsuda, S. Sugita, T. Watanabe, Phys. Rev. B 60, 1377 (1999)
37. A. Matsuda, S. Sugita, T. Watanabe, in Advances in Superconductivity XI, vol. 1, eds. by N. Koshizuka, S. Tajima (Springer-Verlag, Berlin, 1999), vol. 1, p. 151
38. I. Maggio-Aprile, C. Renner, A. Erb, E. Walker, B. Revaz, J.Y. Genoud, K. Kadowaki, O. Fischer, J. Electron. Spectrosc. Relat. Phenom. 109, 147 (2000)
39. C. Renner, B. Revaz, K. Kadowaki, I. Maggio-Aprile, Ø. Fischer, Phys. Rev. Lett. 80, 3606 (1998)
40. T. Cren, D. Roditchev, W. Sacks, K. Klein, J.B. Moussy, C. Deville-Cavellin, M. Laguës, Phys. Rev. Lett. 84, 147 (2000)
41. K. McElroy, D.-H. Lee, J.E. Hoffman, K.M. Lang, J. Lee, E.W. Hudson, H. Eisaki, S. Uchida, J.C. Davis, Phys. Rev. Lett. 94, 197005 (2005)
42. C. Renner, A.D. Kent, P. Niedermann, Ø. Fischer, Phys. Rev. Lett. 67, 1650 (1991)
43. S. Behler, S.H. Pan, P. Jess, A. Baratoff, H.J. Guenterodt, F. Levy, G. Wirth, J. Wiesner, Phys. Rev. Lett. 72, 1750 (1994)
44. A. Kohen, T. Cren, Y. Noat, T. Proslier, F. Giubileo, F. Bobba, A. Cucolo, N. Zhigadlo, S. Kazakov, J. Karpinski, W. Sacks, D. Roditchev, Phys. C: Supercond. 437–438, 145 (2006)
45. I. Guillamon, H. Suderow, S. Vieira, P. Rodiere, Physica C 468, 537 (2008)
46. M. Iavarone, R. Di Capua, G. Karapetrov, A. Koshelev, D. Rosenmann, H. Claus, C. Malliakas, M. Kanatzidis, T. Nishizaki, N. Kobayashi, Phys. Rev. B 78, 174518 (2008)
47. C. Dubois, G. Santi, I. Cuttat, C. Berthod, N. Jenkins, A.P. Petrović, A.A. Manuel, Ø. Fischer, S.M. Kazakov, Z. Bukowski, J. Karpinski, Phys. Rev. Lett. 101, 057004 (2008)
48. M.R. Eskildsen, M. Kugler, S. Tanaka, J. Jun, S.M. Kazakov, J. Karpinski, Ø. Fischer, Phys. Rev. Lett. 89, 187003 (2002)
49. P. de Trey, S. Gygax, J.P. Jan, J. Low Temp. Phys. 11, 421 (1973)
50. H.F. Hess, R.B. Robinson, J.V. Waszczak, Phys. Rev. Lett. 64, 2711 (1990)
51. F. Gygi, M. Schlüter, Phys. Rev. Lett. 65, 1820 (1990)
52. A. Petrovic, Y. Fasano, R. Lortz, C. Senatore, A. Demuer, A. Antunes, A. Par, D. Salloum, P. Gougeon, M. Potel, O. Fischer, arXiv p. 0903.2389v1 (2009)

53. Y. DeWilde, M. Iavarone, U. Welp, V. Metlushko, A.E. Koshelev, I. Aranson, G.W. Crabtree, P.C. Canfield, Phys. Rev. Lett. 78, 4273 (1997)
54. B.W. Hoogenboom, K. Kadowaki, B. Revaz, M. Li, C. Renner, Ø. Fischer, Phys. Rev. Lett. 87, 267001 (2001)
55. U. Hubler, P. Jess, H.P. Lang, A. Erb, E. Walker, M. Regier, D. Schild, J. Halbritter, H.J. Güntherodt, Appl. Phys. A 66, S1219 (1998)
56. K. Shibata, M. Maki, T. Nishizaki, N. Kobayashi, Physica C 392–396, 323 (2003)
57. K. Zhang, D.A. Bonn, S. Kamal, R. Liang, D.J. Baar, W.N. Hardy, D. Basov, T. Timusk, Phys. Rev. Lett. 73, 2484 (1994)
58. B. Keimer, W.Y. Shih, R.W. Erwin, J.W. Lynn, F. Dogan, I.A. Aksay, Phys. Rev. Lett. 73, 3459 (1994)
59. C. Renner, I. Maggio-Aprile, A. Erb, E. Walker, Ø. Fischer, in Spectroscopic Studies of Superconductors, eds. by I. Bozovic, D. van der Marel, Proc. SPIE 2696B (1996) p. 322
60. H. Won, K. Maki, Europhys. Lett. 30, 421 (1995)
61. I. Affleck, M. Franz, M.H. Sharifzadef Amin, Phys. Rev. B 55, R704 (1997)
62. I. Maggio-Aprile, C. Renner, A. Erb, E. Walker, Ø. Fischer, Nature, 390, 487 (1997)
63. K. Matsuba, H. Sakata, N. Kosugi, H. Nishimori, N. Nishida, J. Phys. Soc. Jpn. 72, 2153 (2003)
64. B.W. Hoogenboom, M. Kugler, B. Revaz, I. Maggio-Aprile, Ø. Fischer, C. Renner, Phys. Rev. B 62, 9179 (2000)
65. M. Franz, Z. Tešanović, Phys. Rev. Lett. 80, 4763 (1998)
66. C. Caroli, P.G. de Gennes, J. Matricon, Phys. Lett. 9, 307 (1964)
67. F. Gygi, M. Schlüter, Phys. Rev. B 43, 7609 (1991)
68. H. Nishimori, K. Uchiyama, S.I. Kaneko, A. Tokura, H. Takeya, K. Hirata, N. Nishida, J. Phys. Soc. Jpn. 73, 3247 (2004)
69. Y. Wang, A.H. MacDonald, Phys. Rev. B 52, R3876 (1995)
70. M. Kugler, A sub-kelvin scanning tunneling microscope and its first application of spectroscopy on $Bi_2Sr_2CaCu_2O_8$ and $Bi_2Sr_2Cuo_6$. Ph.D. thesis, University of Geneva (2000)
71. G. Levy, M. Kugler, A.A. Manuel, Ø. Fischer, M. Li, Phys. Rev. Lett. 95, 257005 (2005)
72. B.W. Hoogenboom, C. Renner, B. Revaz, I. Maggio-Aprile, Ø. Fischer, Physica C 332, 440 (2000)
73. P.I. Soininen, C. Kallin, A.J. Berlinsky, Phys. Rev. B 50, 13883 (1994)
74. R.B. Laughlin, Phys. Rev. Lett. 80, 5188 (1998)
75. C. Berthod, B. Giovannini, Phys. Rev. Lett. 87, 277002 (2001)
76. J.C. Phillips, A. Saxena, A.R. Bishop, Rep. Prog. Phys. 66, 2111 (2003)
77. C. Howald, P. Fournier, A. Kapitulnik, Phys. Rev. B 64, 100504(R) (2001)
78. K.M. Lang, V. Madhavan, J.E. Hoffman, E.W. Hudson, H. Eisaki, S. Uchida, J.C. Davis, Nature, 415, 412 (2002)
79. G. Kinoda, T. Hasegawa, S. Nakao, T. Hanaguri, K. Kitazawa, K. Shimizu, J. Shimoyama, K. Kishio, Phys. Rev. B 67, 224509 (2003)
80. K. Matsuba, H. Sakata, T. Mochiku, K. Hirata, N. Nishida, Physica C 388, 281 (2003)
81. M. Kugler, G. Levy de Castro, E. Giannini, A. Piriou, A.A. Manuel, C. Hess, Ø. Fischer, J. Phys. Chem. Solids 67, 353 (2006)
82. K. McElroy, J. Lee, J.A. Slezak, D.H. Lee, H. Eisaki, S. Uchida, J.C. Davis, Science, 309, 1048 (2005)
83. J.A. Slezak, J. Lee, M. Wang, K. McElroy, K. Fujita, B.M. Andersen, P.J. Hirschfeld, H. Eisaki, S. Uchida, J.C. Davis, Proc. Nat. Acad. Sci. USA 105, 3203 (2008)
84. K. Chatterjee, M.C. Boyer, W.D. Wise, T. Kondo, T. Takeuchi, H. Ikuta, E.W. Hudson, Nat. Phys. 4, 108 (2008)
85. Y. Gao, P. Lee, P. Coppens, M.A. Subramanian, A.W. Sleight, Science, 241, 954 (1988)
86. E. Giannini, R. Gladyshevskii, N. Clayton, N. Musolino, V. Garnier, A. Piriou, R. Flükiger, Curr. Appl. Phys. 8, 115 (2008)
87. D.J. Scalapino, E. Loh, J.E. Hirsch, Phys. Rev. B 34(11), 8190 (1986). DOI 10.1103/PhysRevB.34.8190

88. A.V. Chubukov, D. Pines, J. Schmalian, in Conventional and High-T_c Superconductors, The Physics of Superconductors, vol. I, eds. by K. Bennemann, J. Ketterson (Springer, New York, 2003), p. 495
89. J.E. Hoffman, E.W. Hudson, K.M. Lang, V. Madhavan, H. Eisaki, S. Uchida, J.C. Davis, Science, 295, 466 (2002)
90. C. Howald, H. Eisaki, N. Kaneko, M. Greven, A. Kapitulnik, Phys. Rev. B 67, 014533 (2003)
91. J.E. Hoffman, K. McElroy, D.H. Lee, K.M. Lang, H. Eisaki, S. Uchida, J.C. Davis, Science, 297, 1148 (2002)
92. Q.H. Wang, D.H. Lee, Phys. Rev. B 67, 020511(R) (2003)
93. D. Zhang, C.S. Ting, Phys. Rev. B 67, 100506(R) (2003)
94. K. McElroy, R.W. Simmonds, J.E. Hoffman, D.H. Lee, J. Orenstein, H. Eisaki, S. Uchida, J.C. Davis, Nature, 422, 592 (2003)
95. M. Vershinin, A. Misra, S. Ono, Y. Abe, Y. Ando, A. Yazdani, Science, 303, 1995 (2004)
96. K. McElroy, D.H. Lee, J.E. Hoffmann, K.M. Lang, J. Lee, E.W. Hudson, H. Eisaki, S. Uchida, J.C. Davis, Phys. Rev. Lett. 94, 197005 (2005)
97. T. Hanaguri, C. Lupien, Y. Kohsaka, D.H. Lee, M. Azuma, M. Takano, H. Takagi, J.C. Davis, Nature, 430, 1001 (2004)
98. Y. Kohsaka, C. Taylor, K. Fujita, A. Schmidt, C. Lupien, T. Hanaguri, M. Azuma, M. Takano, H. Eisaki, H. Takagi, S. Uchida, J. Davis, Science, 315, 1380 (2007)
99. A. Damascelli, Z. Hussain, Z.X. Shen, Rev. Mod. Phys. 75, 473 (2003)

Chapter 10
Scanning Tunnelling Spectroscopy of Vortices with Normal and Superconducting tips

J.G. Rodrigo, H. Suderow, and S. Vieira

Abstract Scanning tunnelling microscopy and spectroscopy (STM/S) has proved to be a powerful tool to study superconductivity down to atomic level. Vortex lattice studies require characterizing areas of enough size to contain a large number of vortices. On the other hand, it is necessary to combine this capability with high spectroscopic and microscopic resolution. This is a fundamental aspect to measure and detect the subtle changes appearing inside and around a single vortex. We report in this chapter our approach to the use of STM/S, using normal and superconducting tips, to observe the lattice of vortices in several compounds, and the information acquired inside these fascinating entities. The combination of superconducting tips and scanning tunneling spectroscopy, $(ST)^2S$, presents advantages for the study of superconducting samples. It allows to distinguish relevant features of the sample density of states, which manifest itself as small changes in the Josephson coupling between sample and tip condensates, and it has also shown to be very efficient in the study of the ferromagnetic-superconductor transition in the re-entrant superconductor $ErRh_4B_4$.

10.1 Introduction

The invention of the scanning tunnelling microscope by Binning and Rohrer [1] at the beginning of the 1980s was a breakthrough and was a response to the challenge raised by R.P. Feynman on his visionary conference "There is plenty of room at the bottom" [2]. The combination of a pure quantum phenomenon, the electron quantum tunnelling, with a humble, but fascinating, kind of materials, the piezoelectric ceramics, gives us fingers to manipulate objects in the nanoworld. Today, there are many different kinds of probes, such as force or optical detectors, which are scanned at local scale using piezoceramics. Many of these probes are commercially available for biology, chemistry and physics laboratories. Very soon after the invention of the STM, Lozanne et al. [3] built the first STM that could work at low temperatures and made the first experiment of superconducting tunnelling in Nb_3Sn, using a normal platinum tip as counterelectrode. From then to now, the combined microscopic and spectroscopic capabilities of the STM have produced many interesting results, which have signified important advances in our understanding of new superconducting materials and phenomena. Perhaps, one of the most interesting and timely contribution was the one by Hess et al. [4]. They were able for the first time to obtain nice images of the Abrikosov vortex lattice in single crystals of the layered dichalcogenide $NbSe_2$, in which superconducting order coexists with a charge density wave. The internal structure of a single vortex was seen with fine details and explained in terms of previous predictions of Caroli, de Gennes and Matricon [5]. By plotting the zero bias conductance, they obtained a sixfold star around the vortex core, reflecting the spatial behaviour of the localized states of the quasiparticles living in the potential well created by the BCS energy gap. At different bias, changes appears in the star, throwing nice information about the very nature of the vortices. These works were, in our opinion, an illuminating beginning of the relation between STM/S techniques and vortex matter.

It is out of the scope of this chapter to revise the treasure of information on the use STM/S, to study vortices, that has been accumulated in the literature, thanks to the effort of many colleagues. The reviews of [6] and [7], among others, give a rather up-to-date account of many important results. We are going to refer here to the main activities in which our laboratory has been involved in the last years.

We organize the chapter in the following way. First, we discuss some characteristics of our experimental setup, including tip preparation and characterization. Regarding this, we pay special attention to the behaviour under magnetic field, which is especially relevant in the framework of this chapter.

In order to highlight the advantages of the $(ST)^2S$ technique, we will discuss a very well-characterized superconductor $NbSe_2$, in which the coexistence of two band superconductivity and charge density wave produces interesting features on its mixed state. We compare similar studies done with normal (gold) and superconducting (aluminium and lead) tips.

Next, we make a comparative discussion between $NbSe_2$ and its parent compound NbS_2 which does not hold charge density wave order. Significant differences appears between these dichalcogenides, as revealed by $(ST)^2S$. The superconducting properties, including two dimensional vortex lattice behaviour, of a new and promising material, tungsten nanodeposits, fabricated by focused ion beam, will be also summarized in this section.

We finish this contribution with the application of the superconducting tip to the study of a very peculiar phase transition which takes place in $ErRh_4B_4$ close to 0.8 K. This material transits to a superconducting state at 8 K, and this is destroyed close to 0.8 K where a ferromagnetic phase establishes. The use of a superconducting tip was crucial to reveal tiny and relevant differences between the directions, of cooling and heating, in which the transition is passed.

10.2 Experimental: Low Temperature STM with Superconducting tips

10.2.1 Low Temperature STM

The scanning tunneling microscope (STM) [1], as well as related techniques like the atomic force microscope (AFM) [8], are versatile tools to penetrate in the nanoworld realm. The STM allows to study the topography and electronic properties of a conducting surface with atomic spatial resolution. In the little more than 20 years elapsed since its invention, this technique has become widely used. These instruments can be obtained from commercial suppliers, some of them designed to work at low temperatures. However, home-made STM are in use in many laboratories, as they give the required versatility and accuracy for doing specific research.

A cylindrically symetric design is very well suited for low temperatures STMs. In Fig.10.1a, we show a sketch of the STM built and used in the low temperature laboratory of the Universidad Autonoma de Madrid [9], and highlight its original aspects. The coarse approach system is designed to make, if wanted, strong indentation of the tip in the sample surface. With this system, tip and sample can be approached from distances of several millimeters in situ at low temperatures. A piezotube with vertical displacements, at cryogenic temperatures, of several tenths of microns is used for the fine control and scanning. The sample holder stage can be moved in the x–y plane distances in the millimetric range using piezoelectric stacks, allowing access to several samples in the same experimental run [9]. Different configurations can be adopted for the sample holder stage, including a one direction movable holder, with a capacitance sensor to monitor sample displacement with sensitivity of 100 nm, and different types of in situ sample cleavage devices.

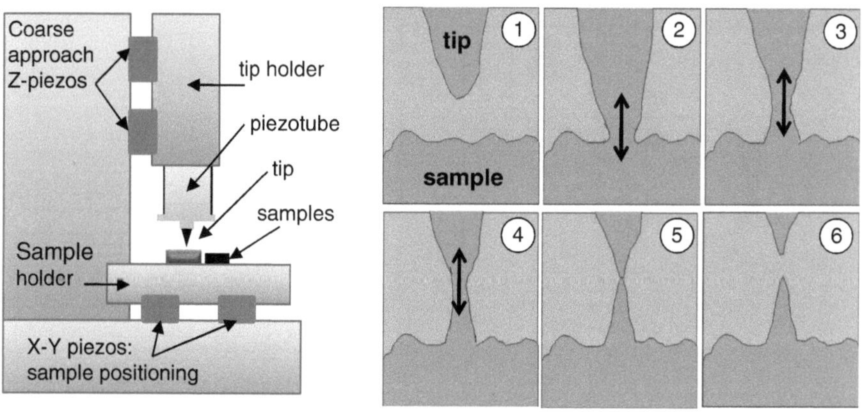

Fig. 10.1 *Left*: Scheme of the STM unit used in the Low Temperature Laboratory of the Universidad Autonoma def Madrid. *Right*: Sketch of the tip fabrication process. Frames 1–6 illustrate different stages of the process

Besides the development of the STM head itself, great care has to be taken in obtaining the desired high level of quality in the different voltage and current signals involved in the measurement. Special filtering, signal conditioning and thermal control procedures have been implemented in our system to allow for the required spectroscopic and thermal energy resolution in these experiments at millikelvin temperatures [10–13].

10.2.2 Tips Preparation and Characterization

The tips used in our experiments are made of normal (Au, Pt) or superconducting (Pb, Al) elements, and the STM itself is used as a tool for its fabrication and tailoring at the nanoscale. The fabrication process allows to obtain large and sharp tips down to the nanoscale with ductile metals such as Au, Pb, Al and Sn [14–17], and it was tested also with semimetals, as Bi [18]. The first step of the fabrication is to crash, in a controlled manner, the tip into a substrate, both of the same material. As the tip is pressed against the substrate, both electrodes deform plastically and then bind by cohesive forces, forming a connective neck. Retraction of the tip, while smaller sized (up to 100 nm) pushing and pulling ramps are added to the z-piezo signal to mechanically anneal the contact region, results in the formation of a neck that elongates plastically and eventually breaks, leading to an atomically sharp tip (see a schematic representation of the process in frames 1–6 in Fig. 10.1).

Current vs. tip displacement ($I - z$) curves are monitored along the fabrication procedure. These $I - z$ curves are staircase-like and strikingly reproducible when the process is repeated many times [16,19]. This reproducibility shows that the resulting nanostructures and tips are clean, without oxides. The results shown in Fig. 10.2a

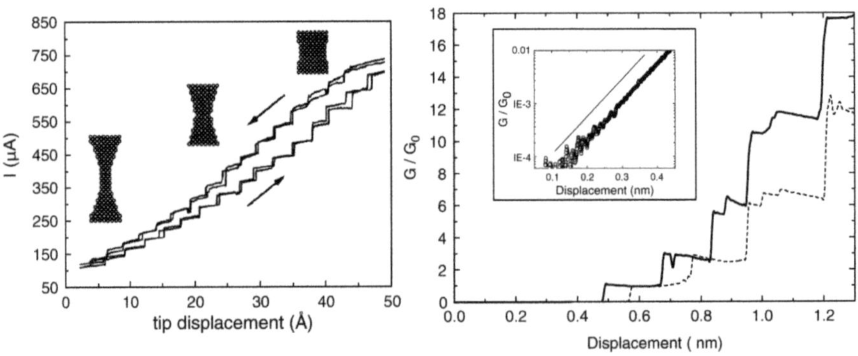

Fig. 10.2 *Left*: Current vs. elongation curves obtained at 4.2 K, using Pb tip and sample, from [19]. The estimated shapes at different stages of the nanobridge along the $I - z$ curves are sketched. *Right*: Typical conductance vs. distance curves close to the atomic contact regime obtained between tip and sample of Al. Inset: exponential conductance vs. distance curve obtained in the tunneling limit. Data from [11]

were the first clear indication that the controlled fabrication of superconducting nanobridges and tips using a STM as a tool and as a probe was possible [19]. In this experiment, made at 4.2 K, high purity lead ($T_c = 7.14$ K) was used for tip and sample.

Figure 10.2b shows the final steps of a typical fabrication and characterization process of atomically sharp aluminum tips. The rupture of the connective neck between aluminum electrodes leads to the characteristic step-like conductance vs. distance curves, $G(z)$. In the tunneling regime, when the contact is fully broken, $G(z)$ has a clear exponential dependence from which a work function of several eV is extracted.

It was soon understood that the staircase shape of the $I - z$ curves reflected the sequence of elastic and plastic deformations followed by the nanobridge [20]. Combined STM–AFM experiments made with lead and gold [14,15,21] established definitively the intimate relation between conductance steps and atomic rearrangements. The conductance observed for the gold atomic contacts is quite close to G_0, where $G_0 = 2e^2/h$ is the value of the quantum of conductance [22, 23]. The force involved in the rupture of these one-atom contacts is also well defined, with a value of 1.5 ± 0.1 nN [21, 24]. Transport experiments in several other elements in the superconducting state (Pb, Al or Nb, and also in Au, made superconducting using the proximity effect), have permitted to establish a clear relationship between the conductance of the last contact and the chemical nature of the atom involved [25–28]. A recent review by Agraït et al. [17] provides a comprehensive vision of this field.

The spectroscopic measurements at zero applied magnetic field not only determine the quality of the superconducting density of states (DOS) of the tip, but also provide the level of sensitivity in the measurements. The curves presented in Fig. 10.3 were obtained in the tunnelling regime, with Pb and Al tips, and represent the optimal energy resolution obtained with the cryogenic STMs in dilution and ^{3}He refrigerators used at LBTUAM [11, 12, 30].

Ideal $I - V$ curves in the S–S tunnelling regime at very low temperature present the well-known features of zero current up to the gap edge at 2Δ, where there is a jump to non-zero current [31, 32]. This appears in the tunneling conductance curves ($\mathrm{d}I/\mathrm{d}V$ vs V) as a divergence at energy 2Δ. Therefore, the measurement of the conductance curves in tunnelling regime is a direct test of the energy resolution of the experimental set-up. This energy resolution was introduced as a narrow Gaussian distribution, with a halfwidth in energy, σ, to simulate the noise in the bias voltage [29].

Results corresponding to aluminum tunneling conductance curves measured in a dilution refrigerator, and the corresponding fitting [29, 30] are shown Fig. 10.3(left). The calculated curve was obtained with the parameters $\Delta = 175\,\mu$ eV, $T = 70$ mK (base temperature of the system) and energy halfwidth $\sigma = 15\,\mu$ eV.

Tuneling experiments using superconducting tips obtained from lead nanobridges fabricated with the STM have been reported elsewhere [29, 30, 33]. Lead is a strong coupling superconductor, and it was found in early tunnelling experiments [34–36] that its gap value is not constant over the Fermi surface. Recent results

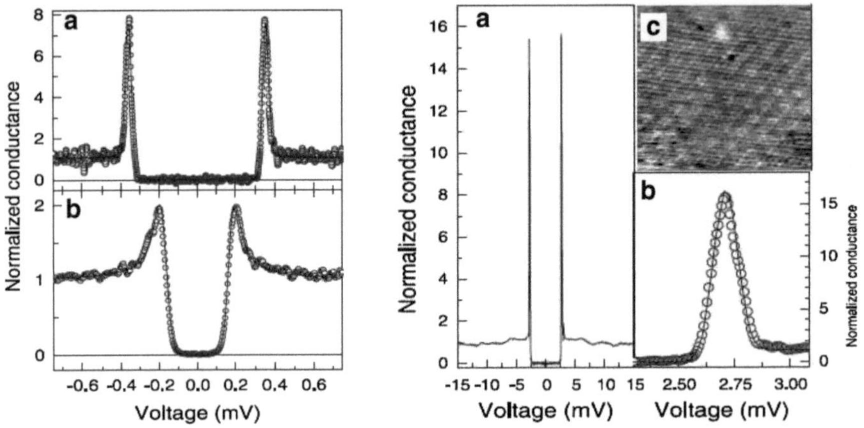

Fig. 10.3 *Left*: Normalized conductance curves obtained in tunnelling regime when both tip and sample are made of Al (**a**), and when the Al tip is located over a gold sample (**b**). The experimental curves (*circles*) were obtained at the lowest temperature of the system (70 mK, see text) and for $R_N = 1$ MΩ. Theoretical curves (*lines*) were calculated as described in the text, using the values $T = 0.150$ K and $\delta = 175$ μeV. *Right*: (**a**) Tunnelling conductance curve obtained in at 0.3 K with tip and sample of Pb ($R_N = 1$ MΩ). The gap edge is zoomed in (**b**). The image (**c**) shows a 10×10 nm^2 area of $NbSe_2$ scanned with atomic resolution using a Pb tip at 0.3 K (from [29])

give new support to this scenario [37]. The tunnelling curves corresponding to Pb tip and sample (Fig. 10.3 (right)) present coherence peaks with a finite width larger than the one expected considering only the energy resolution of the spectroscopic system. This additional width was considered to be a consequence of the above-mentioned different gap values present in the Pb Fermi surface. The experimental spectra were therefore simulated including a Gaussian distribution of gap values (with half-width δ), in addition to the energy resolution of the spectroscopic system. A detailed discussion of this analysis can be found in [30] and [29]. Recently there have been several reports on new superconducting materials which indicate that a single gap in the Fermi surface is the less frequent case [30, 38–43]. Multiband superconductivity and gap anisotropy seem to be more ubiquitous than previously thought. This fact enhances the importance of precise local tunnel measurements to shed light on unsolved problems about the nature of the superconducting state.

10.2.3 Spectroscopic Advantages of Superconducting tips

Soon after the development of first tunneling spectroscopy experiments, it was realized that tunneling spectroscopy on a superconducting sample using a superconducting counter electrode (S–S' tunneling) is more informative than tunneling spectroscopy with a normal electrode. The sharp DOS of the counter electrode allows indeed for a better determination of details in the sample's DOS, even at temperatures where thermal smearing is important [11, 29, 30, 41, 44].

The tunneling current between a superconducting tip with DOS $N_t(E)$ and a sample with $N_s(E)$ can be written as

$$I(V) \propto \int \mathrm{d}E[f(E-eV)-f(E)]N_t(E-eV)N_s(E), \qquad (10.1)$$

where $f(E)$ is the Fermi function. If the temperature and density of states of the tip are known, it is easy to de-convolute the density of states of the sample [11,29,30].

On the other hand, a superconducting tip can be used to probe locally the Josephson coupling between tip and sample [45]. Therefore, the STM with a well-characterized superconducting tip can be used to obtain a direct image of spatial variations of the Cooper pair density of a superconducting sample. This has been proposed as a method to explore the pairing symmetry of the order parameter at atomic level [46]. Mapping of the Josephson coupling around a vortex should provide new fundamental information about vortices.

Another important feature of the superconducting tip is the sensitivity of its DOS to the external magnetic field. This sensitivity can be tuned by means of the sharpness of the tip achieved during the fabrication process, and by the choice of the superconducting material used for the tip. We denominate nanotip (NT), the apical region which remains superconducting above the bulk critical field. Figure 10.4 shows two examples of these situations. The aluminum NT (left) is very sensitive to the magnetic field in the tens of mT range, where its DOS varies in a continuous way until superconductivity fully disappears at 100 mT. Therefore, this NT will be

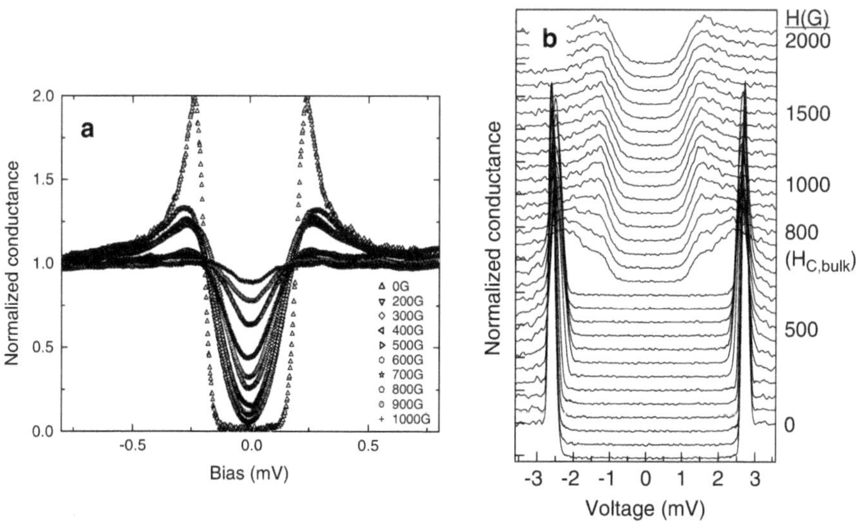

Fig. 10.4 (**a**) The magnetic field behaviour of the tunneling conductance, normalized to the high bias value, of an Al nanotip with a critical field of about 0.1 T (from [11]). (**b**) $Pb - Pb$ conductance curves taken at 300 mK in tunneling regime as the magnetic field is varied from 0 to 2,000 G. This particular nanotip presented a critical field of 9 kG (from [13])

an optimum magnetic field sensor when probing the vortex state in low fields. The magnetic field sensitivity of this NT goes down to the mT range, and the local variations of the magnetic field can be traced down to the nm range. These values, which can be further improved by a more careful measurement of the I–V curves, make this technique very competitive with respect to other local magnetic field probes such as MFM or scanning squid microscopy [47].

Note that the sensitivity to the magnetic field can be largely tuned. For example, in the Pb NT, whose magnetic field behaviour is shown in the right panel of Fig. 10.4, the critical field for destruction of superconductivity is in the range of Tesla, far above the case discussed in previous paragraph. Hence, this NT presents an almost constant DOS in the tens of mT range. Therefore, the local magnetic field variations around vortices will not influence the NT DOS, and all observed variations can be ascribed to the local DOS of the sample.

The fabrication of superconducting tips can be obviously made using different methods. In the first approaches, malleable elements (Pb and Al) and the systematic repeated indentation procedure described above were used for spectroscopic measurements at a single location [16, 45]. Soon after, it was shown that the superconducting condensate can be confined to the tip apex by applying a small magnetic field [27, 33, 48]. Subsequently, it was found that a tip of Nb can be also fabricated by field emission, and atomic resolution images could be achieved [49]. This kind of tips was used to probe superfluid velocity distributions around vortices [50], in a similar manner as the seminal very low temperature experiments with normal tips of [4]. Furthermore, Pb tips were fabricated using an in situ evaporation method on top of sharp Ag tips [51–54]. Thanks to this method, it was possible to make a very careful analysis of Josephson coupling at the local level, complementing earlier work which covered tunneling and contact regime [55]. The feasibility of local Josephson coupling with STM was demonstrated, although no scanning was performed, and the thermal smearing of the Josephson peak was significant. Other important efforts have been made using serendipitously attached small grains, or small single crystals glued on the tip [38, 56, 57]. When using relatively high critical temperature superconductors, such as MgB_2 [56, 57], the Josephson signal was somewhat improved, although thermal smearing was larger than in previous work [51–55].

Contrary to all other approaches, the repeated indentation method developed in [16, 45] allows one to obtain in a systematic way a clean tip, with a control over its geometry and sharpness at atomic level, and which can be carefully characterized before its use in another sample. This allows to cover all the panoply of new possibilities offered in a single experimental set-up.

When all these possibilities are available, the superconducting tip becomes much more than a simple extension of normal tip STM. It gives a truly new kind of nanoprobe, with ultimate spectroscopic capabilities, combining enhanced quasiparticle tunneling resolution, direct probing of the local Josephson coupling and measurement of the local magnetic field, all of them with sub-nanometer scale spatial resolution.

10.3 Vortices Studied by STS

10.3.1 The Vortex Lattice: General Properties and Visualization

The vortex lattice reveals in different ways important aspects of the normal and superconducting state properties. The way the symmetry of the vortex lattice is shaped by the materials' properties has been studied repeatedly in the past using vortex imaging techniques. For example, in the nickel borocarbides, a square to hexagonal transition appears in the vortex lattice when increasing the magnetic field [58]. This transition has been associated with non-local effects and the structure of the Fermi surface [59]. In the heavy fermion superconductor UPt_3, the symmetry of the vortex lattice is closely related to the complex phase diagram showing well-differentiated superconducting phases [60, 61]. The coexistence of magnetism and superconductivity is also expected to shape the symmetry of the vortex lattice in amazing ways. Possibly, the most spectacular example is $ErNi_2B_2C$, where the vortex lattice tilts due to the appearance of weak ferromagnetism below 2.3K with the internal field direction perpendicular to the applied magnetic field [62, 63]. Even more spectacular vortex arrangements are expected in $ErRh_4B_4$ or in the ferromagnetic superconductor URhGe [64–66]. The techniques to measure the symmetry of the vortex lattice are varied, and most of them are sensitive to the local magnetic field [47, 67]. Bitter decoration, scanning squid microscopy and Lorentz microscopy allow access to the lattice at low fields, and small angle neutron scattering as well as STM to higher field lattice studies [47]. In particular, neutron scattering is applied to systems where suitable large single crystals are available [62].

Another important aspect concerns the internal electronic structure of the vortex cores and the superconducting DOS in between cores. Macroscopic techniques such as thermal conductivity, or specific heat have also been used to get information about the DOS in and outside vortices [68, 69]. These give a result averaged over the whole sample [70]. Important information regarding the field dependence of the DOS in between cores has been obtained. Generally, in a material showing a strongly anisotropic or Fermi surface sheet-dependent superconducting gap, the averaged Fermi level DOS increases with the application of a magnetic field [60, 69, 71, 72]. Another technique sensitive to the energy-dependent DOS is photoemission spectroscopy [73]. Its realization at the local scale seems far from state-of-the-art technical possibilities, and the spectral resolution is still generally above 1 K, so that the results appear relatively averaged. Therefore, at present, the only available technique to study locally the DOS inside and around vortex cores, with full energy dependent measurement, and sub-Kelvin spectral resolution is STM.

A single vortex is itself a compendium of the properties of the superconducting material. Mapping the tunnelling conductance curves around a vortex, through STS experiments, it is possible to obtain powerful information on the coherence and penetration lengths, on the superfluid velocity distribution, on the nature of the quasiparticle bound states at the core, on the effect of some impurities on these states, and on the shape of the potential well confining the quasiparticles.

Using a normal metal as the STM probe, it is somewhat difficult to access the way superconductivity is lost when approaching the vortex core. For example, the Cooper pair density variations are hidden in a very complex and indirect way inside the local DOS variations. Thus, the measurement of the quasiparticle excitation spectrum is, necessarily, incomplete. Using a superconducting tip and probing the Josephson coupling between tip and sample, one can obtain the spatial variation of the Cooper pair density. This is crucial to achieve a complete understanding of the very nature of vortices in superconductors.

10.3.2 $NbSe_2$ Studied with Normal and Superconducting tips

As described above, STM/S with normal tips has been extensively used to investigate $NbSe_2$. In this section, we compare these results with those obtained using superconducting tips of lead and aluminum.

Figure 10.5 shows STS measurements on a vortex taken with a normal tip (gold) and a superconducting tip (aluminum), at the same experimental conditions ($T = 0.1$ K, $H = 30$ mT) [11]. When the measurement is done with the Al tip, the superconducting DOS of Al is superposed to the local DOS of $NbSe_2$. Therefore, there is a sharp downturn of the tunneling conductance close to the Fermi level and inside the vortex core (left panels of Fig. 10.5). $(ST)^2S$ images constructed with the tip of Al show also the expected features of the local DOS of $NbSe_2$, namely the star-shaped vortex, although at slightly higher bias voltages due to the gap of Al.

An Al tip whose superconducting DOS shows a stronger field dependence than in the previous example was used to acquire the curves shown in Fig. 10.6 [11]. At the vortex center, the zero bias peak expected with a normal tip is present (top curves of Fig. 10.6). However, when leaving the centre, the superconducting gap of Al appears around the Fermi level (bottom curves of Fig. 10.6 are not flat around zero bias, as expected when using a normal tip). The local magnetic field sensed by the tip is estimated by de-convoluting the Fermi level DOS of the Al tip, $N_T(r_0, E = 0)$, from the conductance curve, $G(r_0, V)$, and using the previously measured dependence of $N_T(E = 0)$ as a function of the magnetic field. This analysis demonstrated the feasibility of measuring variations of the local magnetic field with a superconducting tip as a probe. As mentioned above, the spatial resolution goes down to the nm range, and the magnetic field sensitivity to mT, although both can be significantly improved [11].

The use of a superconducting tip with higher values of both critical temperature and magnetic field allowed to extend the analysis of the details associated with the quasiparticle bound states and the evolution of the superconducting gap around the vortex up to the T_c value of the sample. In [13], a Pb tip was used to investigate the evolution of the conductance patterns in STS measurements of the vortice lattice in $NbSe_2$ at different temperatures. The aim of this study was to detect the effect on

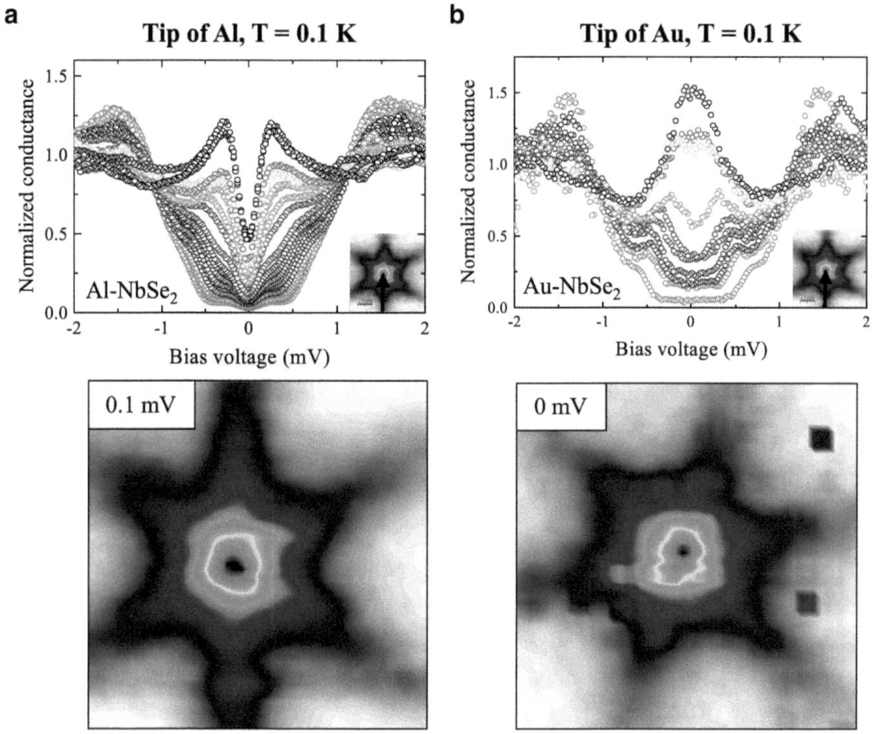

Fig. 10.5 $(ST)^2S$ vortex imaging with a tip of Al (*left*), compared to imaging with a normal tip of Au (*right*), at 0.1 K and 0.03 T. In (**a**), we show tunneling conductance curves obtained along a line of about 60 nm that goes into the vortex core, following one of the rays of the star shown in (**b**) (top to bottom is from the centre to out of the vortex core). Below, $(ST)^2S$ and STS images of a single vortex taken with both tips, at bias voltages corresponding to $E_{F,\text{sample}}$. Images size: 120 × 120 nm^2 (data from [11])

the vortex structure of the two-band character of superconductivity in $NbSe_2$, which presents two main gap features at around 0.6 mV (Δ_S) and 1.2 mV (Δ_L) in the DOS.

$(ST)^2S$ data at low temperature (0.3K) produced the usual star-shaped vortex images corresponding to the bound states at $E_{F,\text{sample}}$, and allowed to detect that the "extra" peak in the conductance curves taken close to the vortex center leads to well-defined spots in the STS images, arranged in a sixfold pattern, at higher energy (Fig. 10.7a–c). These features, which evolve in energy and position even once the peak associated with Δ_L is fully developed, were considered to be caused by the evolution of the smaller of the superconducting gaps present in the multiband superconduct $NbSe_2$ [30, 73, 74].

The sixfold pattern related to the above-mentioned "extra" peak disappears in the $(ST)^2S$ images obtained at 6 K, while the patterns observed at $E_{F,\text{sample}}$ and $E_{F,\text{sample}} + \Delta_{L,\text{sample}}$ remained almost identical to those obtained at 300 mK

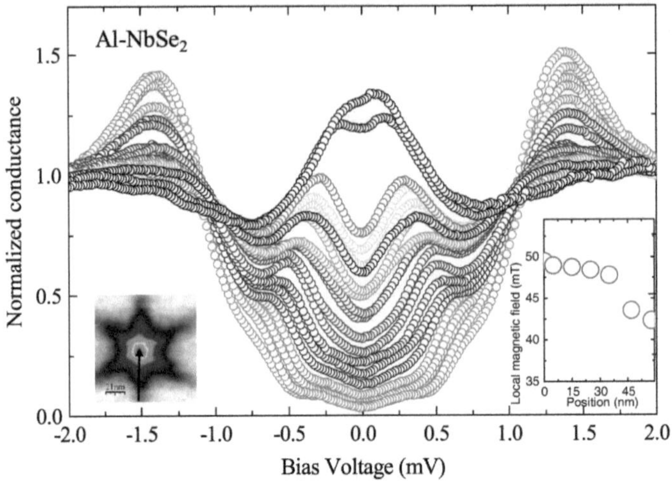

Fig. 10.6 Tunneling conductance as a function of the position taken with an Al nanotip at $T = 0.1$ K and $H = 40$ mT. At the vortex core there is no gap structure corresponding to the tip (top curves of the figure). The gap in the nanotip DOS slowly opens when leaving the vortex core (bottom curves of the figure). (ST)^{2}S curves were taken along a ray of the star shaped vortex. The nanotip acts as a local probe of the magnetic field, whose position dependence is shown in the inset. (Data from [11])

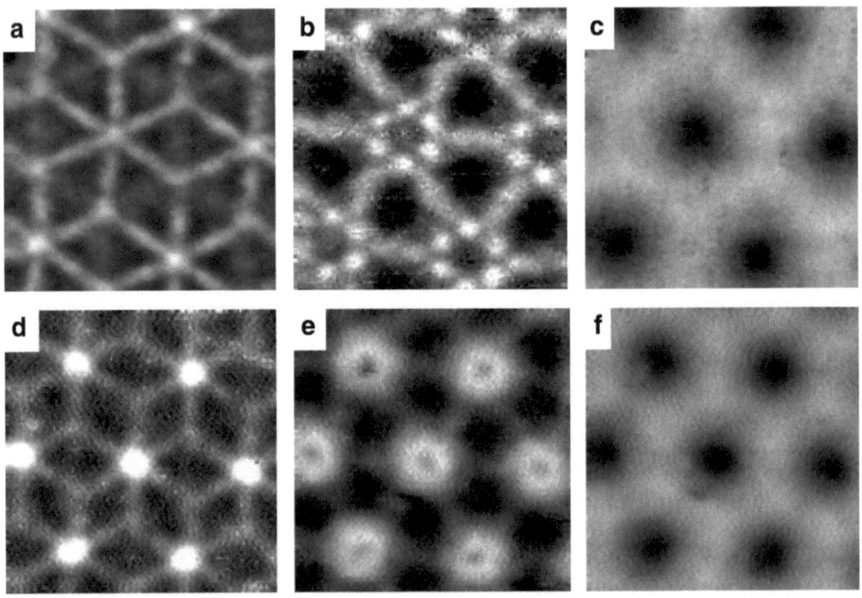

Fig. 10.7 (ST)^{2}S images of the vortex lattice in NbSe$_2$ using a Pb tip ($H = 100$ mT). Measurements were taken at 0.3 K (top row, size: 312 × 312 nm^2) and 6 K (bottom row, size: 364 × 364 nm^2). STS conductance images are composed at different voltages corresponding to the sample energies E_F, $E_F + \Delta_S$ and $E_F + \Delta_L$ (data from [13])

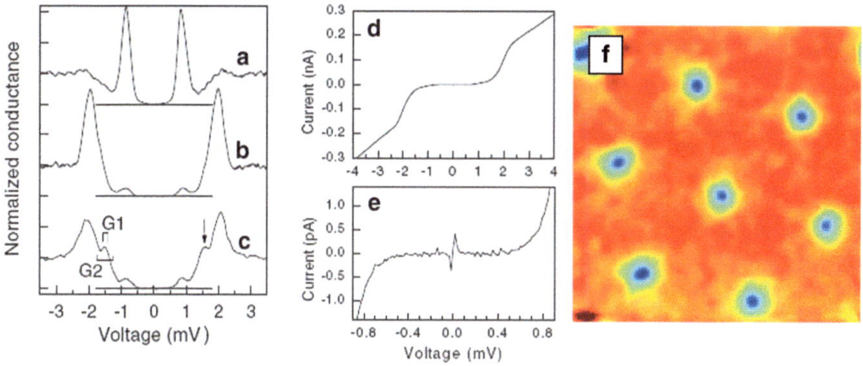

Fig. 10.8 Pb-$NbSe_2$ conductance curves taken at 300 mK in tunneling regime for an applied field of 1 kG: (**a**) at the vortex core, (**b**) at the midpoint among three vortices and (**c**) at an intermediate location. The *arrow* indicates the "extra" peak discussed in the text. Frame (**d**) shows an I-V curve taken at the midpoint among three vortices. The region close to zero bias is zoomed in frame (**e**) showing Josephson current (data from [13]). $(ST)^2S$ data are shown in frame (**f**). Image size: 410 × 410 nm^2 (data from [75]).

(Fig. 10.7d–f). That was interpreted as the consequence of the closing of the smaller gap at about 5K, in agreement with different experimental observations [30, 73, 74].

The atomic sharpness of these tips, and its well-defined DOS, has permitted to observe the complex pattern of electronic bound states in $NbSe_2$, with the sixfold symmetry, in a wide temperature range up to $0.85T_c$, with levels of detail and quality only attainable with normal tips at very low temperatures, below $0.1T_c$.

As observed in Fig. 10.8e, a small Josephson signal appears around zero bias. In Fig. 10.8f, this was plotted as a function of the position, throwing a neat image of the changes of the sample order paramerter in the vortex state.

10.3.3 $NbSe_2$ vs. NbS_2

Recent work on NbS_2 [76] has revealed the presence of a peak at the Fermi level DOS very similar to the one found in $NbSe_2$ (Fig. 10.9), giving strong support to the seminal ideas of Caroli, de Gennes and Matricon, and its subsequent development.

NbS_2 is a material with a very similar critical temperature compared to $NbSe_2$. The a/c ratio of its hexagonal crystal structure is somewhat larger than the one of $NbSe_2$, and it shows no charge density wave order down to 0.1 K [77]. There are very few experiments available in NbS_2. Thus, the comparison between both compounds is best made using the pressure dependence of T_c, which has been studied in detail. In NbS_2, the compressibility is higher than in $NbSe_2$. However, the critical temperature does not depend on pressure up to 2 GPa [78]. This is in striking contrast to the case of $NbSe_2$, where the critical temperature is strongly pressure dependent [79]. On the other hand, the Fermi surface is expected to be very similar

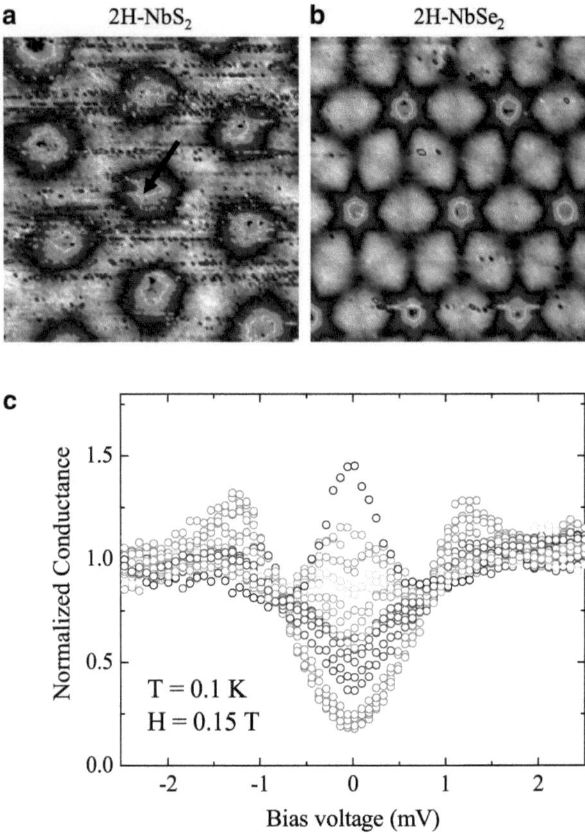

Fig. 10.9 The Fermi level conductance map of the vortex lattice is shown for NbS_2 (**a**) and $NbSe_2$ (**b**). Size of the image is 360 × 360 nm. The colour scale corresponds approximately to the colour used in (**c**) for the full conductance vs. voltage curves. Note that the localized state splits when leaving the vortex cores (see c), following the path given by the arrow in (**a**) (data from [76])

in both compounds [78]. Although there are no Fermi surface measurements available for NbS_2, calculations show that there must be also two Nb-derived cylindrical sheets responsible for the pronounced 2-D character of its properties [78, 80]. The most significant difference should come from the presence of a small Se-derived 3D pocket in $NbSe_2$, which is possibly absent in NbS_2, although this still needs experimental confirmation. Therefore, the striking absence of strong pressure-dependent properties in NbS_2 must come from the absence of CDW order in this material. Zero field superconducting DOS is not far from the one appearing in $NbSe_2$, with a clear two gap structure, which appears to be, however, somewhat more pronounced in NbS_2 than in $NbSe_2$. This would hint at a more smooth gap distribution over the Fermi surface in the latter compound. In both materials, the DOS is modulated at the atomic scale, revealing their intrinsic strongly Fermi surface-dependent superconducting properties [81].

The vortex lattice in NbS_2 has a very similar shape as in $NbSe_2$. The symmetry of the lattice is hexagonal, and vortices appear in the STM images with a somewhat larger radius in NbS_2, in agreement with a larger coherence length and a smaller upper critical field. The DOS inside the core, at the Fermi level and exactly at the centre is very similar in both compounds, a clear peak appears at the Fermi level. However, the peculiar star shape of $NbSe_2$ is absent in NbS_2. The way the localized state peak splits when leaving the vortex core is fully radially symmetric in NbS_2. The same result is found at all bias voltages, and the rich structure appearing in $NbSe_2$ is absent in NbS_2. Clearly, the most important difference in both compounds is the presence of CDW order in $NbSe_2$. The emerging picture is that the CDW order completely shapes the internal electronic structure of the vortex core, "cutting" the radially symmetric core into a sixfold star in the case of $NbSe_2$.

On the other hand, we should note that the appearance of symmetric peak in the Fermi level DOS at the vortex core should be expected, as shown in the initial theoretical treatment of Caroli, de Gennes and Matricon [5], in any superconducting material in the clean limit [82]. However, this was not found again, after its discovery in $NbSe_2$, apart from the presently discussed case of NbS_2. The high T_c superconductors have a very complex DOS inside cores, possibly related to pseudo gap features of the peculiar normal state found in these materials [6, 83]. Boron-doped diamond also shows a peculiar asymmetric DOS inside cores [84]. High quality single crystals of MgB_2 do not show any localized state [85]. In the nickel-borocarbides, the form of the vortices is also star-shaped, with a fourfold symmetry [49, 86], but the tunneling conductance inside the core shows a rather asymmetric structure. Hence, the results in NbS_2 show clearly that the initial predictions of Caroli, de Gennes and Matricon are to be found in clean superconducting materials with an open superconducting gap.

It will be interesting to see what happens in other materials of the series. For instance, in TaS_2, the a/c ratio is smaller and CDW sets in at a much higher temperature than in $NbSe_2$ [78]. The question is whether the internal shape of the vortex core is even more spectacularly shaped than in $NbSe_2$, or if it remains approximately with the same structure. Intercalated TaS_2 compounds give, on the other hand, a very rich playground, where it should be possibly to address wide ranging problems regarding vortex physics, which include pinning or exchange field effects, in a material as TaS_2 which appears to be appropriate for STM studies [87].

10.3.4 The Vortex Lattice in thin Films: A 2D Vortex Lattice

The above-mentioned techniques for vortex lattice imaging have been applied to the vortex lattice in thin films, and, in some cases, also to nanofabricated structures. However, the measurement of thin films or other nanostructures with STM has faced significant technical problems. Often, micro and nanofabrication processes leave residua at the surface, which do not allow for clean tunneling. In other cases, simply, the surface of the thin films is covered with natural oxide, and, unless fabrication is made in situ, it becomes difficult to obtain good tunneling characteristics [88]. Recent spectacular advances have been made in by combining an AFM and STM [89]. In this way, it becomes possible to image and position the tip on top of a given point at the nanostructure, and to take tunneling spectroscopy curves at this point. Although this has allowed to measure fundamental aspects of the DOS in proximity effect nanostructures [89], the tunneling is made through an oxide barrier. Therefore, the experiment should roughly give the spatial dependence of the tunneling DOS at length scales which depend on the tunneling barrier, and may be of the order of some nm. Therefore, local, atomic scale, vacuum tunneling characteristic for STM is lost. For the same reason, scanning also becomes more difficult, so that the observation of the vortex lattice with this technique seems to need further developments.

The physical challenges behind vortex lattice measurements in nanostructures, which could be addressed with clean vacuum tunneling STM, and not with other vortex imaging techniques, are very varied and far-reaching. These challenges cover from the determination of the electronic structure of the vortex cores, down to the study of individual vortices and its arrangement in ultra small structures, with different geometries, and at high magnetic fields, through studying in detail its pinning properties and the temperature, current and field dependence of 2D vortex lattices. Indeed, thin films of extreme type II superconductors are characterized by forming a vortex lattice where the characteristic bending radius are most often very far above the width of the thin film. The vortex lattice appears as a 2D structure, and vortices can be considered as long, rigid entities to a large extent [91, 92].

Recently, studies in a focused-ion-beam fabricated thin film have revealed surprising new insight [90]. These thin films appear to be free of oxides on the surface, and STM images reveal relatively flat regions with a very large number of vortices. Often, STM studies show a small number of vortices as the surface remains flat, and therefore easy to handle with the STM, only in square areas of up to some tens of nm. In the case of these thin films, the vortex lattice can be observed in areas that approach the μm scale, up to fields very close to H_{c2}(T). The DOS in these films can be measured down to atomic level, showing clean s-wave BCS features, which make these films, in addition, particularly simple to model [90].

Work until now has focused on the structure of the vortex lattice, which turns out to be strongly influenced by very small, nm sized surface irregularities. As shown in Fig. 10.10, the lattice arranges beautifully in different hexagonal domains, separated clearly by lines which correspond to features in the topography. Most recently, vortex depinning has been directly observed, leading to different vortex lattice orientations when increasing temperature. 2D vortex lattice melting has been found to occur at temperatures close to T_c. Peculiar hexatic and linear smectic-like vortex arrangements, shedding new light into 2D melting theories, have been found. When approaching the border of the thin film, peculiar vortex arrangements are observed [93]. Remarkably, the vortices are not repelled by the border, but there are many of them located exactly at the border. The thin film is evaporated on top of an Au layer; so possibly the screening currents are also formed in the surrounding gold layer. Another intriguing possibility was recently highlighted in a theoretical work. Under some circumstances, in particular when a vortex is situated close to the border, the

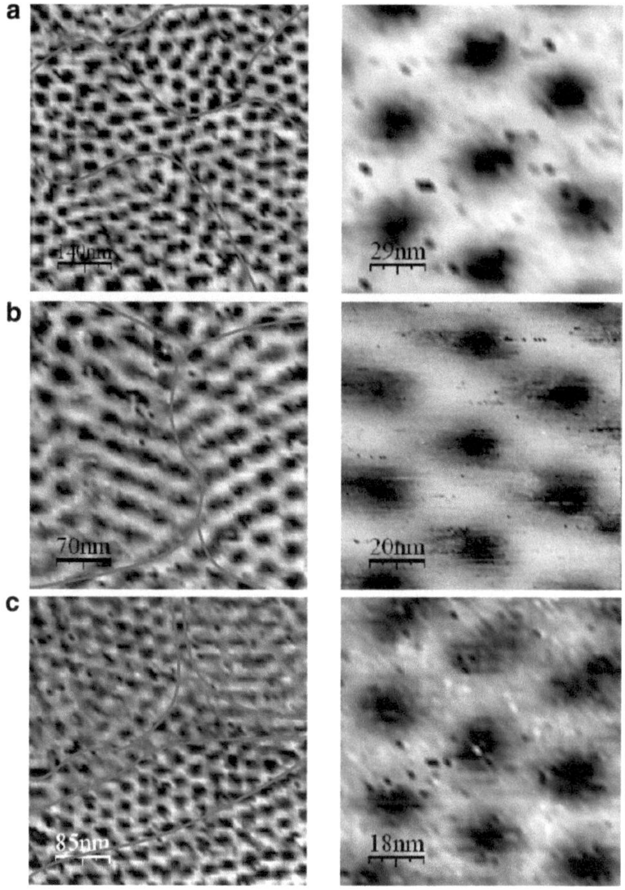

Fig. 10.10 The vortex lattice in an amorphous W deposit is shown in very large scale images involving hundreds of vortices ($H = 1$ T in (**a**), 2 T in (**b**) and 3 T in (**c**)). In the left panels, we show the vortex lattice over large scales, and in the right panels, we show a zoom over individual hexagons (data from [90])

maximum of the Fermi level DOS may not be located at the centre of the vortex core [94]. Further work is being made to try to image vortex arrangements in small islands and other nanostructures.

10.4 Other Scenarios for the Interplay of Magnetism and Superconductivity

We are going to introduce and discuss briefly a problem concerning the relationship between superconductivity and magnetism, namely materials in which a kind of magnetic order, coexists with, or destroys, the superconducting one [64, 65, 95, 96].

A classical example of a superconductor which also exhibits magnetic order is $ErRh_4B_4$ [97]. This is a re-entrant superconductor in which with decreasing temperature a superconducting transition occurs, at $T_{c1} \approx 8$ K, and then at $T_{c2} \approx 0.7$ K, where local Er moments order ferromagnetically, the material becomes normal again. This transition offers a unique case to study how the transition from magnetic to superconducting states, and viceversa, occurs when cooling or heating. Since re-entrant superconductivity was found in this material, in the 1970s of the last century, several macroscopic characterizations of these processes have been done [98–101]. When heating, the transition to the full superconducting state occurs at a temperature, $T_{c2\uparrow}$, which is considerably above (around 100 mK) the transition temperature found when cooling, $T_{c2\downarrow}$.

Neutron scattering experiments [98, 99] have shown that in the superconducting phase, when cooling, a new modulated ferromagnetic structure with a length scale of 10 nm appears around 1 K, and disappears below $T_{c2\downarrow}$. When heating, the same state appears at $T_{c2\uparrow}$ and disappears again around 1 K. Thermal hysteresis is observed both in the normal Bragg peak intensity and the small-angle peaks. For the small-angle peaks, the intensity is higher on cooling than on warming. This is opposite to the behaviour of the regular Bragg peaks from the FM regions. This peculiar magnetic state is supposed to be induced by superconductivity [65], as proposed by Anderson and Suhl [102]. More recently, it has been shown that the modulated state most likely coexists within superconducting domains forming close to ferromagnetism [103]. There are many important open questions raised by these and other studies, which manifest the delicate balance of processes intervening in the transition. Due to this, it is clear that high sample quality is required to give support to the experimental claims.

Recently Crespo et al. [103] characterized, making STM/S experiments, a very high quality single crystal. They use superconducting tips to follow in detail the expected tiny modifications suffered by the sample LDOS when the transition was passed from both sides of the critical temperature. Lead was chosen as the material for the tip. Its advantage has been emphasized previously, mainly considering the temperatures of the transition region, far below the critical temperature of lead.

In the superconducting phase of $ErRh_4B_4$, at 1.1 K, S–S' curves characteristic for tunnelling between two different superconductors were reported (Fig. 10.11). From the analysis of the experimental data, it is found that the LDOS of the sample $N_{ErRh_4B_4}(E)$ has a well open superconducting gap, with a zero DOS up to 0.75 meV, and a rounded quasiparticle peak whose maximum, referred as $\Delta_{ErRh_4B_4}$, is located at 1.2 meV. Most remarkable features in $N_{ErRh_4B_4}(E)$ were found close to the ferromagnetic transition, where the tunnelling conductance curves strongly change its shape. Typical tunnelling conductance curves corresponding to heating and cooling experiments are reproduced in Fig. 10.12a, b, respectively. From these, the sample LDOS was obtained (Fig. 10.12c, d).

When heating, superconductivity in the sample appears abruptly at $T_{c2\uparrow}$, and the tunnelling spectroscopy curves show, within a few mK, the S–S' behavior represented by the lowest curve in Fig. 10.12a. The resulting temperature dependence of $N_{ErRh_4B_4}(E)$ is shown in Fig. 10.12c. Close to $T_{c2\uparrow}$, at $T = 0.86$ K, the curves are

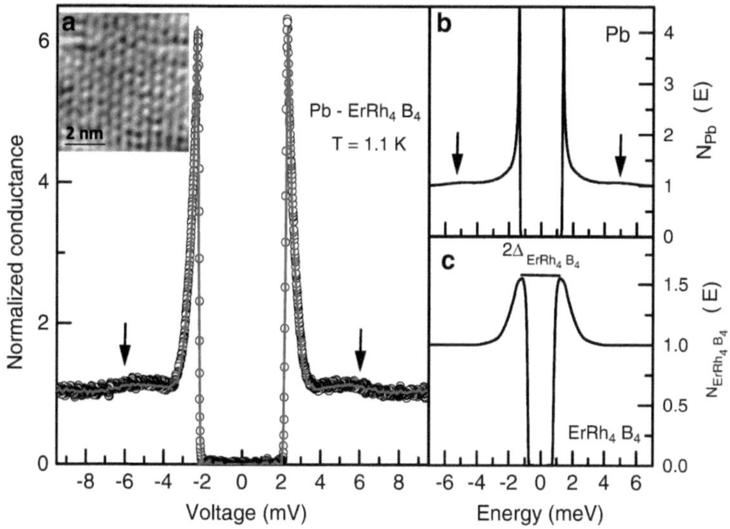

Fig. 10.11 $(ST)^2S$ measurements on $ErRh_4B_4$. (**a**) Local tunneling spectroscopy curve (*circles*) obtained in the superconducting phase of $ErRh_4B_4$ at 1.1 K with a Pb tip, and calculated conductance (*red*) using the densities of states shown in (**b**) and (**c**). Inset: atomic resolution topography. (**b**) $N_{Pb}(E)$, obtained from measurements on Au, used to extract the actual $N_{ErRh_4B_4}(E)$ (shown in (**b**)). *Arrows* in (**a**) and (**b**) mark Pb phonon mode features (data from [103])

significantly different than well within the superconducting phase, at $T = 1.1$ K. When cooling again into the ferromagnetic phase, identical tunneling conductance curves, at, however, lower temperatures than in the heating process, are found. As shown in Fig.10.12b, the superconducting features disappear at $T_{c2\downarrow}$, 90 mK below $T_{c2\uparrow}$. The transition to the normal state occurs again abruptly, within a few mK. Just before the transition, the tunnelling conductance curves appear with most strongly smeared superconducting features (see curves at 0.77 K in Fig. 10.12b, d). Remarkably, the interval with zero $N_{ErRh_4B_4}(E)$ is fully lost close to $T_{c2\downarrow}$, and instead gapless superconductivity with V-shaped increase of $N_{ErRh_4B_4}(E)$ at low energies appears. Such a behaviour is never observed close to $T_{c2\uparrow}$ when heating, as highlighted in Fig. 10.13, where the temperature dependence of the energy interval with zero $N_{ErRh_4B_4}(E)$ has been plotted. This interval is significantly influenced by magnetism up to about 0.2 K above the appearance of the ferromagnetic state, whereas $\Delta_{ErRh_4B_4}$ (inset of Fig. 10.13) remains constant. Small differences in the above-mentioned behaviour were obtained at different locations. Nevertheless, the difference in transition temperatures, $T_{c2\uparrow}$-$T_{c2\downarrow}$, the abrupt nature of the transition, the fact that the position of the quasiparticle peaks remains at $\Delta_{ErRh_4B_4}$, and the gapless V-shaped increase of $N_{ErRh_4B_4}(E)$ close to T_{c2} were found everywhere on the sample's surface.

An important conclusion of this work is that there is no evidence for coexistence between long range ferromagnetic order and superconductivity. If there would be superconducting correlations in extended ferromagnetic regions, these should

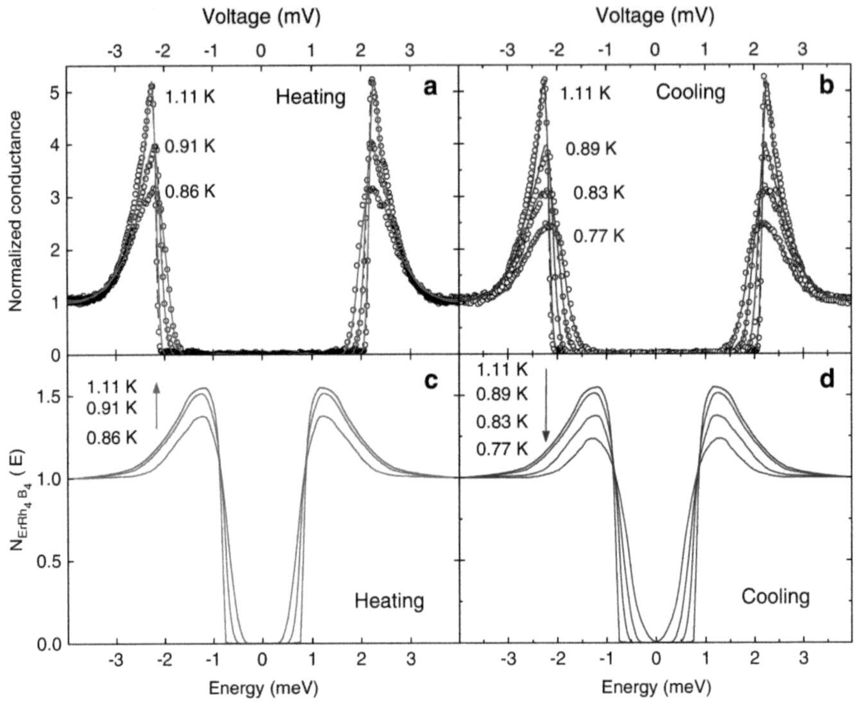

Fig. 10.12 Experimental $(ST)^2S$ curves (*circles*) measured near the ferromagnetic transition during heating (**a**) and cooling (**b**), together with the calculated ones (*lines*). Figure (**c**) and (**d**) show their corresponding $N_{ErRh_4B_4}(E)$ (data from [103])

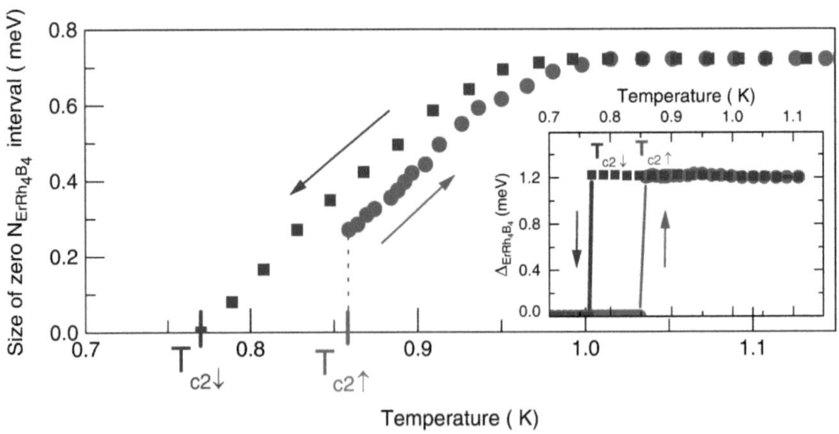

Fig. 10.13 Width of the zero DOS interval when cooling (*blue squares*) and heating (*red circles*) close to the magnetic transition. Inset: Evolution of $\Delta_{ErRh_4B_4}(T)$ at the same temperatures (data from [103])

lead to some signal in the local tunneling conductance curves below $T_{c2\uparrow}$ and $T_{c2\downarrow}$, such as a decrease of $N_{ErRh_4B_4}(E)$ close to the Fermi level, which was not observed by Crespo et al. [103]. The authors conclude that the presence of the magnetically modulated state discovered in [98, 99] best explains the observed smearing in the tunnelling DOS. The peaks satellite to some of the main ferromagnetic Bragg reflections, which correspond to a modulated magnetic moment at (0.042a,0.055c), exist in the superconducting phase on the same temperature range where the changes in the width of the zero $N_{ErRh_4B_4}(E)$(E) interval (Fig. 10.13) were observed. Moreover, the satellite Bragg peaks grow to much higher values on cooling at $T_{c2\downarrow}$ than on heating at $T_{c2\uparrow}$, and disappear abruptly when entering long range ferromagnetic order [102]. Clearly, the smearing of the DOS found here must directly show the effect of this kind of magnetic order on the superconducting DOS. This is strongest when the moments associated with the magnetically modulated state are highest (at $T_{c2\uparrow}$) and where it is observed the peculiar gapless regime. We believe that this is a nice example of the use of STM/S spectroscopy with superconducting tips.

10.5 Summary and Prospects

We have discussed newly developed STM/S techniques to visualize vortex lattices and look into the electronic properties of single vortices. We have shown that the combination of superconducting tip and scanning tunnelling spectroscopy, $(ST)^2S$, is an emerging new and very promising tool. We discuss the main points required to take full advantage of this technique. The tip fabrication and characterization process have to be controlled in situ, at low temperatures. This can be made using a repeated indentation procedure. Tips can be fabricated reproducibly as many times as needed, in particular when it is suspected that the tip has deteriorated. The behaviour of the nanotip DOS under magnetic fields can be modified, according to the problem in hand. For example, the observation of the Josephson effect in the vortex lattice requires well-developed superconducting DOS under high fields, whereas the measurement of the local magnetic field requires a tip DOS which strongly varies as a function of the field. Looking into the near future, we believe that we have two interesting experimental challenges. First, to improve the performances of the superconducting tips as a probe for surface magnetism. Second, and possibly more important, to exploit the new possibilities that $(ST)^2S$ opens through its combined sensitivity to probe quasiparticle, Cooper pair and magnetic field distributions at sub-nanometric scale.

Acknowledgements The Laboratorio de Bajas Temperaturas is associated to the ICMM of the CSIC. This work was supported by the Spanish MICINN (Consolider Ingenio Molecular Nanoscience CSD2007-00010 program, FIS2008-00454 and ACI2009-0905), by the Comunidad de Madrid through programs Citecnomik and Nanobiomagnet and by NES program of the ESF.

References

1. G. Binnig, H. Rohrer, C. Gerber, E. Weibel, Appl. Phys. Lett. 40, 178 (1982)
2. R.P. Feynman, Plenty of Room at the Bottom (American Physical Society, Pasadena, USA, 1959)
3. A.L. de Lozanne, S.A. Elrod, C.F. Quate, Phys. Rev. Lett. 54, 2433 (1985)
4. H.F. Hess, R.B. Robinson, J.V. Waszczak, Phys. Rev. Lett. 64, 2711 (1990)
5. C. Caroli, P.G. de Gennes, J. Matricon, Phys. Lett. 9, 307 (1964)
6. O. Fischer, M. Kugler, I. Maggio-Aprile, C. Berthod, Rev. Modern Phys. 79, 353 (2007)
7. O. Fischer et al., Scanning Tunnelling Spectroscopy of Vortices with Normal and Superconducting tips, ed. by V. Moshchalkov, Nanoscience and Engineering in Superconductivity (Springer, Heidelberg, 2010)
8. G. Binnig, C.F. Quate, C. Gerber, Phys. Rev. Lett. 56, 930 (1986)
9. P. Martínez-Samper, J.G. Rodrigo, N. Agraït, R. Grande, S. Vieira, Physica C 332, 450 (2000)
10. J.G. Rodrigo, H. Suderow, S. Vieira, E. Bascones, F. Guinea, J. Phys.: Condens. Matter. 16, R1151 (2004)
11. I. Guillamón, H. Suderow, S. Vieira, P. Rodiere, Physica C 468, 537 (2008)
12. J.G. Rodrigo, V. Crespo, S. Vieira, Physica C 437, 270 (2006)
13. J.G. Rodrigo, V. Crespo, S. Vieira, Physica C 468, 547 (2008)
14. N. Agraït, J.G. Rodrigo, G. Rubio, C. Sirvent, S. Vieira, Thin Solid Films 253, 199 (1994)
15. N. Agraït, G. Rubio, S. Vieira, Phys. Rev. Lett. 74, 3995 (1995)
16. C. Untiedt, G. Rubio-Bollinger, S. Vieira, N. Agraït, Phys. Rev. B 56, 2154 (1997)
17. N. Agraït, A.L. Yeyati, J.M. van Ruitenbeek, Phys. Rep. 377, 81 (2003)
18. J.G. Rodrigo, A. García-Martín, J.J. Sáenz, S. Vieira, Phys. Rev. Lett. 88 (2002)
19. N. Agraït, J.G. Rodrigo, C. Sirvent, S. Vieira, Phys. Rev. B 48, 8499 (1993)
20. U. Landman, W.D. Luedtke, N.A. Burnham, R.J. Colton, Science, 248, 454 (1990)
21. G. Rubio, N. Agraït, S. Vieira, Phys. Rev. Lett. 76, 2302 (1996)
22. R. Landauer, IBM J. Res. Dev. 1, 223 (1957)
23. D. Wharam, T. Thornton, R. Newbury, M. Pepper, H. Ahmed, J. Frost, D. hasko, D. Peacock, D. Ritchie, G. Jones, J. Phys.: Condens. Matter. 21, L209 (1988)
24. G. Rubio-Bollinger, S.R. Bahn, N. Agraït, K.W. Jacobsen, S. Vieira, Phys. Rev. Lett. 87, 026101 (2001)
25. E. Scheer, P. Joyez, D. Esteve, C. Urbina, M.H. Devoret, Phys. Rev. Lett. 78, 3535 (1997)
26. E. Scheer, N. Agraït, J.C. Cuevas, A. Levy-Leyati, B. Ludolph, A. Martín-Rodero, G. Rubio-Bollinger, J.M. van Ruitenbeek, C. Urbina, Nature, 394, 154 (1998)
27. H. Suderow, E. Bascones, W. Belzig, F. Guinea, S. Vieira, Europhys. Lett. 50, 749 (2000)
28. G. Rubio-Bollinger, C. de las Heras, E. Bascones, N. Agraït, F. Guinea, S. Vieira, Phys. Rev. B 67, 121407 (2003)
29. J.G. Rodrigo, H. Suderow, S. Vieira, Eur. Phys. J. B 40, 483 (2004)
30. J.G. Rodrigo, S. Vieira, Physica C 404, 306 (2004)
31. E.L. Wolf, Principles of Electron Tunneling Spectroscopy (Oxford University Press, Oxford, 1989)
32. R. Mersevey, P.M. Tedrow, Phys. Rep. 238, 173 (1994)
33. H. Suderow, E. Bascones, A. Izquierdo, F. Guinea, S. Vieira, Phys. Rev. B 65 (2002)
34. B.L. Blackford, R.H. March, Phys. Rev. 186, 397 (1969)
35. G.I. Lykken, A.L. Geiger, K.S. Dy, E.N. Mitchell, Phys. Rev. B 4, 1523 (1971)
36. P.G. Tomlinson, J.P. Carbotte, Phys. Rev. B 13, 4738 (1976)
37. J. Short, J. Wolfe, Phys. Rev. Lett. 85, 5198 (2000)
38. P. Martínez-Samper, J.G. Rodrigo, G. Rubio-Bollinger, H. Suderow, S. Vieira, S. Lee, S. Tajima, Physica C 385, 233 (2003)
39. H. Suderow, S. Vieira, J.D. Strand, S. Budko, P.C. Canfield, Phys. Rev. B 69, 060504 (2004)
40. P. Martinez-Samper, H. Suderow, S. Vieira, J.P. Brison, N. Luchier, P. Lejay, P.C. Canfield, Phys. Rev. B 67, 014526 (2003)

41. H. Suderow, V. Crespo, I. Guillamon, S. Vieira, F. Servant, P. Lejay, J.P. Brison, J. Flouquet, arXiV:08.4810 (2009)
42. C. Bersier, A. Florisl, P. Cudazzo, G. Profeta, A. Sanna, F. Bernardini, M. Monni, S. Pittalis, S. Sharma, H. Glawe, A. Continenza, S. Massidda, E.K.U. Gross, J. Phys.: Condens. Matter. 21, 164209 (2009)
43. P. Samuely et al., Scanning Tunnelling Spectroscopy of Vortices with Normal and Superconducting tips, ed. by V. Moshchalkov, Nanoscience and Engineering in Superconductivity (Springer, Heidelberg, 2010)
44. M. Tinkham, Introduction to Superconductivity (McGraw International, Singapore, 1996)
45. J.G. Rodrigo, N. Agraït, S. Vieira, Phys. Rev. B 50, 374 (1994)
46. J. Smakov, I. Martin, A. Balatsky, Phys. Rev. B 64, 212506 (2001)
47. S. Bending, Adv. Phys. 48, 449 (1999)
48. M. Poza, E. Bascones, J. Rodrigo, N. Agraït, S. Vieira, F. Guinea, Phys. Rev. B 58, 11173 (1998)
49. S.H. Pan, E.W. Hudson, A.K. Gupta, K.W. Ng, H. Eisaki, S.U.J.C. Davis, Phys. Rev. Lett. 85, 1536 (2000)
50. A. Kohen, T. Proslier, T. Cren, Y. Noat, W. Sacks, H. Berger, D. Rodritchev, Phys. Rev. Lett. 97, 027001 (2006)
51. O. Naaman, W. Teizer, R.C. Dynes, Phys. Rev. Lett. 87, 97004 (2001)
52. O. Naaman, R.C. Dynes, Solid State Commun. 129, 299 (2004)
53. O. Naaman, R.C. Dynes, E. Bucher, Int. J. Mod. Phys. B 17, 3569 (2003)
54. O. Naaman, W. Teizer, R.C. Dynes, Rev. Sci. Instrum. 72, 1688 (2001)
55. J.G. Rodrigo, N. Agraït, C. Sirvent, S. Vieira, Phys. Rev. B 50, 12788 (1994)
56. T. Proslier, A. Kohen, Y. Noat, T. Cren, D. Roditchev, W. Sacks, Europhys. Lett. 73, 962 (2006)
57. N. Bergeal, Y. Noat, T. Cren, T. Proslier, V. Dubost, F. Debontridder, A. Zimmers, D. Roditchev, W. Sacks, J. Marcus, Phys. Rev. B 78, 140507(R) (2008)
58. P.L. Gammel, D.J. Bishop, M.R. Eskildsen, K. Mortensen, N.H. Andsersen, I.R. Fisher, K.O. Cheon, P.C. Canfield, V.G. Kogan, Phys. Rev. Lett. 82, 4082 (1999)
59. V.G. Kogan, M. Bullock, B. Harmon, P. Miranovic, L.D. Grujic, P. Gammel, D. Bishop, Phys. Rev. B 55, R8693 (1997)
60. R. Joynt, L. Taillefer, Rev. Mod. Phys. 74, 235 (2002)
61. A. Huxley, P. Rodiere, D.M. Paul, N. van Dijk, R. Cubitt, J. Flouquet, Nature, 406, 160 (2000)
62. U. Yaron, P.L. Gammel, A. Ramirez, D.A. Huse, D. Bishop, A.I. Goldman, C. Stassis, P.C. Canfield, K. Mortensen, M.R. Eskildsen, Nature, 382, 236 (1996)
63. S. Bud'ko, P.C. Canfield, Comptes Rendus Phys. 7, 56 (2006)
64. L.N. Bulaevski, A.I. Buzdin, M.L. Kuliç, S.V. Panjukov, Adv. Phys. 34, 175 (1985)
65. J. Flouquet, A.I. Buzdin, Phys. World 15, 9 (2002)
66. D. Aoki, A. Huxley, E. Ressouche, D. Braithwaite, J. Flouquet, J.P. Brison, E. Lhotel, C. Paulsen, Nature, 413, 613 (2001)
67. K. Harada, T. Matsuda, J. Bonevich, M. Igarashi, S. Kondo, G. Pozzi, U. Kawabe, A. Tonomura, Nature, 360, 51 (1992)
68. E. Boaknin, M.A. Tanatar, J. Paglione, D. Hawthorn, F. Ronning, R.W. Hill, M. Sutherland, L. Taillefer, J. Sonier, S.M. Hayden, J.W. Brill, Phy. Rev. Lett. 90, 117003 (2003)
69. G. Seyfarth, J.P. Brison, M.A. Masson, D. Braithwaite, G. Lapertot, J. Flouquet, Phys. Rev. Lett. 97, 236403 (2006)
70. J.P. Emerson, R.A. Fisher, N.E. Phillips, D.A. Wright, E.M. McCarron, Phys. Rev. B 49, 9256 (1994)
71. H. Suderow, J.P. Brison, A.D. Huxley, J. Flouquet, Phys. Rev. Lett. 80, 165 (1998)
72. N. Nakai, P. Miranovic, M. Ichioka, H.F. Hess, K. Uchiyama, H. Nishimori, S. Kaneko, N. Nishida, K. Machida, Phys. Rev. Lett. 97, 147001 (2006)
73. T. Yokoya, T. Kiss, A. Chainani, S. Shin, M. Nohara, H. Takagi, Science, 294, 2518 (2001)
74. E. Boaknin, M.A. Tanatar, J. Paglione, D. Hawthorn, F. Ronning, R.W. Hill, M. Sutherland, L. Taillefer, J. Sonier, S.M. Hayden, J.W. Brill, Phys. Rev. Lett. 90, 117003 (2003)

75. V. Crespo, J.G. Rodrigo, S. Vieira (unpublished)
76. I. Guillamon, H. Suderow, S. Vieira, L. Cario, P. Diener, P. Rodiere, Phys. Rev. Lett. 101, 166407 (2008)
77. A.H. Castro Neto, Phys. Rev. Lett. 86, 4382 (2001)
78. J.A. Wilson, F.J. DiSalvo, S. Mahajan, Adv. Phys. 24, 117 (1975)
79. H. Suderow, V.G. Tissen, J.P. Brison, J.L. Martinez, S. Vieira, Phys. Rev. Lett. **95**, 117006 (2005)
80. M.D. Johannes, I.I. Mazin, C.A. Howells, Phys. Rev. B 73, 205102 (2006)
81. I. Guillamon, H. Suderow, F. Guinea, S. Vieira, Phys. Rev. B 77, 134505 (2008)
82. C. Renner, A.D. Kent, P. Niedermann, O. Fischer, F. Levy, Phys. Rev. Lett. 67, 1650 (1991)
83. K. Matsuba, S. Yoshizawa, Y. Mochizuki, T. Mochiku, K. Hirata, N. Nishida, J. Phys. Soc. Jpn. 76, 063704 (2007)
84. B. Sacépé, C. Chapelier, C. Marcenat, J. Kacmarcik, T. Klein, M. Bernard, E. Bustarret, Phys. Rev. Lett. 96, 097006 (2006)
85. M.R. Eskildsen, M. Kugler, J. Jun, S.M. Kazakov, J. Karpinski, O. Fischer, Phys. Rev. Lett. 89, 187003 (2002)
86. H. Nishimori, K. Uchiyama, S. Kaneko, A. Tokura, H. Takeya, K. Hirata, N. Nishida, J. Phys. Soc. Jpn. 73, 3247 (2004)
87. E.N.C. Marti-Gastaldo, E. Coronado (unpublished)
88. G.J.C. van Baarle, A.M. Troyanovski, P.H. Kes, J. Aarts, Physica C 369, 335 (2002)
89. H. le Sueur, P. Joyez, H. Pothier, C. Urbina, D. Esteve, Phys. Rev. Lett. 100, 197002 (2008)
90. I. Guillamon, H. Suderow, S. Vieira, A. Fernandez-Pacheco, J. Sese, R. Cordoba, J.D. Teresa, M. Ibarra, New J. Phys. 10, 093005 (2008)
91. E.H. Brandt, Rep. Prog. Phys. 58, 1465 (1995)
92. G. Blatter, M.V. Feigel'man, V.B. Geshkenbein, A.I. Larkin, V.M. Vinokur, Rev. Mod. Phys. 66, 1125 (1994)
93. I. Guillamon, H. Suderow, S. Vieira, A. Fernandez-Pacheco, J. Sese, R. Cordoba, J.D. Teresa, M. Ibarra, J. Phys.: Conf. Ser. 150, 052064 (2009)
94. A.A. Mel'nikov, A. Rhyzov, D. Silaev, Phys. Rev. B 79, 134521 (2009)
95. O. Fischer, M.B. Maple, Superconductivity in Ternary Compounds, vols I and II (Springer, New York, 1982)
96. A.I. Buzdin, Rev. Mod. Phys. 77, 935 (2005)
97. W.A. Fertig, D.C. Johnston, L. de Long, Phys. Rev. Lett. 38, 987 (1977)
98. D.E. Moncton, J.D. Axe, F.J. DiSalvo, Phys. Rev. B 16, 801 (1977)
99. S.K. Sinha, H.A. Mook, D.G. Hinks, G.W. Crabtree, Phys. Rev. Lett. 48, 950 (1982)
100. J.M. DePuydt, E.D. Dahlberg, D.G. Hinks, Phys. Rev. Lett. 56, 165 (1986)
101. R. Prozorov, M.D. Vanette, S.A. Law, S.L. Bud'ko, P.C. Canfield, Phys. Rev. B 77, 100503(R) (2008)
102. P.W. Anderson, H. Suhl, Phys. Rev. 116, 898 (1959)
103. V. Crespo, J.G. Rodrigo, H. Suderow, S. Vieira, D.G. Hinks, I.K. Schuller, Phys. Rev. Lett. 102, 237002 (2009)

Chapter 11
Surface Superconductivity Controlled by Electric Field

Pavel Lipavský, Jan Koláček, and Klaus Morawetz

Abstract We discuss an effect of the electrostatic field on superconductivity near the surface. First, we use the microscopic theory of de Gennes to show that the electric field changes the boundary condition for the Ginzburg–Landau function. Second, the effect of the electric field is evaluated in the vicinity of H_{c3}, where the boundary condition plays a crucial role. We predict that the field effect on the surface superconductivity leads to a discontinuity of the magnetocapacitance. We estimate that the predicted discontinuity is accessible for experimental tools and materials nowadays. It is shown that the magnitude of this discontinuity can be used to predict the dependence of the critical temperature on the charge carrier density which can be tailored by doping.

11.1 Introduction

The surface of a superconductor is an important region in which the superconductivity nucleates and which represents a natural barrier for penetrating or escaping vortices. It is desirable to control surface properties so that the nucleation can be stimulated or suppressed. An even more attractive task is to open or shut the penetration barrier for vortices. A promising tool of the surface control is the gate voltage for which we can benefit from the extensive technological experience with field effect transistors.

Unfortunately, the interaction of the electric field applied to the metal surface with the superconducting condensate is very weak. Indeed, the superconducting condensate does not interact with the electrostatic potential as shown by Anderson [1]. The condensate feels only the indirect effects like changes of the local density of states or eventual changes of the surface crystal structure.

It is very likely that it will become possible to enhance the field effect on the superconductivity by a proper surface treatment. To this end, it would be of great advantage to understand how the field interacts with the condensate and to have reliable experimental methods directly aiming to measure the strength of this interaction.

To support the experimental effort in this direction, in this chapter, we provide a phenomenological theory of Ginzburg–Landau (GL) type supplemented with the de Gennes boundary condition derived from the microscopic Bardeen, Copper, and Schrieffer (BCS) theory. It will be shown that the boundary condition captures the field effect on the condensate while the GL equation determines how the condensate responds to the field-affected boundary condition.

The field effect on the superconductivity has been measured under various conditions, nevertheless its actual strength is not yet accurately established. The most pronounced field effects are observed on thin layers, in which it is possible to increase or lower their critical temperature [2–6]. These samples are so thin that the applied field considerably changes the total density of electrons, and the observed effect can be interpreted in terms of the modified bulk properties. With thicker samples, one meets the problem that the potential field effect is restricted to the surface, and the underlying bulk overrides its contribution.

At the end of this chapter, we discuss the field effect on the surface superconductivity near the third critical magnetic field H_{c3}. In this regime, the bulk superconductivity is absent and the surface superconductivity crucially depends on the boundary condition. We will show that the field effect can be observed via the discontinuity in the magnetocapacitance [7, 8].

11.2 Limit of Large Thomas–Fermi Screening Length

To introduce the field effect on the superconductivity, we start from the theory used by Shapiro and Burlachkov [9–14] and by Chen and Yang [15]. It is justified for high-T_c superconductors in which the GL coherence length ξ is very short, while the hole density is low leading to relatively large Thomas–Fermi screening length λ_{TF}. In these materials $\lambda_{TF} \sim \xi$, which allows us to introduce field-induced effects via local changes of the parameters of the GL theory.

Let us assume the jellium model in which the electric charge of electrons is compensated by a smooth positively charged background. Both charges are restricted to the half space $x > 0$. The electric field applied to the surface is exponentially screened $E(x) = E\,\mathrm{e}^{-x/\lambda_{TF}}$ inside the metal. According to the Gauss equation $\epsilon\,\mathrm{div}E = \rho$, the induced electron density $\delta n = \rho/e$ reads

$$\delta n(x) = \frac{\epsilon E}{e\lambda_{TF}}\,\mathrm{e}^{-x/\lambda_{TF}}. \tag{11.1}$$

In the GL equation

$$\frac{1}{2m^*}\left(-i\hbar\nabla - e^*\mathbf{A}\right)^2\psi + \alpha\psi + \beta|\psi|^2\psi = 0, \tag{11.2}$$

the change of the electron density leads to changes in the GL parameters α and β. According to the Gor'kov theory [16], $\beta = \frac{6\pi^2 k_B^2 T_c^2}{7\zeta(3) E_F n}$. It is thus a robust parameter in which δn can create relative changes of the order of $\delta\beta/\beta \sim \delta n/n$. These changes can be neglected. The other parameter $\alpha = \alpha'(T - T_c)$ is a difference of two large constituents. Here, $\alpha' = \frac{6\pi^2 k_B^2 T_c}{7\zeta(3) E_F}$ is modified also only negligibly as $\delta\alpha'/\alpha' \sim \delta n/n$. But, for temperatures close to the critical temperature, $T \to T_c$, even small changes in T_c lead to large relative changes of α eventually even changing its sign. The GL equation with the dominant part of the field effect thus reads

$$\frac{1}{2m^*}\left(-i\hbar\nabla - e^*\mathbf{A}\right)^2\tilde{\psi} + \alpha\tilde{\psi} - \alpha'\frac{\partial T_c}{\partial n}\delta n\tilde{\psi} + \beta|\tilde{\psi}|^2\tilde{\psi} = 0. \tag{11.3}$$

We have denoted the field affected GL function as $\tilde{\psi}$. Initially, we shall use the customary GL condition $\frac{\partial}{\partial x}\tilde{\psi} = 0$ for this equation.

Let us assume now that this theory holds also for conventional superconductors where one always finds a sharp inequality $\lambda_{TF} \ll \xi$. Below, we confirm the result obtained with this unjustified assumption using the well-justified microscopic approach of de Gennes.

We split the GL function according to $\tilde{\psi} = \psi + \delta\psi$, where $\delta\psi(x) = \delta\psi\, e^{-x/\lambda_{TF}}$ is the part of field-induced perturbation which changes on the short scale λ_{TF} and ψ covers the rest changing on the larger scale ξ. Our aim is to establish approximative $\delta\psi$ and to eliminate it, so that in the second step we will be left with the GL equation for the slowly varying function ψ.

The short-scale component has an enormously large space gradient which dominates its contribution to the GL equation,

$$-\frac{\hbar^2}{2m^*}\nabla^2\delta\psi(x) - \alpha'\frac{\partial T_c}{\partial n}\delta n(x)\psi(x) \approx 0. \tag{11.4}$$

Since in this approximation the function $\delta\psi$ is nonzero only in the narrow layer $x \sim \lambda_{TF}$, we can neglect the space dependence of ψ and use its surface value. Performing derivatives, one finds with (11.1) that

$$\delta\psi = -\frac{2m^*\lambda_{TF}^2}{\hbar^2}\alpha'\frac{\partial T_c}{\partial n}\frac{\epsilon E}{e\lambda_{TF}}\psi(0). \tag{11.5}$$

Besides the very local contribution expressed by the function $\delta\psi$, the electric field induces also a perturbation on the scale of the GL coherence length ξ. Indeed, the GL boundary condition demands the zero derivative of the total function

$$\frac{\partial}{\partial x}(\psi + \delta\psi) = 0 \tag{11.6}$$

and the local part has a nonzero derivative

$$\frac{\partial}{\partial x}\delta\psi(x)|_{x=0} = -\frac{\delta\psi}{\lambda_{\mathrm{TF}}}. \tag{11.7}$$

From the GL boundary condition (11.6) and relations (11.5) and (11.7), one thus finds the boundary condition for ψ

$$\left.\frac{\partial\psi}{\partial x}\right|_{x=0} = \frac{2m^*}{\hbar^2}\alpha'\frac{\partial T_{\mathrm{c}}}{\partial n}\frac{\epsilon E}{e}\psi(0). \tag{11.8}$$

This relation can be interpreted as a field-affected GL boundary condition.

In usual problems handled by the GL theory, one ignores the exact gap profile at the surface and focuses on its behavior deeper in the bulk on the scale ξ. In the same spirit, we can ignore the short-scale component $\delta\psi$ using the approximation $\tilde{\psi} \approx \psi$. Doing so, we have to keep in mind that the short-scale component leads to the field-affected boundary condition (11.8). Solving the GL equation (11.2) with the boundary condition (11.8), one obtains the GL function where the field effect manifests itself on the scale ξ.

11.3 de Gennes Approach to the Boundary Condition

The limit of large screening length does not apply to conventional superconductors. In fact, in metals, the Thomas–Fermi screening is smaller than the interatomic distance, and the jellium model is not justified to describe the interaction of the surface with the electric field. Naturally, the gradient correction represented by the kinetic energy of GL theory is not a sufficient approximation of the nonlocal part of the BCS interaction kernel.

To access short screening lengths, Shapiro and Burlachkov [9–14] have used a more sophisticated version of the GL theory in which the 'kinetic energy' is not a mere parabolic function of the gradient but includes all derivatives up to infinite order in a form of the di-gamma function. Since high-order gradients are important only in the short-scale component, the above approach can be easily modified in this way. One merely replaces the kinetic energy in (11.4) by the corresponding di-gamma expression.

Although this high-order gradient correction is elegant and simple, it is likely not sufficient to cover changes on the sub-Ångström scale, i.e., on the scale typical for majority of metals. Apparently, one should use the microscopic approach of Bogoliubov–de Gennes type pioneered by Koyama [17] and other groups [18–25]. Beside microscopic details of the gap near the surface, one has to take into account that the simple exponential decay of the charge from a sharp surface is not a very realistic model of metals, because electrons tunnel out of the metal. Realistic studies of the surface are problematic, however, even in the normal state, namely due to the nontrivial exchange–correlation interaction at the strong gradient of the electron density innate to all surfaces.

Instead of improving the above approach, it is advantageous to formulate the boundary condition directly from the microscopic theory of de Gennes [26]. De Gennes did not assume the electric field explicitly. His result, however, does not specify the forces forming the surface so that the applied field can be included.

Gor'kov has shown that the BCS gap Δ and the GL function are proportional to each other, $\psi = \text{const} \times \Delta$. Using an extrapolation of the BCS gap from the bulk toward the surface, de Gennes has arrived at the boundary condition of the form

$$\frac{1}{\psi(0)} \frac{\partial \psi}{\partial x}\bigg|_{x=0} = \frac{1}{\Delta_0} \frac{\partial \Delta}{\partial x}\bigg|_{x=0} = \frac{1}{b}, \tag{11.9}$$

where b is called the extrapolation length.

Within the BCS theory, the extrapolation length is given by the formula

$$\frac{1}{b} = \frac{1}{\xi^2(0)} \frac{1}{N_0 V} \int_{-\infty}^{\infty} \mathrm{d}x \frac{\Delta(x)}{\Delta_0} \left[1 - \frac{N(x)}{N_0}\right] \tag{11.10}$$

derived by de Gennes ((7–62) in [26]). Here $N(x)$ is the local density of states at the Fermi level and N_0 is its bulk limit. The notation of the gap is different. While $\Delta(x)$ is the true local value of the BCS gap, Δ_0 is the fake surface value after all short-scale components have been removed, i.e., it is a value extrapolated from the near vicinity to the surface. Finally, V is the BCS interaction and $\xi(0)$ is the GL coherence length at 'zero' temperature. In pure metals, it is linked to the BCS coherence length ξ_0 as $\xi(0) = 0.74\,\xi_0$.

De Gennes estimated a typical value of $b \sim 1$ cm at metal surfaces in vacuum. This value is very large on the scale of the GL coherence length; therefore, this contribution is usually neglected. The approximation, $1/b \approx 0$, corresponds to the original GL condition $\frac{\partial}{\partial x}\psi = 0$.

Our aim is to include the effect of electric fields on the extrapolation length b. We denote as b_0 the extrapolation length in the absence of the applied electric field, and $\delta(1/b) = 1/b - 1/b_0$ reflects variation of the inverse length. The electrostatic potential corresponding to the electric field modifies the potential profile near the surface. It results in a change of the density of states $\delta N(x)$. The density of states affects the gap function and creates its deviation $\delta\Delta(x)$. In the linear approximation from (11.10), we find the change of the inverse extrapolation length as

$$\delta\left(\frac{1}{b}\right) = \frac{1}{\xi^2(0)} \frac{1}{N_0 V} \int_{-\infty}^{\infty} \mathrm{d}x \left\{\frac{\delta\Delta(x)}{\Delta_0}\left[1 - \frac{N(x)}{N_0}\right] - \frac{\Delta(x)}{\Delta_0}\frac{\delta N(x)}{N_0}\right\}. \tag{11.11}$$

To estimate this change, we recall the local density approximation in which the local density of states is a function of the local density, $N(x) = N[\rho(x)]$. Since the charge density $\delta\rho(x)$, which screens the applied electric field, spreads over a layer of a few Ångströms near the surface, the perturbed density of state is restricted to this very narrow layer, too. We can thus neglect the x-dependence of $\Delta(x)$ in the second term and write

$$\delta\left(\frac{1}{b}\right)=\frac{1}{\xi^2(0)}\frac{1}{N_0V}\left\{\int\limits_{-\infty}^{\infty}\mathrm{d}x\frac{\delta\Delta(x)}{\Delta_0}\left[1-\frac{N(x)}{N_0}\right]-\frac{\Delta(0)}{\Delta_0}\frac{\delta N^{(2)}}{N_0}\right\}, \tag{11.12}$$

where

$$\delta N^{(2)}=\int\limits_{-\infty}^{\infty}\mathrm{d}x\,\delta N(x) \tag{11.13}$$

is the total change of the density of states per area.

To estimate the second term, we assume that the local density of states achieves the bulk value on the scale of a few Ångströms. Since the gap function changes on the scale of the BCS coherence length which is much larger, we can expect that in the region of non-zero function $1-N(x)/N_0$, the gap function keeps its shape. Assuming that $\delta\Delta(x)\approx\Delta(x)\delta c$, the last x-integration becomes identical to the integral in the de Gennes condition (11.10) so that (11.12) simplifies to

$$\delta\left(\frac{1}{b}\right)=\frac{\delta c}{b}-\frac{1}{\xi^2(0)}\frac{1}{N_0V}\frac{\Delta(0)}{\Delta_0}\frac{\delta N^{(2)}}{N_0}. \tag{11.14}$$

The relative change of the gap δc can be estimated from the GL theory. The GL function obtained with the boundary condition of a large but finite extrapolation length b can be written as a sum of the constant bulk term $\psi_\infty=\sqrt{-\alpha/\beta}$ and a small perturbation ψ', i.e., $\psi=\psi_\infty+\psi'$. For the moment, we ignore the magnetic field and assume a real GL function so that the GL equation reads $\xi^2\nabla^2\psi-\psi+\psi^3/\psi_\infty^2=0$. Since ψ' is proportional to $1/b$, we keep its linear terms only, $\xi^2\nabla^2\psi'+2\psi'=0$. This equation has the exponential solution $\psi'(x)=-\psi_\infty \mathrm{e}^{-\sqrt{2}x/\xi}\xi/(\sqrt{2}b)$. Since the GL coherence length ξ is much smaller than the extrapolation length b, the exponential is small compared to the constant term. Assuming that the GL function provides us with the order of magnitude estimate of the gap function, we find that $\delta c=\delta\Delta(0)/\Delta_0\approx\delta\psi(0)/\psi_0=-(\xi/\sqrt{2})\delta(1/b)$. After substitution of this estimate into (11.14), we obtain

$$\delta\left(\frac{1}{b}\right)=-\frac{1}{1+\frac{\xi}{\sqrt{2}b}}\frac{\Delta(0)}{\Delta_0}\frac{1}{\xi^2(0)}\frac{1}{N_0V}\frac{\delta N^{(2)}}{N_0}. \tag{11.15}$$

We can assume that the induced density of states per area is linearly proportional to the applied electric field,

$$\delta N^{(2)}=E\,g \tag{11.16}$$

so that the modified inverse length is of the form

$$\delta\left(\frac{1}{b}\right) = \frac{E}{U_{\mathrm{s}}}, \tag{11.17}$$

where U_{s} has a dimension of a potential. According to (11.15), this effective potential is given by

$$\frac{1}{U_{\mathrm{s}}} = -\eta \frac{1}{\xi^2(0)} \frac{1}{N_0 V} \frac{g}{N_0}. \tag{11.18}$$

We have introduced a dimensionless parameter

$$\eta = \frac{1}{1 + \frac{\xi}{\sqrt{2}b}} \frac{\Delta(0)}{\Delta_0}, \tag{11.19}$$

which captures the effects of the gap profile near the surface. According to de Gennes' estimate [26], the surface ratio η is of the order of unity.

Let us note that a boundary condition similar to the de Gennes boundary condition described above can be derived also from the minimum free energy principle [27].

11.4 Link to the Limit of Large Screening Length

To draw a link between the de Gennes-type formula (11.18) and the field-affected GL boundary condition (11.8) obtained in the limit of large Thomas–Fermi screening lengths, we evaluate the coefficients of the de Gennes formula in the jellium model. As above we assume that in the absence of the electric field, the extrapolation length diverges, $1/b_0 = 0$.

For zero electric field the density of states is step-like, $N(x) = N_0$ for $x > 0$ and $N(x) = 0$ elsewhere. Now we include the electric field. It is exponentially screened due to the induced electron density given by (11.1). We use the local density approximation and assume that the local density of states is a function of the local density of electrons

$$N(x) = N_0 + \frac{\partial N_0}{\partial n} \delta n(x). \tag{11.20}$$

This approximation yields a simple change of the density of states per area

$$\delta N^{(2)} = \frac{\partial N_0}{\partial n} \int\limits_0^\infty \mathrm{d}x \, \delta n(x). \tag{11.21}$$

The induced density of electrons per area is given by the Gauss law

$$\epsilon E = -e \int\limits_0^\infty \mathrm{d}x \, \delta n(x). \tag{11.22}$$

From relations (11.16), (11.21), and (11.22), we find the coefficient of the density of states

$$g = -\frac{\partial N_0}{\partial n}\frac{\epsilon}{e}. \tag{11.23}$$

To be able to compare the BCS formula with the de Gennes-type one, we have to express both expressions in terms of the same parameters. We thus convert the parameters of de Gennes-type formula to their phenomenological counterparts.

The GL coherence length $\xi = \hbar/\sqrt{-2m^*\alpha}$ depends on the temperature according to $\xi = \xi(0)/\sqrt{1 - T/T_c}$, where $\xi(0)$ is the 'zero' temperature value. From $\alpha = \alpha'(T - T_c)$, one finds

$$\frac{1}{\xi^2(0)} = \frac{2m^*\alpha' T_c}{\hbar^2}. \tag{11.24}$$

Finally, we have to express the BCS interaction potential V in terms of the critical temperature. In the BCS critical temperature

$$T_c = 0.85\,\Theta_D e^{-\frac{1}{N_0 V}}, \tag{11.25}$$

we assume that the Debye temperature Θ_D and the BCS interaction V are independent of the electron density. This corresponds to approximations we have tacitly used above ignoring the electric field effect on the phonon spectrum. The density dependence of the critical temperature follows thus from the density dependence of the density of states

$$\frac{\partial T_c}{\partial n} = 0.85\,\Theta_D e^{-\frac{1}{N_0 V}}\frac{1}{N_0^2 V}\frac{\partial N_0}{\partial n} = T_c \frac{1}{N_0^2 V}\frac{\partial N_0}{\partial n}. \tag{11.26}$$

We will use this relation to express density derivatives of the density of states in terms of the density derivative of the critical temperature.

Now we can rewrite the effective potential in terms of phenomenological parameters. Using $\xi(0)$ from (11.24) in (11.18), we find

$$\frac{1}{U_s} = -\eta \frac{2m^*\alpha'}{\hbar^2} T_c \frac{1}{N_0^2 V} g. \tag{11.27}$$

Next, we substitute g from (11.23)

$$\frac{1}{U_s} = \eta \frac{2m^*\alpha'}{\hbar^2} T_c \frac{1}{N_0^2 V}\frac{\partial N_0}{\partial n}\frac{\epsilon}{e}. \tag{11.28}$$

The group of terms around $\partial N_0/\partial n$ can be substituted with the help of (11.26), so that we obtain

$$\frac{1}{U_s} = \eta \frac{2m^*\alpha'}{\hbar^2}\frac{\partial T_c}{\partial n}\frac{\epsilon}{e}. \tag{11.29}$$

The boundary condition we have obtained now from the de Gennes-type formula

$$\left.\frac{\partial\psi}{\partial x}\right|_{x=0} = \frac{E}{U_{\rm s}}\psi(0) = \eta\,\frac{2m^*\alpha'}{\hbar^2}\frac{\partial T_{\rm c}}{\partial n}\frac{\epsilon E}{e}\psi(0) \tag{11.30}$$

differs from the large screening length limit (11.8) by the factor η. Of course, a heuristic derivation of the field-effect from the GL equation cannot cover the factor η which depends on the gap profile on a scale smaller than the GL coherence length.

In summary, the electric field applied to the surface of the superconductor modifies the GL wave function near the surface. This effect is conveniently described by the GL theory, where the GL equation remains unaffected by the field and the entire electric field effect is covered by a modified boundary condition. In the next section, we discuss an experiment which can be used to measure the predicted field effect on the GL boundary condition.

11.5 Electric Field Effect on Surface Superconductivity

We will investigate now the magneto-capacitance for magnetic fields near the surface critical field B_{c3}. We focus on this region since we expect that the bias voltage affecting only the surface has a relatively large effect on the surface superconductivity. In this section, we show how the electric field affects the nucleation of superconductivity [7].

11.5.1 Nucleation of Surface Superconductivity

At the surface critical field B_{c3}, the superconductivity nucleates in the surface region. At the nucleation point, the GL wave function ψ is infinitely small; therefore, we can work with the linearized GL equation, omitting the cubic term in (11.2)

$$\frac{1}{2m^*}(-i\hbar\nabla - e^*\mathbf{A})^2\psi + \alpha\psi = 0. \tag{11.31}$$

The solution is restricted by the boundary condition (11.9).

The electrode is a superconductor which fills the half space $x > 0$. We assume a homogeneous applied magnetic field $\mathbf{B}_{\rm a} = (0, 0, B_{\rm a})$. Since an 'infinitely' large electrode has translation invariance along the y direction, we use the Landau gauge of the form

$$\mathbf{A} = (0, B_{\rm a}x, 0). \tag{11.32}$$

Nucleation is possible if (11.31) has a nonzero solution, i.e., if the parameter $-\alpha$ becomes equal to an eigenvalue ε of the kinetic energy given by $\frac{1}{2m^*}(-i\hbar\nabla - e^*\mathbf{A})^2\psi = \varepsilon\psi$. Since α changes with temperature, $\alpha = \alpha'(T - T_{\rm c})$, the eigenvalue

ε of the kinetic energy determines the nucleation temperature T^* according to $T^* - T_c = -\varepsilon/\alpha'$. To avoid dual notation for the same quantity, we will treat the equation as an eigenvalue for α. Since α is negative, the nucleation temperature T^* is always below the critical temperature T_c in the absence of the magnetic field. Note that we are looking for maximal α.

Assuming the translation invariance along the y and z axes, we can write the wave function as

$$\psi(x,y,z) = \psi(x)\,\mathrm{e}^{\mathrm{i}ky}\mathrm{e}^{\mathrm{i}qz}. \tag{11.33}$$

Using (11.33) in the GL equation (11.31), we get a one-dimensional equation

$$\frac{\hbar^2}{2m^*}\left(-\left(\frac{\partial}{\partial x}\right)^2 + \left(k - \frac{e^* B_a}{\hbar}x\right)^2 + q^2\right)\psi + \alpha\psi = 0. \tag{11.34}$$

Any non-zero value of q results in the kinetic energy $q^2\hbar^2/2m$, which lowers the value of α reducing the nucleation temperature. The nucleation happens at the first possible occasion, i.e., at the highest allowed temperature. We thus take $q = 0$. Similarly, we have to find the wave vector k from the requirement of the highest nucleation temperature.

11.5.2 Solution in Dimensionless Notation

It is advantageous to express the x-coordinate with the help of the dimensionless coordinate τ

$$x = \tau l + 2l^2 k, \tag{11.35}$$

with the magnetic length

$$l^2 = \frac{\hbar}{2e^* B_a} \tag{11.36}$$

and the momentum $\tau_0 = -2kl$. The wave function is then proportional to the parabolic cylinder function of Whittaker [28]

$$\psi(x) = \mathcal{N} D_{\tilde{\nu}}\left(\frac{x}{l} + \tau_0\right), \tag{11.37}$$

which solves the differential equation (11.31) in the dimensionless notation

$$\frac{\mathrm{d}^2 D_\nu(\tau)}{\mathrm{d}\tau^2} = \left(\frac{\tau^2}{4} - \nu - \frac{1}{2}\right) D_\nu(\tau). \tag{11.38}$$

The dimensionless boundary condition is

$$\left.\frac{D'_\nu(\tau_0)}{D_\nu(\tau_0)}\right|_{\tau_0=-2kl} = \frac{\xi}{b}\sqrt{\nu + \frac{1}{2}}, \tag{11.39}$$

where the prime denotes a derivative with respect to τ_0. The parameter

$$\nu = -\frac{1}{2} - \frac{\alpha m^*}{e^* \hbar B_a} \tag{11.40}$$

plays the role of an eigenenergy.

The boundary condition (11.39) links ν and τ_0. Since we are looking for the minimal ν, we take the solution of (11.39) as a function $\nu(\tau_0)$. Besides the obvious numerical search, we can give directly a nonlinear equation for this desired minimum given by $\nu'(\tau_0) = 0$. For this purpose, we differentiate (11.39) with respect to τ_0 arriving[1] at [7,8]

$$\left.\frac{D_{\tilde{\nu}+1}(\tau_0)}{D_{\tilde{\nu}}(\tau_0)}\right|_{\tau_0=-2\sqrt{(\tilde{\nu}+\frac{1}{2})(1+\frac{\xi^2}{b^2})}} = -\sqrt{\left(\tilde{\nu}+\frac{1}{2}\right)\left(1+\frac{\xi^2}{b^2}\right)} - \frac{\xi}{b}\sqrt{\tilde{\nu}+\frac{1}{2}}, \tag{11.41}$$

where $\tilde{\nu}$ is the minimal value of ν. Solving (11.41), we find the minimal ν to each given τ_0, i.e., $\tilde{\nu}[\tau_0]$. Since $\tau_0 = -2kl$, we find in this way the eigenenergy as a function of momentum k. The GL wave function (11.37) is now specified except for its amplitude $\mathcal{N}$. We discuss this amplitude below. In Fig. 11.1, we see how the shape of the GL wave function evolves with inverse extrapolation length $1/b$, i.e., how it depends on the external bias. For positive electric fields attracting charge carriers to the surface, the superconducting density is pushed from the surface into the bulk while for oppositely directed electric fields the superconductivity is even more squeezed near the surface.

The lowest eigenvalue $\tilde{\nu}$ corresponds to the highest attainable critical magnetic field

$$B_{c3} = \frac{-m^*\tilde{\alpha}}{\hbar e^*(\tilde{\nu}+\frac{1}{2})} \equiv \frac{B_{c2}}{2\tilde{\nu}+1}, \tag{11.42}$$

where B_{c2} is the upper critical field. In Fig. 11.2, we present the result for the surface critical field (11.42) versus the external bias. Without external bias, the known GL

[1] With the minimum condition $\nu' = 0$ at τ_0, the derivative of (11.39) yields

$$\left.\frac{D''_{\tilde{\nu}}}{D_{\tilde{\nu}}}\right|_{\tau_0} = \left.\frac{D'^2_{\tilde{\nu}}}{D^2_{\tilde{\nu}}}\right|_{\tau_0}.$$

The left-hand side follows from (11.38) as $D''_{\tilde{\nu}}/D_{\tilde{\nu}} = \frac{\tau_0^2}{4} - \tilde{\nu} - \frac{1}{2}$. The right-hand side is given by (11.39). The result is

$$\tau_0 = \pm 2\sqrt{\left(1+\frac{\xi^2}{b^2}\right)\left(\tilde{\nu}+\frac{1}{2}\right)}$$

with the negative root being the physical one. Finally, substituting this τ_0 and a general relation $D'_{\tilde{\nu}} = \tau_0 D_{\tilde{\nu}}/2 - D_{\tilde{\nu}+1}$ into (11.39), we obtain (11.41).

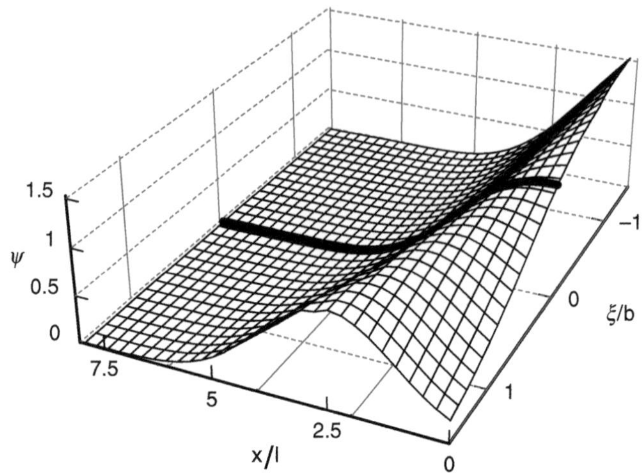

Fig. 11.1 The GL wave function of the condensate vs. the boundary condition (11.9). The normal GL wave function without external bias is marked as *thick line*

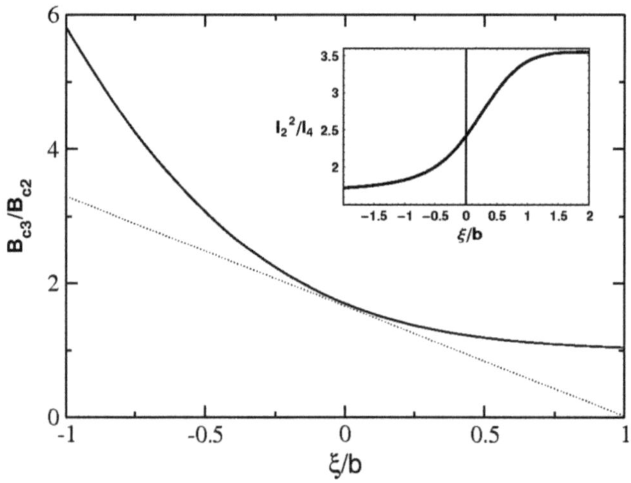

Fig. 11.2 The surface critical field B_{c3} vs. extrapolation length. The exact solution (11.42) is the *solid line*, and its slope at $\xi/b = 0$ is shown by the *dotted line*. The inset shows the ratio of integrals (11.45) and (11.46)

solution [29] with $B_{c3}/B_{c2} = 1.69461$ is reproduced. We see that the external bias can enhance or decrease the surface critical value.

According to the GL wave function (11.33), the current flows only in the y direction. Its net value given by the x integral of the current density equals the k derivative of the eigenenergy. Thus, for the minimal ν the net current is zero. Current distributions for three different boundary conditions with extrapolation lengths $\xi/b = 0, +1$, and -1 are plotted in the Fig. 11.3 as bold, dashed, and dash dot lines.

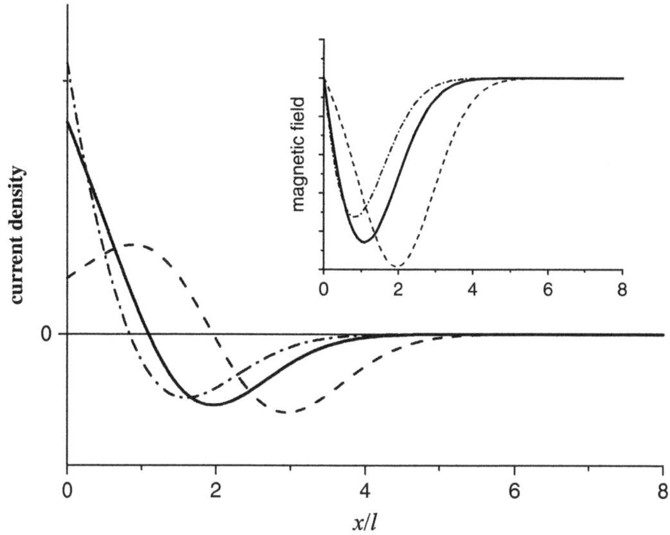

Fig. 11.3 Distribution of the current density for boundary conditions with extrapolation lengths $\xi/b = 0, +1$ and -1 are plotted in *bold*, *dashed*, and *dash dot lines*. The inset shows the corresponding reduction in magnetic field

Since there is no net current circulating around the sample, as is seen in the inset of the Fig. 11.3, the magnetic field is reduced only in the region of nucleation. Each surface thus acts independently.

11.5.3 Surface Energy

With the help of the GL wave function, we can now calculate the surface energy

$$\sigma = \int_0^\infty \mathrm{d}x \left[\alpha|\psi|^2 + \frac{\beta}{2}|\psi|^4 + \frac{[B_\mathrm{a} - B(x)]^2}{2\mu_0} + \frac{|(i\hbar\nabla + e^*\mathbf{A})\psi|^2}{2m^*} \right]. \tag{11.43}$$

Here, we neglect the free energy of the magnetic field, but the term $\sim |\psi|^4$ which determines the amplitude of the GL wave function cannot be omitted even near the nucleation line.

Since the applied field changes the shape of the GL wave function only weakly while its amplitude changes rapidly near the critical point, we fix the shape to be the nucleation function (11.37) and use the amplitude $\mathcal{N}$ as a variational parameter. Briefly, we substitute $\psi = \mathcal{N} D_{\tilde{\nu}}$ into (11.43) which gives

$$\sigma = (\alpha - \tilde{\alpha})\mathcal{N}^2 l\, I_2 + \frac{1}{2}\beta\mathcal{N}^4 l\, I_4, \tag{11.44}$$

where $\tilde{\alpha} = -(\hbar e^* B/2m^*)(2\tilde{\nu}(E, B) + 1)$ is the maximal eigenvalue corresponding to the nucleation temperature under given magnetic and electric field, while $\alpha = \alpha'(T - T_c)$ is the GL parameter given by the actual sample temperature. The quadratic and quartic terms in (11.44) are weighted by dimensionless integrals

$$I_2 = \int_{\tau_0}^{\infty} d\tau D_{\tilde{\nu}}^2(\tau), \tag{11.45}$$

$$I_4 = \int_{\tau_0}^{\infty} d\tau D_{\tilde{\nu}}^4(\tau). \tag{11.46}$$

In the normal state, $\alpha > \tilde{\alpha}$ and the minimum of the surface energy is $\mathcal{N}^2 = 0$ giving $\sigma = 0$. In the superconducting state, $\alpha < \tilde{\alpha}$ and the minimum of σ (11.44) is at $\mathcal{N}^2 = -(\alpha - \tilde{\alpha})I_2/(\beta I_4)$ leading to the surface energy

$$\sigma = -l \frac{I_2^2}{I_4} \frac{(\alpha - \tilde{\alpha})^2}{2\beta}. \tag{11.47}$$

At the critical point, $\alpha = \tilde{\alpha}$ and the surface energy vanishes. The surface energy and its first derivatives with respect to the electric and magnetic field are continuous at the critical point. The second derivatives are discontinuous. In the next chapter, we show how this discontinuity appears in the magneto-capacitance.

11.6 Magneto-capacitance

We assume a capacitor in which the first electrode is a superconducting and the second electrode is a normal metal. Our aim is to evaluate the contribution of the surface superconductivity to the capacitance.

The capacitance of the capacitor with one superconducting electrode reads

$$\frac{1}{C_s} = \frac{1}{C_n} + \frac{1}{\epsilon^2 S} \frac{\partial^2 \sigma}{\partial E^2}, \tag{11.48}$$

where S is the area of the capacitor, C_n is the capacitance when both electrodes are normal, and ϵ is the permittivity of the ionic background in the superconductor. This follows from the inverse capacitance given by the second derivative of the total energy W with respect to the charge Q, $1/C = \partial^2 W/\partial Q^2$. The charge Q is linked to the electric field E at the surface of the superconductor via the Gauss law $\epsilon E = Q/S$. Since energies of normal and superconducting states differ by the surface energy $W_s = W_n + S\sigma$, one arrives at (11.48) for the difference in capacitances.

11.6.1 Discontinuity in Magneto-capacitance

Now, we can evaluate the jump of the capacitance, which appears as the magnetic field B exceeds the critical value B_{c3}. Since $\alpha_E \to \alpha$ for $B \to B_{c3}$, the discontinuity of the inverse capacitance equals

$$\frac{1}{C_s} - \frac{1}{C_n} = -\frac{\hbar^2 e^{*2} B_{c3}^2 l\, I_2^2}{\epsilon^2 m^{*2} S \beta I_4} \left(\frac{\partial \tilde{\nu}}{\partial E}\right)^2, \tag{11.49}$$

where we have used $\partial \alpha_E / \partial E = -(\hbar e^* B / m^*)(\partial \tilde{\nu} / \partial E)$.

To evaluate the slope $\partial \tilde{\nu} / \partial E$, we recall the numerical result shown in Fig. 11.2. The tangential dotted line yields $B_{c3}/B_{c2} = 1/(2\tilde{\nu} + 1) = 1.69 - 1.69\xi/b$, which follows also from an explicit variational calculation, see (20) in [7]. Since $\partial(1/b)/\partial E = 1/U_s$, for $1/b \to 0$ we find $(\partial \tilde{\nu}/\partial E)^2 = 0.087\, \xi^2/U_s^2$. From relation (11.29) thus follows

$$\left(\frac{\partial \tilde{\nu}}{\partial E}\right)^2 = 0.087 \frac{4 m^{*2} \epsilon^2}{\hbar^4 (e^*)^2} \xi^2 \beta^2 \eta^2 \left(\frac{\partial \ln T_c}{\partial \ln n}\right)^2. \tag{11.50}$$

We have used the relation $\beta = \alpha' T_c / n$ which holds for Gor'kov values of GL parameters. Substituting (11.50) into (11.49), we arrive at

$$\frac{1}{C_s} - \frac{1}{C_n} = -0.348 \frac{I_2^2}{I_4} \frac{B_{c3}^2 l}{\hbar^2 S} \xi^2 \beta \eta^2 \left(\frac{\partial \ln T_c}{\partial \ln n}\right)^2. \tag{11.51}$$

The GL coherence length $\xi = \hbar / \sqrt{2 m^* \alpha}$ relates to the surface critical field (see (11.42)). For $1/b \to 0$, it yields $\xi^2 = 1.69\hbar/(e^* B_{c3})$. Moreover, according to (11.36) we can express B_{c3} via the magnetic length; therefore

$$\frac{1}{C_s} - \frac{1}{C_n} = -0.712 \frac{1}{(e^*)^2 S l} \beta \eta^2 \left(\frac{\partial \ln T_c}{\partial \ln n}\right)^2. \tag{11.52}$$

We have used $I_2^2 / I_4 = 2.42$, which is the value at $1/b = 0$.

Since the GL parameter β can be fitted from experimental results, relation (11.52) allows one to establish $\eta\,(\partial \ln T_c / \partial \ln n)$. This material parameter describes the change of the critical temperature with the electron density.

11.6.2 Estimates of Magnitude

For an estimate we assume some typical numbers. The most sensitive measurements of capacitance performed in the $C \sim \mu$F range are capable to monitor changes $\delta C / C \sim 10^{-6}$ with error bars at $\delta C / C \sim 10^{-7}$. From the capacitance $C = \epsilon_d S / L$, one sees that a 1,000 Å-thick dielectric layer with $\epsilon_d = 10^3\epsilon_0$ has an optimal area of $S = 10\,\text{mm}^2$ which is about the usual size of such samples [30].

To estimate β, we use Gor'kov's relation $\beta = 6\pi^2 k_B^2 T_c^2/(7\zeta(3)E_F n)$. For Niobium, $T_c = 9.5\,\text{K}$ and $n = 2.2\times 10^{28}/\text{m}^3$. The free electron model used by Gor'kov then gives the Fermi energy $E_F = 4.6\ 10^{-19}$ J. The corresponding GL parameters then is $\beta = 1.2\ 10^{-53}\ \text{Jm}^3$. The logarithmic derivative is estimated in [31] as $\partial \ln T_c/\partial \ln n = 0.74$. Since η is not known, we take $\eta = 1$ according to the simple theory. Finally, we need the third critical magnetic field B_{c3} to estimate the magnetic length l. From $B_{c3} = 1.69\,B_{c2}$ and the experimental value $B_{c2} = 0.35\,\text{T}$ [32], one finds $B_{c3} = 0.59\text{T}$, which yields $l = 325\,\text{Å}$. With these values from (11.52), we obtain the discontinuity $1/C_s - 1/C_n = -2.8\ 10^{-4}\ \text{F}^{-1}$. Since the capacitance was estimated to be $\sim 10^{-6}$ F, the corresponding relative change $C_s/C_n - 1 \sim 3\times 10^{-10}$ is too small to be observed. A slightly more optimistic estimation can be found in [8].

In high-T_c materials, the GL parameter β is by three to four orders of magnitude larger than in conventional metals due to larger T_c and lower density of holes giving also lower Fermi energy. Moreover, the logarithmic derivative of the critical temperature is about $\partial \ln T_c/\partial \ln n = -4.82$ as estimated in [31]. Finally, higher critical surface magnetic field B_{c3} allows to reduce the magnetic length to values limited rather by experimental facility. For a typical field of 10 T, the magnetic length is 79 Å. These factors together provide an enhancement to values $1/C_s - 1/C_n = -52\ \text{F}^{-1}$ or $C_s/C_n - 1 \sim 5\times 10^{-5}$ which is an experimentally accessible discontinuity.

11.7 Summary

We have shown that the electric field applied to the surface of the superconductor modifies the boundary condition of the GL wave function. Since the surface superconductivity is sensitive to this boundary condition, we have discussed the influence of the electric field. From the surface energy, we predict that a planar capacitor with one normal electrode and the other electrode to be superconducting reveals a discontinuity of the capacitance at the third critical field B_{c3}. This discontinuity is too small for capacitors from conventional superconductors, but it is large enough to be observed in capacitors with ferroelectric dielectric layers of a width of 1,000 Å and nonconventional superconductor electrodes.

Acknowledgements This work was supported in the frame of the Institutional Research Plan AVOZ 10100521, by GAČR project 202/08/0326, DFG-CNPq project 444BRA113/57/0-1 and the DAAD-PPP (BMBF) program. The financial support of the Brazilian Ministry of Science and Technology is also acknowledged.

References

1. P.W. Anderson, J. Phys. Chem. Solids 11, 26 (1959)
2. C.H. Ahn, J.M. Triscone, J. Mannhart, Nature, 424, 1015 (2003)
3. T. Frey, J. Mannhart, J.G. Bednorz, E.J. Williams, Phys. Rev. B 51, 3257 (1995)
4. R.E. Glover, M.D. Sherrill, Phys. Rev. Lett. 5, 248 (1960)
5. D. Matthey, S. Gariglio, J.M. Triscone, Appl. Phys. Lett. 83, 3758 (2003)
6. X.X. Xi, C. Doughty, A. Walkenhorst, C. Kwon, Q. Li, T. Venkatesan, Phys. Rev. Lett. 68, 1240 (1992)
7. K. Morawetz, P. Lipavský, J. Kolácek, E.H. Brandt, Phys. Rev. B 78, 054525 (2008)
8. K. Morawetz, P. Lipavský, J.J. Mares, New J. Phys. 11, 023032 (2008)
9. L. Burlachkov, I.B. Khalfin, B. Ya. Shapiro, Phys. Rev. B 48, 1156 (1993)
10. B. Ya. Shapiro, Phys. Lett. 105A, 374 (1984)
11. B. Ya. Shapiro, Solid State Commun. 53, 673 (1985)
12. B. Ya. Shapiro, Phys. Rev. B 48, 16722 (1993)
13. B. Ya. Shapiro, I.B. Khalfin, Physica C 209, 99 (1993)
14. B. Ya. Shapiro, I.B. Khalfin, L. Burlachkov, Physica B 194, 1893 (1994)
15. J.L. Chen, T.J. Yang, Physica C 231, 91 (1994)
16. L.P. Gor'kov, Zh. Eksper. Teor. Fiz. 36, 1918 (1959) [Sov. Phys. JETP 9, 1364 (1959)]
17. T. Koyama, J. Phys. Soc. Jpn. 70, 2102 (2001)
18. X.-Y. Jin, Z.-Z. Gan, Eur. Phys. J. B 37, 489 (2004)
19. M. Machida, T. Koyama, Phys. C: Supercond. 378–381, 443 (2002)
20. M. Machida, T. Koyama, Phys. Rev. Lett. 90(7), 077003 (2003)
21. M. Machida, T. Koyama, Physica C 388–389, 659 (2003)
22. G.-Q. Zha, S.-P. Zhou, B.-H. Zhu, Y.-M. Shi, Phys. Rev. B 73, 104508 (2006)
23. B.-H. Zhu, S.-P. Zhou, G.-Q. Zha, K. Yang, Phys. Lett. A 338(3–5), 420 (2005)
24. B.-H. Zhu, S.-P. Zhou, Y.-M. Shi, G.-Q. Zha, K. Yang, Phys. Lett. A 355, 237 (2006)
25. B.-H. Zhu, S.-P. Zhou, Y.-M. Shi, G.-Q. Zha, K. Yang, Phys. Rev. B 74, 014501 (2006)
26. P.G. de Gennes, Superconductivity of Metals and Alloys (Benjamin, New York, 1966)
27. J. Koláček, P. Lipavský, K. Morawetz, E.H. Brandt, Phys. Rev. B 79, 174510 (2009)
28. W. Magnus, F. Oberhettinger, R.P. Soni, Formulas and Theorems of the Special Functions of Modern Physics (Springer, Berlin, 1966)
29. D. Saint-James, P.G. Gennes, Phys. Lett. 7, 306 (1963)
30. Ch.S. Hwang, J. Appl. Phys. 92, 432 (2002)
31. P. Lipavský, J. Koláček, K. Morawetz, E.H. Brandt, T.J. Yang, Bernoulli potential in superconductors (Springer, Berlin, 2007) Lecture Notes in Physics 733
32. D.K. Finnemore, T.F. Stromberg, C.A. Swenson, Phys. Rev. 149, 231 (1966)

Chapter 12
Polarity-Dependent Vortex Pinning and Spontaneous Vortex–Antivortex Structures in Superconductor/Ferromagnet Hybrids

Simon J. Bending, Milorad V. Milošević, and Victor V. Moshchalkov

Abstract Hybrid structures composed of superconducting films that are *magnetically coupled* to arrays of nanoscale ferromagnetic dots have attracted enormous interest in recent years. Broadly speaking, such systems fall into one of two distinct regimes. Ferromagnetic dots with weak moments pin free vortices, leading to enhanced superconducting critical currents, particularly when the conditions for commensurability are satisfied. Dots with strong moments spontaneously generate one or more vortex–antivortex (V–AV) pairs which lead to a rich variety of pinning, anti-pinning and annihilation phenomena. We describe high resolution Hall probe microscopy of flux structures in various hybrid samples composed of superconducting Pb films deposited on arrays of ferromagnetic Co or Co/Pt dots with both weak and strong moments. We show directly that dots with very weak perpendicular magnetic moments do not induce vortex–antivortex pairs, but still act as strong polarity-dependent vortex pinning centres for free vortices. In contrast, we have directly observed spontaneous V–AV pairs induced by large moment dots with both in-plane and perpendicular magnetic anisotropy, and studied the rich physical phenomena that arise when they interact with added "free" (anti)fluxons in an applied magnetic field. The interpretation of our imaging results is supported by bulk magnetometry measurements and state-of-the-art Ginzburg–Landau and London theory calculations.

12.1 Introduction

The phenomena of superconductivity and ferromagnetism are characterised by order parameters that are usually mutually exclusive in homogeneous systems. In the rare cases where they have been found to coexist in the same material, it is always accompanied by remarkable new physics. A simple way to overcome the natural antagonism between the two phenomena is to fabricate ferromagnet–superconductor (F–S) hybrid systems in which the two components are spatially separated (see e.g. [1–3] for excellent recent reviews of the field). Two limits of this problem have been extensively studied theoretically and experimentally in recent years:

(1) magnetically coupled hybrids (the type considered here) where the two components are separated by a thin insulating layer [4–10] or (2) electronically coupled hybrids, where no insulating layer is present, and whose structure is designed to minimise magnetic interactions [11–16]. We report here *magnetic* imaging studies of vortex structures in magnetically coupled F–S hybrids composed of relatively dilute arrays of round or rectangular ferromagnetic "dots" with out-of-plane (perpendicular) or in-plane anisotropy, covered by a continuous superconducting film. In the limit of "dots" with weak magnetic moments, vortex–antivortex (V–AV) pairs cannot be created in the superconducting film, but screening currents still flow in response to the local fields of the nanomagnets and the superconducting order parameter is suppressed near their poles. This leads to an additional pinning or anti-pinning potential for free vortices induced by an external magnetic field, depending on the relative direction between the vortex fields and the nanomagnet stray fields. In the strong moment limit, V–AV pairs are spontaneously nucleated in the superconducting film by the stray fields of the nanomagnets. Furthermore, "free" vortices introduced by applying an external magnetic field interact in rather subtle ways with spontaneous V–AV pairs, exhibiting complex pinning, anti-pinning and annihilation phenomena.

The experiments described in this chapter address important outstanding issues in the physics of F–S hybrids via direct magnetic imaging using high-resolution scanning Hall probe microscopy (SHPM) [17]. These studies have generally been performed at temperatures well below T_c, deep in the superconducting state, when vortices have maximum peak fields and are easiest to resolve. Imaging has been complemented by bulk magnetisation measurements, for example, to investigate commensurability phenomena in periodic pinning arrays. Such measurements are, in contrast, generally made close to T_c where the intrinsic critical current of the superconducting films is strongly suppressed. Where appropriate, our results are compared with the predictions of Ginzburg–Landau (G–L) theory or London theory. Technically speaking, G–L theory is only valid close to the normal-superconductor phase boundary where the order parameter is small. However, empirically it appears to have a rather broader range of applications than this and, even deep in the superconducting state (where magnetic imaging is performed), we find excellent qualitative agreement with G–L predictions. London theory has a much broader range of validity, provided we can ignore spatial variations of the amplitude of the order parameter. This is reasonably well justified in our samples since the dimensions of all the structures investigated is considerably larger than the superconducting coherence length, ξ.

12.2 Theoretical Description of F–S Hybrids

12.2.1 Ginzburg–Landau Theory

The properties of F–S hybrids have been extensively studied with G–L theory using the "hard" ferromagnet approximation, i.e. assuming that the magnetisation state of the ferromagnet is not altered due to the presence of supercurrents in the adjacent superconductor or small external applied fields. Under these conditions, the phase transition line is obtained by minimising the free energy functional [1]:

$$G_{\mathrm{sf}} = G_{\mathrm{s0}} + G_{\mathrm{m}} + \int_V \left\{ \alpha |\Psi|^2 + \frac{\beta}{2} |\Psi|^2 + \frac{1}{4m} \left| -i\hbar\vec{\nabla}\Psi - \frac{2e}{c}\vec{A}\Psi \right|^2 \right.$$
$$\left. + \frac{B^2}{8\pi} - \vec{B}\vec{M} - \frac{\vec{B}\vec{H}_{\mathrm{ext}}}{4\pi} \right\} \mathrm{d}V, \tag{12.1}$$

where G_{s0} is a field- and temperature-independent part of the superconductor free energy and G_{m} accounts for the self energy of the ferromagnet. Ψ is the complex superconducting order parameter ($|\Psi|^2$ is the local density of Cooper pairs), $\alpha(\beta)$ is the first(second) G–L parameter in the free energy expansion and $\boldsymbol{A}$ is the total vector potential (containing sources from the nanomagnet and supercurrents).

Minimisation of (12.1) leads to a pair of coupled G–L equations for superconducting films and nanomagnets in a uniform applied magnetic field.

$$-\frac{1}{4m}\left(-i\hbar\vec{\nabla} - \frac{2e}{c}\vec{A}\right)^2 \Psi - \alpha\Psi + \beta|\Psi|^2\Psi = 0, \tag{12.2}$$

$$\vec{\nabla}\times\vec{\nabla}\times\vec{A} = \frac{4\pi}{c}\vec{j}_{\mathrm{s}} + 4\pi\vec{\nabla}\times\vec{M} = \frac{eh}{imc}(\Psi^*\vec{\nabla}\Psi - \Psi\vec{\nabla}\Psi^*)$$
$$-\frac{8\pi e^2}{mc^2}|\Psi|^2\vec{A} + 4\pi\vec{\nabla}\times\vec{M}, \tag{12.3}$$

where j_{s} is the density of superconducting current.

For thin superconductors (thickness, d, smaller than the coherence length and penetration depth) and uniformly magnetised nanomagnets ($\nabla\times M = 0$), Milošević et al. [18] show that the G–L equations can be averaged over the film thickness, d, leading to a simpler two-dimensional form.

$$\left(-i\vec{\nabla}_{2\mathrm{D}} - \vec{A}\right)^2 \Psi - \Psi + |\Psi|^2\Psi = 0, \tag{12.4}$$

$$-\vec{\nabla}^2_{3\mathrm{D}}\vec{A} = \frac{d}{\kappa^2}\delta(z)\vec{j}_{2\mathrm{D}} = \frac{d}{\kappa^2}\delta(z)\left[\frac{1}{2i}\left(\Psi^*\vec{\nabla}_{2\mathrm{D}}\Psi - \Psi\vec{\nabla}_{2\mathrm{D}}\Psi^*\right) - |\Psi|^2\vec{A}\right], \tag{12.5}$$

where $j_{2\mathrm{D}}$ is the supercurrent density, $\kappa = \lambda/\xi$ is the G–L parameter and lengths have been normalised by the coherence length, ξ, the vector potential, $\boldsymbol{A}$, by $c\hbar/2e\xi$ and the magnetic field, $\boldsymbol{H}$, by $H_{\mathrm{c2}} = c\hbar/2e\xi^2$.

Using this formalism, these authors have investigated V–AV nucleation in an infinite superconducting film with a single uniformly magnetised ferromagnetic disk (magnetisation perpendicular to the plane of the film) on top of it with radius, R_{d}, thickness, d_d, and magnetic moment, m. Since the simulation described a single dot on an infinite superconducting sheet, the net flux threading the superconductor was zero and spontaneously generated vortices and antivortices can only appear in pairs. Under these assumptions, Milošević et al. found the ground state flux configurations as a function of magnetic moment (expressed in terms of $m_0 = c\hbar\xi/2e$) and disk radius. For sufficiently small moments, there is no spontaneous nucleation of V–AV pairs corresponding to a Meissner-like screening behaviour. Even in this regime, the ferromagnetic disk represents an additional pinning (or anti-pinning) potential for free vortices introduced by an external magnetic field (this will be discussed further in Sect. 12.2.2). Above a minimum value of magnetic moment (which depends weakly on radius), one or more V–AV pairs are nucleated in the superconductor as illustrated in Fig. 12.1 ($R_\mathrm{d}/\xi = 1$, $d_\mathrm{d}/\xi = 0.5$, $d/\xi = 0.5$). Vortex–antivortex pairs nucleate under the edge of the magnetic disk, where the induced screening currents are maximal. Vortices remain below the ferromagnetic disk in the form of a single giant vortex for small radii splitting up into a multi-vortex structure for large radii. The antivortices are arranged symmetrically in ring-like "shells" outside the ferromagnetic disk (where the stray fields from the lower magnetic pole return back through the superconductor). After each shell has captured its maximum number of antivortices (e.g. 6 for the first shell), a second shell starts to form outside, and so on. Due to the obvious analogy with electrons in atomic shells, these objects are referred to as "vortex molecules".

Figure 12.2 summarises the expected shell structure as a function of m and R_d, where N denotes the number of vortices in the "core" and the Roman numerals indicate the number of antivortex shells. The shaded triangular-shaped region on the right is the regime of multi-vortex cores.

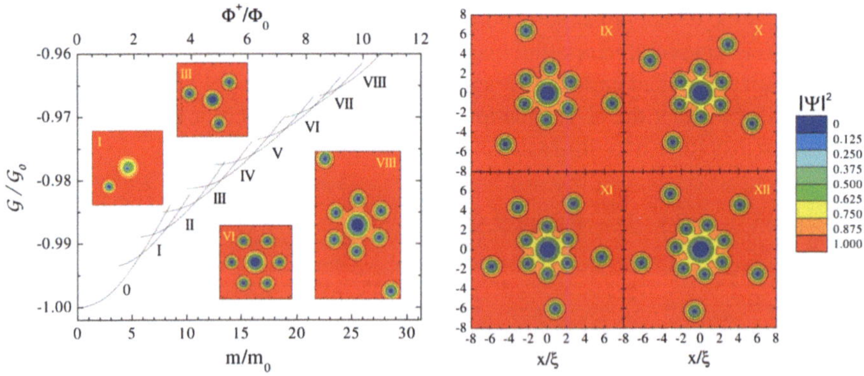

Fig. 12.1 Gibbs free energy as a function of magnetic moment of disk of radius $R_\mathrm{d}/\xi = 1$, $d_\mathrm{d}/\xi = 0.5$ and $d/\xi = 0.5$ placed on top of a superconducting film. Figures on the *right* are contour plots of the Cooper pair density for a few ground state vortex configurations. From [18]

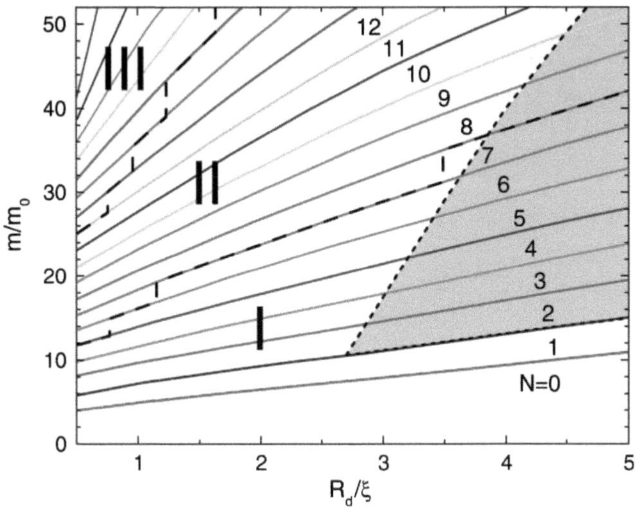

Fig. 12.2 Relationship between the radius of the disk, its magnetic moment, the number of antivortices (N), the number of antivortex rings (*Roman numerals*) and the vortex state under the magnetic disk (*shaded area* = multi-vortex state). From [18]

Other theoretical works have explored flux structures formed in superconducting films with adjacent dense arrays of ferromagnetic dots. In this case, it is found that the vortex–antivortex molecules organise themselves into "ionic" crystals [19, 20].

Milošević and Peeters [21] have used the same G–L formalism described above (12.4, 12.5) to study V–AV nucleation in a superconducting film (thickness d) with arrays of *in-plane* magnetised ferromagnetic dipoles on the surface (separated by an oxide layer of thickness, l). The magnetic dipoles were modelled by uniformly magnetised ferromagnetic rods of dimension w_x, w_y in the basal plane, thickness D and uniform magnetisation M. As in the case for perpendicularly magnetised dots, above a threshold geometry-dependent magnetisation one or more V–AV pairs are spontaneously nucleated in the superconducting film (c.f. Fig. 12.3). Now vortex–antivortex pairs nucleate under the centre of the magnet and arrange in a symmetrical way just below the S and N poles of the nanomagnets. In addition, for the range of moments considered in Fig. 12.3, only single quanta (anti-)vortices are observed.

These authors also demonstrate enhanced pinning of "free vortices" induced by an applied external magnetic field. Free vortices can annihilate spontaneous antivortices at one of the poles of the magnetic dipoles and vice versa. In this way, all flux is strongly pinned at the magnetic centres, and the critical current remains high. Indeed, for a dipole that spontaneously generates two V–AV pairs they found that the critical current is maximum at the second matching field when two free vortices (or antivortices) are associated with each nanomagnet. Moreover, the critical current was symmetrical with field, i.e. independent of whether free vortices or free antivortices were added.

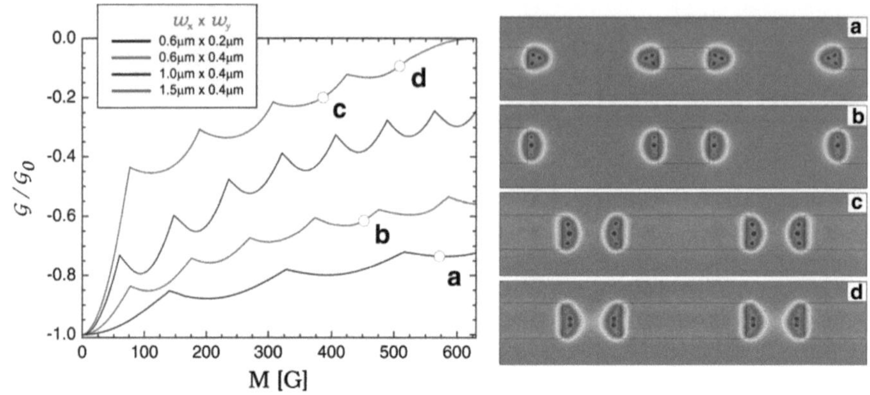

Fig. 12.3 (*Left*) Gibbs free energy of a Pb film [$d = 50\,\text{nm}$, $\xi(0) = 40\,\text{nm}$, $T = 0.9T_c$, $\kappa = 1.2$] with a regular array (period $w_x = 2\,\mu\text{m}$, $w_y = 1\,\mu\text{m}$) of in-plane magnetic dipoles ($D = 100\,\text{nm}$, $l = 20\,\text{nm}$) on *top*. Different curves correspond to different sizes of the magnets ($w_x \times w_y$) and (**a**)–(**d**) show the Cooper-pair density (logarithmic scale, *blue*/*red*-low/high density) of the vortex–antivortex states indicated by *open dots* in the free-energy diagram (*solid lines* denote the edges of the magnets). Adapted from [21]

12.2.2 London Theory

As discussed above, magnetic imaging experiments are generally performed deep in the superconducting state when peak vortex fields are large and easy to resolve. This is technically beyond the regime of validity of G–L theory, although it does still appear to provide a good qualitative description of our experiments. An alternative approach is to use London theory which has a much broader range of validity, but ignores any spatial variation of the amplitude of the order parameter. This is generally a satisfactory approximation for all experiments performed here. At low temperatures, we must explicitly include the magnetic field induced by screening currents or by vortices, which will strongly interact with the ferromagnet. Assuming again "hard" ferromagnets that will not change their magnetisation structure during the performed experiments, the free energy density of the F–S hybrid can be written as [1]:

$$G_{\text{FS}} = G_{\text{M}} + G_{\text{S0}} + \int_V \left\{ \frac{\lambda^2}{8\pi} \left(\vec{\nabla} \times \vec{B} \right)^2 + \frac{B^2}{8\pi} - \vec{B}\vec{M} - \frac{\vec{B}\vec{H}_{\text{ext}}}{4\pi} \right\} \mathrm{d}V, \quad (12.6)$$

where G_{M} and G_{S0} are the self-energies of the ferromagnet and superconductor, respectively. Minimisation of this with respect to the vector potential yields the London–Maxwell equation:

$$\vec{B} + \lambda^2 \vec{\nabla} \times \vec{\nabla} \times \vec{B} = \Phi_0 \sum_i \delta(\vec{r} - \vec{R}_{v,i}) \vec{z}_0 + 4\pi\lambda^2 \vec{\nabla} \times \vec{\nabla} \times \vec{M}, \quad (12.7)$$

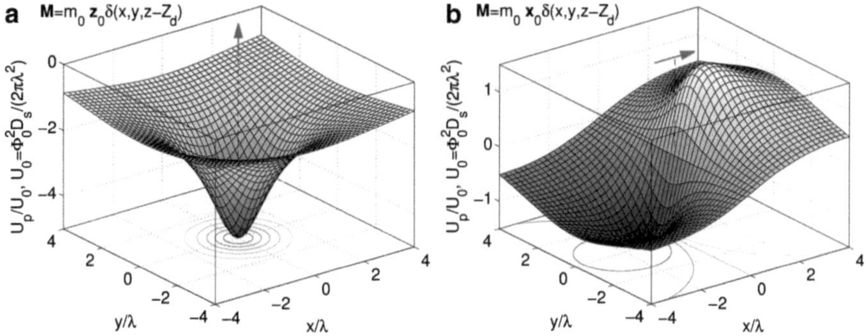

Fig. 12.4 Spatial dependence of the energy of a single vortex in a superconducting film in the field of a point magnetic dipole, $\boldsymbol{M} = \boldsymbol{m_0}(x, y, z\text{-}Z_d)$, calculated according to (12.8) and (12.9) for the height $Z_d/\lambda = 1$ and the dipole strength $m_0/(\Phi_0\lambda) = 10$: (**a**) vertically magnetised dipole, (**b**) horizontally magnetised dipole. From [1]

where the summation should be performed over all the vortices sited at positions $\{R_{v,i}\}$, and $\delta(r)$ is the Dirac delta function. This approach has been used to investigate the flux structures formed for F–S hybrids composed of a single magnetic dot [22].

Milošević et al. [23] have used London theory to calculate the pinning potential for a free vortex near a ferromagnetic dipole with either perpendicular or in-plane magnetisation, assuming that the moment is not sufficiently large to nucleate spontaneous V–AV pairs. They find the following results for perpendicular and in-plane magnetisation, respectively, in the limit that the thickness of the superconductor, d_s, is very much less than the penetration depth [1]

$$U_p^z(\rho) \approx \frac{\Phi_0 d_s}{2\pi\lambda}\left[-\frac{1}{2}\left(\frac{m_0^z}{\Phi_0\lambda}\right)\frac{1}{\sqrt{\rho^2+Z_d^2}}\right], \tag{12.8}$$

$$U_p^x(\rho) \approx \frac{\Phi_0 d_s}{2\pi\lambda}\left[\frac{1}{2}\left(\frac{m_0^x}{\Phi_0\lambda}\right)\frac{\rho\cos\phi}{\sqrt{\rho^2+Z_d^2}\left(Z_d+\sqrt{\rho^2+Z_d^2}\right)}\right], \tag{12.9}$$

where m_0^z and m_0^x are the perpendicular and in-plane dipole moments, respectively, ρ and ϕ are cylindrical polar coordinates of an arbitrary point in the plane of the superconductor measured from the dipole at (0, 0, Z_d). These dependencies are plotted in Fig. 12.4 [1] for a target vortex with field lines pointing upwards, parallel to the z-axis. In this limit, we find that vortices are pinned below the perpendicular moment and near the S pole of the in-plane moment. The N pole of the in-plane moment actually represents an anti-pinning site which repels the vortex. In both

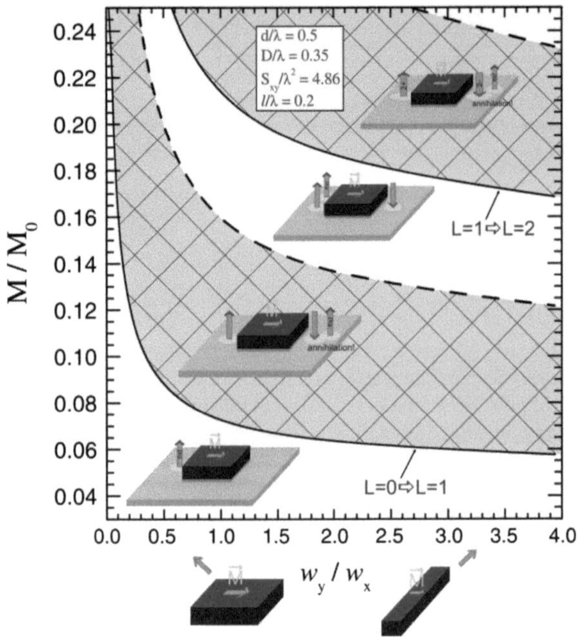

Fig. 12.5 The threshold magnetisation of the magnetic bar necessary for the nucleation of the first and second V–AV pair (*solid lines*) as a function of the aspect ratio for a fixed magnet volume. The *dashed lines* denote changing vortex pinning properties: in the shaded (*white*) regions, an external flux line is pinned at the positive (negative) pole of the magnet. From [24]

cases, pinning can be qualitatively understood in terms of the attraction between a free vortex and the antivortex-like screening currents under the magnetic poles.

Milošević and Peeters [24] have also used London theory to investigate the formation of spontaneous V–AV pairs by in-plane magnetised bars on top of a superconducting thin film as a function of magnetisation and aspect ratio of the magnets. Furthermore, they have studied the pinning of an additional free vortex in the presence of these V–AV pairs. These results are summarised in Fig. 12.5 for a bar uniformly magnetised along the x-axis and of dimensions w_x, w_y in the plane and thickness D. Here, the magnetisation has been normalised by $M_0 = \Phi_0/\lambda^2$. The lower solid line represents the threshold magnetisation for the generation of a single V–AV pair. The shaded region above this represents a regime where free vortices are drawn to the N pole of the nanomagnet by attraction to the spontaneous antivortex there. Remarkably, the white region above the lower dashed line represents a regime at higher moments when the vortex becomes attracted to the S pole of the nanomagnet because the magnet–vortex interaction now dominates over the vortex–antivortex attraction.

12.3 Experimental Results

12.3.1 Scanning Hall Probe Imaging

High resolution SHPM has been used, both to image the magnetisation state of the nanomagnets in our hybrid samples and to visualise vortex and antivortex structures in them in the superconducting state. The SHPM used (Fig. 12.6) is a modified commercial low temperature STM in which the usual tunnelling tip has been replaced by a microfabricated GaAs/AlGaAs heterostructure chip. Electron beam lithography and wet etching were used to define a Hall probe in the two-dimensional electron gas at the intersection of two ~500-nm-wide wires approximately 5 μm from the corner of a deep mesa etch. The latter had been coated with a thin Au layer to act as an integrated STM tip. The sample is first approached towards the sensor until tunnelling is established and then retracted about 100 nm allowing rapid scanning with minimum detectable fields $\sim 1\,\mu T/Hz^{0.5}$. The Hall probe makes an angle of about 1° with the sample plane so that the STM tip is always the closest point to the surface, and the Hall sensor was typically about 300–400 nm above the sample in the images shown here. A more detailed description of the instrument is given elsewhere [17].

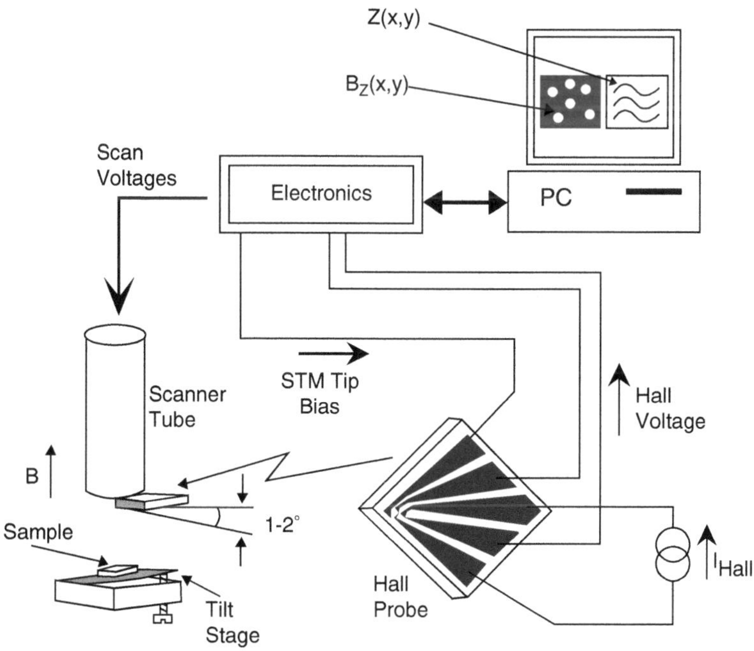

Fig. 12.6 Schematic diagram of the scanning Hall probe microscope. Adapted from [25]

12.3.2 Low Moment Dot Arrays with Perpendicular Magnetisation

This experiment was designed to explore the predictions of London theory (see discussion in Sect. 12.2.2 above) that weak moment magnetic dots with perpendicular magnetisation do not generate spontaneous V–AV pairs, but act as polarity-dependent pinning centres for free vortices [26]. Vortices with fields parallel to the direction of magnetisation should be pinned on-site and those with fields anti-parallel should be repelled by the nanomagnets.

The sample was prepared on a Si single crystal substrate with an amorphous SiO_2 top layer. The magnetic dots consisted of a $[\mathrm{Co}(0.3\,\mathrm{nm})/\mathrm{Pt}(1.1\,\mathrm{nm})]_{10}$ multilayer on a 3 nm Pt buffer. They were fabricated by electron-beam lithography, electron-beam evaporation in ultra-high vacuum and lift-off techniques. They had a square shape with a side length of 400 nm and rounded corners and were arranged in a square lattice with a 1 μm period. To confirm the perpendicular anisotropy of the Co/Pt dots, hysteresis loops of the dot array were measured at room temperature using the magneto-optical Kerr effect (MOKE). With H perpendicular to the substrate, the dot array had a coercive field of 2.3 kOe and showed 100% magnetic remanence. After saturating the dots perpendicular to the substrate plane, H could be swept in the range $-1\,\mathrm{kOe} < H < 1\,\mathrm{kOe}$ without changing the magnetic response of the array, indicating that the magnetic state of the dots remains virtually unchanged. A 10 nm Ge film, a 50 nm superconducting Pb thin film (behaving as a type-II superconductor with critical temperature $T_c = 7.17\,\mathrm{K}$), a 25 nm protective Ge layer and a 50 nm Au film were subsequently evaporated on top of the dot array. The insulating Ge layers prevent proximity effects between the metallic and superconducting layers, while the Au layer facilitates the approach of the SHPM probe in tunnelling control mode. The penetration depth $\lambda(0)$ and the coherence length $\xi(0)$ in the sample were estimated from electrical transport measurements on a 50 nm reference Pb film: $\xi(0) = 37\,\mathrm{nm}$ and $\lambda(0) = 45\,\mathrm{nm}$. Near to T_c when $\lambda(T) \gg d$, where $d = 50\,\mathrm{nm}$ is the thickness of the superconductor, we expect an effective penetration depth $\Lambda(T) = \lambda^2(T)/d = 27.4\,\mathrm{nm}/(1 - T/T_c)$.

The flux pinning properties of our F–S hybrid system were investigated in a Quantum Design SQUID magnetometer. $M(H)$ magnetisation curves in perpendicular field, H, at $T = 7.00\,\mathrm{K}$ and $T = 7.10\,\mathrm{K}$ are shown in Fig. 12.7. The curves were obtained after the dots were magnetised above T_c in (a) $H = +40\,\mathrm{kOe}$ ($\boldsymbol{m} > 0$) and (b) $H = -40\,\mathrm{kOe}$ ($\boldsymbol{m} < 0$) perpendicular to the surface. The field axes have been normalised to the first matching field, H_1, given by $\mu_0 H_1 = \Phi_0/1\,\mu\mathrm{m}^2 = 2.068\,\mathrm{mT}$ at which exactly one superconducting flux quantum Φ_0 is generated per unit cell of the dot array. We observed large $\Delta M(H)$ and enhanced matching effects when H and $\boldsymbol{m}$ have the same polarity, i.e. for $\boldsymbol{m} > 0$ and $H > 0$ and for $\boldsymbol{m} < 0$ and $H < 0$. In these cases, pronounced matching effects were obtained at integer (1, 2 and 3) and several fractional multiples (1/4, 1/3, 1/2, 2/3, 3/4, 5/4, 4/3, 3/2, 5/3 and 7/4) of H_1. These pronounced matching effects indicate commensurate vortex configurations in a strong periodic pinning potential. If

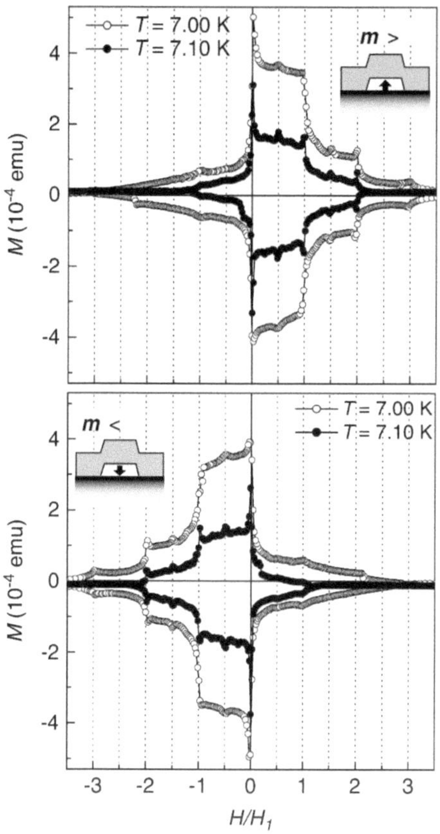

Fig. 12.7 Magnetisation measured at two different temperatures near T_c (7.00 K *open symbols*, 7.10 K *filled symbols*) showing the superconducting response of the Pb layer on top of the Co/Pt dot array with all dots aligned in a positive (*upper panel*) and negative (*lower panel*) sense. $H_1 = 20.68$ Oe is the first matching field. From [26]

H and $\boldsymbol{m}$ have opposite polarity, ΔM is substantially reduced and only rather weak matching features were observed at $|H/H_1| = 1/2$, and 1.

To unambiguously establish the origin of the strong asymmetry observed in the $M(H)$ data, SHPM was used to investigate the microscopic vortex distribution in the sample. Images were first recorded in zero applied field at temperatures slightly above and below T_c. When the dots are magnetised in a large negative field perpendicular to the surface, they appear as a square array of dark spots at $T = 7.4\,\text{K} > T_c$, as is shown in Fig. 12.8a. The peak-to-valley B_z contrast of the image in Fig. 12.8a is 0.30 G. After zero field cooling, the sample to $T = 6.8\,\text{K} < T_c$ the B_z peak-to-valley amplitude decreases by about 20% to 0.25 G, as can be clearly seen from the weaker contrast in Fig. 12.8b. This effect is attributed to Meissner-like screening currents generated in the superconductor in response to the local magnetic stray field of the dots. Below T_c, supercurrents, j_s, appear in the Pb film encircling the dots whose

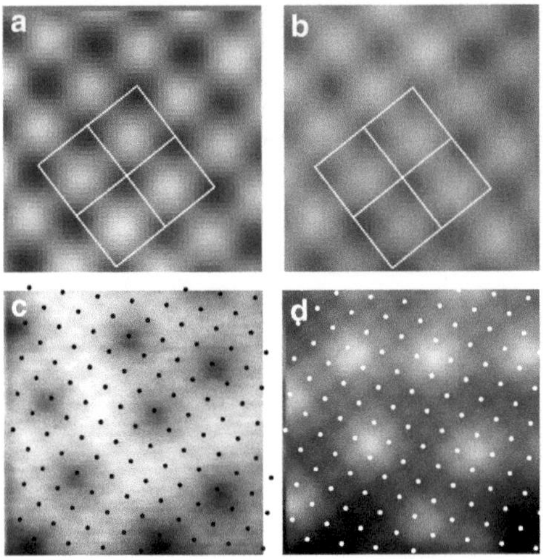

Fig. 12.8 (*Top*) Zero field SHPM images showing the local induction, B_z, above the sample surface at (**a**) $T = 7.4\,\mathrm{K} > T_c$ and (**b**) $6.8\,\mathrm{K} < T_c$ after magnetising the dots in a negative field ($\boldsymbol{m} < 0$). The dots appear as *dark spots*, the *dotted line* indicates the square dot array. (*Bottom*) SHPM images of the sample in (**c**) $H = -1.6\,\mathrm{Oe}$ and (**d**) $H = +1.6\,\mathrm{Oe}$ after field-cooling to $T = 6.8\,\mathrm{K}$. The tiny *black/white* points indicate the positions of the Co/Pt dots, which are all magnetised downwards ($\boldsymbol{m} < 0$). Vortices are visible as diffuse *dark* ($H < 0$) or *bright* ($H > 0$) spots in the SHPM images. Adapted from [26]

direction and magnitude depend on the amount of flux generated by the dot. We estimate that the magnetic moment of these dots is $m \approx 1.9m_0$ (where $m_0 = c\hbar\xi/2e$ as defined in Sect. 12.2.1) and $R_d/\xi \approx 5.4$. Comparison with Fig. 12.2 above places the disks firmly in the Meissner-like screening regime with $N = 0$, where the moment is too small to nucleate V–AV pairs. This calculation is hence fully consistent with our imaging results.

Figure 12.8c, d shows SHPM images obtained after field-cooling the sample in small positive ($H = -1.6\,\mathrm{Oe}$) and negative fields ($H = +1.6\,\mathrm{Oe}$), respectively. This field value corresponds to about 8.5 flux quanta in the scanned area of $(10.5\,\mu\mathrm{m})^2$. Additional vortices appearing in the superconductor can be clearly recognised by their white or black contrast. There are nine negative (dark) vortices in Fig. 12.8c and eight positive (bright) vortices in Fig. 12.8d, in good agreement with these expectations. The square array of dots produces a much weaker contrast and is indicated by black and white dots for clarity. The location of the vortices clearly depends on the field polarity. When the magnetic field of the vortex points in the same direction as the magnetic moments of the dots (Fig. 12.8c, $\boldsymbol{m} < 0$ and $H < 0$), the vortices are positioned on-site. In contrast, vortices with opposite field polarity (Fig. 12.8d, $\boldsymbol{m} < 0$, $H > 0$) are positioned at interstitial positions of the dot array. These local visualisation experiments can be correlated with the global magnetisation experiments depicted in Fig. 12.7. When the applied field is *parallel* to the moment of the dots, the flux lines are pinned by the dots and high critical currents and pronounced matching effects are observed. On the other hand, for the *antiparallel* alignment of applied field and dot magnetisation, the flux lines are positioned at interstitial positions where they have a higher mobility and the pinning is substantially reduced. This scenario is fully consistent with the pinning potential calculated for weak moment magnetic dots in Fig. 12.4. In effect, the screening currents above the magnetic dot are antivortex-like, and a free vortex is attracted to them (on-site pinning) and a free antivortex is repelled by them (anti-pinning).

12.3.3 High Moment Dot Arrays with Perpendicular Magnetisation

This experiment was designed to explore the predictions of Ginzburg–Landau theory (see discussion in Sect. 12.2.1) that strong moment magnetic dots with perpendicular magnetisation generate one or more spontaneous V–AV pairs with vortices pinned on the dots and antivortices organised into shells around the dots [25, 27]. Furthermore, subtle pinning and annihilation phenomena should occur for free vortices and antivortices introduced by an applied magnetic field.

The samples investigated consisted of a *dilute* square array of circular ferromagnetic disks with perpendicular magnetic anisotropy covered with a type II superconducting Pb film. The disks were patterned by electron beam lithography and reactive ion etching of a $[Co(0.5\,nm)/Pt(1\,nm)]_{12}$ multilayer film sputtered on a Si/SO_2 substrate. Four different diameter disks with different magnetic moments were patterned on the corners of a $5\,\mu m \times 5\,\mu m$ square cell which was repeated periodically in a square lattice, allowing the behaviour of dots with different spontaneous V–AV numbers to be compared in the same sample. The design diameters of 522 nm (dot A), 738 nm (D), 808 nm (B) and 902 nm (C) were chosen, corresponding *theoretically* to 1, 3, 3 and 5 spontaneous V–AVs, respectively [18]. Magneto-Optical Kerr Effect (MOKE) measurements of the unpatterned Co/Pt film at 300 K showed very high remanence and a coercive field of ~1,000 Oe. SHPM images of magnetisation reversal in the patterned disks at low temperature indicate a range of coercive fields spanning 700–1,000 Oe and magnetic saturation above $H \cong 1{,}000$ Oe. After saturation, magnetisation disks exhibit almost complete remanence at $H = 0$ and remain in a single domain state with highly uniform out-of-plane moments (c.f. Fig. 12.9). The disks were coated with a 20-nm Ge layer to suppress proximity effects and an 80-nm Pb film deposited using dc magnetron sputtering followed by a 10-nm Mo capping layer to prevent oxidation. Magnetisation measurements on a single Pb film of the same thickness indicate that it is a type II superconductor with $T_c = 6.68$ K, $\lambda_{eff}(5K) \cong 120$ nm and $\xi(5K) \cong 50$ nm. Finally the sample was also coated with 20 nm Ge and 50 nm Au to enhance the stability of the SHPM when in tunnelling contact.

In order to observe clear spontaneous V–AV structures, it was necessary to magnetise the sample to saturation ($H > 3{,}000$ Oe) at low temperature. An undesirable consequence of this was the trapping of a small amount of magnetic flux in the

Fig. 12.9 SHPM image of the magnetised disk array at $T = 77\,\mathrm{K}$ (Co/Pt dots appear *white*). The corners of the dotted white square indicate the positions of the four different diameter magnetic disks

superconducting solenoid (equivalent to $-3.5\,\mathrm{Oe}$), which acts in the opposite direction to the dot magnetisation. To avoid unnecessary confusion arising from this constant offset to the applied field (H_a), the "effective" field values as defined by $H_{eff} = H_a - 3.5\,\mathrm{Oe}$ are always quoted here.

All SHPM images were obtained after field-cooling the sample from above T_c in the indicated applied field. Figure 12.10a shows the direct observation of spontaneous V–AV shell structures in the hybrid sample at $H_{eff} = 0$. Since the amplitude of the (black) antivortices was typically only 10–20% of that of the (white) vortices trapped on the magnetic disks, a strongly non-linear grayscale has been used to enhance their contrast. As theoretically expected, the (black) AVs order in shell-like structures around the magnetic dots, while (white) vortices remain confined above the dots. A line-scan analysis has been used to determine the exact AV locations and these are sketched for clarity in Fig. 12.10b. Figure 12.10c shows the maps of the local magnetic induction obtained from Ginzburg–Landau simulations (using the framework described in Sect. 12.2.1) for the same sample geometry. The observed vorticity is in good agreement with simulations, and broadly speaking increases with the magnetic moment of the disks (subject to flux quantisation).

The vorticity of these V–AV shell structures can be "tuned" by applying an external magnetic field to either add ($H_{eff} < 0$) or annihilate ($H_{eff} > 0$) antivortices. Figure 12.10d–f show such "tuned" V–AV shells containing 2, 3 and 5 antivortices, respectively.

For negative effective fields, free AVs are introduced into the spontaneous V–AV shell system. In this situation there is a subtle competition between the n-fold rotational symmetry of the V–AV "molecules" and the translationally symmetric lattice of magnets. Sometimes spontaneous AVs are observed to detach from their

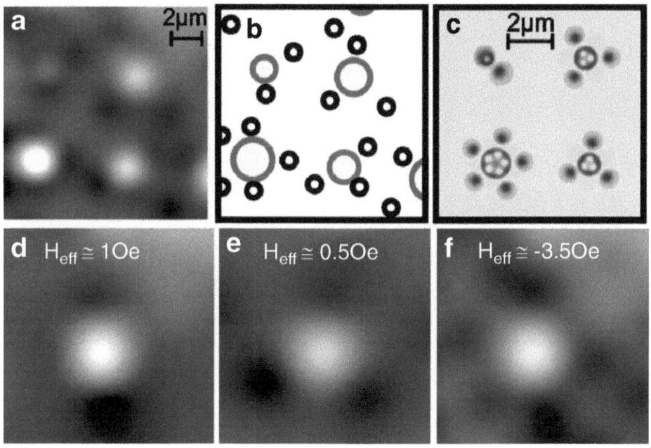

Fig. 12.10 (**a**) SHPM image and (**b**) schematic depiction of spontaneous V–AV configurations at $H_{\text{eff}} \cong 0$ and $T = 5\,\text{K}$. (**c**) Map of magnetic induction across the G–L calculation under the same conditions (*bright colour* – negative field). (**d**)–(**f**) Different vorticity V–AV shell structures at the indicated effective applied fields and $T = 5\,\text{K}$. From [27]

nanomagnet (effectively breaking a V–AV bond) to join the interstitial AV-lattice. At other field values, the reverse occurs; an AV that is not needed in the interstitial lattice approaches a magnetic dot and joins its AV shell, attracted by the "positive" vortex core. In this way, at sufficiently negative effective fields, all the V–AV molecules become net "negative", i.e. they contain more antivortices than vortices.

For positive effective magnetic fields, the general behaviour of the V–AV structures is governed by the annihilation of AV shells, and each of the molecules progressively becomes more "positively" charged. Even after all AVs have become annihilated, the vortex "charge" of individual magnetic dots keeps increasing due to the trapping of additional vortices on the magnetic dots due to the attraction between a magnet and a vortex when their moments are parallel (c.f. the discussion in Sect. 12.2.2).

This process is illustrated in Fig. 12.11 (left), where the number of observed off-site fluxons is shown (circles; antivortices are counted negative and vortices positive) as a function of applied field, as well as the results of G–L calculations (solid line), which are in excellent agreement. Both data sets clearly exhibit a "nulling" field ($H_{\text{a}} \sim 6\,\text{Oe}$, $H_{\text{eff}} \cong 2.5\,\text{Oe}$), when there are no free fluxons. The right side of Fig. 12.11 depicts SHPM images as this state is approached. The scan at $H_{\text{eff}} \cong 1.5\,\text{Oe}$ clearly shows some remaining black AVs in shells around the dots that are annihilated in the next image at $H_{\text{eff}} \cong 2\,\text{Oe}$. The subtracted "difference" image on the right-hand side highlights changes that have occurred between the two scans. White dots in this either represent annihilated antivortices in the first image or added vortices in the second. The difference in applied field between successive images of $\Delta H = 0.5\,\text{Oe}$ corresponds to three added flux quanta on average. While this is consistent with what is seen in the difference image, it is found unexpectedly that a vortex has been removed from above the disk in the lower right-hand corner

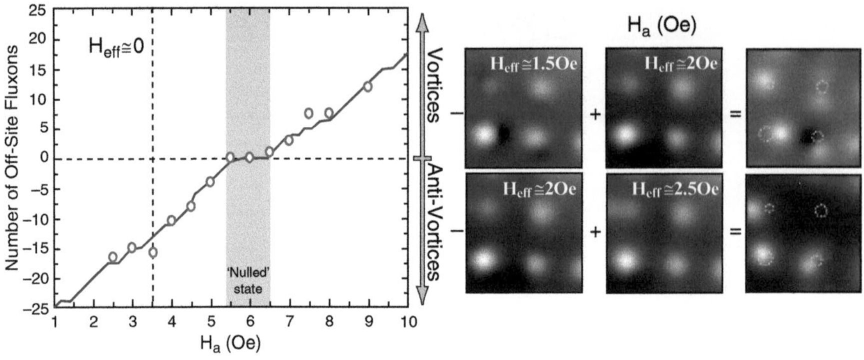

Fig. 12.11 (*Left*) Number of observed (anti)vortices as a function of applied field at $T = 5\,\mathrm{K}$ (*circles*). *Solid line* shows the predictions of G–L theory (shifted by $+4\,\mathrm{Oe}$ on the H-axis to simulate flux trapping in the solenoid). (*Right*) Two pairs of successive SHPM images near the "nulling" state and their "difference images". Adapted from [27]

where a black dot is seen, having annihilated with one of the adjacent antivortices. The image of the vortex state at $H_{\mathrm{eff}} \cong 2.5\,\mathrm{Oe}$ looks superficially similar to that at $2\,\mathrm{Oe}$, but the difference image reveals that the number of trapped on-site vortices has increased at the higher field. Moreover, the asymmetric position of the added vortices in the difference image is suggestive of a multi-vortex state above the dots. For the parameter of this disk ($m/m_0 \sim 10$ and $R_{\mathrm{d}}/\xi \sim 7.4$), this is exactly what is expected theoretically (c.f. Fig. 12.2 in Sect. 12.2.1). The trapping of excess vortices leads to a fairly robust "nulling" state that survives over a reasonable range of applied fields ($\Delta H > 1\,\mathrm{Oe}$). This "locking" behaviour ensures the absence of any off-site fluxons and consequently corresponds to a field-induced enhancement of the critical current of the sample.

Upon increasing the (positive) effective field still further, non-uniform changes in on-site vortex occupation lead to a "dynamic" pinning landscape for off-site vortices. At sufficiently large magnetic fields ($H_{\mathrm{eff}} = 4.5\,\mathrm{Oe}$), a distorted square interstitial vortex lattice is recovered (Fig. 12.12a), mirroring conventional matching phenomena. The influence of multi-quanta vortices at the nanomagnets becomes more evident at the highest resolvable vortex densities. Figure 12.12b shows how the presence of four different repulsive potentials propagating radially from the corners of the square cell, and strong interactions between interstitial vortices, results in their arrangement in shells. The same structure is found in G–L simulations under identical conditions. While AVs form shells around confined vortices due to their mutual attraction, counter-intuitively, vortex shells are formed here by *repulsion*.

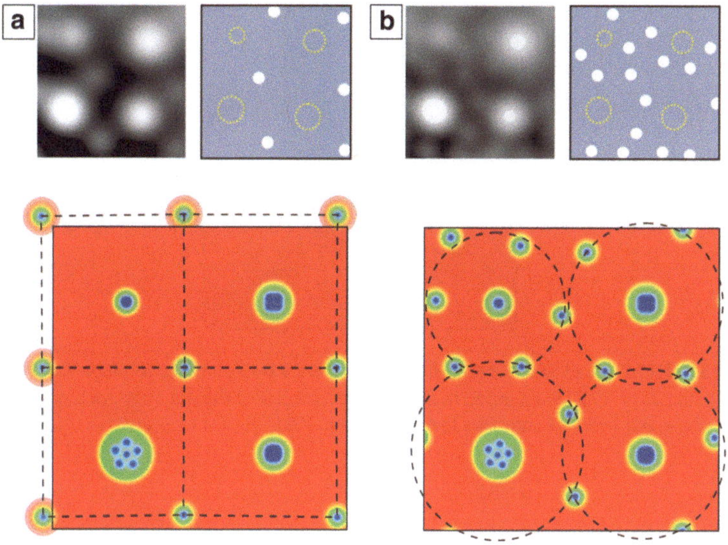

Fig. 12.12 SHPM images at $T = 5\,\mathrm{K}$ and corresponding Cooper-pair density plots illustrating the formation of (**a**) an interstitial vortex lattice ($H_{\mathrm{eff}} = 4.5\,\mathrm{Oe}$) and (**b**) interstitial vortex shells at high vortex densities ($H_{\mathrm{eff}} = 7.5\,\mathrm{Oe}$). From [27]

12.3.4 High Moment Arrays with In-Plane Magnetisation

12.3.4.1 Arrays of Rectangular Nanobars

This experiment was designed to explore vortex pinning and the generation of spontaneous V–AV pairs near the poles of strong moment magnetic nanobars with in-plane magnetisation [28].

Square lattices (period $a = 1.5\,\mu\mathrm{m}$) of rectangular sub-micron polycrystalline magnetic dots, consisting of a Au(7.5 nm)/Co(20 nm)/Au(7.5 nm) trilayer, were fabricated by electron-beam lithography and molecular beam deposition. The dots had lateral dimensions of 540 nm (easy axis) × 360 nm. Magnetic force microscopy (MFM) at room temperature revealed a multi-domain as-grown state. After magnetisation along the easy axis, all dots were in a single-domain remanent state [29]. The dot array was covered with a 50 nm superconducting Pb film, a protective Ge layer (20 nm) and a 10 nm Au layer for the scanning tunnelling microscopy (STM) height control of the SHPM.

SHPM measurements were performed above and below the superconducting critical temperature of the Pb film ($T_c = 7.16\,\mathrm{K}$) with an applied field (H) normal to the sample plane. Before all measurements, the dots were aligned into a single-domain state using a large in-plane magnetic field. Figure 12.13 shows a SHPM image of a region near the centre of the sample at $H = 0$ and temperature

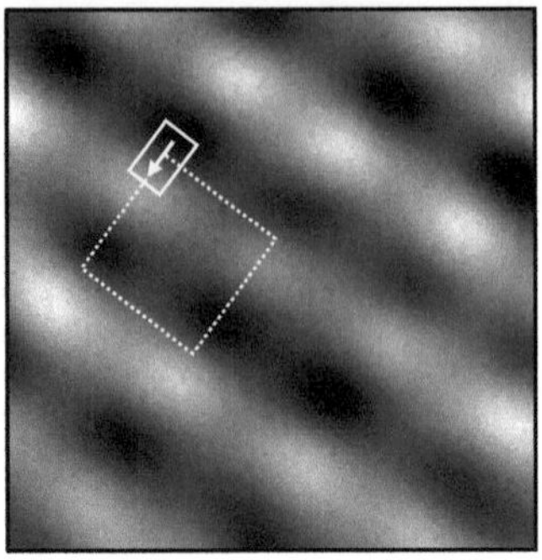

Fig. 12.13 SHPM image of the dipole stray fields from a square lattice of single-domain Co dots at $H = 0$ and $T = 77$ K. As a guide to the eye, the unit cell of the dot array (*dashed line*) and the position of one dot (*full line*) are also indicated. From [28]

$T = 77$ K. The dipole stray fields characteristic of an ordered array of single-domain in-plane magnetised nanobars is clearly visible.

A series of images of the type shown in Fig. 12.14 were captured upon cooling the sample through T_c. Just as the temperature dropped below T_c, we observed a sudden increase (~20%) in the magnetic contrast indicating that a spontaneous V–AV pair was created in the superconducting state. We estimate the normalised magnetisation $M/M_0 \approx 0.12$ ($M_0 = \Phi_0/\lambda^2$ as defined in Sect. 12.2.2) and the aspect ratio $w_y/w_x = 0.67$. This places our sample firmly in the lower shaded region of the phase diagram shown in Fig. 12.5, where a single V–AV pair is spontaneously nucleated at the poles of our nanobars, fully consistent with our imaging results.

Complementary measurements on the system confirm that these magnetic dots act as strong field-symmetric pinning centres for flux lines in a perpendicular applied magnetic field [29]. In order to investigate how additional free vortices interact with the magnetic dipoles, SHPM experiments were performed in small applied magnetic fields. The SHPM results at $T = 6\,\text{K} < T_c$ after field cooling at $H/H_1 = +1/2$ and $-1/2$ are shown in Fig. 12.14a, b, respectively. Here $\mu_0 H_1 \equiv \Phi_0/a^2 = 0.92$ mT is the first matching field defined as the applied field at which exactly one flux quantum Φ_0 is present per unit cell of the pinning array. For comparison, Fig. 12.14a, b also show maps of the magnetic dipole array at the same location of the sample in the same applied field at $T = 7.5\,\text{K} > T_c$. We find that after cooling through T_c at $H/H_1 = +1/2$, the negative (black) poles of some of the dots have disappeared, indicating that a vortex has been added at these locations. This is schematically indicated in the diagram on the left of Fig. 12.14a; at the position of the disappearing dark poles, a vortex (white circle) has been trapped.

In order to identify the free flux line locations more precisely, we subtract the dipole contribution at $T = 7.5\,\text{K} > T_c$ from the image at $T < T_c$ to obtain the difference image on the right-hand side of Fig. 12.14a. This predominantly contains information about the additional free vortices, and white spots are clearly seen that can be associated with vortices pinned at the S poles of the nanomagnets. A similar

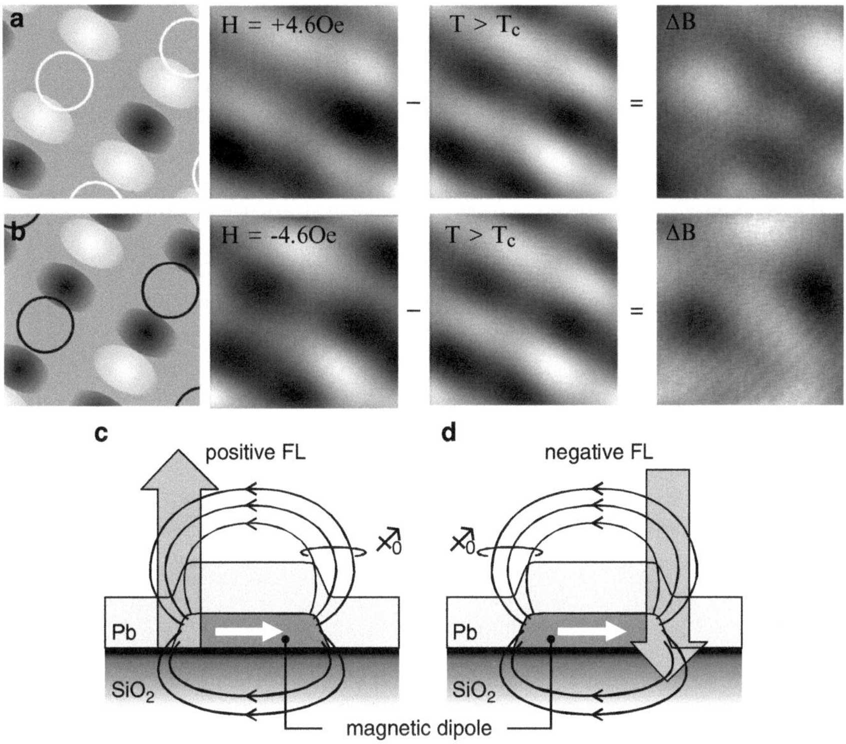

Fig. 12.14 SHPM images at (**a**) $H/H_1 = +1/2$ and (**b**) $H/H_1 = -1/2$. From *left* to *right* panels show a schematic representation of the flux image at $T = 6\,\mathrm{K}$, SHPM image at $T = 6\,\mathrm{K} < T_c$, SHPM image at $T = 7.5\,\mathrm{K} > T_c$ and the SHPM difference image constructed by subtracting the high temperature scan from the low temperature one as shown. The observed dipole fields are indicated in the schematic and the flux lines induced by the positive (negative) applied field are represented as *white* (*black*) circles. (**c**), (**d**) Sketch of the observed polarity-dependent flux pinning process; (**c**) a positive vortex (wide gray arrow) is attached to a dot at the pole where a spontaneous antivortex is induced by the stray field (*black arrows*), and (**b**) an antivortex is pinned at the pole where a vortex is induced by the stray field. Adapted from [28]

scenario occurs at $H/H_1 = -1/2$ (Fig. 12.14b) where now antivortices become pinned at the N poles of the nanomagnets. These observations are in excellent agreement with the London theory calculations summarised in Fig. 12.5 for our sample parameters, provided one takes into account the fact that the nanobars are *underneath* the superconductor in our experiments and *on top* of the superconductor in the theoretical simulation. In both cases, an added free vortex is attracted to a spontaneous antivortex at one of the poles of the magnet. It is indeed fascinating that Fig. 12.5 predicts a reversal of the polarity-selective vortex pinning mechanism for nanobars with slightly larger (>50%) magnetisation. This remarkable result remains to be experimentally verified.

12.3.4.2 Arrays of Nanoscale Ferromagnetic Rings

Recently, Menghini et al. [30] have explored the magnetic moment-dependence of vortex pinning in F–S hybrids using ferromagnetic rings with in-plane magnetisation in which the domain structure, and hence vortex pinning potential, can be controlled via the magnetic history used.

The samples studied in this work consisted of a 4-μm period square array of square Co rings fabricated by electron beam lithography and lift-off techniques. The matching field of the array, when the density of vortices equals that of the Co rings, was $H_1 = 1.3\,\mathrm{Oe}$. The square rings were based on 250-nm-wide "wires", 25-nm thick, and had an overall lateral size of 2 μm. A 10-nm insulating layer of Ge was deposited on the magnetic array to suppress proximity effects, followed by a 50-nm evaporated film of Pb. Transport measurements of the normal-superconductor phase boundary as a function of applied magnetic field revealed that the Pb film was a type II superconductor with a slightly lower T_c than an otherwise identical film without an adjacent Co ring array (due to the influence of the ferromagnet stray fields). Transport data after magnetisation of the entire array into a dipolar "onion" state reveal commensurability effects out to $H = 3H_1$, above which the phase boundaries of the films with and without rings converge. This suggests that three V–AV pairs are produced by each magnetised ring in the "onion" configuration.

Field-cooled Bitter decoration experiments at 4.2 K revealed that about 45% of the as-grown Co rings were in a flux-closure state with the north pole (head) of each magnetised side pointing towards the south pole (tail) of the next side around and vice versa. The remaining rings were nearly all either in a dipolar "onion" state with head-to-head and tail-to-tail poles on diagonal corners of the square, or a "horseshoe" state with head-to-head and tail-to-tail poles on the same edge of the ring. Single decoration spots were also occasionally observed along one side or in one corner of the ring, suggesting that a few magnetic vortices with out-of-plane directed "cores" may also have been present. After magnetisation in a direction parallel to the diagonal across the squares, all rings existed in the "onion" state described above with average magnetic moments directed along the diagonal.

Figure 12.15 shows a set of SHPM images of the hybrid sample after field-cooling to 4.2 K in the indicated applied magnetic fields. In each case, the established location of the Co rings has been superimposed on the images (dotted black or white lines). Figure 12.15a, b shows images obtained after field-cooling an array of as-grown Co rings in fields of $H = +1.5H_1$ and $H = 0$, respectively. The adjacent bright and dark spot towards the centre as well as dark spots near the edge of the field of view in both images correspond to the stray fields generated at magnetic poles in the ferromagnetic rings. The adjacent spots of different polarity arise from a ring in the "horseshoe" domain configuration. A superficial comparison between the two images reveals few differences since the maximum stray fields of the ferromagnets are about five times larger than the peak vortex fields. Figure 12.15c illustrates how vortices can, however, be resolved by numerically subtracting image (b) at $H = 0$ from image (a) at $H = +1.5H_1$ to suppress the approximately constant ferromagnetic stray fields. Each bright spot in Fig. 12.15c corresponds to a single

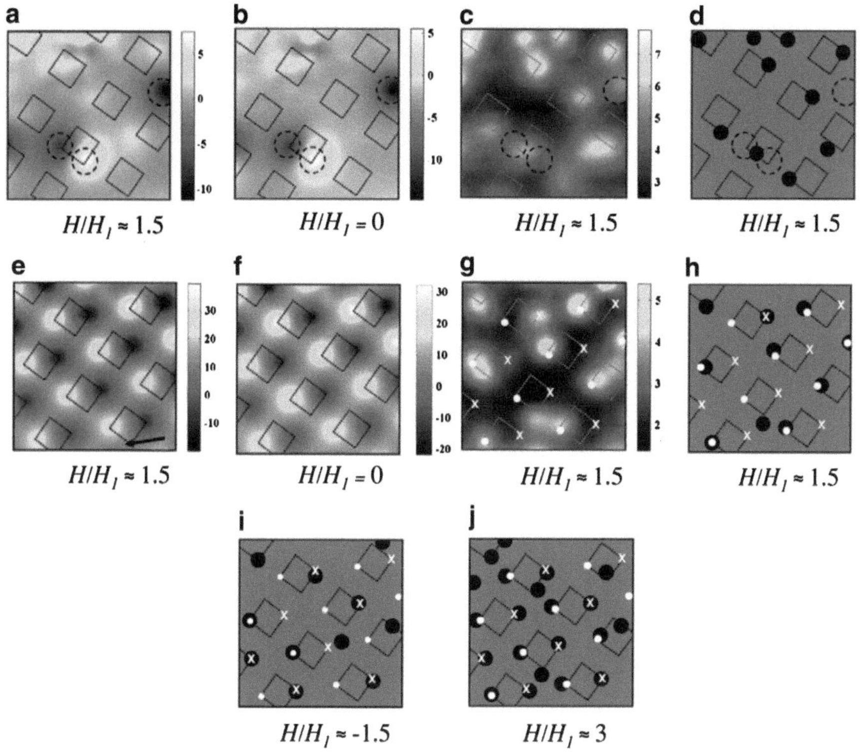

Fig. 12.15 (**a**), (**b**) SHPM images taken after field-cooling to $T = 4.2\,\mathrm{K}$ in a field of $H = +1.5H_1$ and $H = 0$, respectively, for Co rings in the as-grown state. (**c**) Vortex image obtained after subtracting image (**b**) from image (**a**). (**d**) Sketch of vortex positions (*black dots*) in (**c**). (**e**)–(**h**) Same as (**a**)–(**d**) after magnetisation of all rings in the array into a dipolar "onion" state and field-cooling at $H = +1.5H_1$. (**i**), (**j**) Vortex configurations after field-cooling at $H = -1.5H_1$ and $H = +3H_1$, respectively. In all images, the field of view is $12.6 \times 12.6\,\mu\mathrm{m}^2$ and the colorscale bars correspond to the magnetic induction in units of Gauss. From [30]

vortex, and these are sketched as black dots in Fig. 12.15d for clarity. It is evident that vortices sit primarily on the corners of the square rings with flux closure. The one interesting exception is a vortex that sits in the middle of one side between the two poles of the ring with "horseshoe" domain structure. Figure 12.15e–h shows a similar family of images after the entire ring array has been magnetised into the dipolar "onion" state with a 3,000 Oe in-plane field in the direction of the black arrow shown in Fig. 12.15e. The strong dipole fields exhibited in the onion state are now clearly visible in Fig. 12.15e, f. These have been suppressed in the difference image shown in Fig. 12.15g and vortex locations sketched in Fig. 12.15f. We now find that vortices induced by field-cooling in $H = +1.5H_1$ sit primarily on the bright north poles of the magnetised rings, although not exclusively so. One vortex actually sits on one of the south poles, one on an interstitial position and one on a flux-closed corner, possibly due to intrinsic disorder or inhomogeneity in the sample. Figure 12.15i sketches the vortex configuration for field-cooling in $H = -1.5H_1$, when we find that antivortices now sit primarily on the south poles of the rings, although again not exclusively so. Finally, Fig. 12.15j shows the vortex configuration for $H = +3H_1$ where we see that vortices now occupy both north and south poles of the magnetised rings, as well as other flux-closed corners where the stray field is small.

The primary conclusions of this study were that the corners of the rings always represent strong vortex pinning sites and vortices initially appear to prefer to sit on poles where the local stray field is *parallel* to the vortex fields when the rings are in the "onion" state. The latter observation is the exact opposite of the case of ferromagnetic nanobars discussed in Sect. 12.3.4.1 and is superficially consistent with the predictions of [24] for stronger in-plane magnetic moments. Reference [24], however, only considers nanobars that generate a single V–AV pair while the rings discussed here generate up to three V–AV pairs in the "onion" state. Hence, further theoretical work must be done to fully understand these experimental results. At higher applied magnetic fields (vortex densities), it is clear that the balance between vortex–vortex and vortex–magnetic pole interactions must be a crucial factor.

12.4 Conclusions

We have described a number of theoretical and experimental investigations of F–S hybrid structures composed of thin superconducting films magnetically coupled to arrays of ferromagnetic "dots" with both in-plane and out-of-plane anisotropy. A rich variety of pinning, anti-pinning, spontaneous vortex–antivortex nucleation, annihilation and interaction phenomena have been theoretically predicted, and several experiments have been described that are in excellent qualitative agreement with these results. This remains a very active area of research, with recent results taking theory and experiments into new regimes, e.g. investigations of a ferromagnetic dot on a mesoscopic superconducting sample [31–33] and studies of the influence of the superconductor on the magnetisation process in the adjacent ferromagnet [34]. To date, only a very small fraction of hybrid sample phase space (e.g. magnetisation, size and areal density of the "dots") has been mapped out in experiments, and there is considerable potential for more comprehensive investigations of the systematic dependence on these parameters. One challenge here will be to find ferromagnetic materials with appropriate values of magnetisation and magnetic anisotropy to fully map out different regimes. A need also remains for more quantitative comparison with theory, for example, a better understanding of the pinning potential for free vortices is needed. The London theory described in Sect. 12.2.2 ignores many factors, e.g. core pinning, local suppression of T_c and changes in the mean free path [1], and experiments should be designed to carefully identify the contributions of these different factors.

Acknowledgements This work was supported by the ESF VORTEX and NES programs, EPSRC in the UK under grant no. GR/J03077, GR/L96448, GR/D034264/1 and GR/P02707/1, the EU Marie-Curie Intra-European program, the Fund for Scientific Research-Flanders (FWO), the Belgian Science Policy and the Flemish GOA. The work at KULeuven was supported by the FWO, IAP and Methusalem.

References

1. Y.A. Aladyshkin, A.V. Silhanek, W. Gillijns, V.V. Moshchalkov, Nucleation of superconductivity and vortex matter in superconductor-ferromagnet hybrids. Supercond. Sci. Technol. **22**, 053001 (2009)
2. M. Vélez, J.I. Martín, J.E. Villegas, A. Hoffmann, E.M. González, J.L. Vicent, I.K. Schuller, Superconducting vortex pinning with artificial magnetic nanostructures. J. Magn. Magn. Mater. **320**, 2547–2562 (2008)
3. I.F. Lyuksyutov, V.L. Pokrovsky, Ferromagnet-superconductor hybrids. Adv. Phys. **54**, 67–136 (2005)
4. I.F. Lyuksyutov, V. Pokrovsky, Magnetization controlled superconductivity in a film with magnetic dots. Phys. Rev. Lett. **81**, 2344 (1998)
5. I.K. Marmorkos, A. Matulis, F.M. Peeters, Vortex structure around a magnetic dot in planar superconductors. Phys. Rev. B **53**, 2677–2685 (1998)
6. S. Erdin, A.F. Kayali, I.F. Lyuksyutov, V.L. Pokrovsky, Interaction of mesoscopic magnetic textures with superconductors. Phys. Rev. B **66**, 014414 (2002)
7. J.I. Martín, M. Vélez, J. Nogués, I.K. Schuller, Flux pinning in a superconductor by an array of submicrometer magnetic dots. Phys. Rev. Lett. **79**, 1929–1932 (1997)
8. D.J. Morgan, J.B. Ketterson, Asymmetric flux pinning in a regular array of magnetic dipoles. Phys. Rev. Lett. **80**, 3614–3627 (1998)
9. M. Lange, M.J. Van Bael, Y. Bruynseraede, V.V. Moshchalkov, Nanoengineered magnetic-field-induced superconductivity. Phys. Rev. Lett. **90**, 197006 (2003)
10. M. Lange, M.J. Van Bael, A.V. Silhanek, V.V. Moshchalkov, Vortex–antivortex dynamics and field-polarity-dependent flux creep in hybrid superconductor/ferromagnet nanostructures. Phys. Rev. B **72**, 052507 (2005)
11. Prokić, V., A.I. Buzdin, L. Dobrosavljević-Grujić, Theory of the π-junctions formed in atomic-scale superconductor/ferromagnet superlattices. Phys. Rev. B **59**, 587–95 (1999)
12. Y.A. Izyumov, M.G. Khusainov, Y.N. Proshin, Competition between superconductivity and magnetism in ferromagnet/superconductor heterostructures. Phys. Usp. **45**, 109–148 (2002)
13. A.I. Buzdin, Proximity effects in superconductor–ferromagnet heterostructures. Rev. Mod. Phys. **77**, 935–976 (2005)
14. F.S. Bergeret, A.F. Volkov, K.B. Efetov, Odd triplet superconductivity and related phenomena in superconductor–ferromagnet structures. Rev. Mod. Phys. **77**, 1321–1373 (2005)
15. V.V. Ryazanov, V.A. Oboznov, A.Y. Rusanov, A.V. Veretennikov, A.A. Golubov, J. Aarts, Coupling of two superconductors through a ferromagnet: evidence for a π-junction. Phys. Rev. Lett. **86**, 2427–2430 (2001)
16. T. Kontos, M. Aprili, J. Lesueur, F. Genêt, B. Stephanidis, R. Boursier, Josephson junction through a thin ferromagnetic layer: negative coupling. Phys. Rev. Lett. **89**, 137007 (2002)
17. A. Oral, S.J. Bending, M. Henini, Real-time scanning Hall probe microscopy. Appl. Phys. Lett. **69**, 1324–1327 (1996)
18. M.V. Milošević, F.M. Peeters, Superconducting Wigner vortex molecule near a magnetic disk. Phys. Rev. B **68**, 024509 (2003)
19. M.V. Milošević, F.M. Peeters, Vortex–antivortex lattices in superconducting films with magnetic pinning arrays. Phys. Rev. Lett. **93**, 267006 (2004)
20. D.J. Priour Jr, H.A. Fertig, Vortex states of a superconducting film from a magnetic dot array. Phys. Rev. Lett. **93**, 057003 (2004)

21. M.V. Milošević, F.M. Peeters, Vortex-antivortex nucleation in superconducting films with arrays of in-plane dipoles. Physica C **437/438**, 208–212 (2006)
22. S. Erdin, London study of vortex states in a superconducting film due to a magnetic dot. Phys. Rev. B **72**, 014522 (2005)
23. M.V. Milošević, S.V. Yampolskii, F.M. Peeters, Magnetic pinning of vortices in a superconducting film: the (anti)vortex-magnetic dipole interaction energy in the London approximation. Phys. Rev. B **66**, 174519 (2002)
24. M.V. Milošević, F.M. Peeters, Vortex pinning in a superconducting film due to in-plane magnetized ferromagnets of different shapes: the London approximation. Phys. Rev. B **69**, 104522 (2004)
25. S.J. Bending, J.S. Neal, M.V. Milošević, A. Potenza, L. San Emeterio, C.H. Marrows, Competing symmetries in superconducting vortex-antivortex “molecular crystals”. Physica C **468**, 518–522 (2008)
26. M.J. Van Bael, M. Lange, S. Raedts, V.V. Moshchalkov, A.N. Grigorenko, S.J. Bending, Local visualization of asymmetric flux pinning by magnetic dots with perpendicular magnetization. Phys. Rev. B **68**, 014509 (2003)
27. J.S. Neal, M.V. Milošević, S.J. Bending, A. Potenza, L. San Emeterio, C.H. Marrows, Competing symmetries and broken bonds in superconducting vortex–antivortex molecular crystals. Phys. Rev. Lett. **99**, 127001 (2007)
28. M.J. Van Bael, J. Bekaert, K. Temst, L. Van Look, V.V. Moshchalkov, Y. Bruynseraede, G.D. Howells, A.N. Grigorenko, S.J. Bending, G. Borghs, Local observation of field polarity dependent flux pinning by magnetic dipoles. Phys. Rev. Lett. **86**, 155–158 (2001)
29. M.J. Van Bael, K. Temst, V.V. Moshchalkov, Y. Bruynseraede, Magnetic properties of submicron Co islands and their use as artificial pinning centers. Phys. Rev. B **59**, 14674–14679 (1999)
30. M. Menghini, R.B.G. Kramer, A.V. Silhanek, J. Sautner, V. Metlushko, K. De Keyser, J. Fritzsche, N. Verellen, V.V. Moshchalkov, Direct visualisation of magnetic vortex pinning in superconductors. Phys. Rev. B **79**, 144501 (2009)
31. C. Carballeira, V.V. Moshchalkov, L.F. Chibotaru, A. Ceulemans, Multiquanta vortex entry and vortex-antivortex pattern expansion in a superconducting microsquare with a magnetic dot. Phys. Rev. Lett. **95**, 237003 (2005)
32. Q.H. Chen, C. Carballeira, V.V. Moshchalkov, Symmetry-breaking effects and spontaneous generation of vortices in hybrid superconductor-ferromagnet nanostructures. Phys. Rev. B **74**, 214519 (2006)
33. Q.H. Chen, C. Carballeira, V.V. Moshchalkov, Vortex matter in a hybrid superconducting/ferromagnetic nanostructure. Phys. Rev. B **79**, 104520 (2009)
34. J. Fritzsche, R.B.G. Kramer, V.V. Moshchalkov, Visualization of the vortex-mediated pinning of ferromagnetic domains in superconductor-ferromagnet hybrids. Phys. Rev. B **79**, 132501 (2009)

Chapter 13
Superconductor/Ferromagnet Hybrids: Bilayers and Spin Switching

J. Aarts, C. Attanasio, C. Bell, C. Cirillo, M. Flokstra, and J.M.v.d. Knaap

Abstract In research on superconductor (S)/ferromagnet (F) multilayers, a number of issues are at play simultaneously. In phenomena such as interlayer coupling in S/F/S systems or superconducting spin valve effects in F/S/F systems, questions about the interface transparency, the effect of magnetic domains on the F-side of the interface, or the stray fields produced by the F-layer arise all the time. Investigating bilayers of S/F combinations is useful and necessary to discern the effects which can be produced by the bare interface and one F-layer from those which occur when coupling between S- or F-layers also plays a role. In this chapter, we review insights gained from bilayer studies, using both weak and strong ferromagnets, with particular attention to the interface transparency and the issue of inhomogeneous exchange fields; we also review some of the questions pertaining to the superconducting spin valve effect, again for both weak and strong ferromagnets.

13.1 Introduction

Over the last 15 or so years, it has become clear that combining thin film superconductors (S) and thin film ferromagnets (F) can evoke a host of interesting physical phenomena. The interaction of the Cooper pairs of the superconductor with the exchange field of the magnet has led to new concepts such as the π-junction, the spin switch, and odd-frequency triplet pairing [1, 2], while the stray field of the magnet can be used to manipulate vortex matter in, and critical currents of the superconductor [3, 4]. Here, we concentrate on the former type of physics. We focus on what we have learnt from investigating (S/F) bilayers, and we make some observations on the issue of the F/S/F superconducting spin switch device. Throughout, we make a distinction between the case of weak magnets and strong magnets. We start with a short history of the field, reviewing some of the early developments which have led to the present state of research on S/F hybrids. We then present a summary of work on bilayers of Nb/$Cu_{1-x}Ni_x$ and Nb/$Pd_{1-x}Ni_x$ and show that measurements of the variation of T_c with the S-layer thickness d_S and the F-layer thickness d_F can be satisfactorily described with proximity effect theory, yielding values for the interface transparency and the exchange energy of the magnet. Following that, we discuss T_c-enhancement in Nb/$Cu_{41}Ni_{59}$ which occurs when an applied field generates a domain state in the magnetic layer. We then turn to experiments on Nb/$Ni_{80}Fe_{20}$, with $Ni_{80}Fe_{20}$ or Permalloy (Py) as an example of a strong magnet. Similar to the case of weak magnets, domain walls in the Py layer enhance superconductivity, although this can easily be masked by the effects of stray fields, as we will show. The domain wall effects lead to a discussion of the effects observed in spin switch devices. We end with a brief outlook on current and possible future developments, in particular, with respect to long-range proximity effects and odd frequency triplet correlations induced in ferromagnets.

13.2 Some History of the Field

Historically, bilayers actually spearheaded the field. Combinations of Pb with different magnetic materials (ferromagnets, antiferromagnets and dilute magnetic alloys) were investigated by Hauser et al. [5], who constructed a remarkable double cathode sputtering system where the substrate was held at 77 K to avoid contamination of the interfaces and interdiffusion. The results of such investigations of what came to be called the proximity effect were described in terms of the de Gennes–Werthamer theory [6, 7], treating the effect of the ferromagnet by introducing a spin flip scattering time as was done by Abrikosov and Gorkov [8] for the case of dilute paramagnetic impurities. Although strictly speaking not valid, the model yielded a very good estimate for the decay length of Cooper pairs into Fe of 0.6 nm. Another remarkable experiment was performed by Meunier et al., who measured the field dependence of the resistance of a Permalloy (Py)/Pb bilayer as function of applied magnetic field, for a temperature in the resistive transition [9]. Sharp resistance peaks were observed at the values of the coercive field, which were attributed to the stray fields coming out of domain walls.

The second phase with respect to S/N (normal metal) and S/F proximity effects started with the development of advanced vacuum systems which allowed thinner films, different materials and the deposition of multilayers. This shifted the research focus to coupling effects, since the coupling between S layers now could be varied by changing the thickness of the N(F) layers. It became of interest to investigate not only T_c but also the anisotropy in $H_{c2\parallel}$ and $H_{c2\perp}$, the parallel and perpendicular upper critical fields, due to the layering. The coupling is temperature dependent, which gives rise to the so-called dimensional crossover (DCO) phenomenon: for S-layer thicknesses $d_S \leq \xi_S(T)$, the layer system can behave three-dimensionally (3D) near T_c but two-dimensionally (2D) at lower temperatures, which shows up as a changing temperature dependence of $H_{c2\parallel}$, from $H_{c2\parallel} \propto 1 - t$ (3D) to $H_{c2\parallel} \propto \sqrt{1-t}$ (2D) (with $t = T/T_c$ the reduced temperature). DCOs were already known in S/I (insulator) multilayers [10] and now showed up in Nb/Cu, the first metallic system to be investigated exhaustively [11]. Experiments on S/F multilayers in this context were few. The F-layer decouples very effectively and S-layers behave simply in a 2D fashion, although DCOs can be observed [12–14]. Special mention is deserved by the experiments on V/Fe by Wong et al. [12] who reported reentrant behavior of T_c as function of the Fe layer thickness, probably the first observation of what now is called the π-state.

For the description of the critical field behavior, Radovic et al. proposed a method using quasiclassical equations for the Gorkov Green's functions (the Eilenberger/Usadel formalism [15, 16]), in which the exchange energy can be incorporated in a natural way [17]. This led to the celebrated predictions of the existence of oscillations of the superconducting order parameter and the π-phase [18–21]. As a consequence, T_c should oscillate as function of F-layer thickness in S/F multilayers, with an oscillation wavelength $\lambda_F = 2\pi\xi_F$, and ξ_F set by the exchange energy E_{ex} according to $\xi_F = \sqrt{(\hbar D_F)/(E_{ex})}$ (with D_F the electronic diffusion constant in the ferromagnet). The superconducting order parameter Ψ_F in the F-layer then can be described by

$$\Psi_F(x) \propto e^{-x/\xi_F}\cos(x/\xi_F). \qquad (13.1)$$

In this simple approximation, the length scales of the damping of the order parameter and the oscillation are the same. However, when comparing the theory to results for S/F multilayers with strong magnets such as Fe, Co, or Ni, a problem surfaced which has not been adequately dealt with to the present day. In the original version [17], the interface transparency was taken to be ideal, meaning that for the boundary conditions only the ratio of resistivities of the S- and F-layers should play a role. Although measurements of the change of T_c with the F-layer thickness d_F could be readily fitted to the theory, the resulting numbers for this ratio were rather unphysical [22]. Qualitatively, the solution to this problem is now known. Apart from a resistivity mismatch which may lead to a reflection rather than transmission of electrons, the spin polarization of the ferromagnet also leads to a suppression of the Andreev reflections which take care of the particle exchange between both materials [23]. This was demonstrated in an early experiment on the variation of the critical thickness for the onset of superconductivity d_S^{cr} of the V layers inside an V/V_xFe_{1-x} multilayer as function of the magnetic moment μ_F of the V_xFe_{1-x} alloy [24]. The experiment, reproduced in Fig. 13.1, shows that upon lowering the value of μ_F by diluting Fe with V, the value for d_{cr}^S first increases before a decrease with decreasing μ_F sets in. This seems counterintuitive, since a lower moment might lead to less pair breaking and a lower d_{cr}^S. The explanation is that the decrease in the magnetic moment (and supposed decrease of the spin polarization) is counteracted by the increase in interface transparency (less suppression of the Andreev reflections). Quantitatively, the tools to describe this effect are still missing, since for strong magnets the splitting of the spin subbands and the resulting difference in spin-up / down density of states cannot be simply translated into an interface transparency parameter. For weak magnets, where the spin subbands can be taken the same, this is different, since the quasiclassical theory can be simply extended to cover the case of finite transparency of the interface. Using this formalism for the experiments on V/V_xFe_{1-x}, the transparency for the V/Fe interface was found to

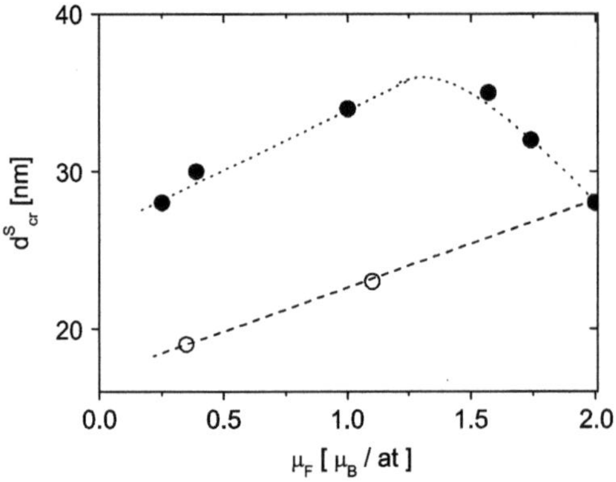

Fig. 13.1 Critical thickness d_S^{cr} of the superconducting layer for the onset of superconductivity in two sets of S/F multilayers, as function of the atomic magnetic moment of the ferromagnetic layer. *Open circles* (right to left): multilayers of V/Fe, V/Co, V/Ni; *closed circles* (right to left) multilayers of V/V_xFe_{1-x} with $x = 1, 0.88, 0.77, 0.53, 0.38, 0.34$. Adapted from [24]

be very low, and linearly increasing with decreasing μ_F. So, although the physics is not treated fully correctly, the effects of suppressed Andreev reflections are clearly evident.

The demonstration of T_c-oscillations in, e.g., Nb/Gd [25] and Nb/Co [26] pushed further developments in the S/F field. It was realized that in weak magnets with small values of E_{ex}, such as CuNi- or PdNi-alloys, the oscillation wavelength of the order parameter can be significantly larger (of the order of a few nm) than in the case of Co or Fe (order 1 nm). Experiments with such ferromagnets resulted in the observations of π-coupling in S/F/S junctions [27, 28]. Also the superconducting spin switch effect in F/S/F sandwiches, predicted by Buzdin [29] and Tagirov [30], was subsequently observed using a CuNi alloy [31]. With strong magnets, π-junctions can also be fabricated [32] and spin switching observed [33]. This is an ongoing research, and some of the pertinent issues still are the role of interface transparency and the difference between weak and strong magnets. For homogeneous exchange fields, quantitatively satisfactory descriptions appear to be possible in the case of weak magnets, but not for strong magnets where the effects of strong pair breaking and the accompanying strong decrease of the order parameter at the interface are counteracted by the suppression of Andreev reflections which tend to relax the order parameter to its bulk value. More recent questions concern the possible consequences of an inhomogeneous exchange field such as can be found in a magnetic domain wall. Here, we are concerned with these issues as far as they can be studied with S/F bilayers. For weak magnets, we show that current theory gives an adequate description of the variation of T_c with layer thickness in terms of the interface transparency and the exchange energy, although there are some caveats. We then turn to the issue of inhomogeneous magnetization and show that the domain state of the ferromagnet leads to a relative increase in T_c, both in the case of weak magnets and strong magnets. This opens the question whether domain wall effects can also play a role in F/S/F devices for spin switching. There are different viewpoints to this question, which also involve the occurrence of the so-called 'inverse spin-switch' effect, and we discuss this in the last section.

13.3 Sample Preparation and Ferromagnet Characteristics

The experiments described here on S/F systems with both weak and strong magnets were mostly performed on samples which were prepared by the group in Leiden. They were always prepared by sputtering in a UHV dc diode magnetron sputtering system with a base pressure less than 10^{-9} mbar, in a sputtering argon pressure of 4×10^{-3} mbar. Standard characterization involved Rutherford backscattering and low-angle X-ray reflectivity. We also make use of the anisotropic magnetoresistance (AMR) effect, and it is useful to give a brief description here. The AMR effect (magnitude and sign) depends on the angle between the (measurement) current I and the magnetization M rather than the applied field H_a. In typical transition metal ferromagnets such as $Ni_{80}Fe_{20}$ (Permalloy, Py), the sign of $\Delta\rho = \rho_{\parallel} - \rho_{\perp}$, with $\rho_{\parallel(\perp)}$ the resistance measured in a configuration $I \parallel (\perp) H_a$ (H_a large, M saturated), is positive. Rotating the magnetization in a thin film bridge with an in-plane field and $I \parallel H_a$ leads to domains with $M \perp I$ and therefore a dip in the resistance at the coercive field H_c, while $I \perp H_a$ leads to domains with $M \parallel I$ and therefore a peak in the resistance at H_c.

In most research involving 'weak magnets', meaning materials with low magnetic ordering or Curie temperature $T_{Curie} < 100$ K or atomic moments $\mu_F < 0.2\mu_B/at$ (μ_B is the Bohr magneton), the workhorses have been the alloys $Cu_{1-x}Ni_x$ (with $x \geq 0.5$) and $Pd_{1-x}Ni_x$ (with $x \leq 0.2$). In all cases, the ferromagnetism (meaning the Curie temperature) can be tuned down to (almost) zero by lowering the concentration of the magnetic element. For $Cu_{1-x}Ni_x$, the critical concentration x_{cr} (where $T_{Curie} \rightarrow 0$) is around $x = 0.43$ [34]. The values used in experiments are above $x = 0.5$, where T_{Curie} is roughly 100 K and $\mu_F \approx 0.1\ \mu_B/at$. [35] A complication with $Cu_{1-x}Ni_x$ around x_{cr} is that Ni-segregation leads to the formation of giant moments and therefore an inhomogeneous magnetic state [36]. Another relevant point for CuNi films is that they can easily show perpendicular magnetic anisotropy as found for CuNi/Cu multilayers [37], which may be of importance in the behavior of such films in Josephson junctions. The AMR behavior of CuNi alloys is the standard one for transition metal alloys: dips (peaks) around H_c for $I \parallel (\perp) H_a$. A final issue in all S/F studies is the spin diffusion or spin-flip scattering length ℓ_{sd} in the magnetic film. Such numbers can be extracted from F/N spin-valve structures; for $Cu_{1-x}Ni_x$-alloys, this was only done for concentrations in the paramagnetic regime, where it was found that ℓ_{sd} decreases from 23 nm at $x = 0.07$–7.5 nm at $x = 0.23$ [38], suggesting an even lower value for $x \approx 0.5$. Also, critical current measurements on $Nb/Cu_{47}Ni_{53}/Nb$ Josephson junctions showed a strong suppression with thickness of the F-layer [39], which

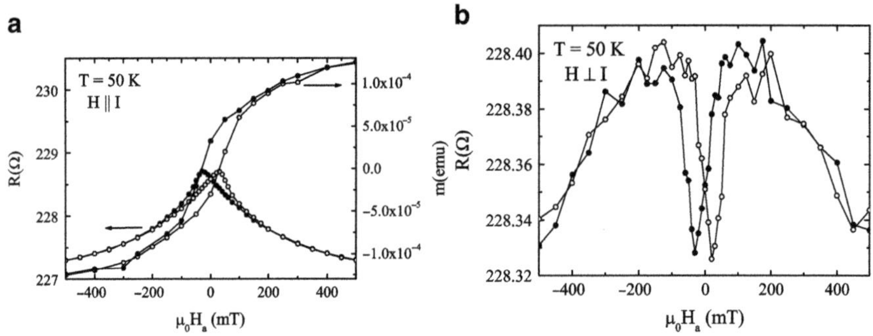

Fig. 13.2 (**a**) Dependence of the magnetic moment m (right-hand scale) and of the resistance R (left-hand scale) on the applied magnetic field H_a for a $Pd_{0.81}Ni_{0.19}$ film. The $R(H_a)$ (magnetoresistance) measurement is performed with the field parallel to the current. (**b**) $R(H_a)$ dependence for the field perpendicular to the current. In both cases, the thickness of the $Pd_{0.81}Ni_{0.19}$ film is 20 nm and the measurements were performed at $T = 50$ K. *Filled symbols* refer to field sweep from positive to negative values, *open* ones to the opposite direction

was interpreted in terms of spin-flip processes in strength equal to the the exchange energy itself, and probably due to the aforementioned inhomogeneities. In that case, ℓ_{sd} would again be of the order of a few nanometers.

For $Pd_{1-x}Ni_x$, there is also a critical concentration of $x = 0.025$, below which the material is not a homogeneous ferromagnet but shows giant magnetic moments [40, 41]. S/F-type experiments are usually performed well above this concentration. For $x = 0.1$, T_{Curie} is roughly 100 K [42], for $x = 0.14$, T_{Curie} is 180 K and $\mu_F \approx 0.17\ \mu_B/at$ [43]. Similar to the CuNi alloys, thin films of PdNi also appear to show perpendicular magnetic anisotropy, as was recently reported for films of $Pd_{0.88}Ni_{0.12}$ [44]. The AMR effect in PdNi is anomalous. In bulk $Pd_{1-x}Ni_x$, $\rho_{\parallel} - \rho_{\perp}$ is *negative* [45, 46], which was ascribed to the formation of local orbital moments on the Ni atoms. In films we find the same behavior, as shown in Fig. 13.2 for a 20 nm film of $Pd_{0.81}Ni_{0.19}$. In the configuration $I \parallel H_a$, the AMR shows a clear peak at H_c (domains with $M \parallel I$), and dips are found for $I \perp H_a$. The spin diffusion length was measured recently for the alloy $Pd_{0.88}Ni_{0.12}$, and again found to be small, $\ell_{sd} \approx 3$–5 nm [47]

By 'strong magnets', we mean ferromagnets where the moment per atom is large (of order 1 μ_B/at), and the exchange splitting between the spin-subbands significant, which also means a significant amount of spin polarization. Here, often used materials are the elemental ferromagnets Fe, Co, Ni, and the alloy $Ni_{80}Fe_{20}$, usually called Permalloy (Py). Much work has been done with these materials, and for bilayer studies we note, e.g., studies on Nb/Fe [48] reporting non-monotonous behavior of $T_c(d_{Fe})$ attributed to the presence of a magnetically dead layer, and a bilayer study on Nb/Ni where a T_c-oscillation was observed [49].

Here, we concentrate on work with Permalloy. Some important properties in the S/F context are the following. First, the coercive or switching fields can be small (order 1 mT) as consequence of the low magnetocrystalline anisotropy [50]. The spin diffusion length $\ell_{sd} \approx 5.5$ nm, similar to that of the weak magnets above, but quite low compared to, e.g., cobalt (60 nm) [51]. Also important, in view of stray fields, is the structure of the domain walls which can be encountered. Two basically different types of wall are the Bloch wall, which separates two domains with in-plane magnetization by an out-of-plane rotation in the wall, and the Néel wall, where the same domains are separated by an in-plane rotation. In Bloch walls, the magnetic charges therefore reside at the surfaces of the film, in Néel walls at the inner surface of the wall. The energy of the Bloch wall (Néel wall) then increases (decreases) with film thickness, and a crossover takes place from the one to the other at some finite thickness. In Permalloy that thickness is around 40 nm [50]. The amount of flux coming out of the wall is different for both types. From numerical simulations, it was found that the magnitude of the stray field above the wall in the Bloch regime slowly increases with decreasing Py thickness until it reaches the order of 10 mT at 40 nm, and then jumps to 40 mT (fourfold increase) at the crossover to the Néel regime, before decreasing to zero upon further decrease of the thickness [52]. The larger amount of stray field over a Néel wall may be found counterintuitive, but it should be rememebered that these walls are broader than Bloch walls. Another point to keep in mind is that the normal component of the field is larger over the Bloch wall.

Below, we use data on bilayers with PdNi to illustrate what can be learnt about interface transparency and the exchange energy of the ferromagnetic layer. Then we use data on bilayers with CuNi and Py to show that the domain state in both cases leads to enhancement of the superconductivity.

13.4 Interface Transparency

The determination of the interface transparency in S/F bilayers with a weak magnet makes use of the particular shape of the dependence of T_c on d_F, which is a consequence of the inhomogeneous order parameter in the F-layer. Opposite to the S/F/S trilayer case, where for certain d_F the coupling leads to a *higher* T_c, destructive interference by the reflection of quasiparticles in the F layer at its vacuum interface leads to a *lower* T_c (a 'dip' in the $T_c(d_F)$ curve), or even to complete disappearance of superconductivity and re-entrance [53, 54]. The depth of the dip in $T_c(d_F)$ then is a measure for the interface transparency. Unfortunately, many more parameters play a role, specifically the superconducting reservoir c.q. the thickness of the S-layer d_S; the bending capability of the order parameter on the S-side, given by the superconducting coherence length ξ_S; and the oscillation wavelength in the F-layer, characterized by $\xi_F = \sqrt{(\hbar D_F)/(E_{ex})}$. In turn, these parameters are determined by the diffusion constants D_X $(X = S, F)$, the resistivities ρ_X, the bulk superconducting transition temperature T_{cS}, and of course the exchange energy E_{ex}. These parameters come together in defining the diffusion of superconducting correlations on both sides of the interface. Apart from that, a boundary resistance (or interface transparency) parameter is needed, called γ_b and defined as $\gamma_b = (R_B A)/(\rho_F \xi_F^*)$.

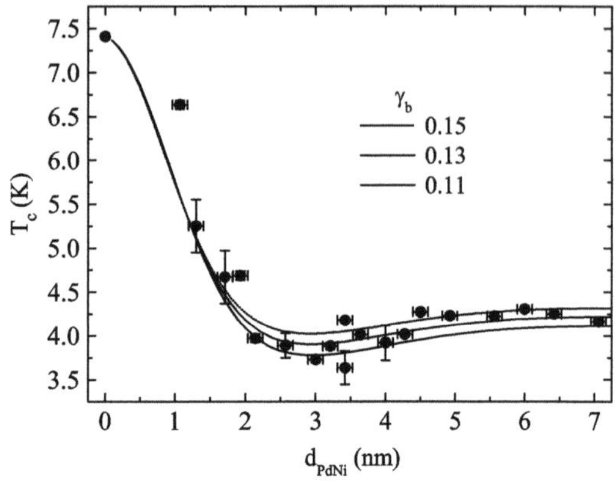

Fig. 13.3 Critical temperature T_c vs. PdNi thickness d_{PdNi} in $Nb/Pd_{0.81}Ni_{0.19}$ bilayers with constant Nb thickness $d_{Nb} = 14$ nm. Different lines (*thick solid, and light solid*) are theoretical fits in the single-mode approximation for $\gamma_b = 0.15$ (*top*), 0.13 (*middle*), 0.11 (*bottom*). Taken from [55]

Here, R_B is the interface resistance, A its area, and $\xi_F^* = \sqrt{(\hbar D_F)(2\pi k_B T_{cS})}$ (k_B is the Boltzmann constant). Note that ξ_F^* has a slightly different meaning than ξ_F. The former measures the above-mentioned basic diffusion, the latter specifically measures the oscillation wavelength. In the full proximity effect model as for instance developed by Fominov et al. [54], ξ_F^* and E_{ex} are used rather than ξ_F. Note also that γ_b is a parameter with values between 0 (perfect transparency) and ∞ (fully reflective), and that spin-flip scattering in the F layer is not taken into account.

To alleviate the problem of many parameters, a point to make here is that, ideally, sets of data on $T_c(d_F)$, taken at constant d_S, and sets of data on $T_c(d_S)$, giving d_S^{cr} and taken at constant d_F, should yield a uniform description. This was demonstrated using $Nb/Pd_{81}Ni_{19}$ bilayers [55]. We shall not repeat the full analysis here, but focus on some pertinent points. Figure 13.3 shows the T_c variation of such a bilayer set for a constant thickness of the Nb layer $d_S = d_{Nb} = 14$ nm with varying $d_F = d_{PdNi}$. The dip in $T_c(d_{PdNi})$, although small, is clearly visible around $d_{PdNi} = 3$ nm. Figure 13.4 shows the T_c variation of a bilayer set for constant thickness of the $Pd_{81}Ni_{19}$ layer $d_{PdNi} = 19$ nm for varying d_{Nb}. Also shown are the variation of T_{cS} with d_{Nb} in single layer Nb films (decreasing with decreasing d_{Nb}), and the variation of $\rho_S = \rho_{Nb}$ with d_{Nb} (increasing with decreasing d_{Nb}). Fitting was performed with the full model of [54], and with a simpler version called the single mode approximation. The fits, given as full lines in Fig. 13.3, show that the behavior is quite sensitive to the choice of γ_b. The best description was obtained for $\gamma_b = 0.13 \pm 0.02$ and $E_{ex} = 230 \pm 20$ K. These same parameters were used to calculate the behavior of $T_c(d_S)$ in Fig. 13.4 which is seen to work very well. Other values can also be extracted. For instance, $\xi_F^* = 6.2$ nm, which is a value comparable to values for ξ_N in dirty normal metals, and from E_{ex} we find $\xi_F = 2.8$ nm (close to the value of d_F at the minimum), which corresponds to an oscillation wavelength $\lambda_F = 2\pi\xi_F = 17.6$ nm, very typical for these weak magnets.

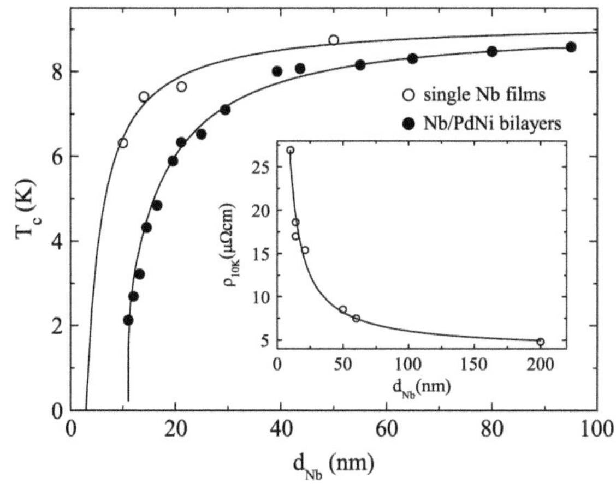

Fig. 13.4 Critical temperature T_c vs. Nb thickness d_{Nb} in Nb/$Pd_{0.86}Ni_{0.19}$ bilayers with $d_{PdNi} = 19$ nm. The *thick line* is the result of the theoretical calculations in the single-mode approximation, using parameter values derived from the fit shown in Fig. 13.3. *Open symbols* refers to single Nb films. The *light line* describes the phenomenological T_c thickness dependence of Nb single films. Inset: thickness dependence of the low-temperature resistivity as a function of the single Nb thickness. The *line* is a guide to the eye. Taken from [55]

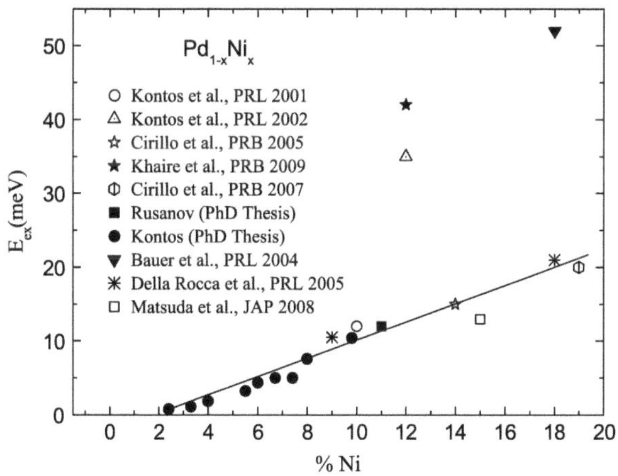

Fig. 13.5 Compilation of values of the exchange energy in the system $Pd_{1-x}Ni_x$ for different values of the Ni-percentage. References are from top to bottom [28, 42–44, 55–60]

The values of γ_b and E_{ex} deserve some more comment. First, the large number of experiments made with $Pd_{1-x}Ni_x$ allows to plot extracted values for E_{ex} as function of x in Fig. 13.5. All values up to 20% Ni fall on a straight line, which is the expected effect for increasing the amount of magnetic ions, and gives some confidence in the procedure. Second, γ_b can be related to an interface transparency T (a dimensionless number between 0 and 1) through the relation

Table 13.1 Values for the boundary resistance parameter γ_b and the transparency T taken from a number of different experiments reported in the literature

Material	γ_b	T	Reference	Remark
$Nb/Pd_{86}Ni_{14}$	0.60	0.39	[43]	
$Nb/Pd_{81}Ni_{19}$	0.13	0.74	[55]	
$Nb/Pd_{89}Ni_{11}$	0.45	0.47	[61]	
$Nb/Pd_{90}Ni_{10}$	7.50	–	[28]	DOS
$Nb/Pd_{88}Ni_{12}$	5.30	–	[42]	J-junctions
$Nb/Cu_{43}Ni_{57}$	0.33	–	[54]	
$Nb/Cu_{48}Ni_{52}$	0.20	0.33	[56]	
$Nb/Cu_{41}Ni_{59}$	0.35	0.25	[56]	
$Nb/Cu_{41}Ni_{59}$	0.25	0.31	[56]	
$Nb/Cu_{50}Ni_{50}$	0.07	0.50	[62]	
$Nb/Cu_{46}Ni_{54}$	1.50	0.22	[63]	
$Nb/Cu_{40}Ni_{60}$	0.57	–	[64]	
$Cu_{47}Ni_{53}/Nb/Cu_{47}Ni_{53}$	0.60	0.29	[65]	Tagirov theory
$Nb/Cu_{50}Ni_{50}$	0.02	–	[66]	DOS
$Nb/Cu_{40}Ni_{60}$	–	0.46	[67]	Tagirov theory
$Nb/Cu_{41}Ni_{59}$	–	0.40	[68]	Tagirov theory

Most are from proximity effect data, one from a Josephson-junction experiment, and two from a density-of-states (DOS) measurement using planar tunnel junctions, as noted. Some values for T were obtained through a version of the quasiclassical theory developed by Tagirov [53], also noted

$$\gamma_b = \frac{2}{3}\frac{\ell_F}{\xi_F^*}\frac{1-T}{T} \qquad (13.2)$$

with ℓ_F the elastic mean free path in the F-metal. In this way, and assuming a thickness limited mean free path of (on average) 3.5 nm, we find $T = 0.74$ for the example above. This is a relatively high but still typical value also for S/N interfaces. However, preparation procedure and materials clearly must play a role. In Table 13.1, we have collected a number of values from different experiments and publications. For the PdNi system, higher values for γ_b (lower T) were found for samples prepared in a different deposition system [43, 61]. For the CuNi system, transparencies are generally lower than found in the PdNi system. A note to make here is that some of the quoted values were obtained from a slightly different version of the quasiclassical theory developed by Tagirov [53]. The transparency parameter T_F in that theory is somewhat differently defined, and corresponds to our definition through $T = T_F/(1 + T_F)$ (see, e.g., [69]). The numbers still suggest that the $Nb/Cu_{40}Ni_{60}$ samples reported in [67, 68] do have highly transparent interfaces, and that that is (part of) the reason that true re-entrance is found for them in $T_c(d_F)$.

Also of interest is the fact that values for both γ_b and E_{ex} extracted from tunnel- or Josephson junctions tend to be different, with less transparent interfaces and higher exchange energies. For PdNi, some uncertainty exists about the value for v_F, as discussed in [44]. There is another caveat to be made, however, which has to do with the fact that spin-flip scattering has not been taken into account. The effect of such magnetic scattering is that the order parameter does not behave as in 13.1, but that the oscillation wavelength *increases* while the damping length *decreases* with increasing magnetic scattering [70], leading to the extraction of different parameter values. This can be illustrated on various experiments involving $Cu_{1-x}Ni_x$ alloys. Proximity effect analysis on bilayers found 11 meV ($x = 0.57$) [54] and 12 meV ($x = 0.59$)[56]. More recently, measurements on Josephson junctions were analyzed taking spin-flip scattering into account. A typical value for γ_b was found of 0.52, but E_{ex} came to 75 meV ($x = 0.53$) [39], which seems an overestimation given the value of T_{Curie} of about 100 K. Proximity effect data have not been analyzed in such a manner, and although their descriptions appear to give consistent and physically reasonable numbers, the possible effects of small spin diffusion lengths (of the order of ξ_F) remain as an issue to be investigated.

13.5 Domain Walls in S/F Bilayers

In the first part of this review, we have assumed that the magnetization and therefore the exchange field in F-layer is homogeneous. In zero applied magnetic field, this will usually not be the case. Cooling down the sample from its virgin state leads to a domain state of the magnetic layer; saturating the magnetization and removing the applied field again lead to the remanent state, which is also a domain state, but a different one than the virgin state. There are several reasons why the question of the influence of this inhomogeneous exchange field on the superconductivity is relevant. One reason comes from the concept of the superconducting spin switch, already mentioned above. The spin switch structure consists of an F_1/S/F_2 sandwich, in which d_S is of the order of ξ_S, and in which the magnetizations of the layers $F_{1,2}$ can be rotated independently to obtain parallel (P) or antiparallel (AP) magnetization configurations [29, 30]. Technically, this usually involves an extra (antiferromagnetic) layer on one side, so that one magnetization direction is pinned while the other can rotate freely. The underlying physics is that the Cooper pair samples the two different exchange fields simultaneously, and that parallel fields break the pair (which is a combination of spin-up and spin-down) more strongly than antiparallel fields. On the same general grounds, it can be expected that the varying exchange fields directions in a domain wall, or the difference which exists between these directions in adjacent domains, leads to a relative enhancement of superconductivity. To find out what the effect of domain walls may be even in the F_1/S/F_2 spin switch sandwich, S/F bilayers can provide an answer. There is a second reason to be interested in domain walls, which is that the inhomogeneous exchange field also can give rise to the so-called odd-frequency pairing phenomenon, which can induce superconducting correlations of a (much) longer range in the ferromagnet than set by ξ_F [71,72]. In S/F/S structures, domain walls therefore could enhance the coupling. A general issue in discussing the role of domain walls is to demonstrate that they are present. The most straightforward way to do this is to measure the AMR effect in the normal state. As explained earlier, AMR is sensitive to the angle between the local magnetization M and the (measurement) current I, and the occurrence of domains will in general lead to a change in resistance which tracks the hysteretic loop measured in magnetization.

13.5.1 Domain Walls in $Nb/Cu_{43}Ni_{57}$

We now illustrate using bilayers of $Nb/Cu_{43}Ni_{57}$ how measurements of $R(H_a)$ above T_c and in the resistive transition can be combined to draw a conclusion of the influence of domains on the emerging superconducting state. In order to have a well-defined measurement geometry, we structured such bilayers into long and narrow bars on which we put Au contacts, as shown in Fig. 13.6 for a bar of 40 μm × 2 μm. Using Au contacts requires a second lithography step, but this is preferred over etching out the bar with contacts included, since stray fields from the contacts then can influence the field dependent measurements in an unwanted way [73]. The length of the bar likewise precludes that stray fields from the end of the bar have a significant influence on the voltage measured between the contacts in the middle. In this case, the bilayer consisted of $Si/Nb(20)/Cu_{43}Ni_{57}(10)/Nb(2)$, with the numbers denoting the thickness of each layer in nm. The capping Nb layer is to prevent oxidation of the underlying CuNi layer. The value of T_c of the sandwich is about 6.5 K, suppressed by about 1 K from the value a single Nb film would have (Fig. 13.3).

Results of $R(H_a)$ on this strip, with H_a oriented along the long axis of the bar, are shown in Fig. 13.7 for different temperatures between 9.0 K and 7.5 K, which is above T_c. The typical AMR signature is visible as dips in the $R(H_a)$ curve with a typical resistance variation of the order 10^{-4} and a minimum value reached around ±24 mT. Domains therefore are apparently present in the bar, and with a dimension smaller than the 2-μm width. This seems reasonable in view of recent results obtained with a Bitter decoration technique on thin films of $Cu_{47}Ni_{53}$ in which a characteristic scale of the domain structure was found to be 100 nm [74]. Next, we show $R(H_a)$ data taken between 6.6 K and 6.4 K, which is in the resistive transition (Fig. 13.8). We find very similar dips as above T_c, with two notable differences. One is that the field of the minimum in $R(H_a)$ is around ±10 mT (-10 mT when

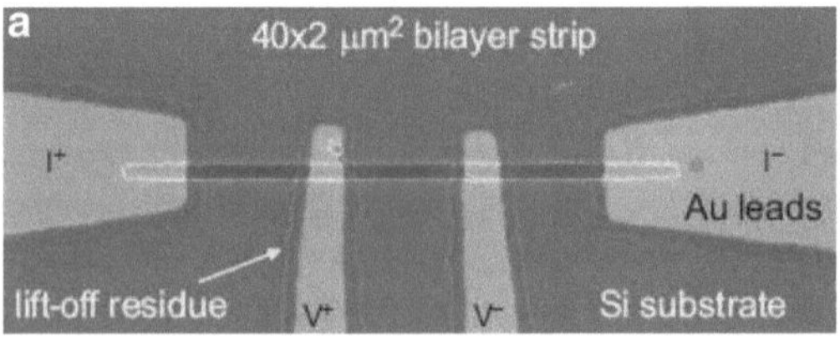

Fig. 13.6 Electron microscope image of a 40 μm × 2 μm bilayer device. The small bar is the $Nb/Cu_{43}Ni_{57}$ bilayer; both current- ($I^{+,-}$) and voltage-contacts ($V^{+,-}$) are made from Au

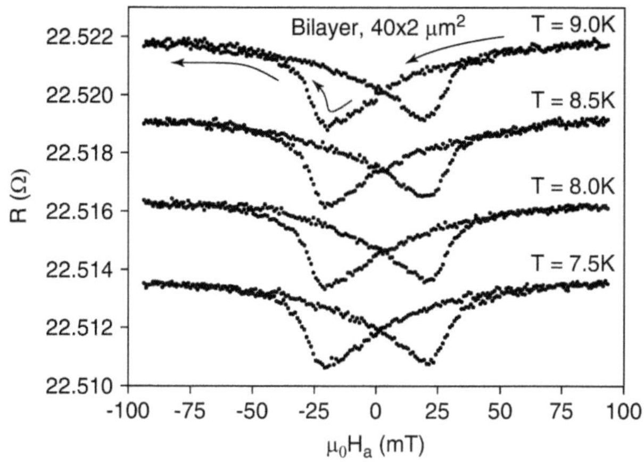

Fig. 13.7 Resistance R vs. in-plane applied field H_a on a 40 μm × 2 μm Nb/$Cu_{43}Ni_{57}$ bilayer strip at several temperatures above the superconducting transition. *Arrows* in the top curve show the direction of the field sweep starting from high fields. Taken from [73]

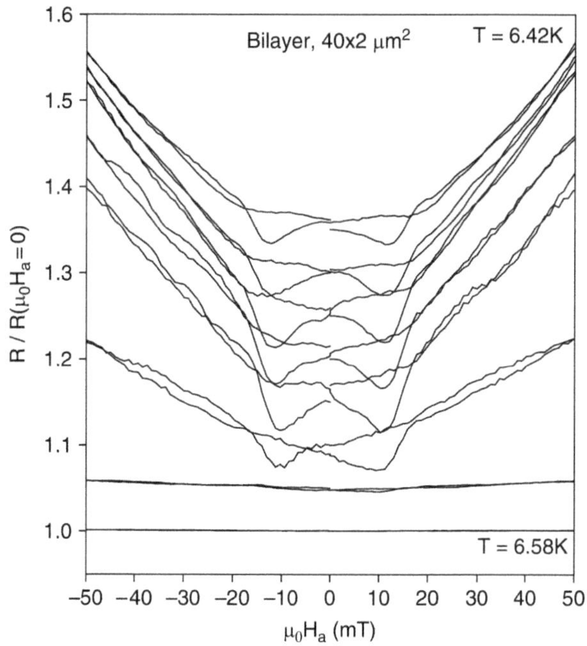

Fig. 13.8 Resistance normalized to the value at zero field of a 40 μm × 2 μm Nb/$Cu_{43}Ni_{57}$ bilayer strip at several temperatures in the transition curve. The *bottom curve* is for $T = 6.58$ K. Each subsequent curve is shifted by +0.05 with respect to the previous one, with corresponding temperatures 6.55, 6.52, 6.50, 6.48, 6.46, 6.44, and 6.42 K (*top curve*). Taken from [73]

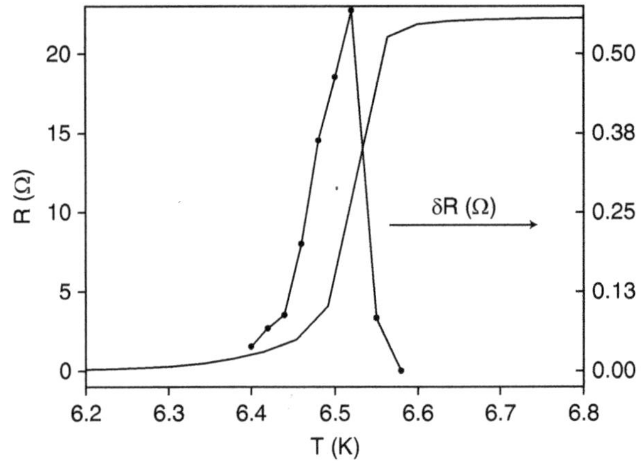

Fig. 13.9 Temperature variation of the resistance of the 40 μm × 2 μm Nb/$Cu_{43}Ni_{57}$ bilayer strip in the transition (left-hand scale). Also shown is the temperature variation of the magnitude of the resistance dip (right-hand scale). 40 μm × 2 μm Nb/$Cu_{43}Ni_{57}$ bilayer strip. Taken from [73]

coming from positive H_a), significantly smaller than found earlier. The other is that the magnitude of the signal is of order 10^{-2} and therefore significantly larger than the AMR signal. The lower resistance in the dip signifies enhanced superconductivity, and the fact that the behavior as a whole reflects the hysteresis loop of the magnet allows to conclude that the domain state enhances the superconductivity. The amount of enhancement in terms of the variation in T_c can be estimated by estimating the resistive change δR due to the domains and comparing that number to the steepness of the $R(T)$-transition. For δR, we take the difference of the value of R in the down-going branch of the field sweep at the field of the dip and the value of R at the same field for the up-going branch, where presumably no influence of domains is yet present. In Fig. 13.9, this difference is plotted together with the $R(T)$-transition. The transition width is of the order of 100 mK, and from $\delta T_c = \delta R / \frac{\partial R}{\partial T}$ it follows that δT_c is a few mK. Very similar values were found in the reports on superconducting spin valve effects with CuNi as magnetic material [31, 65], as will be discussed below.

13.5.2 *Domain Walls in Nb/Py*

Work on the effect of domain walls in Nb/Py preceded the study on Nb/CuNi presented above. Shown in Fig. 13.10 (taken from [75]) is a $R(H_a)$ measurement on a sample *s*/Py(20)/Nb(21), where *s* stands for the (Si) substrate The temperature is just below the middle of the resistive transition and this sample is structured as a large bar of 0.5 mm × 4 mm. Also important here is that no magnetic sample holder was used during the sputter deposition. Nonetheless, magnetic fields

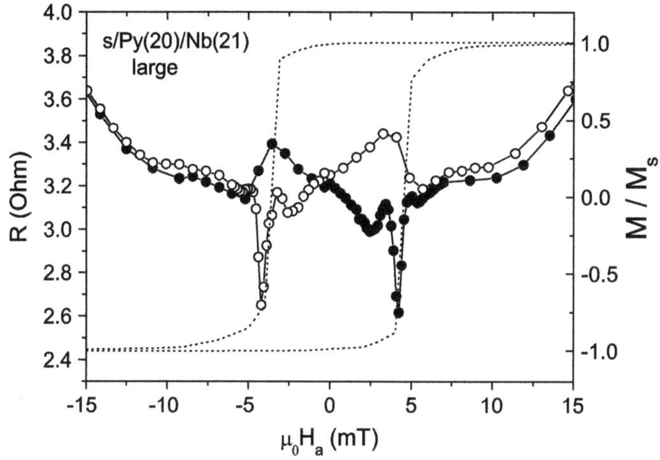

Fig. 13.10 Left-hand scale: resistance R vs. applied field H_a of a 4 mm × 0.5 mm sample s/Py(20)/Nb(21). *Filled symbols* are in the positive field (forward) direction, *open symbols* in the backward direction. Right-hand scale: magnetization M (*dotted lines*) normalized on the saturation magnetization M_s vs. H_a, measured at 8 K. In both cases, $H_a \parallel \hat{e}_e$, the easy axis of magnetization. Taken from [75]

from the sputter guns were present, which induced an easy axis of magnetization $\hat{e}_e$. By mapping out the direction of $\hat{e}_e$ in the Py film with magnetization measurements, the long axis of the bar could be aligned with the easy magnetization axis. Also shown in Fig. 13.10 is a magnetization measurement at 8 K, with H_a along $\hat{e}_e$, which shows a very square loop with sharp switching of the domain state. This switching is fully reflected in $R(H_a)$, which shows deep dips with widths of not more than 1 mT. Similar behavior was found for a sample with the S- and F-layers inverted, *s*/Nb(19)/Py(20). Two points stand out in these measurements. One is that the switching is very sharp. The other is that, apparently, the effect of stray fields is virtually absent, since no resistance increase is seen at the coercive fields. That this can be different is illustrated by another set of measurements. Figure 13.11 shows data on two bilayers *s*/Nb(50)/Py(20)/Nb(2) (called BL20) and *s*/Nb(50)/Py(50)/Nb(2) (called BL50), both above T_c and in the transition, and for large (2 mm × 20 μm) and small (40 μm × 4 μm) structures. These samples were sputtered slightly differently than the earlier set. A protection layer of 2 nm Nb was deposited to prevent the Py layer from oxidizing, and now a magnetic sample holder was used with a defined stray field pattern to induce an easy axis along the long axis of the substrate. No further alignment was done, however, in the lithography step. The data above T_c for the large structures show the usual AMR effect, and make clear that in these samples the switching is not as sharp as in the earlier ones. In the transition, the behavior is also different now. Instead of the sharp dips, there is no signal for BL50, and two peaks for BL20, which in their position and width, look

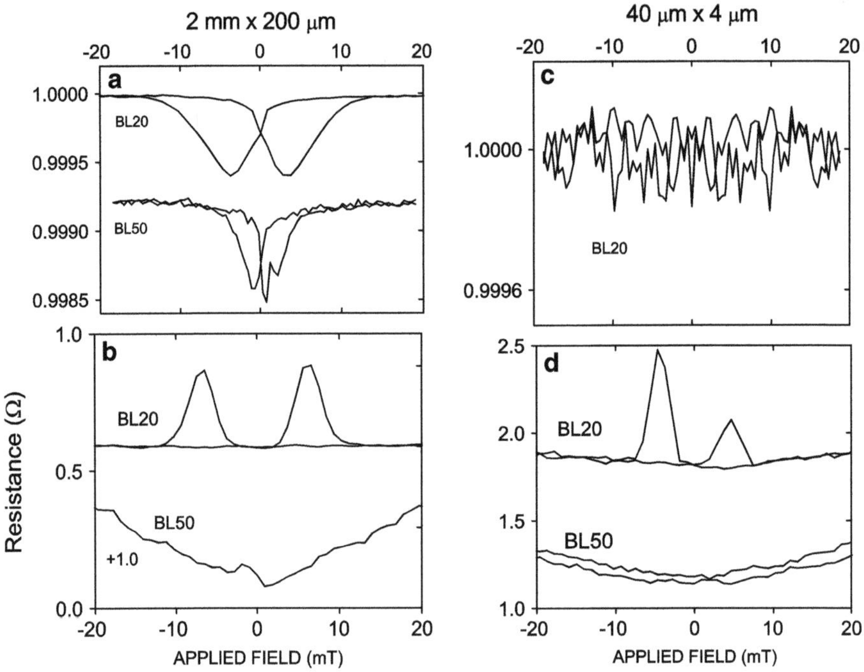

Fig. 13.11 Resistance vs. applied field at temperatures above and in the transition. Shown are two different bilayers (BL20 = s/Nb(50)/Py(20)/Nb(2), BL50 = s/Nb(50)/Py(50)/Nb(2)) structured into two different devices as indicated. (**a**) *large structure*, 10 K (normalized resistance); (**b**) same, in transition; (**c**) *small structure*, 10 K (normalized resistance); (**d**) same, in transition

similar to the dips in the AMR data. The domain state now apparently produces an increase in R rather than a decrease. We assume this is due to stray fields coming from the domain structure. Noting that for this thickness of the Py layer we can expect Néel walls, the stray fields have strong in-plane components which suppress the superconductivity, but do not necessarily form vortices. Comparing the earlier and the later data, it appears that the different preparation and alignment procedures lead to different domain configurations, and that the effects of T_c enhancement can be easily masked by stray fields. For the small structures, no AMR could be measured, but in the transition two peaks are found for BL20, which are very similar to that seen for the large structure. For BL50, no signal is measured in the transition. The explanation here is probably that in this thicker Py layer Bloch walls occur, with a smaller amount of stray fields [52]. It is interesting to compare these observations to recent work of Rusanov et al. [76]. They made another comparison of two bilayer sets Py(20)/Nb(20) and Nb(20)/Py(50), both structured into small bars with contacts included in the structure, and the easy axis of magnetization aligned with the long axis of the bars. The Py(20) sample showed a sharp *negative* peak in the magnetoresistance in the transition, but in the Py(50)-sample a sharp *positive* peak.

They attributed the difference to the normal component of the flux being stronger in the Bloch wall, giving rise to vortices and flux flow. Combining their observations with our own data, it appears that, in the Néel wall regime, the alignment of magnetic field direction, magnetization easy axis, and sample long axis all play a crucial role in determining the domain wall configuration and the amount of stray field. Correct alignment, with a precision of the order of degrees, minimizes stray fields and make it possible to observe the domain wall-enhanced superconductivity. This should therefore be simpler in thin Py layers. What stands out here is that these conclusions still await direct confirmation from measurements of the evolution of the low-temperature magnetic domain structure (for instance, by magnetic force microscopy) in such samples under magnetization reversal.

13.6 On the Superconducting Spin Switch

In the framework of research into S/F hybrids, much attention has been given to the concept of the superconducting spin switch and its experimental realization. The device consists of an F/S/F trilayer with two homogeneous ferromagnets. Both banks suppress superconductivity, but due to the nonlocal nature of the Cooper pair, the total suppression also depends on the mutual orientation of the magnetization in the two F layers [29, 30], with the parallel (P) state having a T_c^P significantly lower than that of the antiparallel (AP) state T_c^{AP}. The effect obviously requires d_S to be of the order of ξ_S and occurs for any thickness of the F-banks. It can be considerably boosted by choosing $d_F \approx \lambda_F/4$ [54], with $\lambda_F = 2\pi\, \xi_F$ the wavelength of the oscillation of the order parameter in the banks. In this way, true re-entrance can be obtained theoretically, in which superconductivity is fully suppressed for the P-case, but present in the AP-case. If the magnetization of one of the layers can be switched with a small magnetic field, this in some sense represents infinite magnetoresistance. A variant on the device was even earlier proposed by Oh et al., who calculated the behavior of an $F_1/N/F_2/S$ sandwich, and also found switching and reentrance effects [77]. Experimentally, there are several hindrances. One is that the condition $d_S \approx \xi_S$ requires good thickness control, since it is almost by definition met only near d_S^{cr}, as can be seen from, e.g., Fig. 13.3. The steepness of $T_c(d_F)$ means that not only thickness but also the roughness now is important; as is the interface transparency, since a low value will be detrimental to the reentrance effect. Also, the magnetization of one of the F-layers needs to be switched while the other remains unchanged. This can be done with an antiferromagnetic pinning layer such as FeMn, or using shape anisotropy. In theory then, and with sufficient control over growth, a value of $\Delta T_c = T_c^{AP} - T_c^P$ of several Kelvin could be possible. Instead, the values reported in a variety of experiments are rather in the range of 10 mK. We discuss these first, making a distinction between weakly magnetic CuNi and strongly magnetic Py.

13.6.1 Spin Switch Effects with CuNi

Successful realization of spin switch effects was reported by Gu et al. [31], and by Potenza and Marrows [65]. They used unstructured samples and alloys with a concentration close to $Cu_{50}Ni_{50}$ (here called CuNi). A problem with CuNi is that in large samples the switching of the magnetization is not sharp. To overcome this problem, two outer layer of Py were added. It is supposed that the fast switching magnetization of the Py layer also rotates the magnetization of the CuNi layer through exchange coupling. One Py layer then was pinned with an FeMn layer. A full structure typically looked like s/F_1/F_2/Nb(18)/F_2/F_1/FeMn(6), with F_1 = Py(4) and F_2 = CuNi(d_{CuNi}), a variable thickness of the CuNi layer d_{CuNi}. Bringing such a device in the P state or in the AP state resulted in a ΔT_c of maximally 3 mK at $d_{CuNi} = 7$ nm [31] (transition temperature 2.8 K). Several comments can be made on this result. Apart from ΔT_c being much lower than predicted, also the maximum is reached at higher d_{CuNi} than expected. It does not appear therefore that the interference effects in the oscillatory order parameter play a role. This is not too surprising, since the vacuum interface which should take care of reflections is also absent, and replaced by a CuNi/Py interface. Cooper pairs arriving at this interface could actually undergo stronger dephasing. Also, in unstructured samples it is unlikely that the thickness variations are small enough to preclude detrimental averaging effects. Furthermore, the value of a few mK is equal to what we found above as the effect of domains in a CuNi layer of similar thickness (10 nm). In spin switch devices, not much attention was yet paid to the question of domain formation. We recently prepared a spin switch device equal to the one mentioned above (s/Py(4)/CuNi(10)/Nb(18)/CuNi(10)/Py(4)/FeMn(6)), and measured magnetization and resistance as shown in Fig. 13.12. The magnetization measured at 10 K behaves fully as required. Coming from high fields, the free F layers switch at less than a mT below zero field, after which a plateau sets in (the AP state), and saturation is only reached beyond 150 mT. In sweeping the field back up, hysteresis is observed and the switching back to the P state is again around zero field. The transition temperature, around 2.7 K, is also very close to what was reported earlier. The inset in Fig. 13.12 shows $R(T)$ measurements in the P state (at 30 mT), and the AP state (at -30 mT), which yield a surprisingly large difference of 30 mK, ten times more than observed previously. Quite probably, small roughness has helped to give this large effect. Equally interesting, however, is the AMR measurement performed at 10 K in a geometry with $H_a \parallel I$ and shown in Fig. 13.12b. Around zero field, a clear downward jump in R is seen, but it is not the spike expected for a sharp and full switching of the magnetization. Instead, R stay low before slowly climbing up. Sweeping back up from below -150 mT reveals clear hysteresis, and again a downward jump at zero field. The difference in resistance in the negative quadrant between the down sweep and the up sweep appears to be due to a domain state which is set up when the device is switched into the AP state. It seems safe to conclude that the observed effects are at least mixtures between domain effects (which can be thought of as in-plane P/AP configurations) and cross-layer spin switch effects,

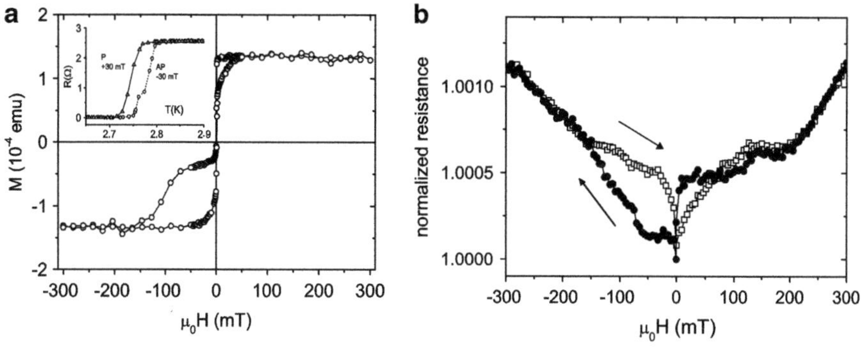

Fig. 13.12 (**a**) Magnetization M vs. applied field H_a at 10 K for a multilayer s/Py(4)/CuNi (10)/Nb(18)/CuNi(10)/Py(4)/FeMn(6). The rotations of the free Py layer around 0 field, and of the pinned Py layer around −100 mT are clearly visible. Inset: resistance R vs. temperature T around the superconducting transition, measured in a field of 30 mT (the Parallel (P) state), and in a field of −30 mT (the Antiparallel (AP) state). (**b**) R vs. H_a at 10 K (above the superconducting transition). The resistance is normalized to the value at zero field

although the observation of a 30 mK difference at 2.7 K seems to be outside what can be expected from domain structures alone.

13.6.2 *Spin Switch Effects with Py*

Spin switch effects also have been reported for devices with strong ferromagnets, in particular with Ni [33] and Py [78], using Nb as the superconductor. In both cases, samples were unstructured, and use was made of an FeMn pinning layer. By tuning the Nb thickness $d_{\rm Nb}$ very close to d_S^{cr}, in the Ni case values for ΔT_c were 41 mK at $d_{\rm Nb} = 17$ nm (transition around 0.34 K), and for the Py case 20 mK at $d_{\rm Nb} = 20.5$ nm (transition around 0.39 K). For these samples, increasing $d_{\rm Nb}$ by only 1 – 2 nm resulted in the decrease of ΔT_c to a few mK, reinforcing the point that the switching effects are very sensitive to the S layer thickness. Magnetic domain effects were not investigated, but it seems clear that stray fields had no effect, even though the F layers with were not particularly thin (7 nm Ni, 8 nm Py). The reason for this must lie in the use of the FeMn pinning layer, notwithstanding the fact that we showed in the CuNi case above that domain structures are probably present.

Without such a pinning layer, a series of reports has shown that magnetization switching yields an opposite effect, namely an *increase* in R when going out of the P state. Without a pinning layer, shape anisotropy has to be used to switch the F layers separately, specifically using two different thicknesses. In this way, for a series of microstructured (3 μm × 20 μm) samples of s/Py(50)/Nb($d_{\rm Nb}$)/Py(20) (in thickness similar to the ones above), very sharp switching to higher resistance was found at the switching field of the 50 nm Py layer (1.5 mT) and sharp switching at that of the

20 nm layer (9.5 mT), appearing as a 'reverse effect' in an inferred AP regime [79]. Subsequent work by different groups on unstructured samples with various strong ferromagnets (combination of Co and Fe [80]; Py [76, 81]; Co [82] indicated that stray field effects do play a role in these experiments. The clearest evidence for the difference between presence and absence of a pinning layer came recently from Zhu et al. [83], who showed, again on the Nb/Py system, that in an unstructured sample including a FeMn pinning layer, both a P state, an AP state, and a 'D' (domain) state could be invoked, with the AP state showing a (mK) higher T_c, and the D state a (mK) lower T_c. They also made the point that the domain walls (Néel walls in their sample) may be coupled, as is known to happen in F/N/F structures [84]. Coupling of domain walls is probably what also gives rise to the block-like switching in s/Py(50)/Nb(d_{Nb})/Py(20). To investigate this further, we prepared a set of small (40 μm × 1 μm) structures of trilayers s/Py(50)/Nb(50)/Py(20) and measured MR above and in the transition as shown in Fig. 13.13, where several measurement traces are shown, taken at the same temperatures.

What is striking here is that, above T_c, the (A)MR shows block-like dips in R. The small structure apparently allows only very few domain states, which are not always exactly equal. The onset is always close to 5 mT (the switching field of the 50 nm layer), and the return to zero between 10 mT and 15 mT, where the 20-nm layer would switch. The first layer switch appears to drive both layers into a domain state which only disappears when the other layer cannot sustain that state any more.

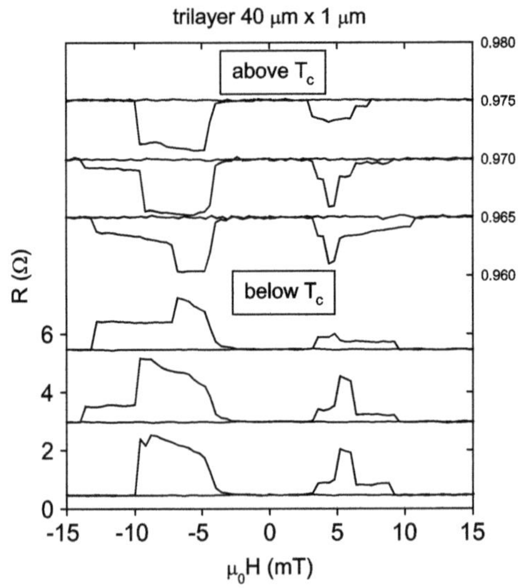

Fig. 13.13 Comparison of the resistance R as function of the in-plane applied field H_a on a 40 μm × 1 μm Py(50)/Nb(50)/Py(20) structure between temperatures on the low-end of the transition curve (Nb in superconducting state) and above the transition temperature (Nb in normal state). All curves have a different magnetic history. Left-hand scale: results for $T < T_c$ with curves repeatedly shifted by +2.5 as indicated. Right-hand scale: results for $T > T_c$

Also remarkable is that the superconducting state mimics the block pattern of the AMR state, but with increased R. Even in this small strip domains are present, and the stray fields from the coupled walls are visible in the superconducting state.

Concluding this section on the spin switch, research in the last few years has made a number of points clear. Both for weak and strong ferromagnets, domain walls can lead to enhancement of superconductivity, with effects up to 10 mK. All spin valve devices show $\Delta T_c = T_c^{AP} - T_c^{P} > 0$ if stray field effects can be avoided. This is effectively done using an antiferromagnetic pinning layer, although domain walls may still be present and their effects on the superconductivity cannot be excluded yet. The value of ΔT_c is still quite low, with the highest values less than 40 mK. The question then arises whether there is still room for improvement. As starting point, it would be optimal to use a system where reentrance at least in the bilayers already was demonstrated, as is the case for the CuNi alloys. As shown by Sidorenko et al., utilizing strips from a wedge-shaped grown sample, clear T_c oscillations could be obtained for the case of Nb/Ni [49] and full re-entrance in Nb/CuNi bilayers [67]. Their experiments demonstrate that there is still gain to be made by thickness control. In unstructured samples with sizes of millimeters as have mostly been used, thickness averaging effects are unavoidable. Another possibility is furnished by the system V/Fe system, where the low miscibility of the two elements leads to very sharp interfaces, and re-entrance behavior was found in trilayers Fe/V/Fe when varying d_V [85]. However, the magnet strongly suppresses Andreev reflections, and the value of $d_S^{cr}/\xi_S \approx 3$ becomes large (see Sect. 13.2), which makes the system less suitable for spin switch effects. It was not tried to prepare a switching geometry. Also, mention should be made of the multilayer device investigated by Westerholt et al. [86]. This was an $S/F_1/N/F_2$ geometry with the N layer such that the interlayer exchange coupling between F_1 and F_2 is antiferromagnetic. Using V as superconductor and an $[Fe_2V_{11}]_{20}$ superlattice block as the antiferromagnetic stack, they showed that the AP state (the zero field state of the device) has a significantly (order 100 mK) higher T_c than what could be extrapolated for the P state. Although the device cannot be switched, it clearly hints at the possibility for higher values of ΔT_c. A final remark concerns the role of spin polarization P_s. Although it is clear now that with ferromagnets such as Py (P_s of order 50%) the AP state has the higher T_c, the argument can still be made that for high P_s this is opposite, since the crossed Andreev reflection process, in which the Cooper pair is broken by simultaneous accommodation of both parts in the F banks, is possible in the AP case but not in the P case. This is still a point to research both theoretically and experimentally.

13.7 Concluding Remarks

In this contribution, we have tried to sketch a part of the current research scene in metallic S/F proximity systems. Much progress has been made, although some questions have not been answered in a fully satisfactory way. In the case of the basic proximity effect description of an S/F bilayer, this certainly pertains to the role of spin-flip scattering. In the case of the superconducting spin switch, there is still the promise of much larger effects than have been reported until now, although it also has to be remarked that these will be confined to a region close to T_c, where the superconducting order parameter is still weakly developed. For strong current switching effects, the device looks less promising. What has been done in the process is to begin distinguishing the effects of inhomogeneous versus homogeneous exchange fields. In this direction, however, much is still to be learned. Especially the theoretical work of Bergeret et al. [71] and Kadigrobov et al. [72] pointed out another consequence of the presence of inhomogeneous exchange fields such as present in domain walls. They can generate and sustain so-called odd-frequency triplet correlations with a long-range character: the equal-spin triplet component, once formed, is not broken in the ferromagnet by the exchange field, so its coherence length is proportional to $\sqrt{(\hbar D_F)/(k_B T)}$ rather than to $\sqrt{(\hbar D_F)/E_{ex}}$. Triplet correlations in or through domain walls have not yet been explicitly demonstrated, partly due to the difficulty in controlling the wall characteristics. Of course, the question is also relevant for the description of the behavior of S/F/S junctions, where the question of the domain configuration is usually not explicitly answered. It may well be that aside from spin-flip scattering, the existence of domain walls in the junctions has effects on the critical currents.

Answering such questions is resulting in the shift of the field to the research of triplet correlations in general, since they may appear not only through inhomogeneous exchange fields but also through differences in spin scattering at the S/F interface [87], or through noncollinear magnetizations between two ferromagnets, as in an S/F_1/F_2/S configuration [88]. The first effect was probably observed in devices made with the halfmetallic ferromagnet CrO_2 [89], where a supercurrent was reported to flow through the magnet between two superconducting contacts up to a μm apart. Around the same time, it was reported that ferromagnetic Holmium wires up to a length of 150 nm could sustain superconducting conductance oscillations in an Andreev interferometer configuration [90]. The reports from 2006 did not yet lead to general acceptance of the basic physics behind odd frequency triplets, or the existence of long-range proximity effects in ferromagnets. During the preparation of this manuscript, however, the experimental scenery has started to change, with several reports demonstrating such long-range effects. One concerned a device in which the F layer is a combination of PdNi and Co, with a supercurrent induced in the Co over more than 20 nm [91]; another showed zero resistance induced in single crystalline Co nanowires with a length of 1.5 μm [92]; one used a Ho/Co/Ho sandwich as the central F-layer and also found supercurrents up to a Co thickness of at least 20 nm [93]; and finally also in CrO_2, supercurrents were found again extending over a μm range [94]. There is little doubt that the next few years will see much progress in resolving the issues around the existence of odd frequency triplet correlations and the resulting long-range proximity effects induced by superconductors in ferromagnetic materials.

Acknowledgments Many colleagues have contributed to discussions involving this work, but we especially want to mention A. Buzdin, A. Golubov, M. Kupriyanov, S. Prischepa, A. Rusanov, V. Ryazanov and L. Tagirov. This research was supported in part by a research program of the Stichting F.O.M., which is financially supported by the Dutch national science foundation NWO. Scientific information exchange was enabled by the ESF program "Nanoscience and Engineering in Superconductivity – NES".

References

1. A.I. Buzdin, Rev. Mod. Phys. 77, 935 (2005)
2. F.S. Bergeret, A.F. Volkov, K.B. Efetov, Rev. Mod. Phys. 77, 1321 (2005)
3. I.F. Lyuksyutov, V.L. Pokrovsky, Adv. Phys. 54, 67 (2005)
4. A.Y. Aladyshkin, A.V. Silhanek, W. Gillijns, V.V. Moshchalkov, Supercond. Sci. Techn. 22, 053001 (2009)
5. J.J. Hauser, H.C. Theuerer, N.R. Werthamer, Phys. Rev. 142, 118 (1966)
6. P.G. De Gennes, E. Guyon, Phys. Lett. 3 168 (1963)
7. N.R. Werthamer, Phys. Rev. 132, 2440 (1963)
8. A.A. Abrikosov, L.P. Gorkov, Sov. Phys. JETP 12, 1243 (1961)
9. F. Meunier, J.P. Burger, G. Deutscher, P.A. van Dalen, Phys. Lett. 24A 722 (1967)
10. S.T. Ruggiero, T.W. Barbee Jr., M.R. Beasley, Phys. Rev. Lett. 45, 1299 (1980)
11. I. Banerjee, Q.S. Yang, C.M. Falco, I.K. Schuller, Phys. Rev. B 28, 5037 (1983)
12. H.K. Wong, B.Y. Jin, H.Q. Yang, J.B. Ketterson, J.E. Hilliard, J. Low Temp. Phys. 63, 307 (1986)
13. H. Homma, C.S.L. Chun, G.-G. Zheng, I.K. Schuller, Phys. Rev. B 33, 3562 (1986)
14. P. Koorevaar R. Coehoorn, J. Aarts, Physica C 248, 61 (1995)
15. G. Eilenberger, Z. Phys. 214, 195 (1968)
16. L. Usadel, Phys. Rev. Lett. 25, 507 (1970)
17. Z. Radović, L. Dobrosavlević-Grujić, A.I. Buzdin, J.R. Clem, Phys. Rev. B 38, 2388 (1988)
18. A.I. Buzdin, M.Y. Kupriyanov, JETP Lett. 52, 487 (1990)
19. A.I. Buzdin, M.Y. Kupriyanov, B. Vujicic, Physica C 185, 2025 (1991)
20. Z. Radović, M. Ledvij, L. Dobrosavljević-Grujić, A.I. Buzdin, J.R. Clem, Phys. Rev. B 44, 759 (1991)
21. A.V. Andreev, A.I. Buzdin, R.M. Osgood, Phys. Rev. B 43, 10124 (1991)
22. P. Koorevaar, Y. Suzuki, R. Coehoorn, J. Aarts, Phys. Rev. B 49, 441 (1994)
23. M.J.M. de Jong, C.W.J. Beenakker, Phys. Rev. Lett. 74, 1657 (1995)
24. J. Aarts, J.M.E. Geers, E. Brück, A.A. Golubov, R. Coehoorn, Phys. Rev. B 56, 2779 (1997)
25. J.S. Jiang, D.Davidović, D.H. Reich, C.L. Chien, Phys. Rev. Lett. 74, 314 (1995). Some uncertainty still exists whether the T_c variation reported here is due to π-coupling or due to onset of ferromagnetism; see the discussion in [24]
26. Y. Obi, M. Ikebe, H. Wakou, H. Fujimori, J. Phys. Soc. Jpn. 68, 2750 (1999)
27. V.V. Ryazanov, V.A. Oboznov, A.Yu. Rusanov, A.V. Veretennikov, A.A. Golubov, J. Aarts, Phys. Rev. Lett. 86, 2427 (2001)
28. T. Kontos, M. Aprili, J. Lesueur, X. Grison, Phys. Rev. Lett. 86, 304 (2001)
29. A.I. Buzdin, A.V. Vedyayev, N.V. Ryzhanova, Europhys. Lett. 48, 686 (1999)
30. L.R. Tagirov, Phys. Rev. Lett. 83, 2058 (1999)
31. J.Y. Gu, C.-Y. You, J.S. Jiang, J. Pearson, Ya.B. Bazaliy, S.D. Bader, Phys. Rev. Lett. 89, 267001 (2002)
32. J.W.A. Robinson, S. Piano, G. Burnell, C. Bell, M.G. Blamire, Phys. Rev. Lett. 97, 177003 (2006)
33. I.C. Moraru, W.P. Pratt, N.O. Birge, Phys. Rev. Lett. 96, 037004 (2006)
34. S.A. Ahern, M.J.C. Martin, W. Sucksmith, Proc. R. Soc. (Lond.) A 248, 145 (1958)

35. A.Yu. Rusanov, R. Boogaard, M. Hesselberth, H. Sellier, J. Aarts, Physica C 369, 300 (2002)
36. C.G. Robbins, H. Claus, P.A. Beck, Phys. Rev. Lett. 22, 1307 (1969)
37. A. Ruotolo, C. Bell, C.W. Leung, M.G. Blamire, J. Appl. Phys. 96, 512 (2004)
38. S.-Y. Hsu, P. Holody, R. Loloee, J.M. Rittner, W.P. Pratt Jr., P.A. Schroeder, Phys. Rev. B 54, 9027 (1996)
39. V.A. Oboznov, V.V. Bolginov, A.K. Feofanov, V.V. Ryazanov, A.I. Buzdin, Phys. Rev. Lett. 96, 197003 (2006)
40. A. Tari, B.R. Coles, J. Phys. F: Met. Phys. 1, L69 (1971)
41. M. Nicklas, M. Brando, G. Knebel, F. Mayr, W. Trinkl, A. Loidl, Phys. Rev. Lett. 82, 4268 (1999), and references therein
42. T. Kontos, M. Aprili, J. Lesueur, F. Genêt, B. Stephanidis, R. Boursier, Phys. Rev. Lett. 89, 137007 (2002)
43. C. Cirillo, S.L. Prischepa, M. Salvato, C. Attanasio, M. Hesselberth, J. Aarts, Phys. Rev. B 72, 144511 (2005)
44. T.S. Khaire, W.P. Pratt Jr., N.O. Birge, Phys. Rev. B 79, 094523 (2009)
45. S. Senoussi, I.A. Campbell, A. Fert, Solid State Commun. 21, 269 (1977)
46. H.P. Kunkel, Z. Wang, G. Williams, J. Phys. F: Met. Phys. 17, L-157 (1987)
47. H.Z. Arham, T.S. Khaire, R. Loloee, W.P. Pratt Jr., N.O. Birge, Phys. Rev. B 80, 174515 (2009)
48. Th. Mühge, K. Theis-Bröhl, K. Westerholt, H. Zabel, N.N. Garifyanov, Yu.V. Goryunov, I.A. Garifullin, G.G. Khaliullin, Phys. Rev. B 57, 5071 (1998)
49. A.S. Sidorenko, V.I. Zdravkov, A.A. Prepelitsa, C. Helbig, Y. Luo, S. Gsell, M. Schreck, S. Klimm, S. Horn, L.R. Tagirov, R. Tidecks, Ann. Physik 12, 37 (2003)
50. R. O'Handley, Modern Magnetic Materials (Wiley, New York, 2000)
51. J. Bass, W.P. Pratt, J. Magn. Magn. Mater. 200, 274 (1999)
52. M. Redjdal, J.H. Giusti, M.F. Ruane, F.B. Humphrey, IEEE Trans. Magn. 39, 2684 (2003)
53. L.R. Tagirov, Physica C 307, 145 (1998)
54. Ya.V. Fominov, N.M. Chtchelkatchev, A.A. Golubov, Phys. Rev. B 66, 014507 (2002)
55. C. Cirillo, A. Rusanov, C. Bell, J. Aarts, Phys. Rev. B 75, 174510 (2007)
56. A. Rusanov, PhD Thesis, Universiteit Leiden (2005)
57. T. Kontos, PhD Thesis, Université Paris XI (2002)
58. A. Bauer, J. Bentner, M. Aprili, M.L. Della Rocca, M. Reinwald, W. Wegscheider, C. Strunk, Phys. Rev. Lett. 92, 217001 (2004)
59. M.L. Della Rocca, M. Aprili, T. Kontos, A. Gomez, P. Spathis, Phys. Rev. Lett. 94, 197003 (2005)
60. K. Matsuda, Y. Akimoto, T. Uemura, M. Yamamoto, J. Appl. Phys. 103, 07C711 (2008)
61. A. Rusanov, J. Aarts, M. Aprili, Nanoscale Devices – Fundamentals and Applications, NATO Science Series, ed. by Gross et al., 189 (2006)
62. H. Sellier, PhD Thesis, Université Grenoble I (2002)
63. A.A. Armenio, C. Cirillo, G. Iannone, S.L. Prischepa, C. Attanasio, Phys. Rev. B 76, 024515 (2007)
64. J. Kim, J.H. Kwon, K. Char, H. Doh, H.-Y. Choi, Phys. Rev. B 72, 014518 (2005)
65. A. Potenza, C.H. Marrows, Phys. Rev. B 71, 180503R (2005)
66. G.P. Pepe, R. Latempa, L. Parlato, A. Ruotolo, G. Ausanio, G. Peluso, A. Barone, A.A. Golubov, Ya.V. Fominov, M.Yu. Kupriyanov, Phys. Rev. B 73, 054506 (2006)
67. V. Zdravkov, A. Sidorenko, G. Obermeier, S. Gsell, M. Schreck, C. Muller, S. Horn, R. Tidecks, L. Tagirov, Phys. Rev. Lett. 97, 057004 (2006)
68. A. Sidorenko, V. Zdravkov, J. Kehrle, R. Morari, G. Obermeier, S. Gsell, M. Schreck, C. Muller, M.Yu. Kupriyanov, V.V. Ryazanov, S. Horn, L.R. Tagirov, R. Tidecks, JETP Lett. 90, 139 (2009)
69. L. Lazar, K. Westerholt, H. Zabel, L.R. Tagirov, Yu.V. Goryunov, N.N. Garifyanov, I.A. Garifullin, Phys. Rev. B 61, 3711 (2000)
70. M. Fauré, A.I. Buzdin, A.A. Golubov, M. Yu. Kupriyanov, Phys. Rev. B 73, 064505 (2006)
71. F.S. Bergeret, A.F. Volkov, K.B. Efetov, Phys. Rev. Lett. 86, 4096 (2001)
72. A. Kadigrobov, R.I. Shekhter, M. Jonson, Europhys. Lett. 54 394 (2001)

73. M. Flokstra, J. Aarts, Phys. Rev. B 80 144513 (2009)
74. I.S. Veshchunov, V.A. Oboznov, A.N. Rossolenko, A.S. Prokofiev, L.Ya. Vinnikov, A.Yu. Rusanov, D.V. Matveev, JETP Lett. 88, 758 (2008)
75. A.Yu. Rusanov, M. Hesselberth, J. Aarts, A.I. Buzdin, Phys. Rev. Lett. 93, 057002 (2004)
76. A.Yu. Rusanov, T.E. Golikova, S.V. Egotrov, JETP Lett. 87, 175 (2008)
77. S. Oh, D. Youm, M.R. Beasley, Appl. Phys. Lett. 71, 2376 (1997)
78. I.C. Moraru, J.W.P. Pratt, N.O. Birge, Phys. Rev. B 74, 220507(R) (2006)
79. A. Yu. Rusanov, S. Habraken, J. Aarts, Phys. Rev. B 73, 060505R (2006)
80. R. Steiner, P. Ziemann, Phys. Rev. B 74, 094504 (2006)
81. D. Stamopoulos, E. Manios, M. Pissas, Phys. Rev. B 75, 184504 (2007)
82. G. Carapella, F. Russo, G. Costabile, Phys. Rev. B 78, 104529 (2008)
83. J. Zhu, X. Cheng, C. Boone, I.N. Krivorotov, Phys. Rev. Lett. 103, 027004 (2009)
84. J.C. Slonczewski, S. Middelhoek, Appl. Phys. Lett. 6, 139 (1965)
85. I.A. Garifullin, D.A. Tikhonov, N.N. Garifyanov, L. Lazar, Yu.V. Goryunov, S.Ya. Khlebnikov, L.R. Tagirov, K. Westerholt, H. Zabel, Phys. Rev. B 66, 020505R (2002)
86. K. Westerholt, D. Sprungmann, H. Zabel, R. Brucas, B. Hjörvarsson, D.A. Tikhonov, I.A. Garifullin, Phys. Rev. Lett. 95, 097003 (2005)
87. M. Eschrig, J. Kopu, J.C. Cuevas, G. Schön, Phys. Rev. Lett. 90, 137003 (2003)
88. M. Houzet, A.I. Buzdin, Phys. Rev. B 76, 060504(R) (2007)
89. R.S. Keizer, S.T.B. Goennenwein, T.M. Klapwijk, G. Miao, G. Xiao, A. Gupta, Nature, 439, 825 (2006)
90. I. Sosnin, H. Cho, V.T. Petrashov, A.F. Volkov, Phys. Rev. Lett. 96, 157002 (2006)
91. T.S. Khaire, M.A. Khasawneh, W.P. Pratt, N.O. Birge, Phys. Rev. Lett. 104, 137002 (2010)
92. J. Wang, M. Singh, M. Tian, N. Kumar, B. Liu, C. Shi, J.K. Jain, N. Samarth, T.E. Mallouk, M.H.W. Chan, Nat. Phys. 6, 389 (2010)
93. J.W.A. Robinson, J.D.S. Witt, M.G. Blamire, Science, 329, 59 (2010)
94. M.S. Anwar, F. Czeschka, M. Hesselberth, M. Porcu, J. Aarts, Phys. Rev. B 82, 100501(R) (2010)

Chapter 14
Interplay Between Ferromagnetism and Superconductivity

Jacob Linder and Asle Sudbø

Abstract This chapter presents results on transport properties of hybrid structures where the interplay between ferromagnetism and superconductivity plays a central role. In particular, the appearance of so-called odd-frequency pairing in such structures is investigated in detail. The basic physics of superconductivity in such structures is presented, and the quasiclassical theory of Greens functions with appropriate boundary conditions is given. Results for superconductor|ferromagnet bilayers as well as magnetic Josephson junctions and spin valves are presented. Further phenomena that are studied include transport in the presence of inhomogenous magnetic textures, spin-Josephon effect, and crossed Andreev reflection. We also investigate the possibility of intrinsic coexistence of ferromagnetism and superconductivity, as reported in a series of uranium-based heavy-fermion compounds. The nature of such a coexistence and the resulting superconducting order parameter is discussed along with relevant experimental results. We present a thermodynamic treatment for a model of a ferromagnetic supercondcutor and moreover suggest ways to experimentally determine the pairing symmetry of the superconducting gap, in particular by means of conductance spectroscopy.

14.1 Introduction

It may seem paradoxical that a research field revolving around placing ferromagnetic and superconducting materials in close proximity to each other has evolved into a highly active research area in condensed matter physics today. After all, ferromagnets tend to align electron spins with each other, a property which at first glance appears completely detrimental to superconductivity for two reasons. First, the main constituent of a superconductor, known as a Cooper pair, consists of two electrons with opposite spins bound together in momentum space in the traditional Bardeen–Cooper–Schrieffer paradigm [1]. This does not resonate very well with the presence of ferromagnetism. Second, a fundamental property of superconducting materials is that they tend to expel magnetic fields from their interior, until the field strength exceeds a critical value and destroys superconductivity all together. The long-range orders of ferromagnetism and superconductivity thus seem to be about as compatible as fire and ice, and one may wonder what the point could possibly be of putting the two together to see what happens. One would expect the story to be quite short: fire melts ice, and that's it.

However, as the reader might suspect, there is more to the story than that. We shall aim in this chapter to give a modest overview of the wealth of interesting physical phenomena that occur in situations where there is an interplay between ferromagnetism and superconductivity. Quite generally, one may envision at least two scenarios where this is possible. One possibility, which is highly relevant in terms of nanostructures engineered with specific applications in mind, is to artificially grow ferromagnet|superconductor (F|S) hybrid structures. Examples of such structures are F|S bilayers, S|F|S Josephson junctions, and F|S|F spin-valves. All of these systems have their own distinct properties which provide interesting opportunities with regard to spin-functionality in superconducting devices. A second possibility is the exotic scenario of materials where ferromagnetism (FM) and superconductivity (SC) appear to coexist intrinsically. This is the case for a series of uranium-based heavy-fermion compounds in which case superconductivity may be triggered by the onset of ferromagnetism, an issue we shall return to later. Such an observation is dramatic in the sense challenges our present understanding of the pairing mechanism giving rise to superconductivity in addition to the manner in which magnetism and superconductivity interacts.

The study of the interplay between ferromagnetism and superconductivity dates back to investigations in the late 1950s and early 1960s [2–8]. Coexistence of superconductivity and long-range magnetic interactions arranged in a domain-like structure, albeit in a narrow temperature range, were also discussed in the 1970s [9–11] for two families of ternary compounds: the Chevrel phases $REMo_nX_n$, $X = S$ and Se, in addition to the rhodium borides $RERh_4B_4$, which both contain a rare-earth (RE) sublattice [12]. In the above mentioned works, possible routes to reconciling the coexistence of magnetism and superconductivity were proposed that will be elaborated upon later. However, it was not until the millenium shift that the research field concerned with the interplay between ferromagnetism and superconductivity would sky-rocket. Two crucial experimental discoveries played a major part in touching off a storm of activity.

The first experiment was presented in a Nature paper by Saxena et al. [13] who reported on the observation of ferromagnetism and superconductivity intrisically coexisting in UGe_2. The abrupt loss of resistivity signaling the transition to a superconducting state was found to be a robust property of all the investigated samples in this experimental study, and the survival of ferromagnetism below the superconducting critical temperature T_c was explicitly verified via elastic neutron scattering [14]. These findings obviously raised a number of questions, challenging in particular our very understanding of how superconductivity is formed and its interplay with ferromagnetic order.

The second experiment was presented in *Physical Review Letters* by Ryazanov et al. [15] who reported on so-called $0-\pi$ oscillations in the critical current of an S|F|S junction. This finding was significant for at least two reasons. On one hand, the authors experimentally verified the existence of an induced Fulde–Ferrel–Larkin–Ovchinnikov [7, 8] state in the ferromagnet, which previously had never been observed in bulk materials. Moreover, the presence of $0-\pi$ oscillations offered potential for using the coexistence of FM and SC order in quantum computing by making use of π-phase shifts in superconducting networks [16, 17].

Since these landmark experiments, a vast literature revolving around the interplay between FM and SC order has transpired. We cannot possibly do justice to the entire literature published on this field of research, but shall attempt to cite the most relevant works in each of the below sections covering a specific scenario where FM and SC coexist. The reader may find it useful to consider the two nice review articles by Bergeret et al. [18] and Buzdin [19], which will also supplement the list of references offered in this chapter. Also, we will not deal with the interesting scenario of a purely electromagnetic interaction in S|F hybrids (see, e.g., [20, 21]). This topic is covered in the chapter "Polarity-dependent Vortex Pinning and Spontaneous Vortex–Antivortex Structures in Superconductor/Ferromagnet Hybrids" of this book.

A twofold motivation for studying the interplay between FM and SC emerges from the above discussion. From the fundamental physics point of view, the study of how superconductivity and ferromagnetism interact with each other under various circumstances simply offers a very rich arena to explore. In terms of practical applications, the combination of spin-polarization and a dissipationless flow of a current suggests exciting prospects in low-temperature nanotechnology. In the following, we shall distinguish between two generic scenarios that feature an interplay between FM and SC order: (1) artifical synthesis obtained in tailored F|S structures and (2) the intrinsic coexistence found in certain uranium-based heavy-fermion compounds. We shall begin with the former.

14.2 Artifical Synthesis: F|S Hybrid Structures

14.2.1 Basic Physics

In order to appreciate the rich physics of F|S heterostructures, it is necessary to first review a few basic concepts that are intimately linked to such systems. We do not aim here to give a detailed introduction to these fundamental concepts, but rather highlight their importance and refer the reader to more comprehensive literature where required.

14.2.1.1 Proximity Effect and Andreev Reflection

In the context of superconducting heterostructures, the *proximity effect* denotes physical phenomena arising when a nonsuperconducting material, such as a paramagnetic metal or a ferromagnet, is placed in contact with a superconductor. In this scenario, Cooper pairs enter the normal metal while electronic excitations leak into the superconducting region. As a result, one may observe both a decrease in the critical temperature T_c for superconductivity in the superconducting region and an induction of superconducting correlations in the normal region. For instance, it is the proximity effect that renders possible the flow of a Josephson current between two superconductors separated by a normal metal of finite width. Also, the superconducting correlations in the normal part are manifested in the density of states, where a so-called minigap is induced in the diffusive regime. The minigap amounts to a substantial suppression of the density of states over an energy interval dictated by the interface transparency and the Thouless energy ε_{Th}, which is the governing energy scale for the proximity effect in normal/superconductor (N/S) junctions. The Thouless energy is defined as $\varepsilon_{\text{Th}} = D/d^2$, where D is the diffusion constant of the non-superconducting region while d is its width.

A key aspect in understanding low-energy quantum transport at the interface of a non-superconducting and superconducting material, e.g., a N/S interface, is the process of Andreev reflection. Although the existence of a gap in the energy spectrum of a superconductor implies that no quasiparticle states may persist inside the superconductor for energies below that gap, physical transport of charge and spin is still possible at a N/S interface in this energy-regime if the incoming electron is reflected as a hole with opposite charge. The remaining charge is then transferred to the superconductor in the form of a Cooper pair at Fermi level.

14.2.1.2 Non-monotonous Decay of Superconductivity

While in an N/S junction the order parameter decays monotonously upon penetrating the normal region, the presence of an exchange splitting h between the majority and minority spin energy bands in a ferromagnet gives rise to a damped oscillatory decay of the superconducting order parameter in an F|S junction (Fig. 14.1). This may be understood physically by realizing that the Cooper pair entering the ferromagnetic region obtains a finite center-of-mass momentum due to the exchange splitting. The majority-spin electron experiences a reduction in energy by h, while the minority spin electron gains an energy h, thus leading to a center-of-mass momentum $q = 2h/v_F$. In addition to the decaying behavior of the superconducting order parameter, the scattering potential at the interface between the F and S regions may lead to a suppression of the order parameter magnitude-wise in the vicinity of the interface. For low-transparency interfaces (high barrier resistances), it is usually a good approximation to neglect this effect. As we shall see in the following sections, the finite center-of-mass momentum acquired by the Cooper pair has a number of implications for experimentally observable quantities.

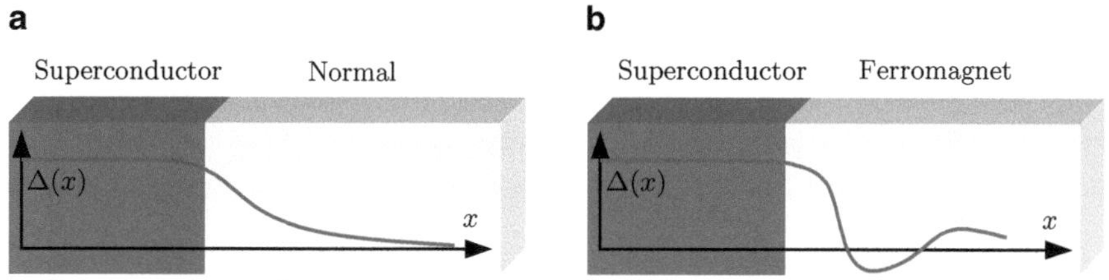

Fig. 14.1 The characteristic spatial dependence of the superconducting order parameter in a (**a**) N|S and (**b**) F|S junction

14.2.1.3 Spin-dependent Interfacial Phase-shifts

An intrinsic property of F|S interfaces is that the particles partaking in the scattering processes undergo spin-dependent interfacial phase-shifts (spin-DIPS) upon transmission and reflection at the interface. This is a result of the energy-split majority and minority spin bands in the F region. Nevertheless, these phase-shifts have not received much attention in the literature, and nonmagnetic boundary conditions [22, 23] have in the vast majority of works been used to characterize F|S interfaces [18, 19]. However, the presence of spin-DIPS may have a strong influence on the system, even qualitatively. The easiest way to capture the effect of spin-DIPS is to start with a Blonder–Tinkham–Klapwijk (BTK) [24] theory for an N/S junction and incorporate a barrier potential at the interface which depends on the spin of the incident electrons, i.e., a potential $V_\sigma = V_0 + \sigma V_\mathrm{M}$, where V_M is the magnetic part of the barrier potential. In this scenario, two effects come into play. First, the transmission amplitudes for spin-$\uparrow$ and spin-$\downarrow$ particles are no longer degenerate, but one of the spin species are favored tunneling-wise. Second, the scattered particles pick up phase-shifts at the interface due to the exchange splitting near the interface. These phase-shifts may actually induce triplet correlations [25, 26]. To see this, consider a singlet correlation function in the superconductor:

$$|\psi\rangle = |\uparrow\rangle_k|\downarrow\rangle_{-k} - |\downarrow\rangle_k|\uparrow\rangle_{-k}. \tag{14.1}$$

Upon scattering at the interface, the spins acquire different phase shifts according to:

$$|\uparrow\rangle_{-k} = \mathrm{e}^{\mathrm{i}\theta_\uparrow}|\uparrow\rangle_k,\ |\downarrow\rangle_{-k} = \mathrm{e}^{\mathrm{i}\theta_\downarrow}|\downarrow\rangle_k, \tag{14.2}$$

which transforms (14.1) into

$$|\psi\rangle = -\cos(\Delta\theta)\Big(|\uparrow\rangle_k|\downarrow\rangle_{-k} - |\downarrow\rangle_k|\uparrow\rangle_{-k}\Big) - \mathrm{i}\sin(\Delta\theta)\Big(|\uparrow\rangle_k|\downarrow\rangle_{-k} + |\downarrow\rangle_k|\uparrow\rangle_{-k}\Big), \tag{14.3}$$

where $\Delta\theta = \theta_\uparrow - \theta_\downarrow$. As seen, the spin-dependent phase-shifts at the interface induce the $S_z = 0$ triplet component which contributes to the total wavefunction $|\psi\rangle$ as long as $\Delta\theta \neq 0$. If there are also spin-flip scattering processes at the interface of the type $|\uparrow\rangle_k \to |\downarrow\rangle_k$ and $|\downarrow\rangle_k \to |\uparrow\rangle_k$, equal-spin pairing $S_z = \pm 1$ components may be generated. Here, we have sketched the situation for a normal|superconductor junction with an explicitly magnetic interface. For a ferromagnet|superconductor junction, however, the barrier potential at the interface should still be modelled as spin-dependent even if the material constituting the barrier is nonmagnetic, due to the close proximity of the F region.

14.2.1.4 Odd-frequency Pairing

The presence of superconducting correlations is in general represented by a non-zero expectation value for the anomalous Green's function

$$\mathcal{F}_{\alpha_1\alpha_2}(x_1, x_2) = -\mathrm{i}T\{\langle\psi_{\alpha_1}(x_1)\psi_{\alpha_2}(x_2)\rangle\}, \tag{14.4}$$

with T as the time-ordering operator, while $x_j = (\boldsymbol{r}_j, t_j)$ are the space- and time-coordinates for particle j while α_j is the spin-coordinate. This correlator must satisfy the Pauli principle at equal times $t_1 = t_2$, meaning that a sign change is due for $\boldsymbol{r}_1 \leftrightarrow \boldsymbol{r}_2, \alpha_1 \leftrightarrow \alpha_2$. This leads to a finite number of possibilities for the allowed symmetries of the space-, time- and spin-part of the Green's function. In the conventional BCS-case, the correlator is odd with respect to spin such that $\mathcal{F}_{\alpha_1\alpha_2}(\boldsymbol{r}_1, \boldsymbol{r}_2, t) = -\mathcal{F}_{\alpha_2\alpha_1}(\boldsymbol{r}_1, \boldsymbol{r}_2, t)$. This is known as an even-frequency, spin-singlet, even-parity symmetry. The wording stems from the Fourier-transformed and mixed representation [27,28], where frequency (or energy) is the Fourier-transform of the relative time-coordinate and the momentum is the Fourier-transform of the relative space-coordinate. In fact, in this Fourier-transformed and mixed representation, the condition $\mathcal{F}_{\alpha_1\alpha_2}(\varepsilon, \boldsymbol{k}) = -\mathcal{F}_{\alpha_2\alpha_1}(-\varepsilon, -\boldsymbol{k})$ must be satisfied in general. As seen, this opens up the possibility that the anomalous Green's function, and hence the superconducting order parameter, is *odd* in frequency while even in spin and momentum.

Such an odd-frequency pairing state was proposed by Berezinskii in 1974 [29], who argued that it could be formed due to retarded paramagnon exchange in the context of ^{3}He, although it was later experimentally established that this was not the case. However, it was realized in 2001 by Bergeret et al. [30, 31] that the odd-frequency pairing state could be obtained by means of the proximity effect between a ferromagnet and a superconductor. Very recently, it was argued in [32, 33] that odd-frequency pairing is actually generated whenever time-reversal symmetry or translational symmetry in space is broken in a superconductor. As a consequence, odd-frequency pairing should be generated under very general conditions even in N|S junctions in the absence of any exchange field.

14.2.2 *Quasiclassical Theory*

14.2.2.1 Green's Functions and Equations of Motion

The quasiclassical theory of superconductivity has proven to be a highly useful tool to study the proximity effect and transport properties of F|S hybrid structures. The central quantity in the quasiclassical theory of superconductivity is the quasiclassical Green's functions $\check{g}(\boldsymbol{p}_\mathrm{F}, \boldsymbol{r}; \varepsilon, t)$, which depends on the momentum at Fermi level $\boldsymbol{p}_\mathrm{F}$, the spatial coordinate $\boldsymbol{r}$, the energy measured from the chemical potential ε, and time t. For a detailed introduction to this theory, see, e.g., [27,28]. The quasiclassical Green's functions $\check{g}(\boldsymbol{p}_\mathrm{F}, \boldsymbol{r}; \varepsilon, t)$ are obtained from the Gor'kov Green's functions $\check{G}(\boldsymbol{p}, \boldsymbol{r}; \varepsilon, t)$ by integrating out the dependence on kinetic energy, assuming that $\check{G}$ is strongly peaked at Fermi level,

$$\check{g}(\boldsymbol{p}_\mathrm{F}, \boldsymbol{r}; \varepsilon, t) = \frac{\mathrm{i}}{\pi} \int \mathrm{d}\xi_{\boldsymbol{p}} \check{G}(\boldsymbol{p}, \boldsymbol{r}; \varepsilon, t). \tag{14.5}$$

This is typically applicable to superconducting systems where the characteristic length scale of the perturbations present, such as mean-free path and magnetic coherence length, is much larger than the Fermi wavelength. Also, the corresponding characteristic energies of such phenomena must be much smaller than the Fermi energy ε_F. The quasiclassical Green's functions may be divided into an advanced (A), retarded (R), and Keldysh (K) component, each of which has a 4×4 matrix structure in the combined particle-hole and spin space. One has that

$$\check{g} = \begin{pmatrix} \hat{g}^\mathrm{R} & \hat{g}^\mathrm{K} \\ 0 & \hat{g}^\mathrm{A} \end{pmatrix}, \tag{14.6}$$

where the elements of $\check{g}(\boldsymbol{p}_\mathrm{F}, \boldsymbol{r}; \varepsilon, t)$ read

$$\hat{g}^\mathrm{R,A} = \begin{pmatrix} \underline{g}^\mathrm{R,A} & \underline{f}^\mathrm{R,A} \\ -\underline{\tilde{f}}^\mathrm{R,A} & -\underline{\tilde{g}}^\mathrm{R,A} \end{pmatrix}, \quad \hat{g}^\mathrm{K} = \begin{pmatrix} \underline{g}^\mathrm{K} & \underline{f}^\mathrm{K} \\ \underline{\tilde{f}}^\mathrm{K} & \underline{\tilde{g}}^\mathrm{K} \end{pmatrix}. \tag{14.7}$$

The quantities $\underline{g}$ and $\underline{f}$ are 2×2 spin matrices, with the structure

$$\underline{g} = \begin{pmatrix} g_{\uparrow\uparrow} & g_{\uparrow\downarrow} \\ g_{\downarrow\uparrow} & g_{\downarrow\downarrow} \end{pmatrix}. \tag{14.8}$$

Due to internal symmetry relations between these Green's functions, all of these quantities are not independent. In particular, the tilde-operation is defined as

$$\tilde{f}(\boldsymbol{p}_\mathrm{F}, \boldsymbol{r}; \varepsilon, t) = [f(-\boldsymbol{p}_\mathrm{F}, \boldsymbol{r}; -\varepsilon, t)]^*. \tag{14.9}$$

The quasiclassical Green's functions $\check{g}(\boldsymbol{p}_{\mathrm{F}}, \boldsymbol{r}; \varepsilon, t)$ may be determined by solving the Eilenberger equation [34], which is derived from the Gor'kov equation at the price of losing information about the physics at length scales comparable to the Fermi wavelength. It reads:

$$[\varepsilon\hat{\rho}_3 - \hat{\Sigma}, \check{g}] + \mathrm{i}\boldsymbol{v}_{\mathrm{F}}\nabla\check{g} = 0, \tag{14.10}$$

where $\hat{\Sigma}$ contains the self-energies in the system such as impurity scattering, the superconducting order parameter, and exchange fields. Above, we have assumed that there is no explicit time-dependence in the problem. The operation $\hat{\rho}_3\check{g}$ inside the commutator should be understood $\hat{\rho}_3\check{g} \equiv \mathrm{diag}\{\hat{\rho}_3, \hat{\rho}_3\}\check{g}$. Pauli-matrices in particle-hole×spin (Nambu) space are denoted as $\hat{\rho}_i$, while Pauli-matrices in spin-space are written as $\underline{\tau}_i$. The Green's functions also satisfy the normalization condition

$$\check{g} \otimes \check{g} = \check{1}. \tag{14.11}$$

The self-energies entering (14.10) should be solved in a self-consistent manner. For instance, a weak-coupling s-wave superconducting order parameter is obtained by

$$\Delta(\boldsymbol{r}; t) = -\frac{\lambda}{4}\int_{-\omega_{\mathrm{c}}}^{\omega_{\mathrm{c}}} \mathrm{d}\varepsilon \langle f^{\mathrm{K}}_{\uparrow\downarrow}(\boldsymbol{p}_{\mathrm{F}}, \boldsymbol{r}; \varepsilon, t)\rangle_{\hat{\mathbf{p}}_{\mathrm{F}}}, \tag{14.12}$$

where λ is the coupling constant and ω_{c} is the cut-off energy, which may be eliminated in favor of the transition temperature. The notation $\langle \ldots \rangle$ is to be understood as an angular averaging over the Fermi surface. In the special case of an equilibrium situation, one may express the Keldysh component in terms of the retarded and advanced Green's function by means of the relation

$$\hat{g}^{\mathrm{K}} = (\hat{g}^{\mathrm{R}} - \hat{g}^{\mathrm{A}}) \tanh(\beta\varepsilon/2), \tag{14.13}$$

where $\beta = T^{-1}$ is inverse temperature. In nonequilibrium situations, one must derive kinetic equations for nonequilbrium distribution functions in order to specify the Keldysh part.

In the dirty limit, the scattering time due to impurities satisfies $\zeta\tau \ll 1$, where τ is the impurity scattering time while ζ is the energy scale of any other self-energy in the problem. The Eilenberger equation is then reduced to the Usadel equation [35], which may be derived by expanding the Green's function in spherical harmonics and averaging the Eilenberger equation over the Fermi surface. The Usadel equation reads

$$D\nabla(\check{g}_s\nabla\check{g}_s) + \mathrm{i}[\varepsilon\mathrm{diag}(\hat{\rho}_3, \hat{\rho}_3) - \check{\Sigma}_s, \check{g}_s] = 0, \tag{14.14}$$

where the subscript 's' indicates that the Green's function and self-energy have been averaged over the Fermi surface, thus rendering them independent of the direction of the momentum $\boldsymbol{p}_{\mathrm{F}}$. The diffusion constant is given by $D = v_{\mathrm{F}}^2\tau/3$. Considering an equilibrium situation, it suffices to solve for the retarded part of (14.14) which has the same form, namely (hereafter dropping the subscript 's')

$$D\nabla(\hat{g}^{\mathrm{R}}\nabla\hat{g}^{\mathrm{R}}) + \mathrm{i}[\varepsilon\hat{\rho}_3 - \hat{\Sigma}, \hat{g}^{\mathrm{R}}] = 0. \tag{14.15}$$

Omitting the superscript 'R' on the Green's function, we expand it around the bulk solution $\hat{g}_0$ as $\hat{g} \simeq \hat{g}_0 + \hat{f}$, where $\hat{g}_0 = \mathrm{diag}(\bar{1}, -\bar{1})$ and

$$\hat{f} = \begin{pmatrix} \bar{0} & \bar{f}(\boldsymbol{r};\varepsilon) \\ -[\bar{f}(\boldsymbol{r};-\varepsilon)]^* & \bar{0} \end{pmatrix},$$
$$\bar{f}(\boldsymbol{r};\varepsilon) = \begin{pmatrix} f_{\uparrow\uparrow}(\boldsymbol{r};\varepsilon) & f_{\mathrm{t}}(\boldsymbol{r};\varepsilon) + f_{\mathrm{s}}(\boldsymbol{r};\varepsilon) \\ f_{\mathrm{t}}(\boldsymbol{r};\varepsilon) - f_{\mathrm{s}}(\boldsymbol{r};\varepsilon) & f_{\downarrow\downarrow}(\boldsymbol{r};\varepsilon). \end{pmatrix} \tag{14.16}$$

One may now multiply out the matrix equation (14.15), only keeping the lowest order terms in the anomalous Green's functions $f_{\alpha\beta}(\boldsymbol{r};\varepsilon)$. For more compact notation, we have defined the quantities $f_{t(s)} = [f_{\uparrow\downarrow} + (-) f_{\downarrow\uparrow}]/2$.

Consider now for concreteness a ferromagnet where $\Sigma = -h\mathrm{diag}(\underline{\tau_3}, \underline{\tau_3})$, one then obtains the Usadel equation:

$$D\partial_x^2 f_\pm + 2\mathrm{i}(\varepsilon \pm h) f_\pm = 0, \;\; f_\pm = f_{\mathrm{t}} \pm f_{\mathrm{s}}, \tag{14.17}$$

which is easily solved by

$$f_\pm = A_\pm \mathrm{e}^{\mathrm{i}k_\pm x} + B_\pm \mathrm{e}^{-\mathrm{i}k_\pm x}, \tag{14.18}$$

where $k_\pm = \sqrt{2\mathrm{i}(\varepsilon \pm h)/D}$ and $\{A_\pm, B_\pm\}$ are unknown coefficients to be determined by the boundary conditions of the problem.

If one is interested in the full-proximity effect regime, it should be noted that the Green's function $\check{g}$ may be conveniently parametrized to facilitate both analytical and numerical calculations. For instance, the Green's function in a F|S bilayer with a homogeneous magnetization may be parametrized as

$$\hat{g}_j = \begin{pmatrix} c_{\uparrow,j} & 0 & 0 & s_{\uparrow,j} \\ 0 & c_{\downarrow,j} & s_{\downarrow,j} & 0 \\ 0 & -s_{\downarrow,j} & -c_{\downarrow,j} & 0 \\ -s_{\uparrow,j} & 0 & 0 & -c_{\uparrow,j} \end{pmatrix}, \;\; j = \{S, F\}, \tag{14.19}$$

where we have introduced $s_{\sigma,j} = \sinh(\theta_{\sigma,j})$ and $c_{\sigma,j} = \cosh(\theta_{\sigma,j})$. Note that $(\hat{g}_j)^2 = \hat{1}$ is satisfied as a direct consequence of the above parametrization. The parameter $\theta_{\sigma,j}$ is a measure of the proximity effect and vanishes in the ferromagnetic region in the absence of a proximity effect. In the superconducting region, the bulk solution for a conventional s-wave BCS superconductor reads $c_\sigma = \cosh(\theta_{\mathrm{BCS}})$ and $s_\sigma = \sigma\sinh(\theta_{\mathrm{BCS}})$, where $\theta_{\mathrm{BCS}} = \mathrm{atanh}(\Delta/\varepsilon)$. The Usadel equation on the ferromagnetic side now reads:

$$D\partial_x\theta_\sigma^2 + 2\mathrm{i}(\varepsilon+\sigma h)\sinh\theta_\sigma = 0. \tag{14.20}$$

The Usadel equation may be augmented to incorporate different types of spin-dependent scattering, stemming from, e.g., magnetic impurities or spin–orbit interactions. It is moreover possible to distinguish between different types of spin–flip scattering, such as uniaxial or isotropic. Assuming an exchange field along the z-direction, we may write:

$$\begin{aligned} &D\partial_x\theta_\sigma^2 + 2\mathrm{i}(\varepsilon+\sigma h)\sinh\theta_\sigma - \frac{\sigma S_{xy}}{2\tau_{\mathrm{sf}}}\sinh(\theta_\uparrow-\theta_\downarrow) \\ &\quad - \frac{S_z}{4\tau_{\mathrm{sf}}}\sinh(2\theta_\sigma) - \frac{1}{2\tau_{\mathrm{so}}}\sinh(\theta_\uparrow+\theta_\downarrow) = 0, \end{aligned} \tag{14.21}$$

where τ_{sf} and τ_{so} are the spin–flip and spin–orbit scattering times. Above, $S_z = S_{xy} = 1$ for isotropic spin–flip scattering, whereas $S_z = 1$ and $S_{xy} = 0$ for uniaxial scattering.

Finally, we mention the most general parametrization available for the Green's function, which may be used for any magnetization texture and heterostructure [36–38]. It reads:

$$\hat{g} = \begin{pmatrix} \underline{\mathcal{N}}(\underline{1}-\underline{\gamma}\underline{\tilde{\gamma}}) & 2\underline{\mathcal{N}}\underline{\gamma} \\ 2\underline{\tilde{\mathcal{N}}}\underline{\tilde{\gamma}} & \underline{\tilde{\mathcal{N}}}(-\underline{1}+\underline{\tilde{\gamma}}\underline{\gamma}) \end{pmatrix}. \tag{14.22}$$

The unknown functions $\underline{\gamma}$ and $\underline{\tilde{\gamma}}$ are key elements in this parametrization of the Green's function and will be solved for below. Here, $\underline{\ldots}$ denotes a 2×2 matrix and

$$\underline{\mathcal{N}} = (1+\underline{\gamma}\underline{\tilde{\gamma}})^{-1}\ \underline{\tilde{\mathcal{N}}} = (1+\underline{\tilde{\gamma}}\underline{\gamma})^{-1}. \tag{14.23}$$

Inserting (14.22) into the Usadel equation, we obtain the transport equation for the unknown function $\underline{\gamma}$ (and hence $\underline{\tilde{\gamma}}$)

$$D[\partial^2\underline{\gamma} + (\partial\underline{\gamma})\underline{\tilde{\mathcal{F}}}(\partial\underline{\gamma})] + \mathrm{i}[2\varepsilon\underline{\gamma} + \boldsymbol{h}\cdot(\underline{\boldsymbol{\sigma}}\underline{\gamma} - \underline{\gamma}\underline{\boldsymbol{\sigma}}^*)] = 0, \tag{14.24}$$

with $\underline{\tilde{\mathcal{F}}} = -2\underline{\tilde{\mathcal{N}}}\underline{\tilde{\gamma}}$, while

$$D[\partial^2\underline{\tilde{\gamma}} + (\partial\underline{\tilde{\gamma}})\underline{\mathcal{F}}(\partial\underline{\tilde{\gamma}})] + \mathrm{i}[2\varepsilon\underline{\tilde{\gamma}} + \boldsymbol{h}\cdot(\underline{\tilde{\gamma}}\underline{\boldsymbol{\sigma}} - \underline{\boldsymbol{\sigma}}^*\underline{\tilde{\gamma}})] = 0, \tag{14.25}$$

with $\underline{\mathcal{F}} = -2\underline{\mathcal{N}}\underline{\gamma}$-

14.2.2.2 Boundary Conditions

The above equations suffice to completely describe, for instance, a bulk superconducting structure, but must be supplemented with boundary conditions when treating heterostructures such as F|S junctions. These boundary conditions take different forms depending on the physical properties of the interface, and we proceed to describe possible scenarios in this respect. Transport across interfaces in heterostructures may in general be characterized according to three particular properties: (1) the transmission of the interface, (2) the resistivity of the compounds separated by the interface, and (3) whether the interface is spin-active or not. Let us clarify the distinction between the two first properties. The transmission of the barrier determines the likelihood of electron transport to occur across the interface. On the other hand, the resistivities of the compounds separated by the interface are in principle unrelated to the transmissivity of the interface, and one may have for instance a tunneling contact with electrodes attached to it that have either a large or small resistance.

Zaitsev [39] derived boundary conditions for a clean N|S interface, while Kupriyanov and Lukichev (KL) [22] worked out simplified boundary conditions in the dirty limit, valid for atomically sharp interfaces in the tunneling regime with a low barrier transparency. In a heterostructure with a region 1 located in the half-space $x < 0$ and a region 2 located in the half-space $x > 0$, the KL boundary conditions may be expressed as follows for the retarded part of the Green's function:

$$2d_1\gamma_1(\hat{g}_1\partial_x\hat{g}_1)\Big|_{x=0} = 2d_2\gamma_2(\hat{g}_2\partial_x\hat{g}_2)\Big|_{x=0} = [\hat{g}_1, \hat{g}_2]\Big|_{x=0}. \tag{14.26}$$

The parameter γ_j models the interfacial transmission properties and is given by $\gamma_j = R_{\mathrm{I}}/R_j$, where R_{I} is the interface resistance while R_j is the resistance in region j. Also, d_j is the width of region j. In the case of an arbitrary interface transparency, the most compact way of writing the boundary conditions for the Green's functions for nonmagnetic interface was introduced by Nazarov [23]:

$$\frac{\gamma_1 d_1}{2}\hat{g}_1\partial_x\hat{g}_1\Bigg|_{x=0} = \frac{[\hat{g}_1, \hat{g}_2]}{4 + T(\{\hat{g}_1, \hat{g}_2\} - 2)}\Bigg|_{x=0}, \tag{14.27}$$

and similarly for $\hat{g}_2$. We have here assumed that all N transmission channels at the interface are characterized by the same transmission probability, i.e., $T_N = T$. As seen, this reduces to the KL boundary condition in the tunneling limit $T \to 0$. It should be noted that the above boundary condition has also been extended to include the scenario of unconventional superconductors, such as p-wave or d-wave, by Tanaka and co-workers [40, 41].

The third property determines to what degree the interface discriminates between incoming quasiparticles with different spins. In all the preceding references, a nonmagnetic (spin-inactive) interface was assumed. The generalized boundary conditions for magnetically active interfaces have also been derived [42] and further studied in [43–45]. As described in a previous section, spin-dependent interfacial phase shifts (spin-DIPS) come into play when one of the materials on each side of the interface is magnetic or if the barrier itself is explicitly magnetic. The spin-DIPS may have an appreciable effect on the physics when the term describing these

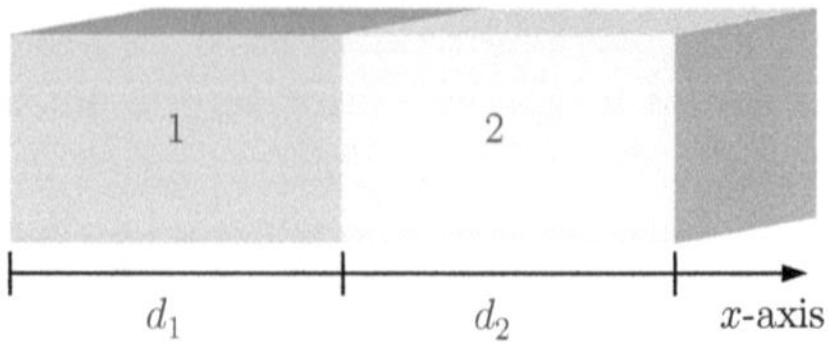

Fig. 14.2 Junction consisting of two regions 1 and 2 with an interface perpendicular to the x-axis

phase-shifts in the boundary conditions is comparable in magnitude to the normal-state tunneling conductance. When the contribution from the spin-dependent phase shifts is much smaller than the tunneling conductance, as may be appropriate for a relatively weak ferromagnet, their effect can often be neglected and the K-L or Nazarov boundary conditions can be employed.

In the context of F|S structures, the experimentally most relevant boundary conditions are the ones derived in [43–45]. To facilitate and encourage use of the spin-active boundary conditions required for an S|F interface, we here write down their explicit form in the diffusive limit for the case of a magnetization in the z-direction. Consider a junction consisting of two regions 1 and 2, as shown in Fig. 14.2. The regions have widths d_j and bulk electrical resistances R_j. The matrices used below are 4×4 matrices in particle-hole$\otimes$spin space, using a basis

$$\psi(\boldsymbol{r},t) = \begin{pmatrix} \psi_\uparrow(\boldsymbol{r},t) \\ \psi_\downarrow(\boldsymbol{r},t) \\ \psi_\uparrow^\dagger(\boldsymbol{r},t) \\ \psi_\downarrow^\dagger(\boldsymbol{r},t) \end{pmatrix}.$$

Introducing $\hat{\alpha} = \mathrm{diag}(1,-1,1,-1) = \mathrm{diag}(\underline{\sigma_3},\underline{\sigma_3})$, where $\underline{\sigma_3}$ is the third Pauli matrix in spin-space, we may write the boundary conditions as follows when the spin-DIPS constitute the dominating effect of the spin-active interface:

$$\begin{aligned} 2(d_1/R_1)\hat{g}_1\partial_x\hat{g}_1 &= G_T[\hat{g}_1,\hat{g}_2] - \mathrm{i}G_{\phi,1}[\hat{\alpha},\hat{g}_1], \\ 2(d_2/R_2)\hat{g}_2\partial_x\hat{g}_2 &= G_T[\hat{g}_1,\hat{g}_2] + \mathrm{i}G_{\phi,2}[\hat{\alpha},\hat{g}_2]. \end{aligned} \tag{14.28}$$

Here, G_T is the conductance of the junction while $G_{\phi,j}$ are the phase-shifts on side j of the interface. The parameters $\{G_T, G_{\phi,j}\}$ may be calculated by relating them to microscopic transmission and reflection probabilities within, e.g., a Blonder–Tinkham–Klapwijk (BTK) [24] framework. Explicitly spin-active barriers were considered in ballistic S|F bilayers using the BTK-approach for both s-wave [46] and d-wave [47] superconductors. In the absence of spin-DIPS ($G_{\phi,j} \to 0$), (14.28) reduce to the Kupriyanov–Lukichev nonmagnetic boundary conditions [22].

Let us make a final remark concerning the treatment of interfaces in the quasi-classical theory of superconductivity. We previously stated that the present theory is valid as long as characteristic energies of various self-energies and perturbations in the system are much smaller than the Fermi energy. At first glance, this might seem to be irreconcilable with the presence of interfaces, which represent strong perturbations varying on atomic length scales. However, this problem may be overcome by including the interfaces as boundary conditions for the Green's functions rather than directly in the Eilenberger equation.

14.2.3 F|S Bilayers

The most basic structure that probes the interplay between FM and SC is probably a F|S bilayer. Unless the effective barrier transparency is identically zero, the proximity effect ensures that ferromagnetic correlations leak into the superconducting region and vice versa. The influence of the proximity effect may be experimentally investigated by probing the local density of states (DOS), as made possible by, e.g., scanning-tunneling microscopy (STM) measurements. The experiment by Kontos et al. [48] pioneered this kind of experiment in F|S bilayers and showed explicitly how the spatially modulated FFLO-like superconducting state [depicted in Fig. 14.1b] induced in the ferromagnet was manifested experimentally by means of an oscillating DOS near Fermi level.

To gain further insight into the microscopic structure of the proximity effect, we revert to the analytical expression (14.18) which expresses the anomalous Green's function induced in the ferromagnetic region. There are two important observations that should be made. First of all, it is seen that the spatial dependence of the superconducting correlations are governed by the term $e^{ik_\sigma x}$, where $k_\sigma = \sqrt{2i(\varepsilon + \sigma h)/D}$. The exchange field h is typically much larger than the quasiparticle energy ε measured from the Fermi level. Considering a geometry where the ferromagnet occupies the region $x > 0$ such that the superconducting correlations must die out as $x \to \infty$, it is readily seen that the decaying solution is proportional to $e^{-(1+i)x/\xi_F}$. Here, we have defined the coherence length $\xi_F = \sqrt{D_F/h}$ where D_F is the diffusion constant in the F region. The solution thus has both an oscillatory and decaying part which both vary on the relevant length scale ξ_F. For weak ferromagnetic alloys such as PdNi, one typically finds that ξ_F lies in the range of a few nanometers. For stronger ferromagnets, such as Fe or Ni, the coherence length is much shorter. Thus, it is clear that the stronger the exchange field h, the shorter the penetration depth of superconductivity into the ferromagnet. The oscillating behavior may be directly inferred from the DOS, which in the weak proximity effect $|f_\sigma| \ll 1$ regime can be evaluated according to

$$N = \sum_\sigma N_\sigma,\ N_\sigma = N_{F,\sigma}\text{Re}\Big\{\sqrt{1 + f_\sigma(\varepsilon) f^*_{-\sigma}(-\varepsilon)}\Big\}, \tag{14.29}$$

where $f_\sigma = f_t + \sigma f_s$ and $N_{F,\sigma}$ is the normal-state (zero proximity effect) spin-resolved DOS.

The second important observation pertains to the symmetry of the proximity-induced superconducting correlations. We restrict first our attention to the experimentally most relevant diffusive regime, i.e., strong impurity scattering. The scattering effectively averages the correlations over the Fermi surface such that only isotropic components (s-wave) survive. This is why impurity scattering in general is detrimental toward higher angular momentum pairing superconductivity such as p-wave or d-wave, since the gap averages to zero. At first glance, one might then be tempted to conclude that the strong impurity scattering should lead to a vanishing triplet component $f_\mathrm{t} \to 0$, since only the s-wave spin-singlet anomalous Green's function f_s should be robust in the diffusive regime. Closer examination reveals that this is not so. Spin-triplet pairing can actually find a way of accomodating itself even in the hostile environment of considerable impurity scattering: namely, by adapting the previously mentioned odd-frequency symmetry in Sect. 14.2.1.4. In this way, it is possible to obtain superconducting correlations which are both isotropic s-wave and with a spin-triplet symmetry. Therefore, both f_t and f_s remain non-zero in the ferromagnetic region.

One might wonder if it is possible to discern between even- and odd-frequency correlations or if they just blend together in the proximity-induced superconductivity without leaving individual fingerprints. It turns out that symmetry of the superconducting correlations actually strongly dictate the behavior of the DOS in the ferromagnetic region. In general, it can be shown analytically [49] that the conventional even-frequency correlations suppress the low-energy DOS (just as they do in the bulk of the superconductor, where the DOS is zero in the range $\varepsilon \in [\Delta, -\Delta]$), whereas the odd-frequency correlations *enhance* the low-energy DOS. The oscillations of the DOS observed by Kontos et al. can thus be reinterpreted in terms of even- and odd-frequency correlations: where the DOS is suppressed compared to its normal-state value, the even-frequency correlations dominate, and vice versa.

There are, however, two problems associated with detecting odd-frequency correlations in F|S bilayers. One is that the odd-frequency state induced in S/F bilayers has a short penetration depth into the ferromagnetic region. It is limited by the magnetic coherence length ξ_F, which is usually much smaller than the superconducting coherence length ξ_S [18]. Another problem is that these odd-frequency correlations are often masked by the simultaneous presence of even-frequency correlations in the same material. Clear-cut signatures of the odd-frequency correlations are therefore only accessible in a limited parameter regime [53].

In the majority of works on superconducting proximity structures, the diffusive limit and spin-inactive interfaces have been considered [22]. For a nonmagnetic bilayer, a minigap appears in the density of states of the normal metal. It scales with the Thouless energy of the normal layer and with the transmission probability of the interface. For a spin-active interface, the transmission properties of spin-$\uparrow$ and spin-$\downarrow$ electrons into a ferromagnetic metal are different, and this gives rise to both spin-dependent conductivities and spin-dependent phase shifts at the interface [42–45, 50–52]. We will now show how a spin-active interface in an S|N bilayer produces clear signatures of purely odd-frequency triplet pairing amplitudes that can be tested experimentally.

Fig. 14.3 Proposed experimental setup for observation of the odd-frequency component in a diffusive normal metal layer|superconductor junction

Let us consider the system shown in Fig. 14.3. The superconductor is conventional (even-frequency s-wave) while the interface is magnetic. There will be a dramatic change in the nature of proximity correlations when the spin-dependent phase shifts exceed the tunneling probability of the interface. The spin-active interface in a superconductor/normal metal (S|N) bilayer causes the even-frequency correlations to vanish at zero excitation energy (Fermi level), while odd-frequency correlations appear. At the same time, the minigap, one of the hallmarks of the conventional proximity effect, is replaced by a low-energy band with enhanced density of states. We focus on the density of states (DOS) in the normal region, which can be probed by tunneling experiments. Since the exchange field is absent in the normal metal, it should be possible to detect the odd-frequency amplitude without any interfering effects of even-frequency correlations. This resolves the two main difficulties associated with the experimental detection of odd-frequency correlations mentioned above.

We adopt the quasiclassical theory of superconductivity [27, 28], where information about the physical properties of the system is embedded in the Green's function. For an equilibrium situation, it suffices to consider the retarded part of the Green's function, here denoted $\hat{g}$. Due to the symmetry properties of $\hat{g}$, one may parametrize it conveniently in the normal (N) region by a parameter θ_σ, which allows for both singlet and triplet correlations [53]. In the superconducting (S) region, we use the bulk solution $\hat{g}_S = c \cdot \underline{\tau_3} \otimes \underline{\sigma_0} + s \cdot \underline{\tau_1} \otimes (\mathrm{i}\underline{\sigma_2})$, $c = \cosh(\theta)$, $s = \sinh(\theta)$, $\theta = \mathrm{atanh}(\Delta/\varepsilon)$, $\underline{\tau_i}$ and $\underline{\sigma_i}$ being Pauli matrices in particle-hole and spin space, respectively.

We use the formalism described in [53] and consider the diffusive limit. Then, the orbital symmetry for all proximity amplitudes is reduced to s-wave, and hence the singlet component always has an even-frequency symmetry while the triplet component has an odd-frequency symmetry. The Green's functions are subject to boundary conditions, which assume at the S|N interface in the tunneling limit the form [43–45]:

$$2\gamma d \hat{g}_N \partial_x \hat{g}_N = [\hat{g}_S, \hat{g}_N] + \mathrm{i}(G_\phi/G_T)[\underline{\tau_0} \otimes \underline{\sigma_3}, \hat{g}_N], \tag{14.30}$$

and at the outer interface read $\partial_x \hat{g}_N = \hat{0}$. Here, $\gamma = R_B/R_N$ where R_B (R_N) is the resistance of the barrier (normal region), and d is the width of the normal region, while G_T is the junction conductance in the normal state. The boundary condition above contains an additional term G_ϕ compared to the usual nonmagnetic boundary conditions in [22]. This term is due to spin-dependent phase shifts of quasiparticles being reflected at the interface. G_ϕ may be non-zero even if the transmission $G_T \to 0$, corresponding to a ferromagnetic insulator [43, 44]. We define the superconducting coherence length $\xi_S = \sqrt{D/\Delta}$ and Thouless energy $\varepsilon_{\text{Th}} = D/d^2$, where D is the diffusion constant, and assume that the inelastic scattering length, l_{in}, is sufficiently large, such that $d \ll l_{\text{in}}$.

The Usadel equation [35] reads $D\partial_x^2\theta_\sigma + 2\mathrm{i}\varepsilon \sinh\theta_\sigma = 0$, with boundary condition $\gamma d\partial_x\theta_\sigma = (cs_\sigma - \sigma sc_\sigma) + \mathrm{i}\sigma(G_\phi/G_T)s_\sigma$ at $x = 0$ and $\partial_x\theta_\sigma = 0$ at $x = d$. Here, $c_\sigma = \cosh(\theta_\sigma)$ and $s_\sigma = \sinh(\theta_\sigma)$. At zero energy, we find that the pairing amplitudes are either purely (odd-frequency) triplet,

$$f_s(0) = 0, \; f_t(0) = \frac{G_T \cdot \text{sgn}(G_\phi)}{\sqrt{G_\phi^2 - G_T^2}} \quad \text{for } |G_\phi|/G_T > 1, \tag{14.31}$$

or purely (even-frequency) singlet

$$f_s(0) = \frac{\mathrm{i} \cdot G_T}{\sqrt{G_T^2 - G_\phi^2}}, \; f_t(0) = 0 \quad \text{for } |G_\phi|/G_T < 1. \tag{14.32}$$

Thus, the presence of G_ϕ induces an odd-frequency component in the normal layer. The remarkable aspect of (14.31) and (14.32) is that they are valid for any value of the width d below the inelastic scattering length, and for any interface parameter γ. Thus, the vanishing of the singlet component is a robust feature in S|N structures with spin-active interfaces, as long as $|G_\phi|/G_T > 1$. Without loss of generality, we focus on positive values of G_ϕ from now on.

The DOS is given as $N(\varepsilon)/N_0 = \sum_\sigma \text{Re}\{c_\sigma\}/2$, yielding

$$N(0)/N_0 = \text{Re}\{G_\phi/\sqrt{G_\phi^2 - G_T^2}\}. \tag{14.33}$$

At zero-energy, the DOS thus vanishes as long as $G_\phi/G_T < 1$, which means that the usual minigap in S|N structures survives in this regime. However, the zero-energy DOS is enhanced for $G_\phi/G_T > 1$ since the singlet component vanishes there.

The ratio G_ϕ/G_T depends on the microscopic barrier properties [45]. In the tunneling limit, one finds that G_ϕ can be considerably larger than G_T. Although not included here, we underline that performing the same analysis in the ballistic limit gives the same result: namely, a complete separation between the even- and odd-frequency proximity amplitudes above a critical value of the interface resistance. This suggests that the effect predicted here should be quite robust, as it occurs both in the clean and dirty limit and is independent of the specific system size and/or junction conductance.

Let us now consider a qualitative explanation for the mechanism behind the separation between even- and odd-frequency correlations. The superconductor induces a minigap proportional to G_T in the normal metal, while the spin-active barrier induces an effective exchange field proportional to G_ϕ. The situation in the normal metal then resembles that of a thin-film conventional superconductor in the presence of an in-plane external magnetic field [54], with the role of the gap and field played by G_T and G_ϕ, respectively. In that case, it is known that superconductivity is destroyed above the Clogston–Chandrasekhar limit [5, 6], due to pair-breaking of a spin-singlet Cooper-pair. In the present case, we observe coexistence of the exchange field and spin-singlet even-frequency superconductivity as long as G_ϕ is below the critical value of $G_\phi = G_T$. However, for $G_\phi > G_T$, spin-singlet pairing is no longer possible at the chemical potential. It is then replaced by spin-triplet pairing, which must be odd in frequency due to the isotropization of the gap in the diffusive limit. Thus, there is a natural separation between even-frequency and odd-frequency pairing in the normal metal at a critical value of the effective exchange field G_ϕ.

The simplest experimental manifestation of the odd-frequency component is probably a zero-energy peak in the DOS [49,55,56]. In S|F layers, where this phenomenon has been discussed previously, a clear zero-energy peak is unfortunately often masked by the simultaneous presence of singlet correlations f_s, which tend to suppress the DOS at low energies. This is not so in the system we consider, provided only that $G_T < |G_\phi|$. This is ideal for a direct observation of the odd-frequency component, manifested as a zero-energy peak in the DOS. The even-frequency correlations vanish completely when the interface transmission is sufficiently low, and the parameter G_ϕ can be increased by increasing the magnetic polarization of the barrier separating the superconducting and normal layers. By fabricating several samples with progressively increasing strength of magnetic moment of the barrier, one should be able to observe an abrupt crossover at the zero-energy DOS above a certain strength of the magnetic moment. Alternatively, one could alter the interface transmission by varying the thickness of the insulating region.

14.2.4 S|F|S Josephson Junctions

14.2.4.1 0–π Oscillations of Critical Current

In a superconducting Josephson junction, two bulk superconductors are connected via a weak link to support a dissipationless flow of electrical current as long as there is a phase difference $\Delta\theta = \theta_L - \theta_R$ between the order parameters $\Delta_j = \Delta_{0,j}e^{i\theta_j}$, $j = L, R$ in each superconducting bank. The relationship between the magnitude of the current that flows through the junction and the corresponding phase difference between the superconducting banks is known as the current–phase relationship (see [57] for a detailed review). The current–phase relationship is under many circumstances given as $I = I_c \sin\Delta\theta$, where I_c is the critical (maximum) value of the current. However, it may in general include contributions from higher harmonics. In order to explain the concept of $0 - \pi$ oscillations, let us nevertheless proceed with the sinusoidal form $I = I_c \sin\Delta\theta$ for the sake of clarity. The energy ground-state

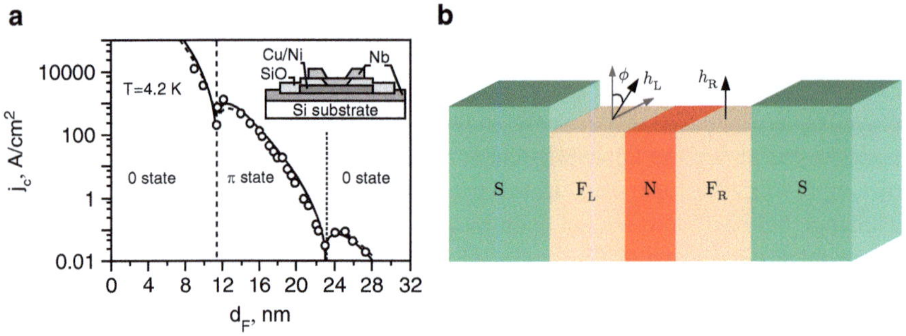

Fig. 14.4 (**a**) Experimental data from the experiment of Oboznov et al. [58]: critical current density for Nb–$Cu_{0.47}Ni_{0.53}$-Nb junctions vs. thickness d_F of the ferromagnetic region at temperature 4.2 K. *Open circles* represent experimental results; *solid* and *dashed lines* show model calculations discussed in the second part of the Letter. The inset shows a schematic cross-section of the S|F|S junctions. The figure is taken from [58]. (**b**) The Josephson current can be actively tuned by the misalignment angle ϕ in a S|F|N|F|S multilayer structure. The normal spacer is required to reduce the exchange coupling between the ferromagnets which tends to lock the misalignment at $\phi = 0$ or $\phi = \pi$

E of the Josephson junction is related to the phase difference as follows [19]:

$$E = \frac{I_c}{2e}(1 - \cos \Delta\theta), \tag{14.34}$$

giving the sinusoidal current–phase relationship $I = (2e)\partial E/\partial \Delta\theta = I_c \sin \Delta\theta$. From the above equation, we see that the minimum energy is obtained at $\Delta\theta = 0$ when $I_c > 0$. However, the minimum energy of the junction is obtained at $\Delta\theta = \pi$ when $I_c < 0$. The constant I_c, and in particular its sign, depends in general on the parameters of the junction, such as width, temperature, and barrier transparency. This means that it is possible to manipulate the preferred ground-state of the system *if* one may control the sign of I_c by altering for instance the width d of the junction. As a result, one should observe an oscillatory behavior of the critical current when plotted against a parameter in the system that may provoke a $0-\pi$ transition. Precisely, such behavior has been observed experimentally both as a function of the interlayer width d_F and the temperature T, as demonstrated, e.g., in the experiment by Oboznov et al. [58] (see Fig. 14.4a).

We here briefly sketch how the Josephson current may be calculated analytically in the quasiclassical framework, assuming a weak proximity effect. The solution of the Usadel equation in the ferromagnet provides $f_\pm = f_t \pm f_s = A_\pm e^{ik_\pm x} + B_\pm e^{-ik_\pm x}$, as discussed previously. The values of $A_\pm$ and $B_\pm$ depend strongly on the boundary conditions used at the interfaces. The general expression for the charge-current density is:

$$I = \frac{N_F e D}{16} \int_{-\infty}^{\infty} d\varepsilon \text{Tr}\{\hat{\rho}_3(\check{g}\partial_x \check{g})^K\}. \tag{14.35}$$

In an equilbrium situation (no applied voltage or thermal gradient), the above expression simplifies to

$$I = \frac{N_F e D}{8} \int_{-\infty}^{\infty} \mathrm{d}\varepsilon \mathrm{Re}\{M_+(\varepsilon) + M_-(\varepsilon)\} \tanh(\beta\varepsilon/2),$$
$$M_\pm(\varepsilon) = [f_\pm(-\varepsilon)]^* \partial_x f_\mp(\varepsilon) - f_\pm(\varepsilon)[f_\mp(-\varepsilon)]^*. \tag{14.36}$$

This is accomplished by expressing the Keldysh component of the Green's function via the retarded and advanced components as described in Sect. 14.2.2, in addition to inserting $f_\pm$ into $\hat{g}^{\mathrm{R}}$. It can be shown analytically that for the S|F|S case with s-wave superconductors one has $I = I_0 \sin \Delta\theta$, since only the first harmonic survives in the weak proximity effect regime. It should also be mentioned that the current–phase relationship acquires a substantial contribution from higher harmonics than $\sin \Delta\theta$ when the interface transparency is high [53, 59], which has some interesting effects for the ground-state properties of the system [60]. In particular, it is possible to obtain ground-state phase differences which are neither 0 or π, but instead some intermediate value θ_0. The presence of higher order harmonics in the current–phase relationship may be experimentally manifested through residual values at the cusps of the critical current corresponding to the $0-\pi$ transition. In the majority of experiments so far on the Josephson current in S|F|S junctions, the residual value of the current at the transition points has been practically vanishing, indicating a sinusoidal current–phase relationship. However, the existence of higher harmonic terms, although small in magnitude, was demonstrated experimentally in [61].

14.2.4.2 Inhomogeneous Magnetization Textures

Most of the works studying S|F|S junctions have taken the exchange field in the ferromagnet to be homogeneous. When the magnetization texture is inhomogeneous, several new effects come into play. One of these effects is the possibility of a superconducting long-range triplet component which, contrary to the case of homogeneous magnetization, penetrates much longer into the ferromagnetic region. Such an inhomogeneous magnetization texture can be realized by artificially engineering several ferromagnetic layers with misaligned magnetizations [62–67]. Alternatively, inhomogeneous magnetization arises naturally in the presence of domain walls or nontrivial patterns for the local ferromagnetic moment. An example of the latter is the conical ferromagnet Ho. Very recently, two theoretical studies have predicted qualitatively new effects in S|F and S|F|S hybrid structures, where F is a conical ferromagnet [68, 69]. Due to the inhomogeneous nature of the magnetization in Ho, the spin properties of the proximity-induced superconducting correlations are expected to undergo a qualitative change compared to the case of homogeneous ferromagnetism.

The study of S|F|S Josephson junctions with inhomogeneous magnetization textures is interesting for at least two reasons. In the case where the inhomogeneity is intrinsic and difficult to alter experimentally, which is the case for domain wall structures, it is important to clarify how the current is altered compared to the homogeneous magnetization case to be able to differentiate between these two scenarios when interpreting experimental data. When the inhomogeneous is artifically created, such as two magnetic layers with misaligned magnetization vectors, the focus is shifted toward active control of the Josephson current by means of rotating the magnetizations relative each other. This may be accomplished by an external magnetic field when one of the magnetic layers has a coercive field that is much smaller than the other layer. In this way, the magnetization of this layer is alterable at low external fields, whereas the latter remains unchanged.

The Josephson current through a ferromagnet with a domain wall has been studied in, e.g., [65, 70–72], where it was shown that the current depends strongly on both how rapidly the magnetization texture in the domain wall varies and its spatial extension. For instance, when the variation is very rapid, the effective magnetization cancels out and the result becomes similar to that of an S|N|S junction. To see this, one may consider the result of [70] where it was derived that the Josephson current in the presence of a spiral magnetization structure acquires an additional term compared to the homogeneous case. This term is proportional to $\sum_{\omega_n}(Ql)^2\mathrm{e}^{-\kappa d_F}$, where Q is the wavevector characterizing the spiral structure, l is the mean free path, and d_F is the thickness of the F layer, while

$$\kappa = \sqrt{2(\omega_n/D) + 4h^2/(DQ)^2}. \tag{14.37}$$

Above ω_n is the fermionic Matsubara frequency. As seen, the wavevector κ above reduces to that found in an S|N|S layer when $Q \to \infty$.

Although the behavior of the Josephson current thus in principle can be tuned by altering Q, this is not feasible experimentally. Turning instead to the case of artificially created consecutive F layers with misaligned magnetizations, the possibility of exerting active control over the Josephson current via the misalignment angle opens up. Several studies, most of them considering the diffusive regime of transport [62–67], studied the influence of the misalignment angle ϕ on the Josephson current. While no simple analytical dependence on ϕ could be identified in general, it was shown how varying ϕ could trigger a long-range Josephson effect [64], strongly modify the $0-\pi$ phase diagram [65], and also generate a very large inverse magnetoresistance effect [66].

It should be mentioned that the majority of works considering S|F|S Josephson junctions with multiple misaligned F layers did not take into account any normal spacer between the F layers. In an experimental situation, this is absolutely necessary to do in order to reduce the exchange coupling between the F layers which would work against any misalignment angle apart from 0 or π, depending on the distance between the layers. Qualitatively, it is probably a reasonable approximation to neglect the normal spacer since it just reduces the proximity effect. In order to obtain quantitatively correct predicitions for the actual magnitude of phenomena such as the long-range triplet Josephson effect, it would, however, be necessary to take into account the normal spacers as shown in Fig. 14.4b.

14.2.4.3 Spin-Josephson Current

Another consequence of inhomogeneous magnetization, be it in the form of multiple misaligned layers or intrinsic nonuniformity within a single ferromagnetic layer, is that the Josephson current should become spin-polarized. This has been noted by several authors in the context of superconductors coexisting with helimagnetic or spiral magnetic order [73, 74] as well as ferromagnetic superconductors [75–77]. In each case, the charge- and spin-currents are proportional to a term $\cos(\phi + \alpha_0)$ where α_0 is a constant offset phase, suggesting that both the charge- and spin-transport can be tuned actively by varying ϕ. There is presently no experimental way of measuring directly a spin-current, and one must find signatures of the spin-current indirectly via some consequence it exerts on a charge-distribution and/or charge-current. Nevertheless, the prospect of a spin-supercurrent is certainly fascinating and may find use in future applications.

14.2.5 F|S|F Spin-valves

In the previous section, we discussed how the misalignment of the magnetization vectors of F layers in contact with superconductors could be used to control the Josephson current. In a related scenario, one can imagine a superconductor sandwiched between two F layers with misaligned magnetization, as shown in Fig. 14.5 for a parallel and antiparallel alignment. It turns out that such a setup offers an arena for the observation of two very interesting phenomena: the spin-switch effect and crossed Andreev reflection.

14.2.5.1 Controlling T_c by a Spin-switch

When a superconductor of finite size d_S is inserted between two ferromagnetic layers, the proximity effect causes the superconducting order parameter, and consequently the critical temperature, to be suppressed. While this seems obvious, it is less obvious that the critical temperature (controlling whether the material is superconducting or not) can be carefully controlled by the relative magnetization orientation of the F layers. For example, if the critical temperature is higher in the antiparallell (AP) alignment ($\phi = \pi$) than in the P alignment, it means that the spin-valve structure in Fig. 14.5a can act as an on–off switch for superconductivity simply by reversing one of the magnetization vectors.

The spin-switch effect has been studied in several papers, but we would here like to make particular mention of the work by Fominov et al. [78, 79], who developed a robust method for accurately calculating the critical temperature in such a spin-valve structure in the diffusive limit. The problem must in general be tackled numerically, although an analytical solution is viable under limiting circumstances [80]. In the majority of experiments, the critical temperature seems to be higher in

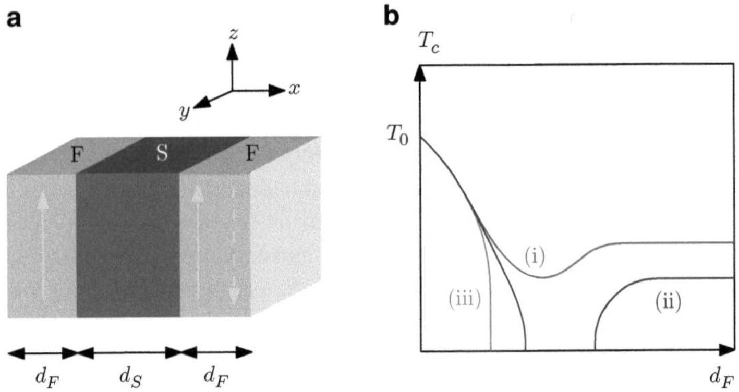

Fig. 14.5 (**a**) F|S|F trilayer where the magnetizations of each F layer are in a parallel (*full line*) or antiparallel (*dashed line*) alignment. The widths of the ferromagnetic and superconducting regions are d_F and d_S, respectively. (**b**) Characteristic behavior of the critical temperature in F|S and F|S|F layers, where T_0 denotes the critical temperature in the absence of any ferromagnetic region. In case (i), the critical temperature T_c decays in a nonmonotonic way and then saturates at a finite value, featuring a minimum at a particular value of d_F. In case (ii), T_c exhibits a reentrant behavior where it vanishes in a finite interval of d_F and then saturates similarly to case (i). In case (iii), T_c decays in a monotonic fashion and vanishes at a critical value of d_F

the AP alignment compared to the P alignment. The physical reason for this may be understood by looking at the interaction of the exchange field with the spin of the Cooper pair. In the AP alignment, the net effective field is reduced in the middle of the S region. The Cooper pair can in this way be more easily be accommodated in the AP alignment. In the P alignment, the pair-breaking becomes stronger and the critical temperature is reduced.

The critical temperature exhibits a rich variety of behavior as a function of the ferromagnetic layer width d_F. This behavior is similar in both F|S and F|S|F junctions. As shown schematically in Fig. 14.5b, there are mainly three types of behavior which we shall explain the origin of below. In case (1), the critical temperature T_c decays in a nonmonotonic way and then saturates at a finite value, featuring a minimum at a particular value of d_F. In case (2), T_c exhibits a reentrant behavior where it vanishes in a finite interval of d_F and then saturates similarly to case (1). Finally, in case (3) T_c decays in a monotonic fashion and vanishes at a critical value of d_F. The key to understand this variety of behavior lies in considerations of the proximity effect. We here follow the arguments put forward in [78]. The saturation of T_c at large d_F can be attributed to the fact that only a region extending a distance λ into the F region will influence the superconductor by means of the proximity effect. Once the F layer thickness exceeds this value, $d_F > \lambda$, T_c remains unaltered compared to its value when the F region has a thickness $\sim \lambda$. The abrupt decrease shown in case (3) is related to the order of the superconducting phase transition changing from an extremely weak first order without the proximity effect (practically speaking second order) to a non-weak first-order transition, as discussed in [81]. Finally, the oscillations of the critical temperature pertain to the FFLO-like oscillating character of the proximity-influenced superconducting order parameter near the interface.

14.2.5.2 Crossed Andreev Reflection and Entanglement

Local superconducting correlations typically extend over a distance ξ, known as the superconducting coherence length. This can also be thought of as the size of a Cooper pair, which is the basic constituent of the superconducting condensate. In a macroscopic superconductor, the correlations can of course extend over much larger distances than ξ as the Cooper pair is not a bound-state in real space. In fact, the electrons continuously find new partners to make up a Cooper pair as they travel through the lattice, in this way mediating the superconducting correlations over macroscopic distances and thus ensuring that the superconducting phase-coherence remains intact.

The size ξ of the Cooper pair plays a pivotal role in proximity structures where a superconductor couples two non-superconducting leads, as in Fig. 14.5a. If the width d_S of the superconductor is of the same order as the coherence length, an incoming electron in the left F region can combine with an electron from the right F region and enter the superconductor as a Cooper pair. The byproduct of this process, known as *crossed Andreev reflection* (CAR), is a hole (missing electron) in the right F region which propagates away from the interface. In effect, CAR is a nonlocal quantum transport process in which an incoming electron in one lead triggers the propagation of a hole in a spatially separated lead. This is made possible by the superconductor sandwiched between the leads. One may also think of CAR in terms of that a voltage difference applied to one part of the system (the left lead and the superconductor) induces an electric current in another spatially distinct part of the system (the superconductor and the right lead) without any applied bias voltage.

We here outline how the transport properties of an F|S|F spin-valve may be investigated, focusing on the ballistic limit. In this case, one may employ a scattering matrix formalism to calculate the current and/or conductance of the system as a function of applied bias voltage, magnitude of the exchange field, width of the superconducting layer, etc. The starting point is the Bogolioubov–de Gennes (BdG) equations in the F and S regions. Following the notation of [82], it may be written as:

$$\begin{pmatrix} -\frac{\nabla^2}{2m} - \mu - \sigma h(z) & \Delta(z) \\ \Delta^*(z) & \frac{\nabla^2}{2m} + \mu - \sigma h(z) \end{pmatrix} \psi_\sigma = E\psi_\sigma. \tag{14.38}$$

Here, m is the electron mass, μ is the chemical potential, h is the exchange field, Δ is the superconducting gap, and $\sigma = \pm 1$ is a spin index. The z-dependence of h and Δ is such that they are non-zero only in the ferromagnetic and in the superconducting region, respectively. Strictly speaking, the gap should be multiplied by a constant σ to account for its bulk spin-singlet symmetry, but this factor can just be gauged away so long as the two sets of BdG equations for $\sigma = \pm 1$ remain uncoupled as here. Note that the BdG equations contain information about processes occurring on the Fermi wave-vector lengthscale, as opposed to the quasiclassical theory outlined in a previous section. From the above BdG equation, one may construct the relevant scattering states in the F and S regions. They read

$$\psi_\sigma^F = \left\{ \begin{pmatrix} 1 \\ 0 \end{pmatrix} \mathrm{e}^{\pm \mathrm{i} p_\sigma^+ z}, \begin{pmatrix} 0 \\ 1 \end{pmatrix} \mathrm{e}^{\pm \mathrm{i} p_\sigma^- z} \right\}. \tag{14.39}$$

Here, we have introduced the notation $p_\sigma^\pm = p_\sigma^\pm(h) = \sqrt{2m[\mu \pm (E + \sigma h)]}$. Note that if the ferromagnets are in the AP alignment, then the substitution $h \to (-h)$ must be performed in one of them. In the S region, we have,

$$\psi^S = \left\{ \begin{pmatrix} u \\ v \end{pmatrix} \mathrm{e}^{\pm \mathrm{i} k^+ z}, \begin{pmatrix} v \\ u \end{pmatrix} \mathrm{e}^{\pm \mathrm{i} k^- z} \right\}, \tag{14.40}$$

with $k^\pm = \sqrt{2m(\mu \pm \sqrt{E^2 - \Delta^2})}$, featuring the usual coherence factors $u_0^2 = 1 - v_0^2 = \frac{1}{2}[1 + \sqrt{E^2 - \Delta^2}/E]$.

Using the above basis set of scattering states, one can write down the wavefunction in the entire system (the left F region, the S region, and the right F region). Consider for concreteness an incoming electron with spin σ from the left F region moving toward the superconductor. In that case, we write

$$\begin{aligned}
\psi_\sigma^{\text{left}} &= \begin{pmatrix} 1 \\ 0 \end{pmatrix} \mathrm{e}^{\mathrm{i} p_\sigma^+ x} + r_\sigma^h \begin{pmatrix} 0 \\ 1 \end{pmatrix} \mathrm{e}^{\mathrm{i} p_\sigma^- x} + r_\sigma^e \begin{pmatrix} 1 \\ 0 \end{pmatrix} \mathrm{e}^{-\mathrm{i} p_\sigma^+ x}, \\
\psi_\sigma^{\text{SC}} &= a_\sigma \begin{pmatrix} u \\ v \end{pmatrix} \mathrm{e}^{\mathrm{i} k^+ x} + b_\sigma \begin{pmatrix} u \\ v \end{pmatrix} \mathrm{e}^{-\mathrm{i} k^+ x} + c_\sigma \begin{pmatrix} v \\ u \end{pmatrix} \mathrm{e}^{\mathrm{i} k^- x} + d_\sigma \begin{pmatrix} v \\ u \end{pmatrix} \mathrm{e}^{-\mathrm{i} k^- x}, \\
\psi_\sigma^{\text{right}} &= t_\sigma^e \begin{pmatrix} 1 \\ 0 \end{pmatrix} \mathrm{e}^{\mathrm{i} q_\sigma^+ x} + t_\sigma^h \begin{pmatrix} 0 \\ 1 \end{pmatrix} \mathrm{e}^{-\mathrm{i} q_\sigma^- x}.
\end{aligned} \tag{14.41}$$

Above, $q = p(h)$ in the P alignment, while $q = p(-h)$ in the AP alignment. The coefficients $\{r_\sigma^e, r_\sigma^h, a_\sigma, b_\sigma, c_\sigma, d_\sigma, t_\sigma^e, t_\sigma^h\}$ denote the scattering coefficients of the problem. For instance, r_σ^e denotes normal reflection for spin σ, whereas t_σ^e denotes transmission of an electron with spin σ. To determine these unknown coefficients, we make use of the continuity of the wavefunction at the interfaces which allows us to eliminate four of the unknown coefficients. The above boundary conditions do not suffice, however, to fully determine all scattering amplitudes. We get an additional set of boundary conditions by integrating the BdG equations over an infinitesimal interval $[\epsilon, -\epsilon]$ across the interface with $\epsilon \to 0$. To account for the increased resistance at the interface, we introduce a barrier potential of the form $V_0\delta(x)$, which effectively breaks the continuity of the derivative of the wavefunctions across the interface. Specifically, we obtain the following set of boundary conditions:

$$(\partial_x \psi_\sigma^{\text{SC}} - \partial_x \psi_\sigma^{\text{left}})|_{x=x_L} = 2mV_0 \psi_\sigma^{\text{left}}(x = x_L),$$
$$(\partial_x \psi_\sigma^{\text{right}} - \partial_x \psi_\sigma^{\text{SC}})|_{x=x_R} = 2mV_0 \psi_\sigma^{\text{right}}(x = x_R). \qquad (14.42)$$

where $x_{L(R)}$ denotes the location of the left (right) interface. Inserting (14.41) into the continuity of the wavefunction at the interfaces and using (14.42) allows for determination of the scattering coefficients. In general, the analytical expressions will be quite cumbersome and a numerical solution is often to be preferred. If the geometry is simple as in a bilayer structure, the analytical expressions simplify greatly and can give considerable insight into the physics of the junction. The calculation of current and conductance is then accomplished using the transmission and reflection coefficients in conjunction with the relevant distribution functions for the incoming and outgoing quasiparticle populations. The specific expressions will differ depending on how the various parts of the system are biased with voltage. It should be noted that if one considers a voltage eV applied symmetrically to the junction, i.e. $eV/2$ on the left lead and $-eV/2$ on the right lead, incoming holes from the right side must be taken into account to yield the correct result for the electric current [83, 84].

14.2.6 Future Prospects

Finally, we would like to point out some possible routes for future research which the present authors believe are both interesting and appropriate for further development in this field. One important point regards a more realistic modeling of the interface properties and the magnetization texture in the ferromagnet. The crucial role played by the spin-discriminating properties of the interface have recently been highlighted, particularly in the context of half-metallic ferromagnet|superconductor junctions [52]. Most works so far on F|S structures have completely neglected the spin-dependent properties of the interface, including the aforementioned spin-DIPS and also magnetoresistance effects associated with different transmission probabilities for the two spin species. Similarly, most works have restricted their attention to the case of homogeneous magnetization, whereas domain wall structures and other intrinsical inhomogeneities such as conical ferromagnetism have received far less attention. Both the interface properties and a proper accounting of the magnetization texture are also important with regard to improved understanding and modeling of experimental data. A second venue which has emerged very recently is the study of proximity-induced FM and SC order in exotic condensed-matter systems such as graphene and topological insulators. The peculiar electronic properties in graphene have been shown to qualitatively alter the behavior of Andreev reflection in both N|S and F|S structures [85, 86]. Similarly, quantum transport in topological insulators under the influence of the proximity effect have been shown to give rise to novel properties in Josephson junctions due to the existence of Majorana fermions [87].

14.3 Intrinsic Coexistence: Ferromagnetic Superconductors

The interplay between FM and SC long-range order microscopically coexisting in the same material has attracted much interest during the last decade due to the discovery of superconductivity in ferromagnetic metals, UGe_2, URhGe, UIr, and UCoGe [13, 88–91]. The internal magnetic fields in these materials exceed strongly the Pauli limit, suggesting that spin-singlet superconductivity should be ruled out. In fact, a careful analysis concluded [92] that the coexistence state of spin-singlet pairing and ferromagnetism always turns out to be energetically unfavorable against the nonmagnetic superconducting state even if a finite-momentum pairing state is considered. Later, it was proposed [93] that the coexistence of metallic ferromagnetism and singlet superconductivity may be realized assuming that the magnetic instability is due to kinetic exchange. However, the coexistence of magnetism and spin-triplet superconductivity appears to be a more promising scenario, since the Cooper pairs may use their spin degree of freedom to align themselves with the internal magnetic field. In what follows, we shall first review some of the main experimental results indicating the coexistence of FM and SC order, proceeding later to a phenomenological treatment of ferromagnetic superconductors. Finally, we present results for the thermodynamic and transport properties of ferromagnetic superconductors with various pairing symmetries. For a much more comprehensive review of the experimental results available for ferromagnetic superconductors than what can be offered here, the reader is refered to [94].

14.3.1 Experimental Results

One of the most intriguing experimental facts is that in all ferromagnetic superconductors, except UCoGe, the SC phase is only observed in the ferromagnetic part of the phase diagram. In effect, it does not appear in the paramagnetic regime. The phase-diagram in the T-p plane is illustrated for UGe_2 in Fig. 14.6. Such an observation indicates that the superconductivity may be dependent on ferromagnetism for its very existence in this scenario. Superconductivity was confirmed in these materials by resistivity measurements, which exhibit a sudden and sharp drop at $T = T_c$. Later, specific heat [96] and flux flow resistance [14] measurements confirmed the presence of bulk superconductivity in UGe_2. The magnetization of the samples could be directly measured down to T_c, but is screened in the SC state due to Meissner currents. To verify explicitly the presence of bulk ferromagnetism also in the superconducting phase, Huxley et al. utilized neutron scattering measurements [14]. In UGe_2, ferromagnetic order appears at $T_c = 52$ K under ambient pressure with an exchange splitting estimated to lie around $h = 70$ meV [13]. As seen from the phase diagram, superconductivity emerges only under high pressure, and the transition temperature T_c is typically two orders of magnitude smaller than the Curie temperature T_M.

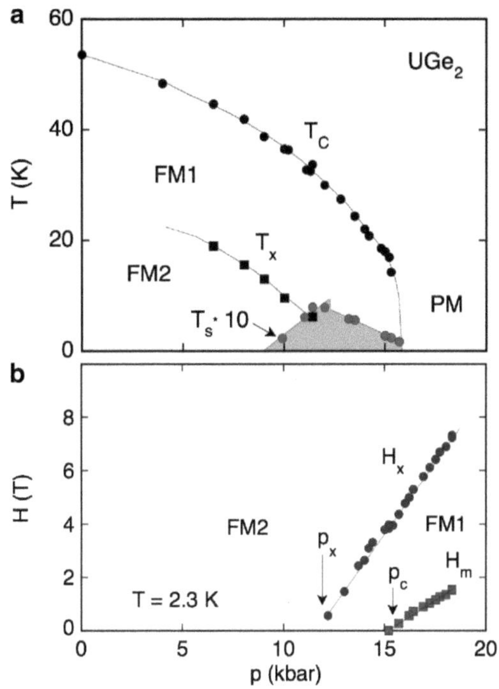

Fig. 14.6 (**a**) Phase diagram in the p-T (pressure–temperature) plane for UGe_2. (**b**) Magnetic field vs. pressure phase diagram of UGe_2. The plot is taken from [95]. Above, T_c denotes the Curie temperature whereas T_S denotes the superconducting transition temperature. Superconductivity is only observed near the ferromagnetic transition. For more details on the phase diagram, the reader is refered to [95]

In URhGe, superconductivity exists deep inside the ferromagnetic regime, just as in UGe_2. These two compounds are also similar in the regard that the exchange splitting is considerable and that experimental data indicate that the ferromagnetic order is itinerant with strongly delocalized 5f-electrons [97]. It may be further noted that the ratio of the Curie temperature T_M to the maximum superconducting transition temperature T_c is very similar in UGe_2 ($T_M/T_c \simeq 30/0.8 = 37.5$) and URhGe ($T_M/T_c \simeq 9.6/0.25 = 38.4$) [94]. The situation is somewhat different for the ferromagnetic superconductors UCoGe and UIr, where superconductivity only appears on the border of the ferromagnetic regime, i.e. not deep inside the part of the phase diagram where FM order is established. As a result, the ordered magnetic moment is much smaller than in UGe_2 and URhGe, and the ratio of the Curie temperature vs. maximum superconducting transition temperature is also substantially reduced. The UCoGe samples studied in [91] were in the ballistic regime of transport $\xi < l$, where ξ is the coherence length and l is the mean free path. For UIr, a qualitatively new aspect compared to the other ferromagnetic superconductors mentioned so far comes into play: namely, it has a non-centrosymmetric lattice structure. This absence of inversion symmetry actually has crucial implications for the allowed symmetries of the order parameter and gives rise to an interesting interplay with ferromagnetic order [98].

The fact that all known ferromagnetic superconductors contain uranium hardly seems like a coincidence. What, then, is the distinctive property of uranium that allows the heavy-fermion compounds UGe_2, UCoGe, UIr, and URhGe to accommodate a ferromagnetic superconducting state? It is interesting to observe that heavy fermion superconductivity is found in these materials at the border of a localization transition of the f-electrons [94]. The accompanying redistribution of charge-density should also influence correlations pertaining to the spin-density, which could be linked to the formation of superconductivity in a ferromagnetic background. A precise answer to the above question nevertheless remains unknown it most likely holds an important clue to understanding the interplay between FM and SC order in these heavy-fermion f-electron compounds.

14.3.2 Phenomenological Framework

As of today, there exists no universally accepted theory for the microscopic mechanism leading to superconductivity in a ferromagnetic phase with a large exchange field such as the one in the aforementioned uranium-based heavy-fermion compounds. Therefore, we shall here use a phenomenological approach where we assume that there exists an attractive interaction responsible for superconductivity without specifying its origin. In doing so, it becomes possible to analyze the thermodynamical and transport properties of the superconducting state for a given symmetry of the order parameter. In particular, we will consider systems where superconductivity arises at a lower temperature than at which ferromagnetic order forms. In all ferromagnetic superconductors discovered thus far, this is the case. The reason for this could simply be the fact that the energy scales for the two phenomena are quite different, with the magnetic exchange energy far exceeding the superconducting gap. Another possibility is that superconductivity intrinsically depends on ferromagnetism to form, as suggested in [99]. The thermodynamics and phase diagrams of ferromagnetic superconductors have been investigated thoroughly via a Ginzburg–Landau approach in [100–102].

One of the most fundamental issues to address in this context is the manner in which FM and SC order coexist. In principle, one may envision at least three ways in which this is possible, still respecting the U(1) broken gauge symmetry in the superconducting state which renders the vector potential massive. One possibility is that a ferromagnetic superconductor accommodates FM order through a spontaneous vortex lattice which exists due to the internal field (i.e., even in the absence of any external magnetic field) [103]. Another possibility is that the magnetism is ordered in a domain structure, effectively canceling out any net magnetic field in the bulk of the superconductor. Finally, homogeneous coexistence of FM and SC order is possible in scenarios where the Meissner effect is suppressed, such as in a thin-film geometry. Actually, this is even possible for spin-singlet superconductivity [5, 6] as long as the exchange field does not exceed the Pauli paramagnetic limitation of

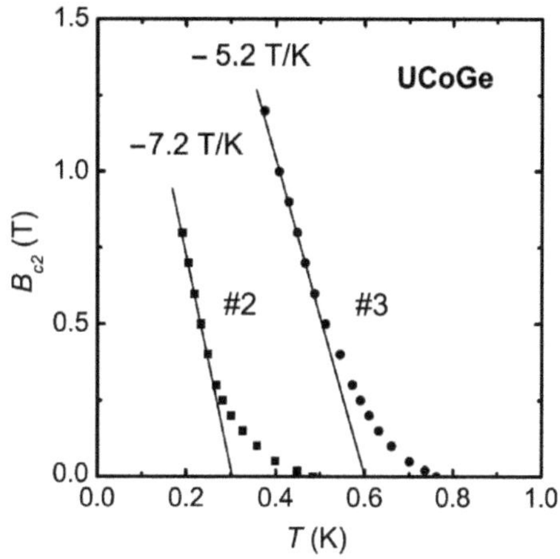

Fig. 14.7 Upper critical field B_{c2} determined by the midpoints of the resistive transitions measured in fixed magnetic fields in the experiment in [91] for their samples #2 and #3. The plot is taken from [91]

$h = \Delta/\sqrt{2}$. For a spin-triplet superconductor, however, once the orbital effect is avoided one need not worry about the paramagnetic limitation since the spin of the Cooper pair can align with the exchange field. In fact, the critical field in UCoGe displays a highly unusual field-dependence with no apparent bounding as $T \to 0$, consistent with the absence of paramagnetic limitation [91] as shown in Fig. 14.7. In a bulk structure where the orbital effect is in full play, the FM and SC order parameters have to be spatially modulated in order to coexist, be it in the form of a vortex lattice or FFLO-like state.

Spin-triplet superconductors are characterized by a multicomponent order parameter, which for the simplest case of the p-wave may be expressed in terms of three independent components of a **d**-vector [104]:

$$\mathbf{d}_k = \Big[\frac{\Delta_{k\downarrow\downarrow} - \Delta_{k\uparrow\uparrow}}{2}, \frac{-\mathrm{i}(\Delta_{k\downarrow\downarrow} + \Delta_{k\uparrow\uparrow})}{2}, \Delta_{k\uparrow\downarrow}\Big]. \tag{14.43}$$

Note that $\mathbf{d}_k$ transforms like a vector under spin rotations. Above, $\Delta_{k\uparrow\downarrow}$ is the $S_z = 0$ triplet component of the gap. In terms of the components of $\mathbf{d}_k$, the order parameter itself is a 2×2 matrix that reads

$$\check{\Delta}_{\alpha\beta}(\boldsymbol{k}) \equiv \langle c_{\boldsymbol{k},\alpha} c_{\boldsymbol{k},\beta} \rangle = [i(\mathbf{d}_k \cdot \boldsymbol{\sigma})\sigma_y]_{\alpha\beta}, \tag{14.44}$$

where $\boldsymbol{\sigma}$ is the vector of Pauli matrices, and $c^\dagger_{\boldsymbol{k},\alpha}, c_{\boldsymbol{k},\alpha}$ are the usual electron creation-annihilation operators for momentum $\boldsymbol{k}$ and spin α.

The superconducting order parameter is characterized as *unitary* if the modulus of the gap is proportional to the unity matrix: $(\check{\Delta} \cdot \check{\Delta}^\dagger) \propto \check{1}$. Written in terms of the vector $\boldsymbol{d}_k$, this condition is equivalent to the requirement that $\langle \mathbf{S}_k \rangle = 0$, where we have introduced the net spin moment of the Cooper pair

$$\langle \mathbf{S}_k \rangle \sim \mathrm{i}(\mathbf{d}_k \times \mathbf{d}_k^*). \tag{14.45}$$

The *unitary* triplet state thus has Cooper pairs with zero spin expectation value, whereas the *nonunitary* state is characterized by non-zero value of $\langle \mathbf{S}_{\boldsymbol{k}} \rangle \neq 0$. The latter effectively means that time-reversal symmetry is spontaneously broken in the spin part of the Cooper pairs [105]. It is thus intuitively clear that having the spin of the Cooper pair aligned with the internal magnetic field of the ferromagnet can lower the energy of the resulting coexistence state. The above argument of an order parameter in ferromagnetic superconductors that must be nonunitary was put forward by Machida and Ohmi [106] and others [107–110]. Distinguishing between unitary and nonunitary states in ferromagnetic superconductors is clearly one of the primary objectives in terms of identifying the correct SC order parameter. To this end, recent studies have focused on calculating transport properties of ferromagnetic superconductors [75, 76, 111–114], an issue we shall return to later.

Inter-subband pairing is expected to be strongly suppressed in the presence of the Zeeman splitting between the $\uparrow, \downarrow$ conduction sub-bands. In other words, only electrons within the same sub-band will form Cooper pairs (the so-called equal-spin pairing) and we shall set $\Delta_{\uparrow\downarrow} = 0$ in what follows. Moreover, the requirement of nonunitarity of the order parameter then reduces to the requirement that the vector $\mathbf{d}_{\boldsymbol{k}}$ in (14.43) should have two non-zero components, i.e., $\Delta_{\uparrow\uparrow} \neq \Delta_{\downarrow\downarrow}$. The spin of the Cooper pair is then $\langle S_z \rangle \sim \frac{1}{2}(|\Delta_{\uparrow\uparrow}|^2 - |\Delta_{\downarrow\downarrow}|^2)$ and is aligned with the magnetic field (z being the spin quantization axis).

Envisioning a nonunitary triplet state in ferromagnetic superconductors, where the spin-$\uparrow$ and spin-$\downarrow$ fermions pair up separately (no interband-pairing), one might ask whether two separate superconducting transitions take place for each spin species. Indeed, in a model where the two spin flavors are decoupled, the superconducting critical temperature is expected to be different for condensation of each spin flavor. This would be manifested, e.g., in two distinct jumps in the specific heat of the sample, each at a temperature $T = T_{\mathrm{c},\sigma}$. In the presence of spin–orbit coupling, however, only there is only one transititon temperature into the superconducting state. In f-electron heavy-fermion compounds, spin–orbit coupling effects are expected to be strong due to the high atomic number, and only one superconducting transition temperature has been observed in the experiments so far.

In order to proceed analytically, we here restrict our attention to a model of a ferromagnetic superconductor described by uniformly coexisting itinerant ferromagnetism and nonunitary, spin-triplet superconductivity. As mentioned previously, such a model holds when the orbital effect is suppressed, e.g., by the geometry of the sample. It may also serve to illustrate the physics on local scale in the case of a domain-wall formation. The main aim is to identify the manner in which FM and SC order mutually influence each other and if a coexistent state indeed is energetically preferred compared to, for instance, a purely ferromagnetic state. We write down a weak-coupling mean-field theory Hamiltonian with equal-spin pairing Cooper pairs and a finite magnetization along the easy-axis similar to the model studied in [[115, 116]], namely,

$$\hat{H} = \sum_{k} \xi_{k} + \frac{INM^2}{2} - \frac{1}{2}\sum_{k\sigma} \Delta^{\dagger}_{k\sigma\sigma} b_{k\sigma\sigma}$$
$$+ \frac{1}{2}\sum_{k\sigma} \begin{pmatrix} \hat{c}^{\dagger}_{k\sigma} \hat{c}_{-k\sigma} \end{pmatrix} \begin{pmatrix} \xi_{k\sigma} & \Delta_{k\sigma\sigma} \\ \Delta^{\dagger}_{k\sigma\sigma} & -\xi_{k\sigma} \end{pmatrix} \begin{pmatrix} \hat{c}_{k\sigma} \\ \hat{c}^{\dagger}_{-k\sigma} \end{pmatrix}, \tag{14.46}$$

where $b_{k\sigma\sigma} = \langle c_{-k\sigma} c_{k\sigma} \rangle$ is the non-zero expectation value of the pair of Bloch states. Applying a standard diagonalization procedure, we arrive at

$$\hat{H} = H_0 + \sum_{k\sigma} E_{k\sigma} \hat{\gamma}^{\dagger}_{k\sigma} \hat{\gamma}_{k\sigma},$$
$$H_0 = \frac{1}{2}\sum_{k\sigma} (\xi_{k\sigma} - E_{k\sigma} - \Delta^{\dagger}_{k\sigma\sigma} b_{k\sigma\sigma}) + \frac{INM^2}{2}, \tag{14.47}$$

where $\{\hat{\gamma}_{k\sigma}, \hat{\gamma}^{\dagger}_{k\sigma}\}$ are new fermion operators and the eigenvalues read

$$E_{k\sigma} = \sqrt{\xi^2_{k\sigma} + |\Delta_{k\sigma\sigma}|^2}. \tag{14.48}$$

It is implicit in our notation that $\xi_{k} = \varepsilon_{k} - E_F$ is measured from the Fermi level, where ε_{k} is the kinetic energy. The free energy is obtained through

$$F = H_0 - \frac{1}{\beta}\sum_{k\sigma} \ln(1 + e^{-\beta E_{k\sigma}}), \tag{14.49}$$

such that the gap equations for the magnetic and superconducting order parameters become [115]

$$M = -\frac{1}{N}\sum_{k\sigma} \frac{\sigma \xi_{k\sigma}}{2E_{k\sigma}} \tanh(\beta E_{k\sigma}/2),$$
$$\Delta_{k\sigma\sigma} = -\frac{1}{N}\sum_{k'} V_{kk'\sigma\sigma} \frac{\Delta_{k'\sigma\sigma}}{2E_{k'\sigma}} \tanh(\beta E_{k'\sigma}/2). \tag{14.50}$$

Specifically, we now consider a model which should be of relevance to the ferromagnetic superconductor UGe_2, and possibly also for UCoGe and URhGe. In [[117]], it was argued that the majority spin (spin-↑ in our notations) fermions were gapped and that the order parameter displayed line nodes, while the minority (spin-↓) fermions remained gapless at the Fermi level in the heavy-fermion compound UGe_2. More specifically, Harada et al. [117] reported on ^{73}Ge nuclear-quadrupole-resonance experiments performed under pressure, in which the nuclear spin-lattice relaxation rate revealed an unconventional nature of superconductivity implying that the majority spin band in UGe_2 was gapped with line nodes, while the minority spin band remained gapless at the Fermi level. An obvious mechanism for suppressing the superconducting instability in the minority-spin channel as compared to the majority-spin channel is the difference in density of states (DOS) at the Fermi level. Indeed, from Fig. 1 in [[115]] (see also Fig. 4 in [[116]]), it is seen that the critical temperature for pairing in the minority-spin sub-band, $T_c^{\downarrow}$, is predicted to be much smaller than the critical $T_c^{\uparrow}$ for the majority-spin sub-band, even for quite weak magnetic exchange splittings. Given the already quite low critical temperature T_c that is observed experimentally in ferromagnetic superconductors ($T_c \lesssim 1$ K), which we associate with $T_c^{\uparrow}$, we therefore conclude that it might indeed be very hard to observe experimentally the even smaller gap in the minority-spin sub-band. Therefore, it is permissible to only consider pairing in the majority-spin channel and neglect a small (if any) pairing between minority-spin electrons. In our notation, this means setting $M \neq 0, \Delta_{\boldsymbol{k}\uparrow\uparrow} \neq 0, \Delta_{\boldsymbol{k}\downarrow\downarrow} = 0$.

We stress that the above statement, although intuitively attractive, may need further justification since we have so far neglected completely the spin–orbit interaction that is expected to be strong in urainium-based compounds. The effect of the latter would be to provide some effective coupling between majority and minority spin sub-bands and would probably lead to induced SC order parameter in minority spin channel. This issue has been addressed in [98].

To model the presence of line nodes in the order parameter, we choose

$$\Delta_{\boldsymbol{k}\uparrow} = \Delta_{\bar{\boldsymbol{k}}_F\uparrow\uparrow} = \Delta_0 \cos\theta, \tag{14.51}$$

where $\bar{\boldsymbol{k}}_F$ is the normalized Fermi wave-vector, such that the gap only depends on the direction of the latter. This is the weak-coupling approximation. The above gap satisfies the correct symmetry requirement dictated by the Pauli principle, namely a sign change under inversion of momentum, $\theta \to \pi - \theta$. Here, θ is the azimuthal angle in the xy-plane. Our choice of this particular symmetry for the p-wave superconducting gap is motivated by the experimental results of Harada et al. [117]. The $\cos\theta$-dependence is also in accord with the results of [[118]], which showed that the majority band at the Fermi level for UGe_2 is strongly anisotropic with a small dispersion along the k_y-direction. In our model, the pairing potential may be written as

$$V(\theta, \theta') = -g \cos\theta \cos\theta', \tag{14.52}$$

where g is the weak-coupling constant. Conversion to integral equations is accomplished by means of the identity

$$\frac{1}{N}\sum_{\boldsymbol{k}} f(\xi_{\boldsymbol{k}\sigma}) = \frac{1}{V_\Omega}\int\int \mathrm{d}\varepsilon \mathrm{d}\Omega N^\sigma(\varepsilon) f(\varepsilon, \Omega), \tag{14.53}$$

where $N^\sigma(\varepsilon)$ is the spin-resolved density of states and Ω denotes the angular dependence such that $V_\Omega = 4\pi$. In three spatial dimensions, this may be calculated from the dispersion relation using the formula:

$$N^{\sigma}(\varepsilon) = \frac{V}{(2\pi)^3} \int_{\varepsilon_{k\sigma}=\text{const}} \frac{\mathrm{d}S_{\varepsilon_{k\sigma}}}{|\hat{\nabla}_{k}\varepsilon_{k\sigma}|}. \tag{14.54}$$

With the dispersion relation $\xi_{k\sigma} = \varepsilon_{k} - \sigma IM - E_F$, one obtains

$$N^{\sigma}(\varepsilon) = \frac{mV\sqrt{2m(\varepsilon + \sigma IM + E_F)}}{2\pi^2}. \tag{14.55}$$

In their integral form, (14.50) for the order parameters read

$$\begin{aligned} M &= -\frac{1}{4\pi}\sum_{\sigma}\sigma\int_0^{2\pi}\int_{-E_F-\sigma IM}^{\infty} \mathrm{d}\theta\mathrm{d}\varepsilon\frac{\varepsilon N^{\sigma}(\varepsilon)}{E_{\sigma}(\varepsilon,\theta)} \\ &\quad\times \tanh[\beta E_{\sigma}(\varepsilon,\theta)/2], \\ 1 &= \frac{g}{4\pi}\int_0^{2\pi}\int_{-\omega_0}^{\omega_0} \mathrm{d}\theta\mathrm{d}\varepsilon\frac{N^{\uparrow}(\varepsilon)\cos^2\theta}{E_{\uparrow}(\varepsilon)}\tanh[\beta E_{\uparrow}(\varepsilon,\theta)/2]. \end{aligned} \tag{14.56}$$

For ease of notation, we also define

$$\begin{aligned} \Delta_{\sigma}(\theta) &= \begin{Bmatrix} \Delta_0\cos\theta \text{ if } \sigma = \uparrow \\ 0 \qquad\quad \text{ if } \sigma = \downarrow \end{Bmatrix}, \\ E_{\sigma}(\varepsilon,\theta) &= \begin{Bmatrix} \sqrt{\varepsilon^2 + \Delta_0^2\cos^2\theta}. \text{ if } \sigma = \uparrow \\ \varepsilon \qquad\qquad\qquad\quad \text{ if } \sigma = \downarrow \end{Bmatrix}. \end{aligned} \tag{14.57}$$

For the following treatment, we define $\widetilde{M} = IM/E_F$, i.e., the exchange energy scaled on the Fermi energy. Moreover, we set $c = gN(0)/2$ to a typical value of 0.2 and $\tilde{\omega}_0 = \omega_0/E_F = 0.01$ as the typical spectral width of the bosons responsible for the attractive pairing potential. Finally, we define the parameter $\tilde{I} = IN(0)$ as a measure of the magnetic exchange coupling. As discussed below, only for $\tilde{I} > 1$ will a spontaneous magnetization appear in our model, in agreement with the Stoner criterion for itinerant ferromagnetism.

At zero temperature, the superconducting gap equation reads

$$1 = \frac{g}{4\pi}\int_0^{2\pi}\int_{-\omega_0}^{\omega_0} \mathrm{d}\theta\mathrm{d}\varepsilon\frac{N^{\uparrow}(\varepsilon)\cos^2\theta}{\sqrt{\varepsilon^2 + \Delta_0^2\cos^2\theta}}. \tag{14.58}$$

Under the assumption that $\omega_0 \gg \Delta_0$, we obtain that

$$\frac{2}{c\sqrt{1+\widetilde{M}}} = \ln\!\left(\frac{2\omega_0}{\Delta_0}\right) - \frac{1}{\pi}\int_0^{2\pi} \mathrm{d}\theta\cos^2\theta\ln|\cos\theta|. \tag{14.59}$$

which may be solved to yield the zero-temperature gap

$$\Delta_0 = 2.426\exp[-2/(c\sqrt{1+\widetilde{M}})]. \tag{14.60}$$

By inserting (14.60) into the gap equation for the magnetization in (14.56), we have managed to decouple the self-consistency equations for M and Δ_0. Numerical evaluation reveals that the gap equation for M is completely unaffected by the presence of Δ_0, which physically means that the magnetization remains unaltered with the onset of superconductivity. This is reasonable in a model where the energy scale for the onset of magnetism is vastly different from the energy scale for superconductivity, such that by the time superconductivity sets in, the ordering of the spins essentially exhausts the maximum possible magnetization.

The onset of a spontaneous magnetization for $\tilde{I} > 1$ is the well-known Stoner criterion for an isotropic electron gas, where the spin susceptibility may be written as [119]

$$\begin{aligned}\chi(\mathbf{q},\omega) &= \frac{\chi_0(\mathbf{q},\omega)}{1 - I\chi_0(\mathbf{q},\omega)},\\ \chi_0(\mathbf{q},\omega) &= N_0\Big(1 - \frac{\mathbf{q}^2}{12k_F^2} + \mathrm{i}\frac{\pi\omega}{2v_F|\mathbf{q}|}\Big),\\ &|\mathbf{q}| \ll 2k_F,\ \ \omega \ll E_F.\end{aligned} \tag{14.61}$$

For a parabolic band, the static susceptibility is maximal for $\mathbf{q} = 0$, where

$$\chi(\mathbf{q} = 0, \omega = 0) = \frac{N_0}{1 - IN(0)} = \frac{N_0}{1 - \tilde{I}}. \tag{14.62}$$

The introduction of a ferromagnetic order is demarcated by the divergence of the susceptibility for $\tilde{I} = 1$, which is precisely Stoner's criterion for itinerant ferromagnetism. In the absence of superconductivity, the self-consistency equation for the magnetization at $T = 0$ reduces to

$$h = -\frac{\tilde{I}}{3\sqrt{E_F}}\sum_\sigma \sigma(E_F - \sigma h)^{3/2}, \tag{14.63}$$

where $h = IM$ is the exchange splitting of the majority and minority bands. Since the energy scales for the magnetic and superconducting order parameter differ so greatly in magnitude, (14.63) is an excellent approximation even in the coexistent state. The critical temperature for the superconducting order parameter is obtained in the standard way [setting $\Delta_0 = 0$ in (14.58)] to yield

$$T_\mathrm{c} = 1.134\omega_0 \exp[-2/(c\sqrt{1 + \tilde{M}})]. \tag{14.64}$$

The present gap equation for the superconducting order parameter has a unique non-trivial solution, which guarantees that the system will prefer to be in the coexistent state of ferromagnetism and superconductivity. This is because the free energy must be bounded from below, and if only one extremal point is found one can be certain that it is a minimum.

14.3.3 Probing the Pairing Symmetry

So far, we have restricted our attention to the nodal gap symmetry $\Delta = \Delta_0 \cos\theta$. Another possibility is that the analogue of an A2-phase in liquid ^{3}He is realized, with a nonunitary state featuring gaps $\Delta_\sigma = \Delta_{\sigma,0}e^{i\theta}$. In the former case, the magnitude of the order parameter is anisotropic whereas in the latter it is isotropic, similar to an s-wave state. A similar mean-field analysis as the one performed above shows that the nonunitary ferromagnetic superconducting state is the thermodynamically preferred ground-state [116].

It would be highly desirable to find experimental signatures that may distinguish between these types of symmetries. To this end, one possibility is to study the tunneling spectra for a N|ferromagnetic superconductor, which reveals clear fingerprints for the symmetry of the order parameter. This geometry is particularly relevant for STM measurements. The normalized tunneling conductance may be written as:

$$\frac{G}{G_0} = \sum_\sigma \int_{-\pi/2}^{\pi/2} d\theta \cos\theta [1 + |r_\sigma^A(eV,\theta)|^2 - |r_\sigma^N(eV,\theta)|^2], \tag{14.65}$$

where G_0 is the normal-state conductance. Above, $r_\sigma^A(eV,\theta)$ and $r_\sigma^N(eV,\theta)$ designate the Andreev- and normal-reflection coefficients for spin-σ carriers, respectively. The conductance spectra of ferromagnetic superconductors were studied in [111, 112, 120] for a variety of order parameter symmetries. The main discriminating factor between a nodal order parameter, such as the one proposed by Harada et al., and an isotropic p-wave order parameter analogous to the A2-phase in ^{3}He, is that the tunneling spectra will be sensitive to the crystallographic interface orientation for a given plane in the former case, whereas they will be insensitive to it in the latter. In both cases, the electron- and hole-like quasiparticles entering the superconductor may experience a constructive phase-interference, which gives rise to the formation of a zero-energy state that is bound to the surface of the superconductor. The resonance condition for the formation of such zero-energy states is $\Delta(\theta) = -\Delta(\pi - \theta)$, [121] and the bound states are manifested as a giant peak in the zero-bias conductance [122]. Note that such states exist even if the spatial depletion of the superconducting order parameter is not taken into account, which may be shown analytically [123]. Taking into account the reduction the gap experiences close to the interface compared to its bulk value, is known to yield the same qualitative features as the usual step-function approximation, with the exception of additional, smaller peaks at finite bias voltages due to non-zero bound states [124]. For a nodal order parameter such as $\sim \cos\theta$, such a zero-energy peak will vanish if the interface orientation is shifted by a rotation of $\pi/2$. For an isotropic order parameter $\sim e^{i\theta}$, it will remain. To see this explicitly, consider tunneling in the crystallographic ab-plane such that the nodal parameter reads $\Delta_0 k_a$, whereas the isotropic one reads $\Delta_0(k_a + ik_b)$. For a tunneling interface normal to the a-axis, it is seen that the order parameter changes sign when $k_a \to (-k_a)$ in both cases (provided $k_b = 0$ in the isotropic case). For a tunneling interface normal to the b-axis, however, only the A2-order parameter changes sign when $k_a = 0$.

While the spin symmetry of the superconducting order parameter seems highly likely to be a spin-triplet one, the orbital symmetry of the superconducting order parameter in ferromagnetic superconductors is a more subtle issue. In [[116]], a mean-field model for isotropic, chiral p-wave gaps in a background of itinerant ferromagnetism was constructed. In that work, pairing was assumed to occur both for majority- and minority-spins, resulting in, for instance, a double-jump structure in the specific heat capacity. An isotropic, chiral p-wave order parameter has a constant magnitude, which is favorable in terms of maximizing the condensation energy gained in the superconducting state. Assuming an isotropic density of states at the Fermi level and a separable pairing potential of the form $V_{\boldsymbol{k}\boldsymbol{k}'} = -g\lambda_{\boldsymbol{k}}\lambda_{\boldsymbol{k}'}$, the condensation energy gained at $T = 0$ in the superconducting state reads

$$E = -\frac{N(0)\Delta_0^2}{2}\langle|\lambda_{\boldsymbol{k}}|^2\rangle, \tag{14.66}$$

where Δ_0 is the maximum value of the gap and $\langle\ldots\rangle$ denotes the angular average over the Fermi surface. This clearly shows the advantage of an isotropic gap $|\lambda_{\boldsymbol{k}}| = 1$. The general principle is wellknown: the system prefers to have the Fermi surface as gapped as possible. However, factors such as spin–orbit pinning energy and lattice structure may conspire to prevent a fully isotropic gap.

The experiments performed so far are indicative of a single gap, or at least a strongly suppressed second gap, in the ferromagnetic superconductors. For instance, no double-jump features have been observed in the specific heat capacity [13] for UGe_2. This warrants the investigation of a single-gap model, possibly with line nodes as suggested by Harada et al. [117] Theoretically, the absence of the SC gap in the minority spin sub-band can be justified by considering the effect of Zeeman splitting on the electronic density of states [115]. In general, it should be possible to discern the presence of two gaps by analyzing specific heat or point-contact spectroscopy measurements, unless one of the gaps is very small.

We note in passing that from an experimental point of view, a complication with UGe_2 is that the superconductivity does not appear at ambient pressure, in contrast to URhGe and UCoGe. The necessity of considerable pressure restricts the use of certain experimental techniques, and this is clearly a challenge in terms of measuring, for instance, conductance spectra of UGe_2. Another experimental quantity which would be of high interest to obtain from, for instance, *ab initio* calculations is the thermal expansion coefficient, which may be directly probed in high-pressure experiments [125].

14.3.4 Future Prospects

Further refinements leading to a more realistic model of a ferromagnetic superconductor should include the presence of spin–orbit coupling, which inevitably is present in heavy-fermion superconductors, in addition to the presence of vortices when studying a spontaneous vortex lattice phase. In particular, experiments on transport properties of ferromagnetic superconductors, such as the Josephson current and point-contact spectroscopy, would be of high interest to further illucidate the pairing symmetry realized in ferromagnetic superconductors. As mentioned previously, such experiments are probably most viable for URhGe and UCoGe where superconductivity appears at ambient pressure. The question of the pairing mechanism from which superconductivity originates in this compounds also remains open. In particular, it would be desirable to shed light on its relation to the presence of ferromagnetic order since the appearance of superconductivity occurs only inside the ferromagnetic regime, except for in UCoGe.

Acknowledgements The authors would like to acknowledge collaborations and fruitful scientific exchanges with Y. Asano, A. Black-Schaffer, A. Cottet, M. Cuoco, M. Eschrig, A. Nevidomskyy, I. B. Sperstad, Y. Tanaka, Z. Tesanovic, T. Yokoyama, and M. Zareyan.

References

1. J. Bardeen, L.N. Cooper, J.R. Schrieffer, Phys. Rev. 108, 1175 (1957)
2. P.W. Anderson, H. Suhl, Phys. Rev. 116, 898 (1959)
3. B.T. Matthias, H. Suhl, Phys. Rev. Lett. 4, 51 (1960)
4. A.A. Abrikosov, L.P. Gorkov, Zh. Eksp. Teor. Fiz. 39, 1781 (1960) [Sov. Phys. JETP 12, 1243 (1961)]
5. A.M. Clogston, Phys. Rev. Lett. 9, 266 (1962)
6. B.S. Chandrasekhar, Appl. Phys. Lett. 1, 7 (1962)
7. P. Fulde, R.A. Ferrel, Phys. Rev. 135, A50 (1964)
8. A.I. Larkin, Y.N. Ovchinnikov, Zh. Eksp. Teor. Fiz. 47, 1136 (1964)
9. Ø. Fischer, A. Treyvaud, R. Chevrel, M. Sergent, Solid State Commun. 17, 721 (1975)
10. R.N. Shelton, R.W. McCallum, H. Adrian, Phys. Lett. 56A, 213 (1976)
11. B.T. Matthias, E. Corenzwit, J.M. Vandenberg, H. Barz, Proc. Natl. Acad. Sci. USA (4th edn.) 74, 1334 (1977)
12. F. Le Berre, O. Pena, C. Hamard, R. Hory, A. Wojakowski, J. Alloys Comp. 262–263, 406 (1997)
13. S.S. Saxena , P. Agarwal, K. Ahilan, F.M. Grosche, R.K.W. Haselwimmer, M.J. Steiner, E. Pugh, I.R. Walker, S.R. Julian, P. Monthoux, G.G. Lonzarich, A. Huxley, I. Sheikin, D. Braithwaite, J. Flouquet, Nature 406, 587 (2000)
14. A. Huxley, I. Sheikin, E. Ressouche, N. Kernavanois, D. Braithwaite, R. Calemczuk, J. Flouquet, Phys. Rev. B 63, 144519 (2001)
15. V.V. Ryazanov, V.A. Oboznov, A. Yu. Rusanov, A.V. Veretennikov, A.A. Golubov, J. Aarts, Phys. Rev. Lett. 86, 2427 (2001)
16. L.B. Ioffe, V.B. Geshkenbein, M.V. Feigelman, A.L. Fauchere, G. Blatter, Nature (Lond.) 398, 679 (1999)
17. J.E. Mooij, T.P. Orlando, L. Levitov, L. Tian, C.H. van der Wal, S. Lloyd, Science 285, 1036 (1999)
18. F.S. Bergeret, A. Volkov, K.B. Efetov, Rev. Mod. Phys. 77, 1321 (2005)
19. A.I. Buzdin, Rev. Mod. Phys. 77, 935 (2005)
20. M. Lange, M.J. Van Bael, Y. Bruynseraede, V.V. Moshchalkov, Phys. Rev. Lett. 90, 197006 (2003)
21. A.Yu. Aladyshkin, A.V. Silhanek, W. Gillijns, V.V. Moshchalkov, Supercond. Sci. Tech. 22, 053001 (2009)

22. M.Yu. Kupriyanov, V.F. Lukichev, Zh. Exp. Teor. Fiz. 94, 139 (1988)
23. Yu. Nazarov, Superlatt. Microstruct. 25, 1221 (1999)
24. G.E. Blonder, M. Tinkham, T.M. Klapwijk, Phys. Rev. B 25, 4515 (1982)
25. M. Eschrig, J. Kopu, J.C. Cuevas, G. Schön, Phys. Rev. Lett. 90, 137003 (2003)
26. M. Eschrig, T. Löfwander, T. Champel, J.C. Cuevas, J. Kopu, G. Schön, J. Low Temp. Phys. 147 457 (2007)
27. J.W. Serene, D. Rainer, Phys. Rep. 101, 221 (1983)
28. We will use the notation and conventions of J.P. Morten, Master Thesis, Norwegian University of Science and Technology, (2003) (unpublished), which in turn is close the notations used in [27]
29. V.L. Berezinskii, JETP Lett. 20, 287 (1974)
30. F.S. Bergeret, A.F. Volkov, K.B. Efetov, Phys. Rev. Lett. 86, 4096 (2001)
31. A. Volkov, F.S. Bergeret, K.B. Efetov, Phys. Rev. Lett. 90, 117006 (2003)
32. Y. Tanaka, A.A. Golubov, S. Kashiwaya, M. Ueda, Phys. Rev. Lett. 99, 037005 (2007)
33. Y. Tanaka, A.A. Golubov, Phys. Rev. Lett. 98, 037003 (2007)
34. G. Eilenberger, Sov. Phys. JETP 214, 195 (1968)
35. K. Usadel, Phys. Rev. Lett. 25, 507 (1970)
36. N. Schopohl, K. Maki, Phys. Rev. B 52, 490 (1995)
37. N. Schopohl, cond-mat/9804064
38. A. Konstandin, J. Kopu, M. Eschrig, Phys. Rev. B 72, 140501 (2005)
39. A.V. Zaitsev, Sov. Phys. JETP 59, 1163 (1984)
40. Y. Tanaka, Yu. V. Nazarov, S. Kashiwaya, Phys. Rev. Lett. 90, 167003 (2003)
41. Y. Tanaka, Yu. V. Nazarov, A.A. Golubov, S. Kashiwaya, Phys. Rev. B 69, 144519 (2004) [see Phys. Rev. B 70, 219907 (2004) for erratum]
42. A. Millis, D. Rainer, J.A. Sauls, Phys. Rev. B 38, 4504 (1988)
43. D.H. Hernando, Yu. V. Nazarov, W. Belzig, Phys. Rev. Lett. 88, 047003 (2002)
44. D.H. Hernando, Yu. V. Nazarov, W. Belzig, arXiv:cond-mat/0204116.
45. A. Cottet, W. Belzig, Phys. Rev. B 72, 180503 (2005)
46. J. Linder, A. Sudbø, Phys. Rev. B 75, 134509 (2007)
47. S. Kashiwaya, Y. Tanaka, N. Yoshida, M.R. Beasley, Phys. Rev. B 60, 3572 (1999)
48. T. Kontos, M. Aprili, J. Lesueur, X. Grison, Phys. Rev. Lett. 86, 304 (2001)
49. T. Yokoyama, Y. Tanaka, A.A. Golubov, Phys. Rev. B 75, 134510 (2007)
50. A. Brataas, Yu. V. Nazarov, G.E. Bauer, Phys. Rev. Lett. 11, 2481 (2000)
51. A. Brataas, G.E.W. Bauer, P.J. Kelly, Phys. Rep. 427, 157 (2006)
52. M. Eschrig, T. Löfwander, Nat. Phys. 4, 138 (2008)
53. J. Linder, T. Yokoyama, A. Sudbø, Phys. Rev. B 77, 174514 (2008)
54. R. Meservey, P.M. Tedrow, Phys. Rep. 238, 173 (1994)
55. Y. Asano, Y. Tanaka, A.A. Golubov, S. Kashiwaya, Phys. Rev. Lett. 98, 107002 (2007)
56. V. Braude, Yu. V. Nazarov, Phys. Rev. Lett. 98, 077003 (2007)
57. A.A. Golubov, M. Yu. Kupriyanov, E. Ilichev, Rev. Mod. Phys. 76, 411 (2004)
58. V.A. Oboznov, V.V. Bolginov, A.K. Feofanov, V.V. Ryazanov, A.I. Buzdin, Phys. Rev. Lett. 96, 197003 (2006)
59. G. Mohammadkhani, M. Zareyan, Phys. Rev. B 73, 134503 (2006)
60. E. Goldobin, D. Koelle, R. Kleiner, A. Buzdin, Phys. Rev. B 76, 224523 (2007)
61. H. Sellier, C. Baraduc, F. Lefloch, R. Calemczuk, Phys. Rev. Lett. 92, 257005 (2004)
62. Yu. S. Barash, I.V. Bobkova, T. Kopp, Phys. Rev. B 66, 140503 (2002)
63. Z. Pajović, M. Bozović, Z. Radović, J. Cayssol, A. Buzdin, Phys. Rev. B 74, 184509 (2006)
64. M. Houzet, A. Buzdin, Phys. Rev. B 76, 060504 (2007)
65. B. Crouzy, S. Tollis, D. Ivanov, Phys. Rev. B 75, 054503 (2007)
66. I.B. Sperstad, J. Linder, A. Sudbø, Phys. Rev. B 78, 104509 (2008)
67. Z. Popović, Z. Radović, arXiv:0907.2042.
68. J. Linder, T. Yokoyama, A. Sudbø, Phys. Rev. B 79, 054523 (2009)
69. G.B. Halasz, J.W.A. Robinson, J.F. Annett, M.G. Blamire, Phys. Rev. B 79, 224505 (2009)
70. F.S. Bergeret, A.F. Volkov, K.B. Efetov, Phys. Rev. B 64, 134506 (2001)
71. A.F. Volkov, K.B. Efetov, Phys. Rev. B 78, 024519 (2008)

72. Y.V. Fominov, A.F. Volkov, K.B. Efetov, Phys. Rev. B 75, 104509 (2007)
73. M.L. Kulic, I.M. Kulic, Phys. Rev. B 63, 104503 (2001)
74. I. Eremin, F.S. Nogueira, R.-J. Tarento, Phys. Rev. B 73, 054507 (2006)
75. M.S. Grønsleth, J. Linder, J.-M. Børven, A. Sudbø, Phys. Rev. Lett. 97, 147002 (2006)
76. J. Linder, M. Grønsleth, A. Sudbø, Phys. Rev. B 75, 024508 (2007)
77. Y. Zhao, R. Shen, Phys. Rev. B 73, 214511 (2006)
78. Ya. V. Fominov, N.M. Chtchelkatchev, A.A. Golubov, Phys. Rev. B 66, 014507 (2002)
79. Ya. V. Fominov, A.A. Golubov, M. Yu. Kupriyanov, JETP Lett. 77, 510 (2003) [Pis'ma v ZhETF 77, 609 (2003)]
80. I. Baladie, A. Buzdin, Phys. Rev. B 67, 014523 (2003)
81. Z. Radovic, M. Ledvij, L. Dobrosavljevic-Grujic, A. Buzdin, J. Clem, Phys. Rev. B 44, 759 (1991)
82. T. Yamashita, H. Imamura, S. Takahashi, S. Maekawa, Phys. Rev. B 67, 094515 (2003)
83. Z.C. Dong, R. Shen, Z.M. Zheng, D.Y. Xing, Z.D. Wang, Phys. Rev. B 67, 134515 (2003)
84. C.J. Lambert, J. Phys. Condens. Matter 3, 6579 (1991)
85. C.W. Beenakker, Phys. Rev. Lett. 97, 067007 (2006)
86. M. Zareyan, H. Mohammadpour, A.G. Moghaddam, Phys. Rev. B 78, 193406 (2008)
87. L. Fu, C.L. Kane, Phys. Rev. B 79, 161408 (2009)
88. D. Aoki, A. Huxley, E. Ressouche, D. Braithwaite, J. Flouquet, J.-P. Brison, E. Lhotel, C. Paulsen, Nature 413, 613 (2001)
89. T. Akazawa, H. Hidaka, T. Fujiwara, T.C. Kobayashi, E. Yamamoto, Y. Haga, R. Settai, Y. Onuki, J. Phys. Condens. Matter 16, L29 (2004)
90. T. Akazawa, H. Hidaka, T.C. Kobayashi, T. Fujiwara, E. Yamamoto, Y. Haga, R. Settai, Y. Onuki, J Phys. Soc. Jpn. 73 3129 (2004)
91. N.T. Huy, A. Gasparini, D.E. de Nijs, Y. Huang, J.C.P. Klaasse, T. Gortenmulder, A. de Visser, A. Hamann, T. Gorlach, H.V. Lohneysen, Phys. Rev. Lett. 99, 067006 (2007)
92. R. Shen, Z.M. Zheng, S. Liu, D.Y. Xing, Phys. Rev. B 67, 024514 (2003)
93. M. Cuoco, P. Gentile, C. Noce, Phys. Rev. Lett. 91, 197003 (2003)
94. C. Pfleiderer, Rev. Mod. Phys. 81, 1551 (2009)
95. C. Pfleiderer, A.D. Huxley, Phys. Rev. Lett. 89, 147005 (2002)
96. N. Tateiwa, T.C. Kobayashi, K. Hanazono, K. Amaya, Y. Haga, R. Settai, Y. Onuki, J. Phys. Condens. Matter 13, L17 (2001)
97. A.D. Huxley, E. Ressouche, B. Grenier, D. Aoki, J. Flouquet, C. Pfleiderer, J. Phys. Condens. Matter 15, S1945 (2003)
98. J. Linder, A. Nevidomskyy, A. Sudbø, Phys. Rev. B 78, 172502 (2008)
99. J. Shi, Q. Niu, `cond-mat/0601531`
100. D.V. Shopova, D.I. Uzunov, Phys. Rev. B 72, 024531 (2005)
101. M.G. Cottam, D.V. Shopova, D.I. Uzunov, Phys. Lett. A 373, 152 (2008)
102. D.V. Shopova, D.I. Uzunov, Phys. Rev. B 79, 064501 (2009)
103. S. Tewari, D. Belitz, T.R. Kirkpatrick, J. Toner, Phys. Rev. Lett. 93, 177002 (2004)
104. A.J. Leggett, Rev. Mod. Phys. 47, 331 (1975)
105. Notice that time-reversal symmetry may be spontaneously broken in the orbital part (angular momentum) of the Cooper pair wavefunction even if the state is unitary.
106. K. Machida, T. Ohmi, Phys. Rev. Lett. 86, 850 (2001)
107. F. Hardy, A.D. Huxley, Phys. Rev. Lett. 94, 247006 (2005)
108. K.V. Samokhin, M.B. Walker, Phys. Rev. B 66, 174501 (2002)
109. K.V. Samokhin, M.B. Walker, Phys. Rev. B 66, 024512 (2002)
110. M.B. Walker, K.V. Samokhin, Phys. Rev. Lett. 88, 207001 (2002)
111. J. Linder, M.S. Grønsleth, A. Sudbø, Phys. Rev. B 75, 054518 (2007)
112. T. Yokoyama, Y. Tanaka, Phys. Rev. B 75, 132503 (2007)
113. A. Brataas, Y. Tserkovnyak, Phys. Rev. Lett. 93, 087201 (2004)
114. Y. Zhao, R. Shen, Phys. Rev. B 73, 214511 (2006)
115. A.H. Nevidomskyy, Phys. Rev. Lett. 94, 097003 (2005)
116. J. Linder, A. Sudbø, Phys. Rev. B 76, 054511 (2007)

117. A. Harada, S. Kawasaki, H. Mukuda, Y. Kitaoka, Y. Haga, E. Yamamoto, Y. Onuki, K.M. Itoh, E.E. Haller, H. Harima, Phys. Rev. B 75, 140502 (2007)
118. A.B. Shick, W.E. Pickett, Phys. Rev. Lett. 86, 300 (2001)
119. M. Sigrist, AIP Conf. Proc. 789, 165 (2006)
120. J. Linder, I.B. Sperstad, A.H. Nevidomskyy, M. Cuoco, A. Sudbø, Phys. Rev. B 77, 184511 (2008)
121. C.-R. Hu, Phys. Rev. Lett. 72, 1526 (1994)
122. Y. Tanaka, S. Kashiwaya, Phys. Rev. Lett. 74, 3451 (1995)
123. T. Lofwander, V.S. Shumeiko, G. Wendin, Supercond. Sci. Technol. 14, R53 (2001)
124. Y.S. Barash, A.A. Svidzinsky, H. Burkhardt, Phys. Rev. B 55, 15282 (1997)
125. N. Sato, private communication.

Index

MIX
Papier aus verantwortungsvollen Quellen
Paper from responsible sources
FSC® C105338

If you have any concerns about our products,
you can contact us on
ProductSafety@springernature.com

In case Publisher is established outside the EU,
the EU authorized representative is:
Springer Nature Customer Service Center GmbH
Europaplatz 3, 69115 Heidelberg, Germany

Printed by Libri Plureos GmbH
in Hamburg, Germany